Resistance to diseases and pests in forest trees

The workshop was organized by the International Union of Forestry Research Organizations and sponsored by the North Atlantic Treaty Organization, the Heidemij (Royal Netherlands Land Development and Reclamation Society) and the Ministry of Agriculture and Fisheries of the Netherlands.

Resistance to diseases and pests in forest trees

Proceedings of the Third International Workshop on the Genetics of Host-Parasite Interactions in Forestry, Wageningen, the Netherlands, 14 – 21 September 1980

Edited by
H.M. Heybroek, B.R. Stephan and K. von Weissenberg

pudoc

Centre for Agricultural Publishing and Documentation
Wageningen – 1982

Preceding Workshops:

Genetic improvement for disease and insect resistance of forest trees, 30 August to 11 September 1964; a NATO and NSF Advanced Study Institute. Pennsylvania State University, University Park, Pennsylvania, USA. Proceedings: H.D. Gerhold, E.J. Schreiner, R.E. McDermott & J.A. Winieski (Eds): Breeding Pest-Resistant Trees. Pergamon Press, Oxford. 1966. 505 p.

Basic biology and international aspects of rust resistance in forest trees, 17-24 August 1969; a NATO-IUFRO Advanced Study Institute. University of Idaho, Moscow, Idaho, USA.
Proceedings: R.T. Bingham, R.J. Hoff & G.I. McDonald (Eds): Biology of Rust Resistance in Forest Trees. USDA Forest Service, Miscellaneous Publication No 1221. 1972. 681 p.

ISBN 90 220 0794 4

Printed in the Netherlands.

Contents

TOPIC 4: RESISTANCE MECHANISMS AND THEIR GENETIC CONTROL

TOPIC 5: BIOCHEMICAL ASPECTS IN RESISTANCE BREEDING

TOPIC 6: GENETIC VARIATION IN FUNGAL PATHOGENICITY

TOPIC 8: PRESENTATION AND ANALYSIS OF CURRENT BREEDING PROGRAMS

TOPIC 9: CONCLUDING REMARKS

Breeding for balance? An introduction

Breeding for resistance to a parasite is breeding against a flexible adversary: one cannot take it for granted that resistance, once obtained, will remain at the same level. The parasite may come up with rare or new genetic variants that circumvent the resistance and render the host susceptible. Therefore, breeding for resistance to a parasite is fundamentally different from breeding for resistance to frost, drought, salt or some other environmental extreme.

Forest tree breeding, at first sight, is particularly vulnerable to changes in the parasite, as a quick change from one tree species to another is difficult to achieve in forest management without big losses. This is caused by the fact that forestry has a long time scale compared to agriculture: the agricultural plant breeder may consider some resistance to be 'stable' if it does not erode within twenty years, which is more than the normal lifetime of a cultivar; for most forest trees, however, this is only half the lifetime of a single crop.

There is no cause for despair, however, for there clearly is a ceiling to the potential of parasites to adapt to 'resistant' hosts. It looks as if not all resistance is overcome by variation of the parasite. Many cases of resistance in agricultural crops have proved to be stable up to now. Nature shows that resistance may hold over much longer periods: poplar rust has not yet adapted to oak, not even to aspen. For the breeder it is most important to find out if there is a pattern to this genetic adaptability: which changes are possible for the parasite, which are not?

When we compare the three international gatherings concerned with forest tree breeding for resistance, some differences stand out. At the first meeting in Pennsylvania in 1964, when the art was still young, the possibility of a break-down of resistance was no more than a far echo from agricultural plant breeding experience. Technical matters such as screening techniques had most interest. The second meeting was held in the den of the breeders for blister rust resistance, in Moscow, Idaho, in 1969. This was one year after Vanderplank's treatise 'Disease Resistance in Plants' appeared. On the basis of experiences gained in agricultural plant breeding, that book predicts that gross adaptation of the parasite can readily be expected in cases of 'vertical' resistance, but is highly unlikely in cases of 'horizontal' resistance. Consequently, participants hotly discussed the danger of ephemeral resistances and how to avoid them. First hand experience with new strains of forest parasites was still lacking, however.

This has changed, alas. In the eleven years before the third meeting in 1980, several changes in pathogenicity of forest tree fungi have been noted:

- a resistance gene in sugar pine was overcome by a new strain in *Cronartium ribicola*;
- new and specific strains of the poplar rust *Melampsora larici-populina* have been found in Australia and New Zealand, soon after the fungus appeared in those new territories;
- in fusiform rust, a certain increase of pathogenicity was observed on host populations that had been selected for higher resistance to the disease;
- in *Ceratocystis ulmi*, a highly aggressive strain (with two variants) was found spreading havoc among the elms in Europe and adjacent Asia;
- a new aggressive strain of *Gremmeniella abietina* seems to pose a big threat to several pines in northeastern North America.

It may be significant that at least the last three examples do not fit the neat pattern of agricultural crops with their stable 'horizontal' and instable 'vertical' resistances. Perhaps in some cases we are witnessing a finer genetic adaptation of a colonizing pioneer-parasite to its new habitat on a new continent.

Where a final victory over the parasite by an absolute and permanent resistance does not seem possible, some sort of balance between host and parasite evidently is the best that can be attained: some situation in which host and parasite are in equilibrium, perhaps by the presence in the host plant of a combination of resistance genes that can be less easily overcome by the parasite, or by the presence in the host population of genetic variability that may diminish chances for parasite change and specialization. Ecological factors can also influence such a balance. The theme of the 1980 Workshop therefore was: 'Breeding for balance'. This turned out to be a stimulating concept which intrigued and provoked the participants. The meeting was structured so that the earlier sessions laid a basis for the day-long session named 'breeding for balance'. It was concluded that a balance between host and parasite, at a level that is acceptable for the forester, may be a vision, an anticipation and an ideal, but that it is an elusive concept too, not sufficiently concrete to serve as a breeding goal. As Professor Dr. J.C. Zadoks of the Agricultural University in Wageningen asked us in his welcome address: Does balance really exist? If so, can you measure it and explain it quantitatively? Is it a static balance or a dynamic equilibrium? From the floor came the question: balance for whom? Apparently several balances or equilibria are possible, but not all are desirable.

We still have much to learn about the interaction between host and parasite, also at a population level. It is difficult to look into the future. The leading concepts of tomorrow may be based on details that are nearly overlooked to-day or they may revive discredited theories of yesterday.

Even if we avoid the word 'balance', some general observations on the current thinking about resistance breeding in forest trees can be given. The search for "the immune genotype" is no longer the sole or final goal of resistance breeding - if it ever was. It is widely realized that even a moderate degree of resistance can suffice to save a crop and that it may be an essential prerequisite for designing an integrated control system, in which adapted spacing, the avoiding of certain sites, species mixture, sanitation, chemical control measures, thinning regime and other silvicultural treatments can all play a part. It is well known that such systems tend to fail if the host is too susceptible. Further, resistance breeding should be an integral part in most general breeding programs, because attention should always be given to the possibility of reducing known susceptibilities and avoiding new ones. It happens that in natural populations of forest tree species, resistance to a certain parasite is the rule, while considerable susceptibility presents itself in selected or new genotypes during genetic improvement of the species. In such cases it seems quite possible to restore the original and stable resistance. Lastly, immunity or absolute resistance, whenever found, is a precious property which should be used and deployed with care and caution. It is not, as some have thought, the conclusion of a resistance breeding program.

The Workshop (72 participants from 21 countries, 65 papers) provided a massive body of information, theory, and ideas. These added to our understanding of hosts and parasites as a basis for future work, both for breeding and for integrated control or pest management.

Without liberal funds, it would have been impossible to bring such a wide-ranging group of distinguished specialists and generalists together, so that a near-complete state-of-the-art could be developed. We are very grateful to our sponsors for their contribution, as well as to our hosts for their massive support: the 'Dorschkamp' Research Institute for Forestry and Landscape Planning, and the International Agricultural Centre, both at Wageningen.

The atmosphere of interest and happy interaction which reigned during the workshop regrettably cannot be transmitted in this book.

The Organizing Committee,

Henry D. Gerhold
Hans M. Heybroek
Robert F. Patton
Bent Søegaard
Richard Stephan
Kim von Weissenberg

The world's need for pest-resistant forest trees

Bruce J. Zobel

Professor Emeritus and Forestry Consultant, School of Forest Resources, North Carolina State University, Raleigh, North Carolina, USA

ABSTRACT

The pattern of pest attack and the genetic flexibility and strong genetic control of resistance to many pests in forest trees, along with the high cost and even impossibility of direct chemical controls and the uncertainty of biological controls, all indicate the necessity for developing pest resistant forest trees.

The commonly cited danger from monocultures resulting from use of genetically improved trees will not be valid for a good forest tree improvement program in which a wide adaptive base has been a major objective of breeding. Genetic control of economically important characteristics is generally independent from that for adaptability or pest resistance; most important characteristics of forest trees are multigenic and are inherited independently from another. Therefore it is highly practical to breed for narrowing the genetic base for economic characteristics such as tree straightness or wood specific gravity while at the same time broadening the base for adaptability to diverse environments or resistance to pests.

Every breeding program involves a struggle between gain and risk; this is especially so when dealing with pest control. We have found that there is considerable genetic resistance to most pests in forest trees although some types of resistance are much easier to breed for than others.

INTRODUCTION

Pests are always a problem in any forest enterprise. As forest management becomes more intensive, pests appear to become more numerous; the same result seems to be evident as rotation ages are made shorter. This apparent increase is partially the result of closer observation as well as a greater concern about pests and their effect on intensive forest management; under this situation, pest attacks that were formerly overlooked or were considered to be a minor nuisance suddenly become important. A good example is spider mites on *Pinus taeda* L. This insect was hardly noticed in natural stands or plantations but is considered to be a serious pest on grafts in seed orchards and to some extent on young trees in progeny tests. Also, the greater apparent incidence of pests is partially a real increase resulting from more intensive management. A good example is *Cronartium fusiforme* Hedgc. & Hunt on *P. taeda*; this disease becomes progressively more serious when management practices such as site preparation are improved or nitrogen fertilizers are used. The damage and wounds that sometimes result from cultivation equipment and other activities related to intensive forest management can become infection courts for pest attack.

Another reason for an increase in pests under intensive forest management, often overemphasized but of importance under some conditions, is the establishment of trees with restricted genotypes over large contiguous acreages. In itself, large plantings of a single species are not serious unless the genotypes used are so uniform that a true monoculture results. There is a special danger when vegetative propagules of restricted genetic background are used for operational planting; vegetative propagation in itself is not bad if a broad enough genetic base is ensured through use of several clones which will discourage wholesale pest attacks.

The true situation related to genetic diversity in forest trees is generally misunderstood. Trees are almost always highly heterozygous genetically, and recent studies have indicated that forest trees have the greatest genetic variability of any organisms. A large adaptability to adverse environments and pests would be expected since a tree must survive many years and reproduce over sometimes drastically changing conditions from day to day and from year to year. Thus, though a forest may be planted as a monoculture with all individuals having similar genotypes, each individual tree is never genetically homozygous in a way similar to some agricultural crops. Each ramet of a clone of a vegetatively reproduced plantation will be quite heterozygous for adaptability and pest resistance. The danger from pests occurs when all ramets of a clone are susceptible to a given pest.

It is often difficult to initiate forest plantation programs, especially in areas where growing forest trees operationally is not a common and accepted practice. It is a real setback, therefore, if the plantations are lost to pests or adverse environments after a program has been initiated with much effort, cost and fanfare. I worked closely with such a program in the three-county area in one of our southern states. Large sums of money and great efforts were expended on propaganda to inform the layman and to get him interested and excited about the importance of planting pines. Everything was favorable and plans went well except that the nurseryman did not pay attention to the provenance or geographic source differences within the pine species he produced in the nursery. The seedlings supplied for planting were not well adapted to the three-county area. A severe drought followed by a *Pales* weevil attack wiped out most of the several hundred thousand acres of plantations. (As I remember, final survival was less than 20 %.) Those few plantings that did survive were later damaged by an ice storm followed by bark beetle attacks. Although this incident occurred over 25 years ago, it still is difficult to this day to get people to plant pines in the area involved. They cite the earlier disaster as 'proof' that pines will not grow there and almost completely ignore the biological explanation of wrong seed source followed by destruction by pests.

Similar incidents are continuously occurring with exotic plantations, and the forestry future of whole countries can be in jeopardy unless pests can be controlled. A prime example occurred in central Brazil. Very large plantings were made of *P. radiata* D. Don; these were later destroyed by *Dothistroma*. Only recently has it been possible to induce the local people to again plant pines in an area ideally suited to grow *P. caribaea* Morelet and *P. oocarpa* Schiede.

In any intensive forestry program there is always a tradeoff between gain achievable and risk encountered to obtain greater gains. This problem is especially evident when pests are involved. Decisions must constantly be made relative to the option of obtaining greater gain from the use of better trees along with more intensive management of the forests, and the possibly increased danger of pest attacks that might result from doing so. Only when pest-resistant strains of trees are used can forestry come close to optimal production of desired products and thus its full contribution to Society.

PESTS AND EXOTICS

Some of the most intensive and widespread forestry operations now being established depend upon plantations of exotic species. Although the major concern and emphasis related to exotics is to the tropical and subtropical regions, there is also considerable usage of exotics in cooler climates; examples are *P. contorta* Dougl. in Sweden or *Pseudotsuga menziesii* (Mirb.) Franco in several European countries. No matter where exotic species are used, a major consideration in their success relates to their susceptibility to attack and their potential as hosts for pests.

Exotics are generally planted in large blocks of single species with the seed for the plantings sometimes being from restricted sources or from small numbers of parents. For some *Eucalyptus* and poplars, vegetative propagules may be used for regeneration, creating an ideal situation for attack by pests if care is not taken to maintain a suitably broad genetic base. In almost all instances the exotic is not too well adjusted to its new environment, with the result that the trees in the forest are growing under considerable stress. Exotics are sometimes used by persons who are uninformed or do not care, so that a large amount of flagrant 'off-site' planting has been done and is still being done. This is especially true for some *P. radiata*, *P. caribaea* and certain *Eucalyptus* plantings where these outstanding forest trees have been placed in quite unsuitable environments because of their fine performance elsewhere under proper conditions.

There are not too many 'absolutes' in biology, but one that comes the closest to certainty is that exotic plantations will be attacked by pests of one kind or another. The situation is especially insidious in that often the exotic grows pest-free in its early years in the new environment, too often leading to a euphoria and false projections about its production potentials, based upon the pest-free growth. But I have never seen it fail - it may take 2 years or it may take 10 - but pests of some type, often serious, will become established in exotic plantations. I have recommended to clients in tropical areas using exotics that they reduce their initial estimates of productive potential by about 30 % in anticipation of pest attacks as the programs become older. Many times pest attack will follow adverse environmental conditions which have placed the trees under severe stress. The poor physiological condition of the exotic trees that results enhances the spread and damage by pests that may have previously either been unknown or which were considered to be of only minor importance.

It is sad and of great concern to me for the future of healthy and profitable forests to view the magnitude and seriousness of losses from pest attacks on exotic forest tree plantings. Massive programs have failed from such pest attacks; examples are the large *P. radiata* plantations destroyed by *Dothistroma* in several regions of the world, including Brazil, Rhodesia and East-central Africa. The loss of confidence in forestry that has occurred as a result of destruction by this disease could have been prevented and millions of dollars could have been saved if those who established the exotic *P. radiata* had heeded the advice of a few pathologists who warned that *Dothistroma* probably would become a killer in radiata pine plantations when planted where there are warm and moist summers. Even when the adverse results of disease attacks resulting from planting in such environments are now so clearcut, there still are instances of planting *P. radiata* in such disease-prone climates. Many persons, including myself, have been puzzled as to how *Dothistroma*, found on *P. radiata* in its indigenous range in California where it is considered to be only a nuisance, has managed to become established in such widely separated regions throughout the world. Although we may not understand, the important fact is that it has spread to these areas and there is no reason to believe it will not occur in other areas with environments suitable for development of the disease.

Exotics are attacked by 3 general types of pests:

1. Those indigenous to the area in which the exotic plant is planted but never before known to attack the exotic. Such pests often adjust to the exotic and can become a major destructive force. This adjustment to the exotic is of rather common occurrence, especially with leaf-eating beetles and worms. Several species of these pests have become major defoliators of eucalypts in Brazil and New Zealand and of *Cupressus lusitanica* Mill. in Colombia.
2. Pests of the exotic in the exotic's native range follow it to the new area. A prime example of this is *Dothistroma* on *P. radiata*. This disease appears to have followed *P. radiata* whenever it was planted in an environment suitable to the development of *Dothistroma*.
3. A pest not native to the area where the exotic is planted but also not indigenous to the home environment of the exotic can develop as a major pest in the forest in its new range. An insect on *P. radiata* serves as a good example. This insect, *Sirex*, seriously damages plantations, especially the suppressed and codominant trees.

Sometimes the pest is harmful, not because it kills the exotic but by deforming it so it is of value only for limited low-quality products. A good example for this is the so-called cypress stem canker on *Cupressus* planted in Kenya and other areas. Stem deformation is so severe that the only suitable use for the tree is for fiber products and they cannot produce the good quality solid wood products for which the plantations were established.

PESTS AND INTENSIVE FOREST MANAGEMENT

As forest management becomes more intensive we must learn to adapt to and reduce the inevitable increased pressures from pest attacks. The high costs of intensive management and the great potential value of the products make it mandatory to prevent pests from seriously reducing the volume and quality, and thus the value, of the forest products grown. It is not possible for an intensive forest management program to operate as some unthinking environmentalists have suggested, i.e., 'let the pests have their share'. Historically, growing forest trees has yielded a low return on the investment and pest attacks can easily turn what would be a profitable venture into a losing enterprise. As the demand for forest products becomes greater it is of paramount importance for Society to make each forested acre more productive. Land on which forest trees can be grown for wood products is becoming less available. Additionally, the best lands are being taken for use in agricultural production, forcing forest production to shift toward the more marginal sites. This in turn results in the trees' being grown under greater stress, and thus they become more susceptible to pest attack.

Certain pests are increased by activities such as fertilization, site preparation or thinning. For example, fusiform rust *(Cronartium fusiforme)* is much worse on fertilized and intensively prepared sites than in plantations that are not intensively managed. Thinning of plantations can result in massive destruction by *Fomes annosus* (Fr.) Cooke because the stumps of the cut trees become an infection court for the disease. Cultivation of plantations cuts roots that are then sometimes attacked by insects and diseases.

If optimal gains through tree improvement are to be obtained, a reduced genetic base for the economically important characteristics will result. On the surface this would appear to develop in the direction of a monoculture that would be ideal for the spread of pests. But greater pest attacks will not occur if care is taken in the development of the tree improvement program. It is most fortunate that in forest trees almost all economically important traits such as tree straightness or wood quality are genetically independent from characteristics of pest resistance. With practically no exceptions, these two sets of characteristics are controlled by multiple genes and are inherited independently and thus are not strongly correlated. It is therefore possible to breed for better economic characteristics by narrowing the genetic base for these while at the same time breeding for greater adaptability and pest resistance by broadening the genetic base. Our program in the southeastern United States has been using such a dual breeding system most effectively for nearly thirty years.

WHY RESISTANT TREES ARE NEEDED AND WHEN SHOULD ONE BREED FOR RESISTANCE

Any breeding program with forest trees is long-term and expensive. Why, therefore, does one spend a lot of time and money on controlling pests through developing resistance rather than using silvicultural, chemical or natural predator controls? The fear is often expressed that 'pests will defeat us' and that we would be best off to do nothing. Such a negative attitude is silly, because any forest, managed or not, will be attacked by pests. When intensive forestry is applied, 'new' pests are found or suddenly become important, not really because they are new but because they had not been closely observed previously. Although intensive site preparation, cultivation, thinning and fertilization sometimes do result in trees becoming more pest-susceptible, sometimes the opposite result occurs where trees become more pest-tolerant. A good example of this relates to pine bark beetles: the healthy, well-tended stands are more resistant to insects than are the off-site, the over-age, the over-dense forests.

Whether breeding for pest resistance should be done depends upon the availability and suitability of other methods of pest control. For example, one would not try to develop a loblolly pine to be resistant to fusiform rust only in the nursery bed; fungicide sprays are so simple and economical in the nursery that this method of spraying is the preferred control. But in forest plantations the situation is reversed. Although actions such as reducing the alternate host can be applied and will help some, no known practical and safe methode is available for efficient control of the rust in large forest plantings. If fusiform rust is to be controlled successfully in large plantations, the only practical way is to breed for resistance. The pine host shows large variability in susceptibility to the disease, and inheritance of resistance is of a magnitude such that it is relatively easy to develop useful resistant strains. The pest also varies genetically but, although this complicates the problem, it appears to be possible to overcome it by resistance breeding. It is unfortunate that the improved silvicultural methods of better site preparation, fertilization and sometimes cultivation currently used to improve tree growth and yield are all conducive to increased rust infection. This indicates that the need for fusiform rust control and thus resistance breeding is constantly increasing as forest management gets 'better'. In the more 'hot spot' rust areas in the southeastern United States, many of us feel that without use of resistant pines the greater growth resulting from site preparation and improved forest management is more than offset by the greater losses from increased fusiform rust.

Opposite to fusiform rust, in my opinion, *Fomes annosus* is an important pest against which breeding for resistance will probably not be of primary value. Although some progress has been made in breeding for resistance to this pest, the major control mechanisms should be through silvicultural and species manipulation. In the southern pines, for example, *Fomes* has generally been kept under reasonable control by good stand management.

Although developing resistant strains is expensive and takes special skills and much time, it has the advantage that, once obtained, resistance is relatively permanent to which additional and better resistance can be added. The long-term cost of control is usually less with breeding than with direct controls because costs of the latter usually reoccur frequently or at least several times throughout the rotation. A good example is *Dothistroma* on radiata pine. Several studies have indicated that there is a reasonable resistance to the disease. Despite this, some organizations have chosen to control *Dothistroma* by making the necessary sprays during the most susceptible period of the plantation's life rather than carry on resistance breeding. If only one generation of plantings is considered, there is no doubt that resistance breeding would be less efficient than sprays for *Dothistroma* control, but I greatly doubt that this will be so for several generations. Additionally, any use of chemicals in the forest environment is not desirable and should be avoided when possible. The advantage of resistance breeding is that no undesirable substances are inserted into the environment. Biological control fills this need when it works but, although much discussed and promoted, effective biological control of forest pests is rare and failures are the norm. A good example is the spruce budworm in northeastern Canada. Much effort has been extended for biological control, with hardly satisfactory results. Large and costly spray control efforts have been undertaken with fair to good success, but the social and environmental objections are strong. Resistance breeding has been essentially ignored but, although I am obviously biased, I feel resistance breeding for spruce budworm should be given a good trial. The pattern of attack, the occasional noninfected tree, the ability to assess attack on young trees, all indicate resistance breeding might be a worthwhile approach.

A common reason given for concern about resistance breeding in forest trees is that new and more virulent strains of diseases will develop or that insects will be evolved that can attack the resistant trees. (This same argument holds for control by sprays or biological control.) This potential, although real, has been emphasized out of all proportion to reality. Because of outstanding examples of developed tolerance, like flies to DDT or rusts on wheat, the lay attitude is that overcoming of resistance by new strains will be regular and normal. This is not so if proper precautions are taken, especially in long-lived, heterozygous forest trees. The concern to the forester would be more real if his management methods and trees planted remained static over consecutive rotations, being the same for succeeding generations of planted trees. But forestry is never static and we have a special advantage so that in our long rotation ages, succeeding crops will be genetically different from the previous crop planted on a given site. Unless the forest managers and geneticists are completely incompetent, the trees planted on a site which had been planted 30 years previously will be genetically improved and different from the original planting. During the 30-year period required for the crop to mature, new and improved strains of trees will have been developed. Therefore, the fear of a pest evolving from the first crop to destroy the succeeding crops may be theoretically correct but does not fit with facts. Such an adaptation and evolving of pests certainly happens on annual crops but should not on long-lived perennials if an active, ongoing breeding program is underway, so the new planting stock will differ from the original.

Often the magnitude and complexity of controls required, whether they be chemical or biological, make it impossible to efficiently use anything other than a resistance program. Although biological control currently is the 'darling' of the environmentalist, and something we would all be delighted to see perfected, it is as of now generally more in the minds of men than a method that is useful on a commercial scale in most forest trees. All of us want to become as free as possible from chemicals for the control of insects and diseases. The conclusion must be:

Breeding for resistance to pests in forestry is absolutely mandatory.

The adoption of agricultural practices for the development of heritable resistance to pests and pathogens in forest crops

I.A.S. Gibson, J. Burley & M.R. Speight

Department of Forestry, Oxford University, South Parks Road, Oxford, UK

ABSTRACT

Objectives, skills and methods of farmer and forester have often conflicted but annual crop breeding provides guidance for perennial crop improvement which is particularly important for the rapidly increasing areas, products and managerial diversity of forest tree plantations. The dangers of reduced genetic variability that are inherent in intensive selection programmes may be offset by the use of clonal mixtures and multiple population breeding but resistance breeding will play an increasingly important part in plantation development and it will call for inter-disciplinary co-operation.

Development of resistance in trees needs long-term knowledge of the biology of the pest including its variation and taxonomy. Prediction and identification of pest problems are made difficult by the lack of pest stability in spreading to new regions. Whereas annual crops favour obligate, host-specific pathogens, trees provide substrate for saprophytic, unspecialized, facultative pathogens and are often grown on marginal sites that reduce their host vigour.

Whereas stability of vertical resistance in trees has been detected, it is bound to decline and mixed cropping should increase. Selection for horizontal resistance is likely to be more profitable and may be extended to generalized field resistance since perennial crops are subject to attack by many pests during a commercial rotation. At both levels, attention must be paid to the effect of the environment on the expression of resistance. While seed orchard development accelerates the application of vertical resistance, vegetative propagation is more suitable for horizontal resistance. Techniques for early detection of resistance are required and may be adopted from agriculture.

INTRODUCTION

Forest management, particularly plantation development, arose largely in response to the erosion of natural resources by agriculture. The objectives, skills and methods of farmer and forester have often conflicted, mainly for reasons entrenched in the genesis of their disciplines, yet the rapid rate of improvement possible with annual crops provides lessons and guidance for the development of perennial, and particularly forest, crops.

The farmer is concerned mainly with the intensive management of single crop species on good sites, yielding one product in which uniformity is at a premium and for which value per unit yield is high. In contrast, the forester commonly acts as an ecologist, maintaining yields of low value products by extensive management of variable crops on poorer sites with several objectives including protection and amenity functions, as well as the production of industrial raw material which, in itself, covers a range of properties for a number of uses.

The evolution, domestication and some 50 years of intensive breeding of many annual crops have reduced their genetic base, attaining a selection plateau beyond which further gain calls for the introduction of genes from wild populations in their natural origin. Forest trees are only 2 or 3 generations removed from the wild type and, with the exception of some intra-specific, natural population variants and some restricted national breeding populations, tree species have considerable resources of genetic variation available. Nevertheless, conservation of genetic resources is essential here as well as in agriculture and, for many species, it is in progress.

THE NEED FOR PLANTATIONS AND THE USE AND IMPROVEMENT OF EXOTIC SPECIES

Around the start of the 20th century the demands of rapidly expanding and increasingly sophisticated world populations called for increases in forest production that have been met largely by the creation of even-aged plantations of single species, which have often been exotics.

These species were selected from natural stands with environments similar to those of the areas under development and subsequent trials led to the selection of 2 or 3 provenances or species for establishment on a large scale (see e.g. Burley & Wood, 1976). Once these species were in use, further selection and testing brought them towards an optimum for the region, largely through culling the worst stock at the nursery stage, selective thinning and seed collection from superior trees.

This has retained a level of genetic versatility in the crop while adapting it to the environment; it has been successful so that it is now common experience to find that such crops are so well adjusted after 2 or 3 generations that their progeny are usually superior to freshly introduced provenances. During the last 3 or 4 decades there has also been an increase in intra-specific improvement for forest plantation crops, similar to that used for other perennial crops, either through selection and vegetative propagation of clonal material of outstanding merit as in poplars, where this includes pest resistance, or Norway spruce (*Picea abies* (L.) Karst.), or by open or controlled pollination in seed orchards between ramets of phenotypically superior trees. The latter has had considerable success for pines and eucalypts and is now being extended to other species.

While these recent lines of development have provided immediate benefits, they may, in the long run, risk the sacrifice of the benefits of the field selection techniques which are associated with plantation management. The danger of reducing the genetic base may be offset by the use of clonal mixtures for vegetatively propagated crops (Schreiner, 1939) and by multiple population breeding (Namkoong et al., 1980).

JUSTIFICATION FOR BREEDING TREES WITH RESISTANCE

The development of pests in agriculture and forestry

The development of even-aged crops in agriculture provided ideal conditions for many pests that have appeared, flourished and declined for reasons that, until recently, were little understood. Now, however, there is an array of expedients for their control, including crop varieties with high and stable levels of resistance to attack which are now important components of integrated field control programmes, especially where pesticide use is not acceptable.

The intensive practices adopted by foresters during the last 80 years have been accompanied by similar threats form pests of all kinds. Before this, noxious insects and pathogens occupied a place in the balanced dynamic complex of the natural forest where they seldom emerged as a threat to the survival of their host species. While these agencies may have caused some serious economic losses, the cost of special control measures and the practical problems that these raised made them generally unacceptable. However, this situation has now changed radically and, with the widespread adoption of forest monocultures, ideal conditions have been created for the establishment and spread of pests of all kinds.

Introduction of exotic pests

While these recent changes in the emphasis of forest practice took place, rapid advances in the volume and speed of international transport greatly increased the risks of introduction of new pests into forests that had previously been protected by natural barriers, threatening natural stands and plantations alike, despite phytosanitary precautions. The reality of this threat is exemplified by the damage caused by white pine blister rust (*Cronartium ribicola* J.C. Fischer), chestnut blight (*Endothia parasitica* (Murrell) Anderson & Anderson) among other introduced pathogens, and the winter moth (*Operophtera brumata* L.) or the gypsy moth (*Lymantria dispar* L.) in the insects. This threat continues to expand as the area of new forest plantations based on exotics extends.

As we shall see below, this factor, in particular, not only establishes a high level of need for forest protection against pests, it also creates an important area of constraint on its implementation.

The need for application of agricultural techniques in forestry and their limitation

It is because forest plantations and natural stands now represent a major part of the resources of many countries, as well as their increased vulnerability, that there is an urgent need to use agricultural technology to protect as well as improve them. However, the degree to which this is possible depends on available funds and the nature of the crop. Both these factors reduce the scope for supplementary protection measures in the natural forest and pose smaller, but still significant limits on their use in plantations. The accounting methods necessary for crops with terminal yields and the difficulties posed by the extent of forest areas, combined with the size and longevity of trees, reduce the extent to which agricultural protection methods can be applied under forest conditions.

There are, therefore, considerable attractions in the use of heritable resistance to protect forests from damage by biotic agencies.

THE DEVELOPMENT OF HERITABLE RESISTANCE

The relation to resistance of other traits

While the breeding of pest resistant trees calls for a high level of interdisciplinary co-operation (Schreiner, 1966; Gerhold, 1970), the practical difficulties in providing a balanced combination of pest resistance and other desirable features may be insuperable. Priorities are allocated to the objectives of a crop improvement programme, based on cost-benefit analysis and selection index weighting, where protection may rank low initially and where genetic weights are unknown.

Further, pest resistance may not be correlated with other desirable crop features; on the simplest level, while host vigour may increase resistance to facultative pests, this may not apply to more specialised organisms. Pests are not a stable feature of any crop, as their impact depends on the environment as well as the genetic nature of the host; they may also take time to emerge and often show a much higher genetic versatility than their hosts. For these reasons, the search for pest resistance tends to become secondary to that for other aspects of crop improvement.

Constraints on developing resistance

Any development of pest resistance in plants on more than an empirical basis requires long term knowledge of the biology and life-cycle of the organisms concerned.

While some progress has been made in the prognosis of forest pests for a few insects, such as the gypsy moth, *Lymantria dispar* (Houston & Valentine, 1977) and pine looper moth, *Bupalus piniarius* (L.) (Bevan, 1974), our information on forest pests in general is quite inadequate to anticipate variations in their impact, once they are established, or to support their prognosis for new regions.

This is largely because research into forest pests has been very limited until recently so that, even now, there is even doubt on the proper indentity of some of our commonest forest pathogens. For example, Minter et al. (1978) have recently shown that many of the cases of pine needle disease previously attributed to *Lophodermium pinastri* (Schrad. ex Hook.) Chev. are due to a new species, *L. seditiosum* Minter, Staley & Millar, and that other records refer to saprophytic species e.g. *L. australe* Dearness (see also Martinsson, 1979). There are many other examples where the taxonomic position, as well as the biology, of important forest pathogens is still poorly understood. While there is not the same lack of basic information on forest insects in the western world, this is certainly the case elsewhere, and particularly in the case of insect parasitoids and the habitat interactions of specialised taxa. There is therefore considerable need to improve our knowledge of forest insects as well as pathogens. A useful review of progress that has been achieved in this field has been given by Hanover (1980).

Our difficulties in the prediction and identification of future forest pest problems are compounded by the fact that their spread to new regions is far from stabilised. While this applies as much to natural forests as to plantations, species such as pines and eucalypts, which are now planted in regions often far from their native areas, may be expected to continue to be at risk for many years to come.

The spread of dothistroma blight (*Scirrhia pini* Funk & Parker; imperfect state, *Dothistroma septosporum* (Doroguine) Morelet) from the northern hemisphere, where in many places it was economically insignificant, to the south, where it has proved extremely damaging to certain pine species, is an example of this threat (Gibson, 1972). Similar problems can be found where forest insects have been introduced to new areas and assumed quite different behaviour patterns from those in their native region. Bacteria, nematodes, viruses, parasitic higher plants and mites may all be expected to pose unpredictable threats to forest crops in due course and in these cases we shall prove to be even less well prepared to deal with them.

In addition to these limitations, the differences between the types of pathogens, and to a less extent, insect pests, that are predominantly important to annual and perennial crops respectively impose still further restrictions on the use of agricultural technology by the forester seeking to protect his crops through heritable resistance.

While it is unrealistic to draw too sharp a distinction between obligate and facultative parasitism, it is nevertheless true that annual plants, largely composed of living tissues, include a very high proportion of obligate, host-specialised pathogens in their pests. Trees, on the other hand, are perennials consisting of a large and increasing proportion of dead tissues (heartwood, dead roots, bark, etc.) after they have passed the juvenile stages. This provides a substrate for the saprophytic build-up of inoculum of unspecialised facultative pathogens in close proximity to living tissues which can become sufficient to permit infection in the course of time in a way impossible in annuals. In addition to this, forest crops are often grown on marginal sites which may serve to reduce their vigour and render them even more susceptible to attack by facultative organisms. Thus, while trees may be subject to a similar range of specialised pathogens as are annuals, a much higher proportion of their diseases are likely to be caused by less specialised organisms.

The situation for insect pests is rather different. These show a similar range of specialisation and host dependence but have little in common with the physiologically close relationships that often occur between fungi, bacteria and viruses (broadly speaking) and their hosts. Nevertheless, stress in the host caused by defoliation, drought or other environmental factors can predispose to attack by such insects as bark beetles (*Scolytidae*), longhorn beetles (*Cerambycidae*), ambrosia beetles (*Platypodidae)* and woodwasps (*Siricidae*), at a level that has no parallel in annual agricultural crops.

Finally, there is a large an economically important group of forest pest which have little in common with agricultural problems and where heritable reactions of host tissues leading directly to resistance are particularly hard to find. This includes the insects, decay fungi and other organisms that destroy the heartwood of trees, often with little external symptoms on the host and with little selection pressure to reduce the potential of the host to survive under natural conditions. As we shall see, the prospect of adapting agricultural technology to provide control through resistance to this class of disease is small indeed.

THE PLACE OF RESISTANCE BREEDING

Types of resistance

VERTICAL RESISTANCE

The outstanding successes in agriculture during the present century that have gained remarkably high levels of pest resistance have largely involved annual species and highly specialised pathogens, where a combination of short generation time in the host and a physiologically close host-parasite relationship (permitting direct genetic control of the diseased state) very largely contributed to these achievements.

This has generally involved a type of resistance now termed 'vertical' (Vanderplank, 1963), which is based on one or a few major genes and is distinguished from horizontal resistance (which we shall discuss below) derived from the interactions of a minor gene complex.

In the majority of cases, vertical resistance, while providing a high level of protection, is effective only against some races of the pest. In the case of fungal pathogens, in particular, which often show a high level of genetic versatility (Buxton, 1961), this has often led to the 'breakdown' of protection with the emergence of races of the pest able to attack the new crop variety. This weakness is aggravated further by the relative simplicity of the genetics of the host-parasite interactions where vertical resistance is effective, many of which conform to the gene-for-gene concept propounded by Flor (1953).

Where vertical resistance has been lost in important annual crops the generation time has often been short enough to counteract this by a flow of new resistant varieties which, while it has led to a cyclic situation, has often proved acceptable. In perennials, however this is nog possible, even where the generation time can be reduced to a minimum. Instead, other expedients, such as the use of pesticides, have proved more cost-effective. Examples can be found in blister blight of tea (*Exobasidium vexans* Massee) and coffee rust (*Hemileia vastatrix* Berk. & Br.).

Although 'breakdown' is not an inevitable result of the use of vertical resistance for crop protection, as Russell (1978) has pointed out, the dangers are enough to justify its avoidance for forest crops, particularly as the economic hazard increases with the interval between the introduction of the new variety and its eventual failure with the appearance of a new race of pathogen. The extent and uniformity of industrial forest crops, and their lenghty rotations, make them particularly vulnerable to late failure of vertical resistance, although little evidence of this has been seen on a field scale.

Fungal forest pathogens as diverse as *Gremmeniella abietina* (Lagerberg) Morelet, *Ceratocystis ulmi* (Buisman) Moreau and *Melampsora larici-populina* Kleb. (Gibbs, 1978; Skilling, 1977; van Vloten, 1949) have all formed races of varying pathogenity, so there seems to be no lack of adaptability in this respect. It is possible, therefore, that the stability of resistance that has been found to forest pests may be through the complexity of the host-parasite interaction. This may be further stabilised by the relatively low populations of existing tree crops, compared to those of cereals and other annuals where selection pressures for new pathogen strains is very high and the emergence of new races is almost commonplace. The physiological complexity of resistance to certain rusts, such as *Cronartium ribicola*, may also contribute significantly to the stability of many tree systems of defence against highly specialised pests (Bingham et al., 1971).

Despite the present situation, there is no room for complacency and, with the increasing extent of forest monocrops, there is no doubt that the stability of vertical resistance to forest pests will be bound to decline. Safeguards against this may be adopted through the use of mixtures of individuals with different inherent resistance factors, either in separate plantations or, better, as mixtures in single stands. Mixed cropping of this kind usually carries with it some sacrifice of crop quality but this can now be avoided by the use of multiline cultivars where a uniform genetic base is combined with a range of alternative resistance genes to a given disease. At present this expedient is expensive but it has potential value for forest protection (Browning & Frey, 1969).

Parallels with insect pests are hard to find although highly specialised host relations and high levels of resistance can be found. Marked variation occurs in the colonisation of individual plants in uniform fields of brussels sprouts by *Brevicoryne brassicae* L. but it is only effective in relation to certain aphid ecotypes (Dunn & Kempton, 1972) giving an analogous situation to major gene resistance to rusts. Similar relationships are known in forest insect pests (i.e. *Cryptococcus fagi* Baerensprung and *Nuculaspis californica* (Coleman)). While these have the appearance of vertical resistance, this has only been shown in a few cases, notably the Hessian fly (*Mayetiola destructor* (Say)) (Hatchett & Gallun, 1970).

HORIZONTAL RESISTANCE

While the distinction between horizontal and vertical resistance was at one time regarded as quite clear-cut, there is a growing body of opinion now that regards the two as extremes of a graded series of genetic states ranging from simple relationships governed by a single major gene to complex states derived from a number of genes, many of which may be minor. Indeed, some doubt has been cast on the validity of the distinction between major and minor genes in this respect.

Whatever the current opinion, the view that horizontal resistance arises from genetic complexes that form part of the genetic base of the host is still tenable (Vanderplank, 1963). The protection afforded by horizontal resistance is never at the same high level as that from vertical resistance but it has the advantage of stability. In general the protection that this kind of resistance offers covers a wider range of pests, including those weaker facultative parasites that have importance in forestry.

The search for horizontal resistance is usually by selection of individuals for desirable characters from large field populations. These are then screened by progeny or clonal tests to ensure that their special features are truly inherent before field propagation. Vegetative propagation is desirable to avoid the risk of dispersal of desirable combinations of genetic fasctor, which might occur through sexual reproduction. Field selection on these lines has been described for resistance to *Heterobasidion annosum* (Fr.) Bref. (Driver & Ginns, 1966).

All this is fully compatible with forest plantation practice which is often indirectly selective for increased crop quality and pest resistance. The balance existing between pests and their hosts in natural forest communities is largely based on horizontal resistance factors and it has been advocated often that lasting solutions to pest problems of plantations may be found in an understanding of the ecological balance existing in the communities where plantation species are native (Dinus, 1974).

Possibly the most important feature of the use of field selection techniques is that they permit selection towards the control of more than one pest at the same time. The majority of crops are at risk from 3 or 4 major pests in the course of the crop cycles and perennials, especially forest trees, are exposed to an even greater number. Practical crop improvement should aim, therefore, at useful protection levels against all the major pests in a given crop situation and it is in this respect that field selection towards horizontal resistance provides much better opportunities than the use of vertical resistance. It must be conceded, however, that the combination of vertical resistance factors to provide protection from an array of pests has been successful in an annual crop such as rice (Khush, 1977) and that some progress seems to have even been achieved in this direction for pines (see Goddard & Rockwood, 1980).

Tolerance

Hitherto, it has been assumed that there is an approximate relation between symptom expression and loss of value of crop from pest attacks (Browning et al., 1977). This is not necessarily so and examples are known where symptom expression may be severe but pest impact is relatively slight, notably in cereal rusts (Caldwell et al., 1958).

FUTURE NEEDS FOR RESISTANCE BREEDING IN FORESTRY

The role of the environment

While agricultural science has much to offer for the improvement of forest crops, much of this calls for development within the context of the special circumstances of forestry.

Forest plantations are characteristically confined to markedly sub-optimal sites and the search for horizontal pest resistance by field selection automatically takes this into account. However, vertical resistance also needs to be sought at times and here allowance must be made for the effects of adverse environmental influences in the screening to select these varieties. Schoeneweiss (1975, 1978) has pointed out that to conduct the screening of forest varieties for pest resistance under optimal growing conditions, and thus follow some agricultural practices, could give dangerously misleading results where the crop may ultimately be raised on sites far from ideal.

It must be accepted in principle, therefore, that due to the circumstances of forestry, much more attention needs to be paid to the interaction between environmental factors and host-pest reaction patterns than is usual in agriculture. Field screening, particularly for vertical resistance, in species, provenances and intra-specific selections, should be made under the future growing conditions for the crop.

Maintenance of genetic resources

The relatively recent development of monocrops and the later initiation of selective breeding in forestry has meant that there have been no major declines at the global level in the genetic resources of the most important forest species. Nevertheless, the warning on the dangers of genetic over-refinement is clear from agriculture and foresters are beginning to heed it.

Excessive exploitation in natural forests, particularly in tropical rain forests, has threatened loss of genetic resources and even loss of species before their utility has become fully known (Kemp, 1978; Kemp & Burley, 1978; Burley & Namkoong, 1980) and for some commercial plantation species important natural populations are at risk even if the species as a whole is not.

For this reason, international programmes have been established by institutes in developed countries, often whith support from multilateral and bilateral agencies, to explore, conserve, evaluate and use to best advantage forest resources of all kinds. However, to date, it has not proved possible to initiate international breeding programmes for pines (Kemp et al., 1972; FAO, 1979) although activities within IUFRO facilitate exchange and testing of natural and improved populations in a range of species.

Accelerated breeding and resistance testing

There is outstanding need to reduce the disadvantages posed by the size and longevity of trees in all aspects of their genetic improvement, and particularly in resistance breeding. In particular, techniques are needed to speed the testing of selections for combinations of desirable qualities and, once found, for their rapid reproduction, without risk of their dispersal.

The development of seed orchards for seed production (particularly if artificial pollination is used) has gone far to accelerate seed production with reliable heritable characters and is a suitable method for handling vertical resistance factors. However, vegetative propagation is much more suitable for horizontal resistance and techniques for this, providing quick utilisation of desirable genetic material, are already well established for many forest species.

Much also needs to be done to reduce the difficulties posed by the search for resistance in forest species and the testing of the results. Many of these drawbacks relate to changes in host reaction with its age and the need to achieve presentation of the pest to the host in tests which correspond reasonably with field conditions. Some advances have already been made to solve these problems, using techniques developed for agricultural crops using either excised tissues or indicators (often empirical) based on the chemistry or morphology of the host. Future problems will call for special research to meet the requirements of forestry. For example, diseases caused by nematodes, such as *Bursaphelenchus lignicolus* Mamiya & Kiyohara (Mamiya, 1972), are only now being recognised outside the nursery and damage from viruses and virus-like organisms to forest trees has scarcely been recognised outside Europe and North America.

Indeed, not only as commercial forest plantations expend, but as their value per unit area increases, so they will be the more closely scrutinised for pest damage and a wider variety of these problems will be revealed at an economic level making them worthy of control. In many cases the most advantageous approach to this will be through the use of inherent resistance mechanisms. Techniques to exploit this may already exist in agricultural science in some respects, but further research is always required to adapt these for use in forestry.

REFERENCES

Bevan, D., 1974. Control of forest insects: there is a porpoise close behind us. In: D. Price-Jones & M.E. Solomon (Eds): Biology in pest and disease control. Proc. 13th Symp. British Ecol. Soc. p. 302-312.

Bingham, R.T., R.J. Hoff & G.I. McDonald, 1971. Disease resistance in forest trees. Ann. Rev. Phytopath. 9: 433-452.
Browning, J.A. & K.J. Frey, 1969. Multiline cultivars as a means of disease control. Ann. Rev. Phytopath. 7: 355-382.
Browning, J.A., M.D. Simons & E. Torres, 1977. Managing host genes; epidemiologic and genetic concepts. In: J.G. Horsfall & E.B. Cowling (Eds): Plant disease. An advanced treatise. Vol. I. How disease is managed. Academic Press, London: p. 191-212.
Burley, J. & G. Namkoong, 1980. Conservation of forest genetic resources. Invited Paper 11th Commonwealth Forestry Conference, Trinidad. CFI, Oxford. 25 p.
Burley, J. & P.J. Wood, 1976. A manual on species and provenance research with particular reference to the tropics. Trop. For. Pap. No. 10, CFI, Oxford. 226 p.
Buxton, E.W., 1961. Mechanisms of variation in the pathogenicity of Fusarium oxysporum. Recent Advances in Botany 1, University of Toronto Press. p. 502-507.
Caldwell, P.S., J.F. Schafer, L.E. Compton & F.L. Patterson, 1958. Tolerance to cereal leaf rusts. Science 128: 714-715.
Dinus, R.J., 1974. Knowledge about ecosystems as a guide to disease control in managed forest. Proc. Amer. Phytopath. Soc. 1: 189-190.
Driver, C.H. & J.H. Ginns, 1966. A method of mass screening of southern pines for resistance to a root-rot induced by Fomes annosus (Fr.) Cooke. In: H.D. Gerhold, E.J. Schreiner, R.E. McDermott & J.A. Winieski (Eds): Breeding pest resistant trees. Pergamon Press, Oxford. p. 421-422.
Dunn, Z.A. & D.P.H. Kempton, 1972. Resistance to attack by Brevicoryne brassicae among plants of brussels sprouts. Ann. Appl. Biol. 72: 1-11.
FAO, 1979. Genetic improvement of tropical lowland conifers. FO: MISC/79/15, Rome. 86 p.
Flor, H.H., 1953. Epidemiology of flax rust in the North Central States. Phytopathology 43: 624-628.
Gerhold, H.D., 1970. A decade of progress in breeding disease-resistant forest trees. Unasylva 24(2-3): 37-44.
Gibbs, J.N., 1978. Intercontinental epidemiology of Dutch elm disease. Ann. Rev. Phytopath. 16: 287-307.
Gibson, I.A.S., 1972. Dothistroma blight of Pinus radiata. Ann. Rev. Phytopath. 10: 51-72.
Goddard, R.E. & D.L. Rockwood, 1980. Co-operative forest genetics research program. Twenty-second Progress Report. Univ. Fla. Schl. For. Res. Conservn., Res. Rept. No. 30. 15 p.
Hanover, J.W., 1980. Breeding forest trees resistant to insects. In: F.G. Maxwell & P.R. Jennings (Eds): Breeding plants resistant to insects. Wiley, New York. p. 487-511.
Hatchett, J.H. & R.L. Gallun, 1970. Genetics of the ability of the Hessian fly Mayetiola destructor to survive on wheats having different genes for resistance. Ann. Entomol. Soc. Amer. 63: 1406-1407.
Houston, D.R. & H.T. Valentine, 1977. Comparing and predicting forest stand susceptibility to gypsy moth. Can. J. For. Res. 7: 447-461.
Kemp, R.H., 1978. Exploration, utilization and conservation of forest genetic resources. Unasylva 30: 10-16.
Kemp, R.H. & J. Burley, 1978. Depletion and conservation of forest genetic resources. In: J.C. Hawkes (Ed.): Conservation and agriculture. Duckworth, London. p. 171-185.
Kemp, R.H., J. Burley, H. Keiding & D.G. Nikles, 1972. International co-operation in the exploration, conservation and development of tropical and sub-tropical forest gene resources. Invited paper 7th World For. Congr. 7 CFM/C:V/49, Buenos Aires, Argentina. 16 p.
Khush, G.S., 1977. Disease and insect resistance in rice. Adv. Agron. 29: 265-341.
Mamiya, Y., 1972. Pine wood nematode Bursaphelenchus lignicolus Mamiya and Kiyohara, as a causal agent of pine wilting disease. Rev. Plant Protect. Res. 5: 46-50.
Martinsson, O., 1979. Testing Scots pine for resistance to Lophodermium needle cast. Stud. For. Suec. Nr. 150. 63 p.
Minter, D.W., J.M. Staley & C.S. Millar, 1978. Four species of Lophodermium on Pinus sylvestris. Trans. British Mycol. Soc. 71: 295-301.
Namkoong, G., R.D. Barnes & J. Burley, 1980. A philosophy of breeding strategy. Trop. For. Paper No. 16. Commonw. For. Inst., Oxford. 67 p.
Russel, G.E., 1978. Plant breeding for pest and disease resistance. Butterworth, London. 485 p.
Schoeneweiss, D.F., 1975. Predisposition, stress and plant disease. Ann. Rev. Phytopathology 13: 193-211.
Schoeneweiss, D.F., 1978. Water stress as a predisposing factor in plant disease. In: T.T. Kozlowski (Ed.): Water deficits and plant growth. Vol. 5. Academic Press, London. p. 61-99.

Schreiner, E., 1939. The possibilities of the clone in forestry. J. For. 37: 61-62.
Schreiner, E., 1966. Future needs for maximum progress in genetic improvement of disease resistance in forest trees. In: H.D. Gerhold, E.J. Schreiner, R.E. McDermott & J.A. Winieski (Eds): Breeding pest-resistant trees. Pergamon Press, Oxford. p. 455-466.
Skilling, D.D., 1977. The development of a more virulent strain of Scleroderris lagerbergii in New York State. European J. For. Path. 7: 297-302.
Vanderplank, J.E., 1963. Plant diseases; epidemics and their control. Academic Press, London. 349 p.
Vloten, H. van, 1949. Kruisingsproeven met rassen van Melampsora larici-populina Klebahn. Tijdschr. PlZiekt. 55: 196-209.

The role of the environment in host-insect interactions

W.H. Parry

Department of Forestry, University of Aberdeen, St. Machar Drive, Old Aberdeen, AB9 2UU, UK

ABSTRACT

Plant host/insect interactions can be extensively modified by changes in the environment. Climate is one major factor involved in such changes, temperature and rainfall being of particular potential importance in the breakdown of host resistance mechanisms. Parasites and predators play an important role in maintaining low pest levels. The survival of pest levels conducive to the continued existence of this complex is considered to be of primary importance during the adoption of resistance strategies. Host nutritional levels are important but their instability in response to environmental changes minimises their effectiveness as resistance factors. Induced density-dependent resistance as demonstrated in birch in Finland provides a useful mechanism meriting further investigation in other tree species. It is concluded that the avoidance of host/insect synchrony by the adoption of a complex of resistance mechanisms aimed also at the maintenance of a non-economic pest level would ensure long-term stability in monocultural forestry as it is now practiced.

INTRODUCTION

Plant host/insect interactions have been reviewed by several authors in recent years: e.g. McNeill & Southwood (1978), Feeny (1976) and Levin (1976). The role of the environment in the context of resistance to forest insects has been neglected, although both Hanover (1975) and Stark (1965) have reviewed the physiology of tree resistance to insects.

The environment of an individual insect can be regarded as a combination of all factors which either directly or indirectly influence the life of that individual. Milne (1962) defines the effective environment as 'everything in the Universe which helps or hinders fulfilment of the individual'. Milne placed the main factors acting on an insect population into

Table 1. Classification of the environment according to Milne (1962).

A. Density-independent factors	1.	Physical circumstances, i.e. weather
	2.	Indiscriminate actions of other species
B. Imperfect density-dependent	1.	Interspecific competition
	2.	Predators, parasites, pathogens
C. Perfect density-dependent	1.	Intraspecific competition

3 groups which were dependent on their relationship to density (Table 1).

Long-term co-evolutionary synchronization of host and insect ideally leads to the occurrence of an apparently stable plant/insect relationship in which neither of the two associates becomes dominant. This stability could be destroyed by the appearance of any of a variety of disruptive factors. One such factor lies in the large-scale planting of artificial monocultures selected for factors other than resistance.

The use of resistant trees has frequently been suggested as being the ultimate means for pest control in both agriculture and forestry. The use of genetically resistant hosts is frequently doomed to failure as genetic resistance will be expressed in different ways under differing environmental conditions. The aim of this presentation is to illustrate the possible effects of environmental conditions on the expression of plant resistance such that genetic resistance can be tailored to meet environmental conditions or, in specific instances, the environment could be modified to meet the requirements of the resistant host.

CLIMATE

Elton (1958) was the first of many observers to associate the absence of pest outbreaks in tropical forests with climatic stability. Gray (1972) concluded that pest problems in tropical situations were triggered by climatic instability. However, it is in the climatically more unstable temperate regions that many of the pest problems associated with afforestation are now encountered.

The 2 main climatic factors associated with insect outbreaks are rainfall and temperature. Rainfall affects the herbivorous insect indirectly through its host and rainfall conditions deviating strongly from normality will have the greatest effect on resistance. Such deviations are particularly important in tree crops subjected to repeated exposure to variable seasonal and annual rainfall.

White (1969) related 2 outbreaks of the psyllid *Cardiaspina densitexta*

Taylor feeding on the leaves of *Eucalyptus fasciculosa* F.v.M. to stress caused by changes in the rainfall pattern. Both outbreak periods were characterized by unusually wet winters and unusually dry summers. The former caused the death by suffocation of feeding plant roots. This reduced ability to take up water was exacerbated by the low availability of water during the succeeding dry summers.

In order to obtain a measure of the stress imposed on the trees under such conditions White calculated a 'stress index' (White, 1969). In the periods of psyllid outbreak a highly positive series of stress indexes associated with dry summers and wet winters were recorded. Outbreaks of the looper caterpillar, *Selidosema suavis* Butler feeding on shallow-rooted *Pinus radiata* D. Don in New Zealand, *Bupalus piniarius* L. in the Netherlands (Klomp, 1968) and *Choristoneura fumiferana* (Clem.) populations in Canada (Morris, 1963) were also characterized by positive stress indexes (White, 1974). White (1974) considered that increases in total and soluble nitrogen and amino acid levels in the foliage of trees under stress was the main factor changing suitability for herbivorous insects.

Of particular interest in Britain is the recent outbreak of the pine beauty moth, *Panolis flammea* D. & S., on lodgepole pine, *Pinus contorta* Dougl., in various areas of Scotland. In Naver forest, Sutherland the outbreak occurred in 1976 and 1977 (Stoakley, 1979). Calculation of annual stress indexes using rainfall data from the nearby Davina Lodge meteorological station indicated that highly positive values occurred in both 1975 and 1976 (Fig. 1). Both these years were characterised by wet winters, the summer of 1976 being particularly dry (Fig. 1). Stoakley (1977) indicated that the outbreak was associated with pine grown on deep unflushed peat rather than on freely drained ground. Substantial root mortality by suffocation would have occurred on these waterlogged, peaty sites during the winter preceding the outbreak.

The conifer oleoresin system is a potent resistance mechanism against many insect species (Hanover, 1975; Stark, 1965). In its hindering role to the entrance of various bark beetles the level of this resistance is closely related to the water status of the trees. E.g., the resistance of *Pinus ponderosa* Laws. to bark beetle infestation is lower following periods of deficient rainfall, injury or other changes in the environment (Vité, 1961). Water stress causes susceptibility by reducing the oleoresin exudation pressure (oep) which is dependent on the turgidity of the tree, specifically the osmotic pressure of the epithelial cells lining the resin ducts (Münch (1921) in Stark, 1965).

Vité (1961) found that oep fluctuated diurnally and was affected during the summer by temperature, light intensity and humidity. However, seasonal fluctuation was primarily influenced by soil water and declined as the season progressed. The variability of the pressure and its dependence on a variety of environmental factors makes it unlikely that oep could be uti-

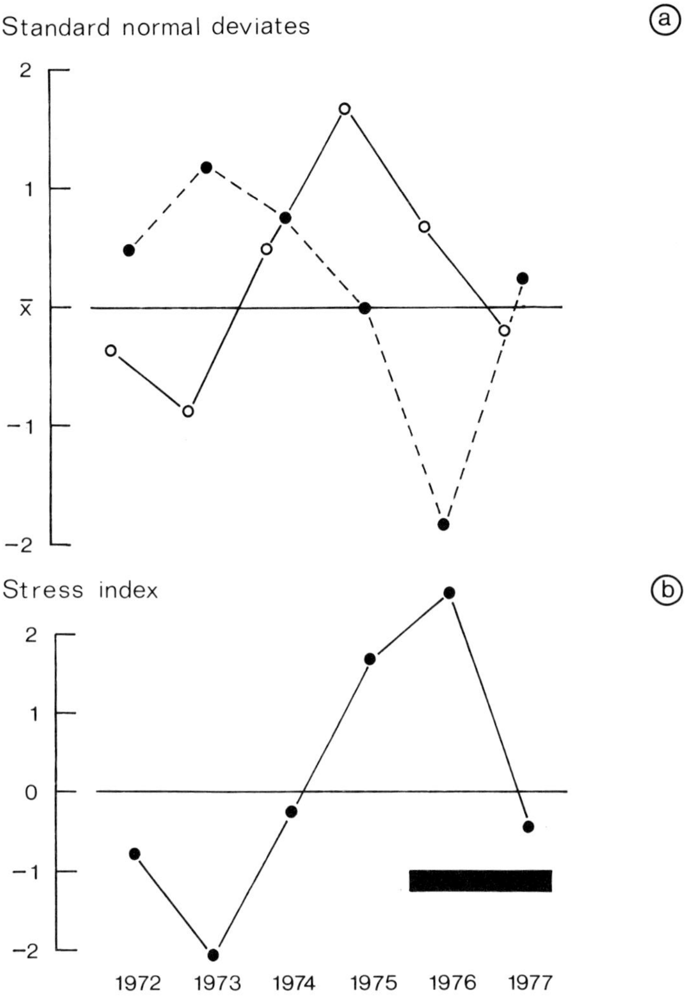

Figure 1. Stress indices associated with outbreaks of *Panolis flammea* in Sutherland. a: Standard normal deviates, summer (●) and winter (o) rainfall. b: Stress index (●).

lised as a predictable genetic resistance mechanism.

Both low and high temperatures can affect herbivorous insect populations by killing insects, slowing their rate of metabolism, modifying reproductive behavior or, changing the susceptibility of the host to the insect pest. The effect of low temperature was considered by Parry (this book, page 32).

Temperature has a marked effect on the duration of larval stages. Watt (1968) correlated warm temperatures with outbreaks of the gypsy moth, *Porthetria dispar* in New York City, the greatest increase in numbers occurring during unusually warm summers. Under such conditions increased rates of fecundity and development minimised the time of exposure to predators and parasites. Such conditions, in combination with the dryness usually associated with periods of high temperature, would put the trees under stress and increase the probability of resistance breakdown.

PARASITES AND PREDATORS

It is clear that the numbers and oscillation range of an insect species in a given habitat is relevant to the regulatory ability of its parasites and predators. The planting of trees with genotypes giving an increased resistance to a pest insect will decrease the level of that pest and, in turn, affect its parasite/predator complex.

Führer (1975, 1978) considered that the survival of a spectrum of insectivorous insects was related to a basal level of prey, the 'Eiserner Bestand', sufficiently high to maintain the complex without causing unacceptable damage. A reduction of the basal level could cause a local extinction of those specialised predatory or parasitic species operating at low prey density levels. At higher prey density levels, caused either by environmental or seasonal changes in the resistance of the host or by the absence of the regulatory effect of the low density complex, unspecialised, polyphagous insects are attracted. The sheer number of such forms may further reduce the level of the low density complex, if it still exists, through competition and thus contribute further to the eventual breakdown of resistance.

Following resistant tree stand establishment a heavier selection pressure will be exerted on the pest population. We have, on the one hand, an enormously plastic insect gene pool, on the other, a static tree gene pool. The eventual result of pest selection will be the emergence of aggressive races adapted to exist under the new conditions e.g. individual tree demes of the pine-leaf scale, *Nuculaspis californica* (Coleman) on ponderosa pine (Edmonds & Alstad, 1978). Under such conditions the existence of specialised low density predators and parasites may be crucial.

NUTRITION AS A RESISTANCE FACTOR

The amount and availability of food is frequently the main factor involved in intraspecific competition. The plant can restrict food availability to insect herbivores in a number of ways e.g. reducing protein digestibility with tannins. Competition for food is normally density-dependent and according to Milne (1957) such a relationship is the only truly density-dependent one. Despite the hypothesis of Schwenke (1962, 1963) that the number of leaf and needle eating insects was dependent on the sugar content of the foliage it is now accepted that dietary nitrogen, in the form of an adequate and balanced intake of amino acids and proteins, is a critical factor in diet suitability for herbivorous insects (McNeill & Southwood, 1978). The use of resistant plant strains containing naturally low or unbalanced proportions of amino acids would appear to represent an ideal protective measure. Many factors can contribute to the breakdown of such a resistance mechanism e.g. climate, site variation, fertiliser application or intraspecific and interspecific competition.

Nitrogen is occasionally applied in forest management e.g. application of urea to boost plant growth under certain conditions. Huber & Watson (1974) indicated that the form of nitrogen taken up by plants influenced disease incidence and could also be critical with regard to insect resistance. The effect of the form of N on disease severity was dependent on the specific host-parasite combination, rate and timing of application, level of residual soil N, previous cropping and the physical environment. The level of attack by the spruce aphid, *Elatobium abietinum* (Walker) in a young crop of Sitka spruce (*Picea sitchensis* (Bong.) Carr.) emerging from heather (*Calluna vulgaris* L.) check was considerably lower in checked plants than in others. Spruce aphid populations are sensitive to plant nitrogen levels (Parry, 1974, 1976, 1977) and analysis of these plants indicated that N levels were much higher in the free growing plants (c. 1.2 % N of dry weight) in comparison with plants in check (c. 0.8 % N). Consequently, changes in N levels in forest crops may result in increased tree susceptibility.

Little is known concerning the effects of insect feeding on the nutritional acceptability of host trees although bark beetle attack is frequently initiated by pioneer individuals which discriminate between resistant and susceptible trees. This leads to the eventual colonization of suitable trees by large numbers of beetles (Coulson, 1979). The modification of feeding site nutritional quality is known to occur in e.g. *Brevicoryne brassicae* L. where nutrient 'sinks' are formed by alteration of the normal flow of nutrients in the host (Way, 1968; Way & Cammell, 1970). Uninfested rice-leaf amino acid levels averaged 13.4 $mg.g^{-1}$ while infested leaves averaged 79.3 $mg.g^{-1}$ (Cagampang et al., 1974). Similar reactions may occur in trees, particularly in relation to gall formation where nutritional conditions are changed by the feeding activities of the insect e.g. *Adelgidae* on conifers, *Pemphigus* spp. on poplars.

The classification of plants into apparent and non-apparent by Feeny (1975, 1976) marked a useful step forward in the examination of the co-evolution of plants and insects. Forest trees are classed as apparent plants i.e. they are long-lived and grown in association with large numbers of their own kind. While conifers are generally protected by resins most deciduous trees utilise high leaf tannin levels. Tannins act by progressively binding protein and lowering its availability to sap feeders and interfere with the digestive processes of chewing insects (Feeny, 1969). Tannin levels in young growing tissues with high nitrogen levels are generally much inferior to those in mature tissues with lower nitrogen levels. Consequently, many insects feed and mature quickly on young foliage. By contrast, many sawfly species feed on mature foliage in late summer when the rate of development is decreased in response to the lowered digestibility of the food (Haukioja & Koponen, 1975; Haukioja & Niemalä, 1974).

The forest practice of planting large areas with single clones of trees allows an insect pest to quickly synchronize its life cycle with the phenology of the host with respect to periods of maximum growth potential. Where tree phenology varies selection for synchronization can not occur to such a marked extent. Feeny (1975) indicated that in mature stands of oak, bud-burst may range over 2 weeks. This limited the ability of the herbivores to synchronize precisely with the trees. Such small differences in phenophase of oak stands may have large effects on the population levels of winter moth on oak (Feeny, 1975; Varley et al., 1973). The moth may have responded by attempting to synchronize with individual trees by developing apterous females more inclined to re-colonise trees on which they fed as larvae. It is therefore an essential part of any resistance control programme to promote phenological variability in the host.

INDUCED AND CONSTITUTIVE RESISTANCE

Induced resistance is the result of a response of the plant to the presence of a feeding insect to the detriment of that insect. It is usually a density dependent phenomenon by contrast with constitutive resistance which is independent of pest levels.

The production of large quantities of chemicals in order to reduce the digestibility of leaves imposes a 'strain' on the physiological budget of a plant. The proportion of the total metabolic resources allocated to defence reflects the selective pressures imposed on that plant species by insect pests during its evolution. Heavy selection would result in high defensive compound levels. The most suitable sompromise would be that situation where the chemicals are only produced in response to attack. Such an induced response could be 'delayed' i.e. damage by one insect generation gives plant resistance to subsequent generations; 'immediate' i.e. the plant response decreases its susceptibility to the insect inflicting the damage, or 'indirect' where the damage by one insect species increases tree resistance to others.

Induced resistance to *Oporinia autumnata* (Bkh.) has been demonstrated in *Betula pubescens* ssp. *tortuosa* by Haukioja and his co-workers in Finland where, in addition, seasonal changes in leaf quality, including decreasing water and nitrogen content and increasing phenolic and tannin content, are associated with decreasing suitability and digestibility of birch leaves (Haukioja et al., 1978). *O. autumnata* responds to these constitutive changes by feeding on young leaves prior to development of indigestibility

(Haukioja & Niemalä, 1974).

The birch response to leaf damage retards the growth rate of *O. autumnata* larvae feeding on young foliage both in the same year and in subsequent years, i.e. delayed and immediate induced resistance (Haukioja & Niemalä, 1977). Mechanical damage to leaves also resulted in a decreased growth rate of larvae feeding on adjacent leaves (Haukioja & Niemalä, 1979), a form of indirect induced resistance. Although birches are able to increase the amount of leaf phenolics in response to mechanical damage, a process which appears not to occur in mature leaves where levels are already high (Haukioja & Niemalä, 1979), it is suggested that other processes may also be involved e.g., changes in proteinase inhibitors (Haukioja & Niemalä, 1977).

The level of constitutive resistance may be connected to climatic conditions e.g., in warm summers the leaves mature earlier (Haukioja et al., 1978) and curtail the growth period of *O. autumnata* by the early formation of high phenolic levels. Whether this was critical with regard to population levels remains unclear.

The combination of induced and constitutive resistance mechanisms decreases the probability of the occurrence of an optimally adapted herbivore. The use of birch clones with highly developed density-dependent inductive responses offers an immediate defence mechanism responsive to increased pest numbers caused by environmental variation, such as high temperature.

CONCLUSION

The traditional concept of the utilisation of plant resistance as a simple means of minimising pest damage pays little attention to the effects of the prolonged selection imposed on a series of forest insect pest generations feeding on one host generation. Selection leads to an inevitable diminution of the efficacy of that resistance mechanism. Consequently, it is suggested that resistance mechanisms aiming at the complete local elimination of forest insect pest species should be rejected under forest plantation conditions in favour of the adoption of a complex of weaker resistance mechanisms leading to the formation of a balanced plant-herbivore relationship where neither partner becomes dominant. In this context Mattson et al.'s (this book, page 000) concept of organism trait vectors could prove invaluable as it describes how organisms interact with each other in relation to the manner prescribed by the set of traits that each possesses. Variation in these traits would change the plant/insect relationship in a predictable manner.

The development of suitable resistant varieties should be coupled with environmental studies aimed at obtaining information on the effects of various factors on the expression of these genetic characteristics. Kalkstein (1976) has correlated *Dendroctonus frontalis* Zimmermann activity with various climatic variables in an attempt to develop a predictive technique capable of evaluating future insect activity based upon past climatic conditions. Similar climate-insect evaluations in conjunction with the use of resistant trees could provide practising foresters with the background information necessary to employ the most efficient silvicultural techniques in particular areas.

REFERENCES

Cagampang, G.G., M.D. Pathak & B.O. Juliano, 1974. Metabolic changes in the rice plant during infestation by the brown plant-hopper, Nilaparvata legens Stål (Hemiptera, Delphacidae). Applied Entomolgy and Zoology 9: 174-184.

Coulson, R.N., 1979. Population dynamics of bark beetles. Annual Review of Entomology 24: 417-447.

Edmonds, G.F. & D.N. Alstad, 1978. Co-evolution in insect herbivores and conifers. Science 199: 941-945.

Elton, C.S., 1958. The ecology of invasion by animals and plants. Methuen, London. 181 p.

Feeny, P.P., 1969. Inhibitory effect of oak leaf tannins on the hydrolysis of proteins by trypsin. Phytochemistry 8: 2119-2126.

Feeny, P.P., 1975. Biochemical co-evolution between plants and their insect herbivores. In: L.E. Gilbert & P.H. Raven (Eds): Co-evolution of animals and plants. University of Texas, Austin/London. p. 3-19.

Feeny, P.P., 1976. Plant apparency and chemical defence. In: J.W. Wallace & R.L. Mansell (Eds): Biochemical interaction between plants and insects. Recent Advances in Phytochemistry 10. Plenum, New York/London. p. 1-40.

Führer, E., 1975. Überlegungen zur Wirkung resistenzsteigernder Massnahmen im Wald auf den Massenwechsel forstlicher Schadinsekten. Forstarchiv 46: 228-233.

Führer, E., 1978. Forstzoologie zwischen Theorie und Praxis. II. Abschied von vertrauten Vorstellungen. Forstarchiv 49: 66-69.

Gray, B., 1972. Economic tropical forest entomology. Annual Review of Entomology 17: 315-334.

Hanover, J.W., 1975. Physiology of tree resistance to insects. Annual Review of Entomology 20: 75-95.

Haukioja, E. & S. Koponen, 1975. Birch herbivores and herbivory at Kevo. In: F.E. Wielgolaski (Ed.): Fennoscandian tundra ecosystems Part 2. Springer, Berlin/Heidelberg/New York. Ecological Studies 17: 181-188.

Haukioja, E. & P. Niemalä, 1974. Growth and energy requirements of the larvae of Dineura virididorsata (Retz.) (Hym. Tenthredinidae) and Oporinia autumnata (Bkh.) (Lep., Geometridae) feeding on birch. Annales Zoologici Fennici 11: 207-211.

Haukioja, E. & P. Niemalä, 1977. Retarded growth of a geometrid larva after mechanical damage to leaves of its host tree. Annales Zoologici Fennici 14: 48-52.

Haukioja, E. & P. Niemalä, 1979. Birch leaves as a resource for herbivores: seasonal occurrence of increased resistance in foliage after mechanical damage of adjacent leaves. Oecologia 39: 151-159.

Haukioja, E., P. Niemalä, L. Iso-Iivari, H. Ojala & E.M. Aro, 1978. Birch leaves as a resource for herbivores. Report of the Kevo subarctic Research Station 14: 5-12.

Huber, D.M. & R.D. Watson, 1974. Nitrogen form and plant disease. Annual Review of Phytopathology 12: 139-165.

Kalkstein, L.S., 1976. Effects of climatic stress upon outbreaks of the southern pine beetle. Environmental Entomology 5: 653-658.

Klomp, H., 1968. A seventeen year study of the abundance of the Pine Looper, Bupalus piniarius L. (Lepidoptera: Geometridae). In: T.R.E. Southwood (Ed.): Insect abundance. Royal Entomological Society of London Symposium 4: 98-108.

Levin, D.A., 1976. The chemical defenses of plants to pathogens and herbivores. Annual Review of Ecology and Systematics 7: 121-159.

McNeill, S. & T.R.E. Southwood, 1978. The role of nitrogen in the development of insect/plant relationships. In: J.B. Harbourne (Ed.): Biochemical aspects of plant and animal coevolution. Academic Press, New York/London. p. 77-98.

Milne, A., 1957. The natural control of insect populations. Canadian Entomologist 89: 193-213.

Milne, A., 1962. On a theory of natural control of insect populations. Journal of Theoretical Biology 3: 19-50.
Morris, R.F., 1963. The dynamics of epidemic spruce budworm populations. Memoirs of the Entomological Society of Canada 31: 1-322.
Parry, W.H., 1974. The effects of nitrogen levels in Sitka spruce needles on Elatobium abietinum (Walker) populations in North-eastern Scotland. Oecologia 15: 305-320.
Parry, W.H., 1976. The effect of needle age on the acceptability of Sitka spruce needles to the aphid, Elatobium abietinum (Walker). Oecologia 23: 297-313.
Parry, W.H., 1977. The effects of nutrition and density on the production of alate Elatobium abietinum on Sitka spruce. Oecologia 30: 367-375.
Schwenke, W., 1962. Neue Erkenntnisse über Entstehung und Begegnung von Massenvermehrungen an Kiefern und Fichtennadeln fressender Schadinsekten. Zeitschrift für angewandte Entomologie 50: 137-142.
Schwenke, W., 1963. Über die Beziehung zwischen dem Wasserhaushalt von Bäumen und der Vermehrung blattfressender Insekten. Zeitschrift für angewandte Entomologie 51: 371-376.
Stark, R.W., 1965. Recent trends in forest entomology. Annual Review of Entomology 10: 303-324.
Stoakley, J.T., 1977. A severe outbreak of the pine beauty moth on lodgepole pine in Sutherland. Scottish Forestry 31: 113-125.
Stoakley, J.T., 1979. Pine beauty moth. Forestry Commission, London, Forest Record 120. 11 p.
Varley, G.C., G.R. Gradwell & M.P. Hassell, 1973. Inspect population ecology; an analytical approach. Blackwell Scientific, Oxford. 212 p.
Vité, J.P., 1961. The influence of water supply on oleoresin exudation pressure and resistance to bark beetle attack in Pinus ponderosa. Contributions from Boyce Thompson Institute 21: 37-66.
Watt, K.E.F., 1968. Ecology and resource management. McGraw Hill, New York. p. 62-65.
Way, M.J., 1968. Intraspecific mechanisms with special reference to aphid populations. In: T.R.E. Southwood (Ed.): Insect abundance. Royal Entomological Society of London Symposium 4: 18-36.
Way, M.J. & M.E. Cammell, 1970. Aggregation behaviour in relation to food utilization by aphids. In: A. Watson (Ed.): Animal populations in relation to their food resources. Symposium of the British Ecological Society 10: 229-247.
White, T.C.R., 1969. An index to measure weather-induced stress of trees associated with outbreaks of psyllids in Australia. Ecology 50: 905-909.
White, T.C.R., 1974. A hypothesis to explain outbreaks of looper caterpillars, with special reference to populations of Selidosema suavis in a plantation of Pinus radiata in New Zealand. Oecologia 16: 279-301.

Variability in conifer needle freezing as a resistance factor to aphids

W.H. Parry

Department of Forestry, University of Aberdeen, St. Machar Drive, Old Aberdeen, AB9 2UU, UK

ABSTRACT

A review of the literature on overwintering mortality of the green spruce aphid indicated that a close relationship existed between spruce needle freezing temperatures and aphid freezing temperatures causing mortality. Resistant trees were found to be those where needles froze at higher temperatures than the remainder, such trees exhibiting higher aphid mortality in many winters in N.E. Scotland. As defoliation and aphid population levels are correlated with overwintering aphid survival this mechanism is very important in preventing spruce aphid outbreaks in all except mild winters. The possibilities of incorporating this mechanism into the genetic pool of forest trees is discussed as well as its possible existence in the relationship between other plant-sucking insects and their host.

INTRODUCTION

The existence of Sitka spruce, *Picea sitchensis* (Bong.) Carr. resistant to the green spruce aphid, *Elatobium abietinum* (Walker), has been observed by several authors including Matthews (1950), Behrndt (1961) and Parry & Powell (1977). These trees exhibited the classical resistance characteristic of foliage retention in periods when adjacent trees, frequently with intermingling foliage, exhibited a heavy defoliation. Such trees have been encountered frequently in N.E. Scotland. Long-term observations showed that this apparent resistance was occasionally lost in years following mild winters only to reappear in subsequent years.

Studies on this phenomenon indicated that growth rates of aphid populations in spring were similar on all trees and no differences in the rates of either symptom development or defoliation were evident (Parry & Powell, 1977). The main difference between trees retaining needles (= resistant) and trees showing regular defoliation (= susceptible) lay in the higher number of aphids on the latter in early spring, this being particularly marked on first year needles exposed to greater temperature extremes. The number of aphids present at peak population in late June/early July, and consequently the amount of defoliation, was correlated with the number of aphids surviving winter (Powell & Parry, 1976). Therefore, overwintering survival was the critical factor in the damaging potential of the aphid.

OVERWINTERING SURVIVAL OF E. ABIETINUM

In oceanic regions such as Britain the aphid primarily overwinters as apterous, viviparous, parthenogenetic females. In the continental climate of Central Europe overwintering eggs predominate (Scheller, 1963; Kloft et al., 1961).

Freezing mortality of active, overwintering *E. abietinum* has been frequently observed (Bevan, 1966 and Fig. 1). E.g., in N.E. Scotland a temperature of -5 °C was recorded on the night of March 12/13, 1978 which caused 63 % mortality of adults and 83 % mortality of immature stages in one Sitka spruce stand. The distribution of the live and dead aphids indicated that complete aphid mortality occurred on some individual needles, complete survival on others (Fig. 1). Such a non-random distribution associates individual aphid mortality with the death of neighbouring aphids on the same needle.

Figure 1. Non-random mortality of aphids on Sitka spruce needles following exposure to -5 °C (a = needles with live aphids, b = needles with dead aphids).

INITIATION OF FREEZING IN E. ABIETINUM

As freezing mortality of *E. abietinum* is dependent not only on the temperature of the surrounding air but also on that in the needle feeding site, then variation in the latter could result in differential freezing mortality. This provides a basis for a resistance mechanism.

In isolation from its spruce host the ability of *E. abietinum* to withstand low temperatures without freezing varies in relation to its feeding stage (Table 1). In the field heavy aphid mortality occurs at temperatures well above those required to kill *E. abietinum* in isolation from the needles. E.g., Carter (1972) observed a heavy aphid mortality at -8 °C while Powell & Parry (1976) found that -7 °C greatly reduced overwintering populations. This field mortality at relatively high temperatures must be a corollary of attachment to the needles. A comparison over 4 hours of the time related mortality at -12 °C of two groups of aphids, one remaining undisturbed on the needles, the other detached, showed that detached aphids survived while attached aphids exhibited a 90 % mortality (Powell, 1974).

Carter (1972) indicated that surface water had no effect on the mortality of *E. abietinum* and direct inoculation from external ice through the aphid cuticular pores (Salt, 1963), can, therefore, be eliminated. Powell (1974) considered aphid freezing to be the result of direct inoculation of the aphid from the frozen tissues of Sitka spruce via the sap in the stylet food canal. In *Myzus persicae* (Sulzer), which is roughly comparable in size to *E. abietinum*, the diameter of the stylet food canal is approximately 0.5 µm (Hoof, 1958). This would have a minimal effect on the freezing point depression of liquids within the canal, and would not restrict ice forma-

Table 1. The effect of gut content on the resistance to freezing of (a) *E. abietinum* and (b) *Myzus persicae*.

Aphid instar	Gut	Mean supercooling (°C)	Author
(a) Instar I	Empty	-22.70 ± 1.11	Parry (1979)
Instar I	Empty	-26.97 ± 1.02	Parry (1979)
Instar I	Empty	-24.7	Powell (1974)
Instar I	Empty	<-19	Carter (1972)
Instar II to adult	Full	-15.3 ± 0.32	Powell (1974)
Adults	Full	-11.0 ± 1.0	Carter (1972)
Adults	Full	c. -9 to -10	Parry (1979)
(b) Instar I	Empty	-22.67 ± 1.00	Parry (1978)
Adults	Full	-12.00 ± 0.87	Parry (1978)

tion. Salt & Kaku (1967) showed that the initial nucleation of needles of Colorado blue spruce, *Picea pungens* Engelmann, occurred in the stele, probably in the xylem, the endodermis acting as a barrier to impede ice propagation in the mesophyll.

Observations of Sitka spruce needles during cooling indicated that freezing initially occurred within the vascular bundle and was visually observable as a sudden lightening of the bundle colour following the formation of ice. The mesophyll remained dark green during subsequent cooling, probably due to the impedance to ice propagation by the endodermis, until similar colour changes occurred in this tissue at lower temperatures. Visual observations of aphids cooled while remaining undisturbed on the needles indicated that freezing occurred at the same time as ice became visually apparent in the needle stele. Disturbed aphids cooled on similar needles remained unfrozen on the needle surface for some time following stele, and frequently mesophyll, ice formation.

The supercooling points of spruce needles (Table 2) are considerably higher than those recorded for *E. abietinum* in isolation from its host (Table 1), and are comparable to the -7 °C to -8 °C required to cause heavy aphid field mortality.

FREEZING AS A RESISTANCE FACTOR

In order to test the hypothesis that variation in the ability of needles to supercool could be utilized as a resistance factor two groups of trees were selected, one group showing high aphid overwintering survival, the second a low winter survival in each of 3 successive winters. In order to minimise environmental variation tree pairs were situated as physically close together as possible and frequently possessed intermingling branches and root systems. The temperature at which ice formed was significantly higher in those trees showing low aphid overwintering survival than in

Table 2. The temperatures at which ice formation commences in the needles of various *Picea* species.

Species of *Picea*	Temperature of ice formation commencement (°C)	Author
P. abies	-4 to -8	Pisek (1960)
P. sitchensis	-9 to -12	Powell (1974)
P. sitchensis	-5 to -9	Parry & Powell (1977)
P. sitchensis	-4 to -6	Parry (1979)
P. pungens	-7 to -13	Salt & Kaku (1967)

those trees showing high aphid overwintering survival (Parry & Powell, 1977).

Site variation in the supercooling ability of Sitka spruce needles can occur (Table 2). The high supercooling ability of such needles on one site (Powell, 1974) contrasts with the lower level at the second site (Parry & Powell, 1977), outbreaks being more severe and more frequent at the first.

EFFECTS ON A STAND OF SITKA SPRUCE

High levels of spruce aphid can cause a marked fall in het leader growth of Sitka spruce, particularly young spruce during and immediately following the establishment period (Carter, 1977; G. de Britt, pers. comm.).

Under the winter temperature regimes common in Scotland (Parry & Powell, 1977) susceptible Sitka spruce would be subjected to repeated spring attacks causing a substantial depression in the height increment of these trees. Therefore, as the plantation matures its susceptibility to attack decreases as these susceptible spruce are suppressed and either die or are physically removed. In Britain the traditional practice of selective thinning resulted in the removal of a high proportion of such trees. In the modern practice of line thinning the susceptible trees remain and the stand retains its original characteristics.

There is some evidence of an increased resistance to *E. abietinum* in maturing Sitka spruce crops. Bevan (1966) indicated that young crops of Sitka spruce were prone to attack by *E. abietinum* up to the thicket stage. Powell (pers. comm.) surveyed Sitka spruce stands in Scotland and concluded that pole stage trees were less susceptible than younger crops. Scheller (1963) indicated that trees of lower growth incurred the higher population density, although here the effects of shelter were operative. By contrast Ohnesorge (1959) indicated that >20 year old spruce was more susceptible than <10 year old spruce. More work is needed on the effects of tree age on susceptibility in relation to thinning technique, site characteristics and genetic variability.

SELECTION OF RESISTANT SITKA SPRUCE

In the selection of resistant trees the critical parameter is the inability of Sitka spruce needles to supercool much below the depressed freezing point of the extracellular water, this being determined by a number of environmental and other variables. The mechanism of plant tissue freezing has been reviewed in detail by Mazur (1969), Alden & Hermann (1971) and Burke et al. (1976).

By selecting Sitka spruce trees in which extra-cellular water freezes at relatively high temperatures plantation susceptibility to *E. abietinum* could be significantly decreased, particularly during mild winters when attacks normally develop. Such trees would be found in selectively thinned, mature British plantations which have withstood several aphid attacks. The use of trees grown from imported seed or seed orchards containing trees selected for other characteristics would be of little value in this context.

The selection of Sitka spruce with high temperature extra-cellular water freezing properties offers a resistance mechanism which is unlikely to be broken by genetic selection of resistant strains of aphids. However, selection could occur in favour of holocyclic races of *E. abietinum* overwintering in the egg stage. This latter mechanism removes the positive effects of mild winters on overwintering aphid numbers and considerably decreases the possibility of spring outbreaks.

OVERWINTERING MORTALITY IN OTHER SPECIES

The vulnerability of the active, feeding aphid to low temperature mortality is illustrated by the frequency with which the egg serves as the overwintering stage. In the Adelgidae overwintering normally occurs in the form of diapausing first or later instar sistentes capable of surviving temperatures down below -30 °C (Parry, 1980). When sap imbibition commences these sistentes become susceptible to temperatures as high as -5 °C. There is no relationshop between mortality and the freezing characteristics of the host while the Adelgidae remain in diapause (Parry, 1980). As the termination of diapause and consequent commencement of sap uptake varies within the Adelgidae those species emerging from diapause and commencing feeding at an earlier time should be less resistant to low temperature. *Adelges prelli* (Grosmann), which is considered to be particularly sensitive to low temperature, commences sap uptake in mid-January (Merker et al., 1957). By comparison *A. cooleyi*, in which sap uptake begins in mid-February, is exposed to low temperatures for a much shorter time (Parry, 1980). Consequently there is a greater possibility of modifying Adelges mortality by selection of conifers in which the sap imbibed by the feeding *Adelgidae* freezes at higher temperatures than normal in those species which commence feeding at an earlier time.

In order to avoid ice nucleation within the gut contents most insect species overwinter in various non-feeding states, e.g., eggs, pupae, or adults with empty guts. Very little can be done to modify the low temperature susceptibility of such forms. Therefore, the manipulation of overwintering mortality by selection within the host trees is restricted to a narrow spectrum of sap-sucking conifer aphids overwintering as active, feeding forms.

REFERENCES

Alden, J. & R.K. Hermann, 1971. Aspects of the cold-hardiness mechanism in plants. Botanical Review 37: 37-142.

Behrndt, V.G., 1961. Auftreten und Bekämpfung der Sitkalaus im Frühjahr 1967 im Privatwald des Weser-Ems-Raumes. Allgemeine Forstzeitschrift 16: 659-661.

Bevan, D., 1966. The green spruce aphis Elatobium (Neomyzaphis) abietinum Walker. Scottish Forestry 20: 193-201.

Burke, M.J., L.V. Gusta, H.A. Quamme, C.J. Weiser & P.H. Li, 1976. Freezing and injury in plants. Annual Review of Plant Physiology 27: 507-28.

Carter, C.I., 1972. Winter temperatures and survival of the green spruce aphid, Elatobium abietinum (Walker). Forest Record No. 84, Forestry Commission, London. 10 p.

Carter, C.I., 1977. Impact of green spruce aphid on growth. Research and Development Paper 116, Forestry Commission, London. 8 p.

Hoof, H.A. van, 1958. Onderzoekingen over de biologische overdracht van een non-persistent virus. Mededeling No. 161, Instituut Planteziektenkundig Onderzoek, Wageningen, the Netherlands. 96 p.

Kloft, W., P. Erhardt & H. Kunkel, 1961. Die Fichtenröhrenlaus, Elatobium abietinum (Walk. 1849), ein endemisches Insekt der Rotfichte Picea excelsa in Europa. Waldhygiene 4: 121-125.

Matthews, J.D., 1950. Genetics. Report on Forest Research for the year ending March 1949. Forestry Commission, London. p. 26-29.

Mazur, P., 1969. Freezing injury in plants. Annual Review of Plant Physiology 20: 419-448.

Merker, E., O. Eichhorn & I. von Kleist, 1957. Die vernichtende Wirkung von klimatischen Kälteeinbrüchen auf Tannenläuse der Gattung Dreyfusia. Zeitschrift für angewandte Entomologie 41: 333-352.

Ohnesorge, B., 1959. Die Massenvermehrung der Sitkalaus in Nordwestdeutschland. Forstarchiv 30: 73-78.

Parry, W.H., 1978. Supercooling of Myzus persicae in relation to gut content. Annals of applied Biology 90: 27-34.

Parry, W.H., 1979. Acclimatisation in the green spruce aphid, Elatobium abietinum. Annals of applied Biology 92: 299-306.

Parry, W.H., 1980. Studies on the factors affecting the population levels of Adelges cooleyi (Gillette) on Douglas fir. 3. Low temperature mortality. Zeitschrift für angewandte Entomologie 90: 133-141.

Parry, W.H. & W. Powell, 1977. A comparison of Elatobium abietinum populations on Sitka spruce trees differing in needle retention during aphid outbreaks. Oecologia 27: 239-252.

Pisek, A., 1960. Die Photosynthetischen Leistungen der immergrünen Pflanzen (einschl. Koniferen). In: W. Ruhland (Ed.): Handbuch der Pflanzenphysiologie Bd. V. Teil 2. 439 p.

Powell, W., 1974. Supercooling and the low temperature survival of the green spruce aphid, Elatobium abietinum. Annals of applied Biology 78: 27-37.

Powell, W. & W.H. Parry, 1976. Effects of temperature on overwintering populations of the green spruce aphid, Elatobium abietinum. Annals of applied Biology 82: 209-219.

Salt, R.W., 1963. Delayed inoculative freezing of insects. Canadian Entomologist 95: 1190-1202.

Salt, R.W. & S. Kaku, 1967. Ice nucleation and propagation in spruce needles. Canadian Journal of Botany 45: 1335-1346.

Scheller, H.D. von, 1963. Zur Biologie und Schadwirkung der Nadelholzspinnmilbe, Oligonychus ununguis (Jacobi), und der Sitkafichtenlaus, Liosomaphis abietina Walker (Hom. Aphid.). Zeitschrift für angewandte Entomologie 51: 258-284.

Trees, fungi, and environment

P. Schütt

Lehrstuhl für Forstbotanik, Universität München, Amalienstrasse 52, D-8000 München 40, BRD

ABSTRACT

Using pathological events in the rhizosphere and in the phyllosphere as examples, it is pointed out that a given pathological situation often cannot be explained by means of the established model which says that genetically controlled defense traits are modified within certain limits by environmental influences. There are numerous additional effects like competition, allelopathy, antagonisms, synergisms and secondary plant substances, which are themselves influenced by the environment and which obscure the interrelationships between genetic and environmental factors.

INTRODUCTION

In introductory papers presented at scientific meetings it is customary and useful to provide a fairly broad background dealing with new results in related fields, in an attempt to compare and link them with current aspects of research of more direct relevance to the main topics.

The title of this paper, 'Trees, fungi, and environment', is sufficiently broad to permit such an approach. At first sight, however, this simultaneously creates both an attractive and a confusing feeling. Pathologists and geneticists may well feel that it is far too general a title since it embraces an almost limitless number of deep and far-reaching problems, each of which interact and so lead to a near hopeless situation from which to establish links having general validity.

Central European pathologists may recall the well-known relationship between climate and disease boldly expressed by:

good wine years - good years for insects
bad wine years - good years for fungi

In generalities like these, environment is given equal weight to climate, an over-simplification which is untrue. If we define the term 'environmental influences' as 'parameters which act on trees externally' then we have to include in addition to climate the many biogenic and non-biogenic factors. This leads us into fields which are affected by problems of competition, of allelopathy, and of symbiosis, activities which are often regarded as peripheral disciplines of pathology. Observations and research in these fields, however, often contribute towards a better understanding of the processes of disease outbreak and disease progression.

Before I try to demonstrate the complexities of estimating the influence of single environmental factors on host, parasite, and disease progression, perhaps I should remind you of several facts which may be trivial but none-the-less important biological keywords for the following observations:

First, it is the predisposition for a trait, not the trait itself, which is under genetic control. The expression of a trait, however, is often greatly influenced by the environment. Because the same environmental factor can, in principle, affect different components of an ecosystem by different degrees and ways, the influences of the environment are difficult to analyse and are even more difficult to predict.

Second, the term 'resistance' is understood as the genetically controlled aspect of stability. Disposition or predisposition, however, are regarded as alterations of stability, due to environmental factors.

Third, trees are exposed to varying environmental factors during their long life and are, therefore, different from most agricultural and horticultural crops. Consequently the same host can be affected in different years by different environmental influences. Moreover, long-lived plants have developed an additional method of survival, namely, an ability to recover.

Perhaps you would keep these 3 rules and definitions in mind when trying to understand the interactions between environmental and genetic factors during the development of certain tree diseases.

EXAMPLE 1: ROOT-ROT CAUSED BY FOMES OR ARMILLARIA

Those who have had long experience in research with *Fomes* or *Armillaria* will probably agree when I say that these fungi are still full of mystery for pathologists with regard to: certain parts of the infection biology; the mechanism of predisposition of the host; and, last but not least, to the details and specialization of pathogen reactions against the diverse biotic and non-biotic factors in the substrate.

Our factual knowledge about environmental dependencies of root rots is scarce and is little more than the essence of long practical experience. This means, for *Fomes*, higher infection rates under neutral or alkaline soil conditions; and for *Armillaria*, increased attack after pronounced changes of soil moisture.

Beyond this, can we really draw a more precise picture of environmental influences on fungal aggressiveness in spite of greater research efforts, better methods and improved facilities, and of an increasing demand for control methods? Unfortunately we can't, and it is my feeling that the reasons for these failures are not too difficult to determine.

In Norway spruce, *F. annosus* (Fr.) Cooke is, in economic terms, the most important tree disease of Central Europe - the 'Fichten-Rotfäule'. Differing from Rishbeth's findings in Scots pine, the fungus usually invades through the root and not the stump surfaces (Dimitri, 1969; Käärik, 1975). It is, however, very difficult to artificially copy the infection processes, to develop methods of artificial inoculation and to get a realistic impression of the natural defence mechanisms of the host-tree.

One possible reason for these difficulties appears to be that *Fomes* offers very little competition against other members of the soil microflora, so that its mycelium has very little chance to survive. This seems to be much more the case in acid rather than neutral or alkaline substrates. The fungus itself is largely pH-tolerant (Courtois, 1972). Its most important antagonists, however, usually prefer a lower soil pH, and so the hydrogenion concentration in the soil is an important environmental parameter. However, pH acts not directly, but in a secondary way by influencing the competition with other soil-borne fungi.

The vitality and virulence of a soil-borne fungus such as *Fomes* is influenced by, among other factors, the amount and the quality of nutrients available in the soil, especially in the root region. In turn the availability of these nutrients is to a certain degree dependent upon the quantity, the mineral composition, and the state of decomposition of the litter. Thus the environment directly affects the situation because the mineral nutrient supply of the soil, as well as climatic factors controlling photosynthesis and metabolism, interact and affect both the amount and quality of the litter.

Schuck & Lechler (1980) for the first time proved that aqueous litter-extracts can have a rather dramatic effect on the vegetative development of *F. annosus*. Depending on several variables, either a stimulation or an inhibition of fungus development occurred. The following parameters had relatively pronounced effects:

- species litter: in general conifers promote, deciduous trees inhibit,
- different stands of the same species: site effect,
- duration of extraction and concentration of extract: amount of chemical components dissolved,
- time of extraction: autumn less than spring.

Surely, these results cannot only be explained by differences in the amount of nutrients available to the fungus. Allelopathic effects are also involved since conifers often produce a continuous and permanent litter layer the effects of which seem to be of considerable importance to the soil microflora. Moreover the leaching of chemical compounds has been described for several plant communities including forest trees; it is known to be highly important from an ecological point of view.

To reiterate: in our case study of *Fomes*, important environmental factors affect the development of metabolic compounds which are liberated during the microbiological decomposition of the litter; these compounds act in an allelopathic way. Naturally the processes are affected by temperature and moisture. But for the first time we have to comment also that genetic influences are embodied in this interaction of numerous modifying factors. In essence it is the ability of a tree to:

- absorb a special spectrum of mineral nutrients;
- produce, distribute, and store a special spectrum of secondary plant substances.

There is no doubt that both these abilities are affected by the genetic variability of each tree. If this assumption is correct, then the allelopathic and nutritional components affecting the viability and vitality of the fungus in the soil are, to a certain degree, both qualitatively and quantitatively predetermined by genetic differences in their later host plants.

Allelopathy may also play a direct role during the course of infection by *Fomes* and *Armillaria*. This happens in the rhizosphere, a region of high microbiological activity, caused by the exudation of fluid and volatile substances through root surfaces. Amino-acids, sugars, terpenes, phenolics, and growth substances may be liberated in a species-specific composition. They will obviously affect the environment close to the root surface and therefore affect directly the phytopathological activities of soil-borne fungi.

We know that rhizomorph formation of *Armillaria* is stimulated by glucose in the presence of IAA (Garraway, 1975) and also by 1 - asparagin (Rykowski, 1976). The same fungus exhibits pronounced morphological, physiological, and pathological variations depending on the biochemical complex which is presented. Moreover, there is considerable variation between different isolates when kept under the same environmental conditions. Evidently *Armillaria* differs from other forest pathogens and is manifested in many biotypes each with a different ecological demand. Finally, every single isolate is able to react to changes in the environment in a very plastic manner.

Thus we have to record that *Armillaria* has the puzzling ability to react to environmental factors in a very complicated way. Furthermore, the fungus has additional and rather unusual features which cannot be ignored. For many years we have recognised the ability of *Armillaria* to live as a saprophyte as well as a parasite, and occasionally even the possibility of a symbiotic existence as a mycorrhizal partner has been discussed.

From results of some limited pilot studies in the laboratory, *Armillaria* has been observed to have the capability of reacting allelopathically on tree seedlings and, vice versa, there are growth differences of the fungus with respect to root exudates of various conifer seedlings: 15 ml beakers containing a 3 cm long piece of rhizomorph at the bottom and covered with quartz sand were planted with 10-day-old seedlings of different conifer species. After 6 weeks, the following features were observed:

- the presence of rhizomorphs promoted the formation of secondary roots in *Picea abies* (L.) Karst. and *Pseudotsuga menziesii* (Mirb.) Franco, but not in *Pinus sylvestris* (L.) and *Larix decidua* Mill.;
- root length remained unaffected;
- rhizomorph yield varied with the conifer species planted in close proximity (pine > douglas-fir > spruce).

Although these preliminary trials have not always produced reproducible results with different sources and ages of rhizomorphs, we dare to express some preliminary suppositions:

Obviously there were mutual, species-specific effects between fungus and seedling even though direct contact between roots and rhizomorphs did not occur. So questions arise as to whether or not these relationships are the rule and if they are the first stage of a pathological event. In any case, this kind of experience widens and deepens our picture of the biological roles of *Armillaria*. Parasite - saprophyte - competitor - symbiont: all these functions may be fulfilled by *Armillaria*. What, however, are the conditions which influence a change of role?

Returning to *Fomes*, which is morphologically far less variable than *Armillaria* and much more sensitive to antagonists, we have some evidence that this fungus too could act as a mycorrhizal partner of tree roots - at least temporarily and under favourable conditions. Kowalski (1970) discovered such relationships in Scots pine roots and other examples have been recorded. Especially in *Picea abies*, the roots of which are regularly associated with mycorrhizal fungi, *Fomes* does not seem to play a role as a symbiont (Bücking, 1979), thus differing from *Armillaria* which possibly starts its pathological activities after a symbiotic stage.

In spite of this, mycorrhizae also seem to have some role in the infection biology of *Fomes*. But this is in a more passive way by affecting the pre-infectional stage of 'invading resistance' called *axenie* by Gäumann, Krupa & Nylund (1972). Also Krupa et al. (1973) pointed out that short-roots (roots associated with mycorrhizal fungi) of *Pinus sylvestris* and *P. echinata* Mill. liberated increased amounts of volatile mono- and sesqui-terpenes when compared with roots without mycorrhizae. Apparently the production of the particular group of compounds was markedly promoted, several of which are quite toxic to *Fomes* and some other root pathogens. This is confirmed by many in vitro investigations (Hintikka, 1970; Schuck, 1977). Thus it seems that a good mantle of mycorrhizae around a root system can reduce *Fomes* infection by fungicidal chemical exudates.

Thus, volatile terpenes in the rhizosphere are very relevant environmental factors affecting *F. annosus*. In contrast, the same components, looked at from the host point of view, are metabolic compounds forming part of the defence system.

Again, the formation of terpenes and their composition depend on environmental conditions. As we have seen before, they are a mixture of secondary plant substances, the biogenesis of which is only guaranteed under high photosynthetic activity. Also the rate of liberation of chemical compounds having differing boiling points is a function of temperature.

Volatile terpenes are components of conifer resin. They are not only liberated into the environment through root, stem, and leaf surfaces, but they also fill the space of intercellular systems in many tissues. This is the reason why they are part of the host-specific environment for every invading pathogen. In the case of 'Fichten-Rotfäule', the monoterpenes are said to have resistance-forming abilities, at least in the sense of 'spreading-resistance' after Gäumann.

The composition of the monoterpene fraction is, to some extent, individually fixed; certainly it is species-determined. Individual patterns of monoterpene-composition are, however, modified by non-random differences within the stem which are difficult to explain (Schuck & Schütt, 1975). Consequently for invading pathogens, we have to contend, a priori, with the existence of regions of varying toxicity. Additionally there are alterations to the individual monoterpene composition which may be induced by the parasite itself, or which are to be explained by a traumatic reaction after mechanical wounding. Madziara-Borusiewicz & Strzelecka (1977) found a marked decrease of the limonene and bornyl acetate portion in the monoterpene-fraction of *Picea abies* following infection by *Armillaria*. And Schuck (unpublished) reported that the monoterpenes originating from traumatic resin ducts following wounding have a different composition in comparison with resin collected from undamaged parts of the same tree.

This indicates that the monoterpene fraction, which affects the development of root and wood pathogenic fungi, is not absolutely stable within the same host. It varies for internal reasons between different parts of the same stem, it varies after the invasion of the pathogen, and it is modified after traumatic reactions. Thus again we observe so many interactions between external and internal influences that it is difficult to use the single term 'environmental factor' in its original sense. Also, as a consequence of the interaction of factors, we sense some erosion of the boundary between disposition and resistance-caused differences to disease attack.

EXAMPLE 2: FOLIAGE PATHOGENS

It may be useful to discuss further whether this eroded meaning of proved terms is specific to pathological events in the rhizosphere, or whether it can be discovered elsewhere, for instance in the phyllosphere.

Like the rhizosphere, the phyllosphere represents a borderline between the plant and the environment. It is characterized by high biological activity (Schütt, 1974). Also typical is the rather complicated chemistry of the plant's surface layer of cuticular waxes. They are built up of esters of unsaturated fatty acids with higher aliphatic alcohols, consisting of paraffins, alcohols, esters, ketones, aldehydes, and fatty acids, each of the fractions being represented by 10-15 single components. The net result is an incalculable number of combinations.

Cuticular waxes are chemically and physically heterogenous. They crystallize as plates, rods, grains, and intermediate forms which provide an uneven, microclimatically variable leaf surface. The chemical composition of the waxes, together with their differences in structure, seem to be species-determined and both are genetically controlled.

Because we have evidence that foliar pathogens react partly negatively and partly positively to a given chemical composition of foliar waxes, there is reason to believe that the chemistry of cuticular waxes eventually has an axenic effect on leaf-inhabiting fungi (Schütt, 1971). In principle this association is not affected by those volatile terpenes which fill the intercellular spaces within conifer needles but which are soluble in wax and therefore may be incorporated as true, species-specific components in the chemistry of the cuticular waxes.

It is probable that this predetermined endogenous situation can be modified by environmental factors such as light intensity, which can lead to considerable changes in the wax structure. The pathological situation, however, is extremely influenced by problems arising from the activities of micro-organisms of the phyllosphere. They make it highly complex and very, very difficult to analyze. Micro-organisms, mainly yeasts and bacteria, occur in immense numbers and in many forms; for example, 13 million bacteria can be found per cm^2 of leaf surface.

We know the composition of the flora of the phyllosphere is by no means constant, and successions in the foliar colonization can be observed. Thus the spectrum of species varies with leaf age, which is paralleled by an alteration in the chemical composition of its waxes. Antagonistic effects are also involved in these processes. Some yeasts are able to decompose the leaf waxes, thus considerably changing the chemical situation. Other species produce growth-substances, secrete amino-acids or carbohydrates, and change the ecological condition on the leaf surfaces and inevitably alter their suitability for the establishment of pathogens. This, however, leads to an almost radical change in our thinking. At first it seems to accurately follow our established model, according to which genetically controlled defence traits (resistance) are modified within certain limits by environmental influences.

But as we learned, in practice, this model is not valid under all conditions. Both in the phyllosphere and rhizosphere there are borderline interactive systems affected by high microbiological activities where the inter-relationships between genetic and environmental factors are stratified and obscure. For a proper estimation of a given pathological situation, much more is necessary than the knowledge of simple cause-and-effect relationships between environment and host or pathogen. Thus it is necessary to have a deeper understanding of the complex ecological factors and their interactions, an aim which can only be met through familiarity with recent studies and findings in related disciplines. Pathology can no longer be regarded in isolation but rather as an integral part of a complex ecosystem in which countless abiotic and biotic factors are interwoven one with another: a single division into genetic and environmental components is no longer a wholly acceptable concept.

REFERENCES

Bücking, E., 1979. Fichten-Mykorrhizen auf Standorten der Schwäbischen Alb und ihre Beziehung zum Befall durch Fomes annosus. European Journal of Forest Pathology 9: 19-35.

Courtois, H., 1972. Das Keimverhalten der Konidiosporen von Fomes annosus (Fr.) Cke. bei Einwirkung verschiedener Standortfaktoren. European Journal of Forest Pathology 2: 152-171.

Dimitri, L., 1969. Untersuchungen über die unterirdischen Eintrittspforten der wichtigsten Rotfäuleerreger bei der Fichte (Picea abies Karst.). Forstwissenschaftliches Centralblatt 88: 281-308.

Garraway, M.O., 1975. Stimulation of Armillaria mellea growth by plant hormones in relation to the concentration and type of carbohydrate. European Journal of Forest Pathology 5: 35-43.

Hintikka, V., 1970. Selective effect of terpenes on wood-decomposing Hymenomycetes. Karstenia 11: 28-32.

Käärik, A., 1975. Succession of microorganisms during wood decay. In: W. Liese (Ed.): Biological transformation of wood by microorganisms. Springer-Verlag, Berlin-Heidelberg-New York. p. 39-51.

Kowalski, S., 1970. Mycorrhizal synthesis of Pinus sylvestris L. and Fomes annosus. Bulletin de l'Académie Polonaise des Sciences V 18: 405-408.

Krupa, S. & J.E. Nylund, 1972. Studies on ectomycorrhizae of pine. III. Growth inhibition of two root pathogenic fungi by volatile organic constituents of ectomycorrhizal root systems of Pinus sylvestris L. European Journal of Forest Pathology 2: 88-94.

Krupa, S., J. Andersson & D.H. Marx, 1973. Studies on ectomycorrhizae of pine IV. Volatile organic compounds in mycorrhizal and non-mycorrhizal root systems of Pinus echinata Mill. European Journal of Forest Pathology 3: 194-200.

Madziara-Borusiewicz, K. & H. Strzelecka, 1977. Conditions of spruce (Picea excelsa LK.) infestation by the engraver beetle (Ips typographus L.) in mountains of Poland. I. Chemical composition of volatile oils from healthy trees and those infested with the honey fungus (Armillaria mellea [Vahl.] Québ.). Zeitschrift für Angewandte Entomologie 83: 409-415.

Rykowski, K., 1976. Recherche sur la nutrition azotée de plusieurs souches de l'Armillaria mellea. I. L'utilization de diverses sources d'azote. European Journal of Forest Pathology 6: 211-221.

Schuck, H.J., 1977. Die Wirkung von Monoterpenen auf das Mycelwachstum von Fomes annosus (Fr.) Cooke. European Journal of Forest Pathology 7: 374-384.

Schuck, H.J. & H.H. Lechler, 1980. Die Wirkung wäßricher Streuextrakte verschiedener Waldbäume auf Mycelwachstum und Konidiosporenkeimung von Fomes annosus (Fr.) Cke. European Journal of Forest Pathology 10: 36-47.
Schuck, H.J. & P. Schütt, 1975. Die Verteilung flüchtiger Terpene im Fichtenstamm und ihre Bedeutung für die Ausbreitungsresistenz gegenüber Fomes annosus. Biochemie und Physiologie der Pflanzen 167: 65-77.
Schütt, P., 1971. Untersuchungen über den Einfluß von Cuticularwachsen auf die Infektionsfähigkeit pathogener Pilze. I. Lophodermium pinastri und Botrytis cinerea. European Journal of Forest Pathology 1: 32-50.
Schütt, P., 1974. Die Phyllosphäre, eine Zone hoher biologischer Aktivität. Forstwissenschaftliches Centralblatt 93: 1-10.

Environmental factors affecting host-bacteria interactions in bacterial diseases of forest trees and shrubs

M. Ridé

Station de Pathologie Végétale et Phytobactériologie INRA, Route de Saint-Clément, Beaucouze, 49 000 Angers, France

ABSTRACT

Research achievements are reviewed of the last decennial period about environmental and epidemiological factors affecting host-phytopathogenic bacteria interactions, particularly bacterial tree diseases caused by *Erwinia*, *Pseudomonas*, *Xanthomonas*, and *Agrobacterium* species. Various aspects are illustrated: influence of inoculum sources, temperature, and rains on forecasting the potential activity of fire blight; *Pseudomonas* ice-nucleation activity; effect of edaphic properties upon the development of forest bacterial diseases; and effects of epiphytic microflora and leachates of aerial structures on *X. populi*.

INTRODUCTION

Damages caused by phytopathogenic bacteria to forest trees and shrubs are not so frequent as those induced by fungi. Nevertheless, bacterial diseases increasingly constitute a potential risk which must be taken into account in forest planting choices and breeding programmes for resistance. The evolution of forestry in the recent past is the reason for this increased risk, e.g. use of monoclonal and same age plantings *(Populus, Salix)*, the active interest of sylviculturists in some noble trees (cherry, walnut, ash) and their vegetative micropropagation, and the more and more important intervention of foresters in choosing certain trees and shrubs to integrate into landscape planning and wind protection *(Crataegus, Cotoneaster, Sorbus)*.

The bacteria damaging forest trees are exposed, as are all pathogenic microorganisms, to various interactions between climatic and edaphic factors, genetic host constitution and its physiological development, and evolution of the structure of pathogenic populations inside the complex microflora of aerial structures of trees and rhizosphere. It is the purpose of this review to synthesize our main findings of the last decennial period about some factors affecting host-phytopathogenic bacteria interactions, limited to the environmental and epidemiological aspects, and excluding cellular and molecular host-pathogen relationships.

PRINCIPAL BACTERIAL DISEASES OF FOREST TREES AND SHRUBS

Environmental host-bacteria interactions will be illustrated pertaining to the bacterial diseases mentioned in Table 1, which are of interest for tree breeders and for their economic impacts. Nevertheless, let it be remembered that some pathogenic bacteria localized at the moment in a very restricted area, such as *Erwinia nigrifluens* Wilson, Starr & Berger on walnut in USA (Wilson et al., 1957) or *Xanthomonas populi* subsp. *salicis* on *Salix dasyclada* Wimm. (De Kam, 1978), constitute potential risks for the future. Bacteria which are present in complex infection chains or associated with insects (Ridé & Prunier, 1963; Danilewicz & Tomaczewski, 1970) or with nematodes, such as toxinogenic bacteria associated with the pinewood nematode (*Bursaphelenchus lignicolus* Mamiya & Kiyohara) on *Pinus densiflora* S. & Z. and *P. thunbergii* Parlat (Oku et al., 1980), will not be taken into

Table 1. Principal bacterial diseases of forest trees and shrubs.

Causal agent	Disease	Hosts
Agrobacterium radiobacter var. *tumefaciens*	Crown gall	Various deciduous trees and conifers
Corynebacterium halepensoïdes	Tumors	*Pinus halepensis*
Erwinia amylovora	Fire blight	*Pyrus*, *Malus*, *Crataegus*, *Cotoneaster*, *Pyracantha*, *Sorbus*, and other *Pomoideae*
Erwinia salicis	Watermark disease	*Salix*
Pseudomonas syringae	Twig blight, blossom blight, shoot wilt, canker	*Cerasus*, *Citrus*, *Populus*, *Pyrus*, *Prunus*, *Salix*, *Syringa*, etc.
Pseudomonas savastanoï f. sp. *fraxini*	European ash canker	*Fraxinus*
Pseudomonas mors-prunorum	Bacterial canker, dieback, shoot hole	*Prunus avium*
Xanthomonas corylina	Filbert blight	*Corylus*
Xanthomonas juglandis	Walnut blight	*Juglans*
Xanthomonas populi	Bacterial canker	*Populus*

account in this review, in spite of their evident ecological interest. Also diseases induced by mycoplasmalike bodies and other forms of Mollicutes will not be evoked here.

INOCULUM SOURCES, DYNAMICS OF PATHOGEN POPULATIONS AND CLIMATIC CONDITIONS

Fire blight

The etiology of Pomoideae fire blight was demonstrated for the first time 100 years ago (Burrill, 1882). Fire blight caused by *Erwinia amylovora* (Burrill) Winslow et al. is now present in all the apple and pear growing regions of North America. It was introduced in northwestern Europe probably about 1955, and during the past 20 years it has become the most serious bacterial disease in Belgium, Denmark, England, France, West Germany, and East Germany. Its spread seems stopped in Poland. Though considered to be the most important disease of orchards, fire blight is also of interest for the forester, insofar as he is concerned with landscape appointments and protection mainly with *Crataegus* spp., *Pyracantha* spp., *Cotoneaster* spp., *Sorbus* spp., etc.; these trees and shrubs are frequently dangerous inoculum sources.

E. amylovora is probably the phytopathogenic bacterium on which the influence of environmental conditions has been explored most. It overwinters in bark tissues along the edges of cankers initiated by infections of the previous year. In spring the bacteria multiply at the canker margins producing the primary inoculum (ooze or slime) disseminated by rain, wind, and flying insects. The first infection takes place on blossoms or vegetative shoot tips.

As mentioned by Aldwinckle & Beer (1978), blossom infection occurs first in most regions, 'but in a cooler area such as southern Ontario in Canada, blossoms often escape infection and vegetative shoot tips become infected well after bloom (Dueck & Quamme, 1973). Dissemination of the bacteria in a warm, generally dry area such as California is mainly by insects; but in more humid areas, dissemination by rain can be more important'.

After *E. amylovora* penetrates through wounds or natural openings such as nectaries, hydathodes, or lenticels, the infected tissues frequently exude droplets of ooze consisting of a matrix of polysaccharides. Exudate production is often considered to be related to high temperature and relative humidity. Nevertheless, in climatic chambers exudation takes place within a wide range of temperature and RH, between 15 °C and 28 °C and between 60 and 90 % RH, respectively (Paulin & Lachaud, 1978). Apparently, other factors such as plant nutrition and water supply could also be important for exudate production. Bacteria contained in ooze function as secondary inoculum sources for late primary and secondary blossoms, vegetative shoots, and fruit.

The relationship between inoculum dose and infection development has been studied only recently. The rate of symptom development and the proportion of inoculated blossoms seem directly proportional to the number of *E. amylovora* cells which are introduced into the floral cups (Beer & Norelli, 1975). Californian workers tried to monitor the presence of *E. amylovora* on pear and apple blossoms in an attempt to correlate bacteria populations with weather conditions (Thomson et al., 1975). The presence of the pathogen was demonstrated in blossoms and on the surface of cankers prior to development of symptoms of flower infection. Moreover it was found that *E. amylovora* multiplied in healthy flowers and was detected 14 days before symptoms appeared.

Based on 7 years of monitoring, a formula has been proposed to predict when *E. amylovora* is most likely to be detected in California conditions, only after daily mean temperature exceeded a line drawn from 16.7 °C on March 1 to 14.4 °C in May. Good control of the disease is generally obtained after spraying streptomycin only when the mean temperature of the prediction line is exceeded.

In contrast with observations in California, large epiphytic populations have not been detected in pome fruit blossoms in eastern USA. The same situation seems to exist now in Europe, particularly in the Netherlands (Miller & van Diepern, 1978) but perhaps also in southern France, where the varietal structure of the orchards and tree training techniques are very different from California. Results recently obtained with very late secondary blossoms on Passe Crassane variety would emphasize the possible role of very late and latent infections for the next spring. So blossom monitoring techniques used in California seem difficult to transfer to European climatic conditions.

For this reason Billing (1974, 1980) tried to relate the growth rate of *E. amylovora* and the rain requirement in developing a method for assessing potential fire blight activity in the field in south-eastern England. It is interesting to note that this method takes into account the potential doubling of the bacteria related to temperature. This method is now being tested in different countries in Europe, based on the following system. When the rain is taken into account according to the following score: R = 0, no rain; R = 0.5, rain $\leqq$ 2.5 mm; R = 1, rain > 2.5 mm; the incubation period (I) in days is ended when:

$$\text{P.D.} \geqq 36 \frac{I}{R} - 6,$$

when P.D. = potential doubling per day defined by daily maximum and minimum temperature. This system makes it possible to:
- forecast the potential activity of fire blight in a given area using simple and available data,
- know the most probable appearance of symptoms, by drawing the graphs

daily, and

- assess the risks following a rain or an abnormally high temperature during the presence of primary or secondary blossoms.

But it will be necessary to adapt Billing's system to areas where fire blight is particularly dangerous, such as south-western France. It will be especially important to take into account other factors which could influence initiation to infection, i.e. heavy dews or mists, activity and density of flying insects, frequency of late secondary blossoms, etc.

Pathogenic pseudomonads

In contrast to *E. amylovora* which seems to have temporary epiphytic stages, *Pseudomonas syringae* van Hall and other pathogenic pseudomonads of fruit and forest trees are widely distributed as epiphytic bacteria. But it is necessary to distinguish between *P. syringae*, a collective species which probably includes various ecotypes adapted to different genera such as *Prunus*, *Pyrus*, *Juglans*, *Populus* and various herbaceous species, living and surviving upon nutrients which leach from aerial parts of plants; and other pseudomonads such as *P. mors-prunorum* Wormald on cherry or *P. mors-prunorum* f. sp. *persicae* Prunier, Luisetti & Gardan (= *P. persicae*) on peach, which seem more specific and detectable only in diseased orchards.

By different methods (specific media, antibiotic resistant strains, immunofluorescence techniques) it is possible now to follow the dynamics and structure of *Pseudomonas* populations on leaf, petiole, and twig surfaces, and inside host-tissues. The evolution of *P. syringae* populations depends on tree varieties or species, and organs, as one can notice on graphs drawn according to Luisetti & Paulin (1972). *P. syringae* is always present on the leaf surface without symptoms during the vegetative period. It disappears only when high temperatures occur. The same situation exists on different species of *Populus* such as *P. nigra*, *P. maximowiczii*, and various hybrids. The influence of high temperature and dryness on the contraction of *P. syringae* populations is more contrasted and drastic on *Citrus* leaf surfaces in Greece as mentioned by Panagopoulos & Psallidas (1973). In the case of *P. mors-prunorum* on sweet-cherries the level of populations is maintained by leaf spot infections during the season. Nevertheless, Freigoun (1973) found a positive relationship between bacterial counts and leaf wetness and a negative relationship with sunshine duration. The same scheme is related to *P. persicae*, the causal agent of peach bacterial blight (Gardan et al., 1972); this bacterium maintained a high inoculum level on the leaf surface even in autumn, a period without any symptoms macroscopically perceptible on leaves. This level (10^4-10^5 bacteria per leaf) is high enough to infect fresh leaf scars during the fall.

The influence of low temperatures and chilling is particularly remarkable on the development of peach bacterial blight. Frequently in south-eastern France, negative temperatures interfere with the dormancy period of peach trees and one can observe the first symptoms after leaf fall infections during the first fortnight of December. The influence of chilling is also important when cambial activity of the trees starts again (Vigouroux, 1974).

Ice nucleation activity induced by epiphytic populations of *P. syringae*, a bacterium isolated from a wide variety of plant species, offers another curious and interesting aspect of low temperature-pathogenic bacteria-host tissue interactions.

Fifteen years ago Panagopoulos & Crosse (1964) in England, and Durand et al. (1967) in France, demonstrated that *P. syringae* could develop a rapid infection after the appearance of hypodermic cell microlesions induced by light frost damage on flowers, twigs, leaves, and fruitlets of pear. More recently Arny et al. (1976) showed on maize that *P. syringae* cells had a well defined role in the induction of ice nucleation in host tissues. The same results were obtained by Paulin & Luisetti (1978) on tobacco and on pear twigs. It is possible to observe the same phenomena on poplar shoots. Among phytopathogenic and saprophytic bacteria living on leaf surfaces, only *P. syringae* and sometimes some *Erwinia herbicola* strains seem able to induce ice nucleation activity (Lindow et al., 1978).

This property is dependent on bacterial strains, inoculum dose, temperature, and nature of culture medium on which bacteria are maintained (Paulin & Luisetti, 1978; Hirano et al., 1978). This new aspect of the temperature-bacteria-host-interactions could modify the control of some bacterial diseases such as blossom blight of pear in the future.

Agrobacterium spp. and soil

Unlike phytopathogenic bacteria whose ecology is related to aerial plant structures, *Agrobacterium* species are generally associated with root collars and root systems of conifers and deciduous trees: proximate environments of roots (rhizoplan and rhizosphere) and galls induced by virulent strains. However, some *Agrobacterium* strains are able to give rise to galls on trunks or branches, for instance on *Populus grandidentata* in USA, or *P. euramericana* cv. Robusta or *P. alba* in France, yet this situation is rare (Dochinger, 1969; Ridé & Lopez, 1976).

Tumors which are disaggregating in soil constitute excellent sources of pathogenic *Agrobacterium tumefaciens* (Smith & Townsend) Conn. but also saprophytic *A. radiobacter*. These bacteria have ecological niches comparable with those of *Rhizobium* species, which one can sometimes isolate at the same time as *A. tumefaciens* (Gaur & Sen, 1976).

Agrobacterium species are remarkably concentrated in the rhizosphere, and it is not rare to find there population levels 10 000 times higher than in the soil. Schroth et al. (1971) have shown that *Agrobacterium* has a positive chemiotachism for actively growing roots.

These bacteria overwinter in the soil and temperatures higher than

34 ° C are inhibiting for them; that could explain the low frequency of crown-gall in tropical zones. Furthermore the relative proportions of pathogenic to saprophytic Agrobacteria are variable according to the plant root system and soil conditions: thus on 28 soil samples in California, Schroth et al. (1971) found 17 of them with an appreciable proportion of pathogenic strains varying from 1/13 to 1/500. This balance between pathogenic and saprophytic Agrobacteria probably should be taken into account in the biological control of crown-gall in the field by using the antagonistic bacteriocinogen *A. radiobacter* K 84 strain proposed by Kerr (1972) in Australia.

SOIL, NUTRITION, AND PHYTOPATHOGENIC BACTERIA

Information about the effect of edaphic properties upon the development of forest bacterial diseases and the conservation of phytopathogenic bacteria in soil is very meager. Still, relationships between water supply, nutrient level, succulence of host tissues, tree vigour, and susceptibility to bacterial diseases have been known through empirical observations for a long time.

For example in the Therain Valley of northern France, numerous poplar plantings are generally distributed on 2 types of soil. One of them consists of a thin layer of arable soil on a mixture of clay and turf, with brown turf deeper and incompletely decayed organic matter - plant associations *Cirsium*, *Eupatorium*, *Juncus*, and *Molinia* - pH 6. The other is a deeper arable soil (60 cm) on a black turf and pH varying between 6.5 and 7 - plant association *Phragmites*, *Epilobium*, *Consolida*, and *Spiraea* (Ridé & Viart, 1966). Poplar growth is very poor on the first soil which is associated with low activity of *X. populi*. In contrast, canker expression is excellent on the second type of soil.

Concerning *Pseudomonas* diseases of fruit and forest trees and frequently also fire blight, heavy, poorly drained, or highly acidic soils sustain more extensive and severe damage. Relationships between tree nutrition and shoot succulence particularly with respect to fire blight susceptibility were perceived by Parker et al. (1961). Extra applications of nitrogen induce an increased susceptibility to fire blight, often exacerbated by excesses of potassium or phosphorus. The best theoretical work was probably done by Lewis & Kenworthy (1962): various combinations of nutrients were supplied to potted trees growing in quartz sand in a greenhouse and artificially inoculated with *E. amylovora*. Just as Parker et al. (1961) observed, these authors noted that trees supplied with an excess of nitrogen were more susceptible than control trees which received balanced nutrients. But nitrogen deficiency also increased susceptibility to fire blight resulting from deficiencies of various cations: calcium, copper, iron, magnesium, manganese, molybdenum, or zinc. When a five-fold excess of calcium was supplied, tree shoots could not be infected. Unfortunately, nutrient interactions in natural soils would limit the transfer of results obtained in sandy soil to the orchard.

Similar results were obtained more recently by Vigouroux & Huguet (1977) on the bacterial blight of peach caused by *Pseudomonas persicae*. This bacterial disease is mainly distributed on more or less acidic soils and granitic arena or diluvium sand, generally poor in organic matter. Trials on various sites and potted plants demonstrated the beneficial effect of calcium on the decrease of susceptibility to bacterial blight. Ca^{++} effects on membranes, enzymes, cell walls, and Ca-phytochrome interactions are now considered important, and they probably interfere with the host-bacteria relationship. Nevertheless, the influence of nutrition and Ca balance with other nutrients remain always difficult to interpret in nature.

INTERACTIONS OF HOST, EPIPHYTIC MICROFLORA, LEACHED SUBSTANCES, ENVIRONMENT AND PARASITE

During spring, bacterial poplar canker caused by *Xanthomonas populi* Ridé is characterized by production of bacterial slime which is mainly dispersed by rain, wind, and insects and constitutes an infection source on the new shoots. Infection success depends on the concordance between tissue receptivity, entry ways open to *X. populi*, and quantity of inoculum available on aerial structures of poplar (Ridé et al., 1978). Many factors may contribute to eliminate *X. populi* during its temporary epiphytic stage, and these have implications in the epidemiology of the disease. It is important, for example, to know if the bacterium is present for a long time and at a population level high enough to instigate an effective infection when host tissues become physiologically receptive.

Investigations were undertaken on the evolution of an artificially introduced population of *X. populi* on a susceptible clone (S. 6-2) and on a resistant one (*P. nigra* cv. Italica) during 2 years, 1975 and 1976, the latter year being characterized by drought and very high temperatures during spring and summer. Groups of saprophytic and epiphytic bacteria, yeasts, and yeast-like organisms were evaluated periodically and their role considered in limiting the evolution of *X. populi* (Table 2).

In 1975, *X. populi* rapidly decreased on *P. nigra* after 30 days (it disappeared completely after 60 days); it remained approximately at 10^6 per leaf on the susceptible clone S. 6-2 (its population was still high, 10^4, at the end of September). During the same period, populations of fluorescent pseudomonads (mainly *P. syringae*) were always very low on S. 6-2, and not even detectable at the end of June. In contrast, they varied from 10^5 in June to 10^6 in July on the resistant clone (in vitro this *Pseudomonas* exerted a clear activity against *X. populi*). In 1976, in the drastic conditions mentioned, *X. populi* decreased as slowly on the resistant clone as on

Table 2. Evolution of the numbers (×1000) of epiphytic bacteria (resp. yeast cells) per leaf on a susceptible and a resistant clone of poplar after spraying with *X. populi* on 28.5.1975 (upper table) and 24.5.1976 (lower table).

Fluo. pseudo. = fluorescent pseudomonads including *Ps. syringae*, *Ps. viridiflave* and *Ps. fluorescens* types. Y. like = yeast-like microorganisms (mainly including *Aureobasidium pullulans*). S = susceptible clone (S. 6-2). R = resistant clone (cv. Italica).

Date of sampling	*X. populi*		Fluo. pseudo.		Yellow bact.		Yeasts + Y. like	
	S	R	S	R	S	R	S	R
1975								
28-5*			3.3	110	0	5.5	24	56
29-5	4700	2300	2.1	830	0	130	33	200
30-5	12000	5900	22	250	9.4	11	55	150
2-6	5700	1700	0	55	0	48	30	100
5-6	2200	600	6.6	150	0	130	15	72
12-6	860	320	3.3	490	25	60	150	180
18-6	1100	300	0	880	0	110	150	330
25-6	1000	350	0.55	840	0.55	<0.1	93	310
9-7	1300	5.6	0	1300	0	150	140	260
24-7	1200	1.0	0	2300	11	260	1100	490
1976								
24-5*			0	0	0	0.34	3.8	25
25-5	35000	22000	0	3.0	0	0.7	4.4	5.9
26-5	25000	12000	0	0.34	0.50	0.34	5.0	8.3
28-5	46000	21000	0	0	<0.1	0.26	7.0	9.6
1-6	24000	10000	0	3.6	0.33	0.52	3.7	17
8-6	40000	15000	0	22	0.40	0.83	9.0	22
21-6	14000	3000	0	2.7	1.9	0.34	24	86
8-7	8000	5300	0	0.26	1.6	2.6	130	480
27-7	1400	1400	0	2.6	0	34	130	340

* Counted before spraying.

the susceptible one. During the same period, pseudomonads were not detectable on S. 6-2 and their populations were very poor on *P. nigra* and not high enough to exert an antagonistic effect against *X. populi*; yellow bacteria were also affected by drought and temperature on both clones, and yeasts and yeast-like organisms to a lesser degree. The lack of rains during the last experimental period could partially explain why *X. populi* was not eliminated from the leaf surfaces. Moreover, it is possible that the mucoïd polysaccharide of *X. populi* exerts a protective effect against drought conditions, similar to other xanthomonads. Although the role of some other components of leaf microflora is probably not negligible in the regulation of *X. populi*, in 1976 pseudomonad populations were not high enough to exert an antagonistic effect against *X. populi*.

The qualitative evolution of microflora also depends on other factors which influence quality and quantity of substances leached from foliage; genetic constitution, climatic conditions, but also physiological age of the leaves, soil, and fertilizers may influence leaf surface nutrients and canopy microclimate, and thereby may alter phylloplane microflora. Initial investigations undertaken on balsams produced by epidermic cells of stipules and leaf exudates during the first stage of the new shoots show these substances contain flavonoid aglycones and terpenoids and are likely to inhibit *X. populi* on *P. nigra*. In addition to inhibiting compounds localized on the young shoots during the fast growing period, some nutrients are excreted at the same time which are favourable to various components of microflora: carbohydrates, growth substances, and aromatic acids. *Pseudomonas* spp. are able to use some aromatic compounds as carbon sources in degradation in a mixture of phenolic substances. Accordingly these bacteria could have a complex role in the regulation of epiphytic microflora and could be of decisive importance to the natural control of pathogenic populations.

CONCLUSION

Ecological and epidemiological data concerning phytopathogenic bacteria generally require a lot of effort to obtain. One of the main difficulties of a bacteriologist's work is probably the absence of a macroscopically detectable reference stage of the bacterial pathogens, unlike fungi possessing some perfectly defined stage and form of fructification and ripening: ascospores, conidial forms, uredospores, aeciospores, etc.

Twenty years ago the evolution of bacteriological techniques using replication on specific culture media and more sophisticated methods, such as immunofluorescence or immunoenzymology, made it possible to detect pathogenic bacteria in various complex microbial ecosystems: evolving lesions, microflora of aerial structures of trees, and the rhizosphere.

After Crosse (1963), many research workers have been interested in phytopathogenic bacteria population dynamics. Impacts of climatic and edaphic factors on host-bacteria relationships are just beginning to be understood. Genetic background of the host and components of phenotypic expression of the resistance to bacterial diseases of forest trees will be explored more and more in the near future, requiring good cooperation between phytopathologists, physiologists, climatologists, and geneticists.

REFERENCES

Aldwinckle, H.S. & S.V. Beer, 1978. Fire blight and its control. Hort-Reviews 1: 423-474.

Arny, D.E., Lindow, S.E. & C.D. Upper, 1976. Frost sensitivity of Zea mays increased by application of Pseudomonas syringae. Nature 262: 282-284.

Beer, S.V. & J.C. Norelli, 1975. Factors affecting fire blight infections and Erwinia amylovora populations in pome fruit blossoms. (Abstract.) Amer. Phytopath. Soc. Proc. 2: 95.

Beer, S.V. & J.C. Norelli, 1977. Fire blight epidemiology: factors affecting release of Erwinia amylovora by cankers. Phytopathology 67: 1119-1125.

Billing, E., 1974. The effect of temperature on the growth of the fire blight pathogen, Erwinia amylovora. J. Appl. Bact. 37: 643-648.

Billing, E., 1980. Fire blight in Kent, England in relation to weather (1955-1976). Ann. Appl. Biol. 95: 341-364. [And] Fire blight (Erwinia amylovora) and weather: a comparison of warning systems. Ann. Appl. Biol. 95: 365-377.

Burrill, T.J., 1882. The bacteria: an account of their nature and effects, together with a systematic description of the species. Illinois Ind. Univ. Ann. Rpt. 11: 93-157.

Crosse, J.E., 1963. Bacterial canker of stone fruits. V. A comparison of leaf surface populations of Pseudomonas mors-prunorum in autumn on two cherry varieties. Ann. Appl. Biol. 52: 97-104.

Danilewicz, K. & M. Tomaszewski, 1970. Degradation of lignin, lignole and phenolic acids in poplars wood with Pseudomonas, an intestinale bacteria from the larvae of Paranthrene tabaniformis Rott. Acta Microbiologica Polonica, Ser. B, 2: 4-19.

Dueck, J. & H.A. Quamme, 1973. Fire blight in southern Ontario in 1972. Canad. Plant Dis. Survey 53: 101-104.

De Kam, M., 1978. Xanthomonas populi subsp. salicis, cause of bacterial canker in Salix dasyclada. Eur. J. of Forest Pathology 8: 334-337.

Dochinger, L.S., 1969. Agrobacterium gall of hybrid poplar trees in Iowa. Phytopathology 59: 1024.

Durand, R., J. Luisetti & M. Ridé, 1967. Die Beziehungen zwischen den Frühjahrsfrösten und der durch Pseudomonas syringae auf Birnbäumen verursachten Bakteriose. X. Agrarmeteorolog. Dreiländertreffen (Abstract).

Freigoun, S.O., 1973. Relationship between environmental factors and inoculum potential on two cherry varieties. Proc. of Working party on "Pseudomonas syringae group". Angers. p. 79-90.

Gardan, L., J. Luisetti & J.P. Prunier, 1972. Etudes sur les bactérioses des arbres fruitiers. IV. Recherche et étude des variations de Pseudomonas mors-prunorum f. sp. persicae à la surface des feuilles de Pêcher. Ann. Phytopathol. 4: 229-244.

Gaur, Y.D. & A.N. Sen, 1976. The utility of manganese in lactose medium to differentiate rhizobia from agrobacteria. Current Science 45: 865.

Hirano, S.S., E.A. Maher, A. Kelman & C.D. Upper, 1978. Ice nucleation activity of-fluorescent plant pathogenic bacteria. Proc. 4th Int. Conf. Plant Path. Bact. Angers 2: 717-724.

Kerr, A., 1972. Biological control of crown-gall: seed inoculation. J. Appl. Bact. 35: 493-497.

Lewis, L.N. & A.L. Kenworthy, 1962. Nutritional balance as related to leaf composition and fire blight susceptibility in the Bartlett pear. Proc. Amer. Hort. Sci. 81: 108-115.

Lindow, S.E., D.C. Arny & C.D. Upper, 1978. Erwinia herbicola a bacterial ice nucleus active in increasing frost injury to corn. Phytopathology 68: 523-527.

Luisetti, J. & J.P. Paulin, 1972. Etudes sur les bactérioses des arbres fruitiers. III. Recherches sur Pseudomonas syringae (Van Hall) à la surface des organes aériens du Poirier et étude de ses variations quantitatives. Ann. Phytopath. 4: 215-227.

Miller, H.S. & H. van Diepen, 1978. Monitoring of epiphytic populations of Erwinia amylovora in the Netherlands. Acta Horticulturae 86: 57-63.

Oku, H., T. Shiraishi, S. Ouchi, S. Kurozumi & H. Ohta, 1980. Pine wilt toxin, the metabolite of a bacterium associated with a nematode. Naturwissenschaften 67: 198-199.
Panagopoulos, C.G. & J.E. Crosse, 1964. Frost injury as a predisposing factor in blossom blight of pear caused by Pseudomonas syringae. Nature 202: 1352.
Panagopoulos, C.G. & P.G. Psallidas, 1973. Studies on the epiphyllous pseudomonads and other bacteria isolated from Citrus. Proc. Working party on "Pseudomonas syringae group". Angers. p. 67-78.
Parker, K.G., N.S. Luepschen & E.G. Fischer, 1961. Tree nutritions and fire blight development. Phytopathology 51: 557-560.
Paulin, J.P. & J. Luisetti, 1978. Ice nucleation activity among phytopathogenic bacteria. Proc. 4th Int. Conf. Plant Path. Bact. Angers. p. 725-731.
Paulin, J.P. & G. Lachaud, 1978. Inoculations with Erwinia amylovora under controlled environmental conditions. Acta Horticulturae 86: 31-38.
Ridé, M. & M. Lopez, 1976. Crown gall et peupliers d'ornement. Groupe de travail des Maladies du Peuplier. FAO Bordeaux-Angers. 5 p.
Ridé, M. & J.P. Prunier, 1963. Manifestations pathologiques consécutives aux attaques de Dendromyza sp. sur Peuplier. C.R. Congrès de l'Association Française pour l'Avancement des Sciences. p. 156-158.
Ridé, M. & M. Viart, 1966. Etude de la contamination d'une peupleraie par le chancre bactérien. Bull.Service cult. Peuplier Saule SEITA (1-2): 43-61.
Ridé, M., S. Ridé & C. Poutier, 1978. Factors affecting the survival of Xanthomonas populi on aerial structures of poplar. Proc. 4th Int. Conf. Plant Path. Bact. Angers. p. 803-814.
Schroth, M.N., A.R. Weinhold, A.H. Mc Cain & D.C. Hildebrand, 1971. Biology and control of Agrobacterium tumefaciens. Hilgardia 40: 537-552.
Thomson, S.V., M.N. Schroth, W.J. Moller & W.O. Reil, 1975. Occurrence of fire blight of pears in relation to weather and epiphytic populations of Erwinia amylovora. Phytopathology 65: 353-358.
Vigouroux, A., 1974. Obtentions de symptômes de bactériose du Pêcher (Pseudomonas mors-prunorum f. sp. persicae) sur rameaux de Pêcher détachés et conservés en survie: Effet du froid. Ann. Phytopathol. 6: 95-98.
Vigouroux, A. & C. Huguet, 1977. Influence du substrat de culture sur la sensibilité du Pêcher au dépérissment bactérien. C.R. Acad. Agric. Fr. 63: 1095-1103.
Wilson, E.E., M.P. Starr & J.A. Berger, 1957. Bark canker, a bacterial disease of the Persian walnut tree. Phytopathology 47: 669-673.

Biological and physical modifications of the environment and the resulting effect upon the host-parasite interactions in short-rotation tree crops[1]

Harold S. McNabb Jr., Richard B. Hall & Michael E. Ostry[2]

ABSTRACT

Environmental factors of the atmosphere and soil are presented in relation to their effects upon incidence of disease in short-rotation, intensive plantation culture of *Populus*. Environmental modifications inherent for this cultural system and added for pest management schemes are discussed. Tree density, a measure of spatial relationships within stands, is a recurring stand characteristic in the discussion. The term, spatial resistance, is proposed for the summation of the complicated interactions of the many environmental factors associated with stand density that affect disease expression toward decreased disease incidence.

INTRODUCTION

The continued increasing demand for wood fiber in North America has caused silviculturists to study the management of short-rotation intensive forest systems (Larson & Gordon, 1969) that approach the intensity of present-day agricultural systems. The pronounced reduction in rotation age with an accompanying increase in annual wood fiber yield per land unit allows increased management inputs that can make the short-rotation, intensively managed forest ecosystems profitable for industries utilizing wood fiber (Isebrands et al., 1979); i.e., paper and fiberboard, or for energy production.

1. Journal Paper No. J-10120 of the Iowa Agriculture and Home Economics Experiment Station, Ames, IA 50011 USA. Projects No. 2210, 2294, 2295. Also contribution from the North Central Forest Experiment Station, USDA Forest Service, 1992 Folwell Ave., St. Paul, MN 55108, USA.
2. Professor of Forest Pathology, Associate Professor of Forest Genetics, Iowa State University, Ames, IA 50011, and Associate Plant Pathologist, USDA-Forest Service, North Central Forest Experiment Station, St. Paul, MN 55108, respectively.

Selections and hybrids of the genus *Populus* have been considered ideal for short-rotation, intensively managed systems because of their rapid growth, ease of establishment through stem or root cuttings, apparent ease of coppice regeneration, and wide genetic diversity within the genus. *Populus* spp. are considered pioneer species in indigenous forest ecosystems, developing dense, even-aged, pure species stands during the early years of the stand. During the first decade of natural stand development, at least 2 overlapping periods of natural thinning occur in which pathogens and insects are involved (Graham et al., 1963). Our paper considers the relationship of environment to these host-parasite interactions as this environment is modified in the short-rotation, intensive culture of *Populus* spp. and how this environment can be further modified in the development of integrated pest management systems.

We will divide the environment into two broad categories, atmospheric and soil environments, fully realizing that these two environments interact closely, but their separate categorization will allow more critical discussion of the individual factors involved. The discussion will include both biological and physical modifications of the environment.

THE ATMOSPHERIC ENVIRONMENT

Natural reproduction of an aspen (*Populus tremuloides* Michx.) stand after clear-cutting or fire is by root sprouts, producing an initially dense stand of relatively few genotypes. Seedling reproduction of *P. deltoides* Bartr. in open, moist, bottomlands also is dense initially. During the 2 decades after reproduction, natural thinning drastically reduces stand density. Short-rotation, intensively managed systems of *Populus* selections and hybrids are based upon the premise of how efficient, in terms of time and economic benefits, the silviculturist can manage the photosynthesizing area of trees to produce maximum quantity of wood fiber per unit area of land. Thus density of such intensively cultured stands can approach initial densities of naturally reproduced aspen and cottonwood stands. Little research directly relating stand density of poplars to disease incidence is available, but many studies have been conducted relating environmental conditions to production and germination of propagules of poplar leaf and stem pathogens. These latter studies can be used in development of models relating stand density to pathogen incidence and disease development.

In Schipper's (1976a) study with 4 clones of *P.* × *euramericana* Guinier considered to be moderately to highly resistant to leaf rust (*Melampsora medusae* Thum.), 3 tree densities were used: 5 000 (1.4-m spacing), 10 000 (1.0-m spacing), and 15 000 (0.8-m spacing) trees/ha at 2 geographical locations. Ratings of rust severity and resulting recommendation for clonal use were based on Schreiner's (1959) work. At Rhinelander, Wisconsin, where the alternate host, *Larix laricina* (Du Roi) K. Koch, is present, rust ratings for 2 clones were high enough at all densities to exclude them from future plantings at that location. The other 2 clones would have been excluded from future plantings only on their rust ratings at the 15 000 trees/ha density, trees at 10 000 and 5 000/ha densities giving ratings below the arbitrary limit set by Schreiner. Results at the Ames, Iowa, location, where *Larix* spp. are scarce, were similar but less severe for rust disease. In addition, observations on incidence of *Marssonina brunnea* (Ell. & Ev.) P. Magn. paralleled rust observations. Altered disease severity at different planting densities was related to leaves remaining wet for longer periods after rain, dew, or irrigation. This longer period of leaf wetness resulted from reduction of air movement in denser stands with high canopy density. Thus, although actual inherent tree resistance may not be altered, leaves on trees grown at narrow spacing are exposed to longer periods of ideal infection conditions than are trees grown at wider spacing. Further, such interactions may be different for different clones. These observations were made before first coppice cut in these poplar stands. Our observations on sprout growth after first coppice cut indicate that the first coppice generation is even more dense than the original planted density because of multiple stem sprouting.

This example illustrates the relationship of canopy and stand densities with air movement within the stand and resulting disease incidence. In attempting to use results of other research that consider individual environmental factors and their effects upon pathogen growth and reproduction, these individual factors within poplar stands must be separated and related to canopy and stand density. Moisture and temperature are the 2 factors most involved with production and germination of propagules of pathogens and, thereby, intensity of diseases in stands (Anselmi & Cellerino, 1980; Spiers, 1978; Heather & Sharma, 1977; Pinon & Poissonnier, 1975; Castellani, 1966).

Moisture is in 2 forms, free water on leaf and stem surfaces and relative humidity of the surrounding atmosphere. Rainfall, dew, guttation, and irrigation all contribute to leaf wetness. Duration of this free water depends upon a series of complicated interactions among the vapor pressures within plant tissue, the boundary layer, and the surrounding air. In general terms, air movement, temperature, and relative humidity directly affect the duration of leaf wetness. For most pathogens, free water is considered necessary for germination of spores and host penetration by the germ tube. For certain pathogens, relative humidity values above 80 % will allow spore production while relative humidities of 98 % or greater are needed for spore germination.

Temperature directly affects the metabolism of the pathogen, thereby allowing us to determine lower and higher limits for spore production and germination. In addition, for *Marssonina brunnea*, Castellani (1966) has shown an interaction between temperature and rainfall. At higher temperatures, more rainfall is necessary for serious disease incidence. We believe this interaction to be directly related to duration of leaf wetness. Both light and air movement directly affect temperature, but under stand conditions, tend to counteract each other. Denser stands allow less light, thereby less radiant energy making for cooler conditions. Dense stands, however, have less air movement and moisture evaporation and, thus, reduced heat loss to the atmosphere. Under usual growing conditions in these dense *Populus* stands, moisture relationships thereby affect the host-parasite interactions to a greater degree than temperature.

When discussing the characteristics of stand density, mention should be made of clonal and species mixtures. An example of a species mixture will be discussed later under soil environment. Clonal mixtures could modify the atmospheric environment by using mixtures representing different crown architectures. The intensive culture system using *Populus* could violate one of the important forest pest management principles: maintenance of genetic diversity within stands. On the other hand, by judicial choice of clones or selections, we can incorporate a wider range of genotypes than would be present in natural stands that arise by root sprouting or seed dissemination from a restricted local population.

Although much research has been done on measuring atmospheric environmental factors in crop fields and relating them to disease incidence, little has been done in intensive forest culture for the determination of relationships among environmental effects and disease intensity. New instrumentation is now available that records leaf wetness, air temperature and movement, and relative humidity that will greatly aid such research. We are beginning studies of these relationships in our short-rotation, intensively managed *Populus* stands (Young et al., 1980) so that we will be able to better understand the effects of stand density as it is manipulated by tree spacing, clonal crown architecture and pest resistance, pruning (natural and artificial), coppice treatments, species and clonal mixture, rotation age, and the artificial addition of physical, chemical, and biological factors (fertilization, irrigation, pesticides, and biological control agents) on disease incidence.

Before leaving the physical atmospheric environment, further mention should be made of moisture, air movement, and light. Either overhead or ground irrigation may be used to supplement natural rainfall in these intensive-culture systems, especially if pulping effluent is available. Overhead irrigation will maintain longer periods of leaf wetness; however, it may have one benefit in a pest relationship. The cottonwood leaf beetle (*Chrysomela scripta* F.), a serious pest in plantation establishment and clonal nurseries, is effectively managed by periodic rains that physically dislodge larvae and adults from leaves (Caldbeck et al., 1978). Overhead irrigation has the same effect. Any irrigation practice will need to be integrated into pest management systems that consider all pests.

Air movement also is involved with movement of propagules of pathogens. Both spore movement from within and from without the stand are of concern. Primary inoculum may be initiated within the stand from overwintering pathogen on dead leaves (usually sexual reproduction) or from cankers or lesions on stems (usually asexual reproduction) and from outside the stand from natural stands in the locality or from stands some distance away (Kam, 1973, 1975; Kobayashi & Chiba, 1962; Chiba & Zinno, 1960; Waterman, 1954). Convection air currents within stands contribute greatly to movement of spores, especially in the elevation of spores into the horizontal air stream. Air movement is especially important in poplar culture because of the possibility of geographical isolation of specific pests. We find that *Melampsora medusae*, in areas where the alternate hosts, *Larix* spp., are scarce, may not become serious until late season. Thus, for Iowa, where *Larix* is not native, this disease is not significant, but for northern Wisconsin, growth loss may reach 20 % (Widin & Schipper, 1976). Wind movement of urediospores northward from southern cottonwood areas and southward brings the primary inoculum into Iowa.

Thirteen poplar planting sites containing 33 clones replicated three times at each site were planted throughout the north-central United States in an attempt to determine the geographical distribution of poplar pathogens and insects (Ostry, 1979). Initial disease readings indicated specific area distribution of certain leaf diseases (Schipper, 1976b). Iowa plantings were most seriously diseased with *Septoria musiva* Peck and *Marssonina brunnea*. *Septoria musiva* was found for the first time in Rhinelander, Wisconsin, plantings in 1980 even though this pathogen is common on leaves of *Populus balsamifera* L. in natural stands in northern areas of Wisconsin and Michigan. *Marssonina brunnea* was not common in the St. Paul, Minnesota, plantings from 1975 through 1978, but in 1980, this leaf disease has become serious at this location. This situation is typical of many of the diseases that we found initially only at a few locations. As poplars are grown in an area over time, the more serious the poplar leaf and stem diseases become. Some of this spread into new areas probably is associated with distribution of diseased poplar cuttings, but we believe that the spread also is by aerial movement of spores from native and locally grown *Populus* spp. in ornamental plantings. We should mention, though, that 2 leaf pathogens have remained somewhat localized, *Phyllosticta* sp. in the southern sites and *Septotinia populiperda* Waterman & Cash in the northern sites (Ostry, 1979). We believe that leaf rust is the only serious disease for which geographical distribution of the pathogen can be used as a factor in development of pest management systems. For example, moderately susceptible clones can be safely planted in Iowa.

Light, besides interacting with temperature, also has direct and indirect effects upon disease resistance. High light intensities, such as those found at high elevations, caused a breakdown of resistance in oats to *Puccinia graminis avenae* (Browning et al., 1964). Higher incidence of the serious canker of *Populus tremuloides* caused by *Hypoxylon mammatum* (Wahl.) Miller has been associated with stand openings and edges (Anderson & Anderson, 1968) and unpruned trees in these locations (Ostry & Anderson, 1979). Aspen is shade intolerant and, in dense stands, self-prunes, reducing the number of lower branches that otherwise might serve as infection sites for the pathogen. Recently, 2 insects that desire high light conditions for oviposition sites on trees (*Saperda inornata* Say and *Magicicada septendecim* L.) were found associated with the incidence of *Hypoxylon* canker (Anderson et al., 1979). Such oviposition wounds and possible bird feeding at these oviposition sites seem to be entrance points for the pathogen. Thus, light and, therefore, an open stand may cause trees of the stand to be more susceptible to specific diseases associated with light-requiring insects.

Light interacting with temperature also may predispose trees to stem canker and decay pathogens. We observed for the first time in 1980, in exposed situations within poplar stands, serious sunscald wounds on certain poplar clones. This sunscald damage was colonized by canker fungi (i.e., *Cytospora chrysosperma* Pers.) and decay fungi (i.e., *Armillaria mellea* (Vahl.) Quel.). Thus, when making pest management decisions as well as silvicultural and economic recommendations for a short-rotation, intensively managed system, limitations on both ends of the stand-density spectrum are present.

We have discussed physical effects of the atmosphere on pathogen activity. There are chemical and biological factors that should be included. Among the chemical factors are pesticides. Under field and nursery conditions, leaf diseases have been managed with fungicides (Sheridan et al., 1975; Carlson, 1974; Cellerino & Freccero, 1970). Preliminary results indicate that the canker disease caused by *Septoria musiva* in an experimental poplar plantation in central Iowa planted to a density of 10 000 trees/ha was managed with the fungicide captifol (McNabb et al., 1980). A resistant poplar clone was included with 3 moderately to highly susceptible clones. Captifol reduced the incidence of canker in these susceptible clones to the level of the resistant clone. In addition, when a surfactant was added to another fungicide, a slight indication of increased susceptibility was noted for all clones, including the resistant one. The use of fungicides in modifying the chemical atmospheric environment is being investigated further for these *Populus* culture systems.

Relatively little is known about the above-ground biological environment. Both insects and small animals are known to predispose trees to disease. Mice and rabbit feeding wounds are entrance points for canker and decay fungi. In addition, animal feeding can cause tree stress that, in turn, increases the incidence of stress-related cankers caused by *Cytospora* chrysosperma, Dothichiza populea Sacc. & Br., and *Phomopsis macrospora* Kobayashi & Chiba. One insect-breeding predisposition, that for *Hypoxylon* canker, was discussed earlier. Other types of insect predisposition include feeding on leaves and stems that range from puncture wounds to tissue loss (Palmberg, 1977), including branch and stem breakage. Wound pathogens regularly colonize such feeding and breeding sites but have not been important except for *Hypoxylon mammatum*.

Other fungi and bacteria also are part of this biological environment. A number of workers have suggested using leaf fungi that parasitize poplar leaf-rust urediospores as biological control agents (Omar & Heather, 1979). A current study in Poland associates specific leaf bacteria with rust-resistant poplar clones (Danilewicz, 1980). This approach in modifying the biological atmospheric environment should be pursued.

THE SOIL ENVIRONMENT

As with atmospheric environment, stand density has an important influence on the effects that soil environment has on disease incidence. During a drought period from 1975-77 in central Iowa, stress-related canker fungi, *Dothichiza populea*, *Phomopsis macrospora*, and *Cytospora chrysosperma*, caused serious disease in intensive culture poplar plantings (Schipper et al., 1977). Observations made in 1976 in poplar stands containing 4 *P.* × *euramericana* clones planted at 3 different initial densities (5 000, 10 000, and 15 000 trees/ha) indicated greater stem death from girdling cankers for 2 of the clones as stand density increased. The plantings were of 2 types, pure clonal blocks randomly distributed within each density plot, and alternating clonal rows (only 2 of the 4 clones used) within each density plot. The same pattern of clonal susceptibility as influenced by spacing was observed in both cases. Therefore, we concluded that 2 clones had inherent resistance to stress-related cankers.

Previous studies with these stress-related cankers have indicated a close relationship between bark moisture levels and resistance to these fungi (Butin, 1956). Bark moisture was related to wound periderm development and, therefore, to host ability to isolate pathogens and to callus injured tissue.

Further observations on the alternate-row plantations during the next 2 years (1977 and 1978) differed from the 1976 data. The clonal differences of 1976 disappeared (Haywood & McNabb, 1979). Greenhouse and growth-chamber studies with these 2 clones indicated no differences in bark moisture relationships. The only possible known differences between these 2 clones that could account for our observations are root geometry and amount of root growth. Simply stated, the seemingly more resistant clone in 1976 had deeper roots. As drought conditions persisted during the second year, 1976-77, soil moisture continued to be depleted to a lower depth. Therefore, when all roots of the seemingly resistant trees became associated with moisture-deficient soils, stress-related pathogens were able to cause disease.

Although differences among densities continued during 1977 and 1978, such differences became less significant over this period. Prolonged drought conditions, especially those that cause continued fall and early winter soil-moisture deficiency, seemed to even the density and clonal effects with our *P. × euramericana* clones, with the result being serious stem kill of all clones by canker fungi.

As discussed earlier, irrigation may become an alternative in these intensive-culture systems. Under irrigation, soil-moisture deficiencies are eliminated during periods of drought and on drier planting sites. Again, caution is needed so that adverse effects of higher moisture on leaf and stem diseases are minimized.

With the high densities and coppice rotations that are characteristic of intensive culture, the potential for tree-to-tree root grafting is substantial. This may have a significant role in disease transmission in such stands. Alternatively, root grafts might be beneficial in this type of forestry if they protect the original spacing and composition of the stand by allowing cross-feeding between root systems. Our investigations have demonstrated the possibility of such cross-feeding but have been inconclusive on the possibility of wetwood bacteria transmission (Paarmann, 1980), and have demonstrated genetic barriers to root graft formation in *Populus* hybrids.

As would be expected, stand density interacts with other soil factors involved in predisposition of trees to pests. Obvious factors are the many soil nutrients absorbed by plant roots. Drain on such nutrients becomes greater as stand density increases. Although literature is available on nutrient requirements for growth of poplars (Burg & Schoenfeld, 1978; Zoltán, 1978), there are few confirmed examples of disease modifying effects of specific nutrients with *Populus* (Bruck & Manion, 1980; Garbaye & Pinon, 1973; Suzuki, 1973; Gremmen, 1964).

Reports concerning the effects of nitrogen on another forest tree suggest possible modifying effects that nutrients may have in host-parasite interactions in *Populus* plantations. High nitrate levels were associated with a decrease of *Poria weirii* root rot in Douglas-fir in Oregon (Shea, 1970). When red alder, a nitrogen-fixing (actinorhizal) species, occurred within Douglas-fir stands, this root rot was relatively scarce (Trappe, 1971). Thus, besides applications of nitrogen fertilizer, another option is indicated; biological modification of the soil chemical environment with nitrogen-fixing organisms.

Manipulation of soil microorganisms, especially those associated with tree roots, actinorhizal bacteria, and mycorrhizal fungi, is an approach that we believe to be promising for modifying nutrient relationships in intensive, short-rotation systems (Hall et al., 1979). In addition, the use of actinorhizal species in mixed plantings may have beneficial modifying effects on spatial interactions within the stand atmosphere. Tests have begun using European black alder (*Alnus glutinosa* (L.) Gaertn.) in mixtures with *Populus* (Borders, 1980). Careful monitoring of both soil and atmosphere will be essential for critical assessment of the effects of such stand mixtures.

During the past decade, knowledge of mycorrhizal fungi and the relationships that they form with trees has increased to a level that allows possible manipulation of these organisms under nursery and stand conditions. Poplars develop both ectomycorrhizae and endomycorrhizae (Walker, 1979). Ectomycorrhizae may prevent infection of roots by certain root-rot organisms. The endomycorrhizal relationships also seem to modify disease interactions. Interpretation of studies with endomycorrhizae are difficult because these structures increase the efficiency of phosphorus uptake by trees.

Interactions among soil organisms indicate other ways that these root pathogens can be managed. In soils below pH 6 in East Anglia, England, *Fomes annosus* (Fr.) Cke. is not able to compete successfully with other soil fungi, thus causing only minor root-rot incidence in Scots pine plantations on such sites (Rishbeth, 1950). Once an understanding of such soil interactions is available, specific soil treatments may be possible for the manipulation of soil microbiological populations. Indeed, we may be doing such now without realizing it. Although direct herbicidal injury to poplars has been noted (Anselmi, 1978), effect of herbicides on soil organisms in poplar-plantation site preparation has not been examined thoroughly (Walker, 1979).

CONCLUSION

Our discussion of the atmospheric and soil environments within the intensive short-rotation culture systems of *Populus* has indicated a number of environmental modifications resulting from the system itself and others suggested for stand management. One stand characteristic has recurred in our discussion, tree density. This characteristic is a measure of the spatial relationships within the stand. Practical modification of stand environment is directly or indirectly associated with these spatial relationships. The fact that tree density has been associated with reported changes in host-parasite interactions in *Populus* stands indicates the importance of density modifications in the development of pest management schemes.

The complicated interactions among the many environmental factors associated with the spatial relationships of stand density present a challenge for present and future research efforts. If the basic factors are chosen properly, the summation of these interactions affect disease expression producing a type of stand resistance. We propose the term, spatial resistance, for this stand effect. Spatial resistance should not be overlooked when evaluating genetically controlled host-parasite interactions. In addition, the concept of spatial resistance will help to organize observations on stand-host-parasite relationships found in other forest stands. Such a systems approach will promote a better understanding of one of the many complex phenomena associated with forest and plantation ecosystems.

REFERENCES[3]

Anderson, G.W. & R.L. Anderson, 1968. Relationship between density of quaking aspen and incidence of Hypoxylon canker. Forest Science 14: 107-112.

Anderson, N.A., M.E. Ostry & G.W. Anderson, 1979. Insect wounds as infection sites for Hypoxylon mammatum on trembling aspen. Phytopathology 69: 476-479.

Anselmi, N., 1978. Fitotossita verso le salicacee di prodotti ormonici utilizzati nel diserbo del riso. Cellulosa e Carta 29: 17-36.

Anselmi, N. & G.P. Cellerino, 1980. Relationship between the resistance of poplar to Marssonina brunnea and the periods of incubation and of presporification of the parasite. In: Proc. F.A.O. International Poplar Commission Working Party on Diseases - I.U.F.R.O. Working Party S2.05.03 meeting, Kórnik, Poland, 1-5 September. 10 p.

Borders, B.E., 1980. First year growth of two Populus clones grown in mixture with Alnus glutinosa. Unpublished M.S. Thesis, Iowa State University, Ames. 44 p.

Browning, J.A., E. Bustamante, J. Orjuela & H.D. Thurston, 1964. The effect of light intensity on the reactions of oat seedlings to Puccinia graminis avenae. (Abstract). In: Agronomy abstracts, American Society Agronomy meeting, Kansas City, Missouri, 15-19 November. p. 62.

Bruck, R.I. & P.D. Manion, 1980. Interacting environmental factors associated with the incidence of Hypoxylon canker on trembling aspen. Canadian Journal Forest Research 10: 17-24.

Burg, J. van den & P.H. Schoenfeld, 1978. Populierenteelt, bodemvruchtbaarheid en grondonderzoek. Populier 15: 75-80.

Butin, H., 1956. Untersuchungen über Resistenz und Krankheitsanfälligkeit der Pappel genenüber Dothichiza populea Sacc. et Br. Phytopathologische Zeitschrift 28: 353-374.

Caldbeck, E.S., H.S. McNabb, Jr. & E.R. Hart, 1978. Poplar clonal preferences of the cottonwood leaf beetle. Journal Economic Entomology 71: 518-520.

Carlson, L.W., 1974. Fungicidal control of poplar leaf spots. Canadian Plant Disease Survey 54: 81-85.

Castellani, E., 1966. Influence des facteurs climatiques sur les infections des Peupliers euramericains par Marssonina brunnea (Ell. et Ev.) P. Magn. Phytopathologia Mediterranea 5: 41-52.

Cellerino, G.P. & V. Freccero, 1970. Control of Marssonina brunnea scheduled according to the flushing of poplars. In: Proc. F.A.O. International Poplar Commission Working Group on Diseases meeting, Farnham, U.K., 27 September-3 October. 11 p.

Chiba, O. & Y. Zinno, 1960. Uredospores of the poplar leaf rust, Melampsora larici-populina Kleb., as a source of primary infection. Journal Japanese Forestry Society 42: 406-409.

Danilewicz, K., 1980. Erwinia herbicola and Pseudomonas fluorescens as indicators for examination of the resistance of poplar clones to Melampsora larici-populina Kleb. In: Proc. F.A.O. International Poplar Commission Working Party on Diseases-I.U.F.R.O. Working Party S2.05.03 meeting, Kórnik, Poland, 1-5 September. 2 p.

Garbaye, J. & J. Pinon, 1973. Nutrition minérale et sensibilité à Marssonina brunnea (Ell. et Ev.) P. Magn. de Populus X euramericana (Dode) Guinier, cv. 'I-214': étude préliminaire sur jeunes plants en milieu controlé. Annales des Sciences Forestières 30: 423-431.

3. Although 117 references were used in preparing this review, space limitations prevented their inclusion in this section. A complete list is available from the senior author.

Graham, S.A., R.P. Harrison, Jr. & C.E. Westell, Jr., 1963. Aspens - Phoenix trees of the Great Lakes Region. The University of Michigan Press, Ann Arbor. 272 p.
Gremmen, J., 1964. The Marssonina disease of poplar, 2. Inoculation experiments on leaf discs with ascospores and conidia. Nederlands Boschbouwtijdschrift 36: 149-157.
Hall, R.B., H.S. McNabb, Jr., C.A. Maynard & T.L. Green, 1979. Toward development of optimal Alnus glutinosa symbioses. Botanical Gazette 140 (Suppl.): S120-126.
Haywood, W.F. & H.S. McNabb, Jr., 1979. Moisture stress and incidence of stem cankers in intensively grown poplars. In: Proc. First North Central Tree Improvement Conference, Madison, Wisconsin, 21-23 August. p. 123-132.
Heather, W.A. & J.K. Sharma, 1977. Some aspects of poplar rust research in Australia. Australian Forestry 40: 28-43.
Isebrands, J.G., J.A. Sturos & J.B. Crist, 1979. Integrated utilization of biomass. TAPPI 62: 67-70.
Kam, M. de, 1973. Life history, host range and distribution of Septotinia podophyllina. European Journal of Forest Pathology 3: 1-6.
Kam, M. de, 1975. Ascospore discharge in Drepanopeziza punctiformis in relation to infection of some poplar clones. European Journal of Forest Pathology 5: 304-309.
Kobayashi, T. & O. Chiba, 1962. Overwintering of some leaf-attacking fungi of poplars. Journal Japanese Forestry Society 44: 19-24.
Larson, P.R. & J.C. Gordon, 1969. Photosynthesis and wood yield. Agricultural Science Review 7: 7-14.
McNabb, H.S., Jr., M.E. Ostry, P.A. Tipton, R.S. Sonnelitter, J.M. Zito & E.S. Caldbeck, 1980. Fungicidal management of Septoria canker of Populus grown under short-rotation intensive culture. (Abstract 534). In: Program and abstract booklet, American Phytopathological Society-Canadian Phytopathological Society meeting, Minneapolis, Minnesota, 24-28 August. p. 243.
Omar, M. & W.A. Heather, 1979. Effect of saprophytic phylloplane fungi on germination and development of Melampsora larici-populina. Transactions British Mycological Society 72: 225-231.
Ostry, M.E., 1979. Disease research of poplars grown under intensive culture in the north central region of the United States. In: Proc. North American Poplar Council 16th meeting, Thompsonville, Michigan, 14-16 August. p. 83-91.
Ostry, M.E. & G.W. Anderson, 1979. Hypoxylon canker incidence on pruned and unpruned aspen. Canadian Journal Forest Research 9: 290-291.
Paarmann, S.C., 1980. Root grafts in Populus: Growth relations and wetwood disease transmission between trees. Unpublished M.S. Thesis, Iowa State University, Ames. 42 p.
Palmberg, C., 1977. Introduction and improvement of poplars in Australia. Australian Forestry 40: 20-27.
Pinon, J. & M. Poissonnier, 1975. Etude épidemiologique du Marssonina brunnea (Ell. et Ev.) P. Magn. European Journal of Forest Pathology 5: 97-111.
Rishbeth, J., 1950. Observations on the biology of Fomes annosus, with particular reference to East Anglian pine plantations. I. The outbreaks of disease and ecological status of the fungus. Annals of Botany 14: 365-383.
Schipper, A.L., Jr., 1976a. Poplar plantation density influences foliage disease. In: Intensive plantation culture - five years research. USDA Forest Service, General Technical Report NC-21, North Central Forest Experiment Station, St. Paul, Minnesota. p. 81-84.
Schipper, A.L., Jr., 1976b. Hybrid poplar diseases and disease resistance. In: Intensive plantation culture - five years research. USDA Forest Service, General Technical Report NC-21, North Central Forest Experiment Station, St. Paul, Minnesota. p. 75-80.
Schipper, A.L., H.S. McNabb, Jr. & W.F. Haywood, 1977. Dothichiza and Phomopsis cankers of hybrid poplar in Iowa. (Abstract 128). In: Proc. American Phytopathological Society 4: 109-110.
Schreiner, E.J., 1959. Rating poplars for Melampsora leaf rust infection. USDA Forest Service, North-eastern Forest Experiment Station Research Note NE-90. 3 p.
Shea, K.R., 1970. Poria root rot: Problems and progress in the Pacific Northwest. In: T.A. Toussoun, R.V. Bega & P.E. Nelson (Eds): Root diseases and soil-borne pathogens, University of California Press, Berkeley. p. 164-166.
Sheridan, J.E., J.E. Harper & G. Stevenson, 1975. Note on epidemiology and control of poplar leaf rust. New Zealand Journal Science 18: 211-216.
Spiers, A.G., 1978. Effects of light, temperature, and relative humidity on germination of urediniospores of, and infection of poplars by, Melampsora larici-populina and M. medusae. New Zealand Journal Science 21: 393-400.
Suzuki, K., 1973. Studies on the susceptibility to poplar leaf rust influenced by

different nutrient conditions. (I) Changes of susceptibility induced by nutrient deficiency. Journal Japanese Forestry Society 55: 29-34.
Trappe, J.M., 1971. Biological control - forest diseases. In: Proc. Western Forest Pest Committee, Western Forestry and Conservation Association meeting, Spokane, Washington, 30 November. 2 p.
Walker, C., 1979. Hybrid poplar mycorrhizae and endogonaceous spores in Iowa. Unpublished Ph. D. Dissertation, Iowa State University, Ames. 258 p.
Waterman, A.M., 1954. Septoria canker of poplars in the United States. USDA Circular No. 947. 24 p.
Widin, K.D. & A.L. Schipper, Jr., 1976. Epidemiology and impact of Melampsora medusae leaf rust on hybrid poplars. In: Intensive plantation culture - five years research. USDA Forest Service, General Technical Report NC-21, North Central Forest Experiment Station, St. Paul, Minnesota. p. 63-74.
Young, K.R., H.S. McNabb, Jr. & M.E. Ostry, 1980. Aerial pathogens in a central Iowa poplar plantation. (Abstract G-14). Iowa Academy of Science Proceedings 87: centerfold 6.
Zoltán, T., 1978. Anyagforgalmi vizsgálatok nyárfával végzett szennyvízöntözéses tenyészedény kísérletekben. Agrokémia és Talajtan 27: 399-415.

Genes for resistance to insects in agriculture with a discussion of host-parasite interactions in Carya

Marvin K. Harris

Department of Entomology, Texas A & M University, College Station, Texas 77843, USA

ABSTRACT

Information concerning the genetics of resistance in insect/plant interactions in agriculture is available for only some plants and a minuscule number of insects. The scanty data presently available allows numerous interpretations, one of which is that agriculturally useful resistance to insects is usually an artifact of plant domestication rather than a capitalization on a coevolved product of insect and plant. Closer study of natural defense mechanisms in indigenous systems, particularly systems undergoing domestication, can provide new evidence to develop a holistic theory of host plant resistance to insects. Greater emphasis on insect/plant genetics and insect/plant biogeography would help provide the necessary framework for intra- as well as interdisciplinary communication of ideas concerning host plant resistance; and, assist in designing strategies for using this tool in pest management.

ORIGIN AND PERMANENCE OF RESISTANCES

A knowledge of the genetics of host-plant resistance to arthropods and the reciprocal (virulence of arthropods to plants) is important because of the role it can play in discovering, incorporating, deploying and maintaining resistant plants for use in agriculture. Numerous recent publications provide significant space to this subject (Day, 1977; Harris, 1980a; Maxwell & Jennings, 1980; Russell, 1978).

A close inspection of Table 1 and the plant/insect interactions suggests some generalities (also noted by Russell, 1978; Gallun & Khush, 1980). First, quantitatively inherited resistance is stable. Second, biotype development, if it occurs, will be primarily against antibiosis mechanisms when arthropods are exposed to monogenic resistance. Third, monogenic resistance has been quite stable in some plants against some insects. Fourth, agriculturally useful resistance may in general be found in any of

Table 1. Inheritance, numbers of genes involved in resistance, and prior relationship of resistance, of crops to insects (compiled from reviews by Carpenter et al. (1979), Gallun & Khush (1980), Harris (1975), and Russell (1978)).

Crop	Insect	Resistance[1], number of genes			Biotypes known	Prior relationship[2]
		dominant	recessive	quantitative		
Rice						
	brown planthopper	2	2		+	S?
	green leafhopper	4	1			A
	white-backed planthopper	1				U
	gall midge	X	X?		+	U
Sorghum						
	corn leaf aphid	X			+	U
	greenbug	1*			+	A
	shoot fly		X			U
	chinch bug	X				A
Wheat						
	hessian fly	7	1		+	A
	greenbug	1	1			S?
	cereal leaf beetle	?	?	X		S?
	stem sawfly	1*	2			A
Barley						
	hessian fly	1				S?
	cereal leaf beetle		1			S?
	greenbug	2			+'	S?
Corn						
	european corn borer			X		A
	corn earworm			X		U
	fall armyworm			X		S?
Cotton						
	boll weevil	3	1			U
	thrips	3*				U
	tobacco budworm	2				U
	Empoasca spp.	1*	X			U
Apple						
	Dysaphis devecta	1				A
	rosy apple aphid	1				A
	woody apple aphid	1				A
Raspberry						
	Amphoraphora rubi	X			+	A
	Amphoraphora agathonica	X				A

Table 1. Continued

Muskmelon					
melon aphid	1				U
red pumpkin beetle	1				S?
Squash					
striped cucumber beetle	X				S?
squash bug			X		S?
Sweet clover					
sweetclover aphid	1				U
Alfalfa					
pea aphid	1			+	U
spotted alfalfa aphid	X?	X?		+	U
Pear					
pear psylla	X				A
Pecan					
pecan twig phylloxera	X?	X?		+?	S

1. X = number of genes unclear; * = incomplete dominance; ? = status questionable.
2. S = sympatric; A = allopatric; U = unknown.

the categories of inheritance. Fifth, genes for resistance may be obtained both from plant germ plasm which has coevolved with the arthropod as well as from that which has not coevolved with the arthropod.

The stability of quantitatively inherited resistance has been an indication to some, most notably Robinson (1980), that resistance breeding efforts should be concentrated primarily in this direction due to the instability, with occasionally catastrophic consequences, of some monogenic resistance (exhibited primarily in diseases and in insects with parthenogenetic reproduction).

Russell (1978) suggests that pest resistance based on monogenically inherited morphological characters such as pubescence or solid stems confers durability, citing wheat stem sawfly, cereal leaf beetle and *Empoasca* sp. as examples; and also argues that other types of monogenic resistance may have been insufficiently exposed to pest attack to allow biotype expression to occur, or even if biotypes have occurred, their ability to disseminate may be low and thus they have so far escaped discovery. Agreement with Russell (1978) would allow including morphologically based resistant monogenes to the proposed plant archetype. Even this seemingly straightforward generality requires some qualification in that many, perhaps most, morphological characters are polygenically controlled in the plant.

Data in Table 1 summarize most of what has been reported regarding genetics of resistance to arthropods in agriculture. There are 3 major problems with interpretation of these data.

1. Quantity: The genetic bases for most agricultural resistances to arthropods have yet to be described and subsequent studies may change present trends. There are 33 insects attacking 14 crops included in Table 1. The total number of insect species known is about 3/4 million of which only a few thousand or so are considered to be crop pests (Metcalf et al., 1962). Total plant species known number a few hundred thousand of which only a few thousand are considered to be economically important (i.e. Hedrick, 1919). Thus, present knowledge regarding genetics of resistance to insects in agricultural plants has been obtained from less than 1 % of the species in each category.

2. Quality: Furthermore, no complete characterization of the genetics of resistance has been conducted for even 1 insect or plant so that even the best studies of insects on the most well investigated crops are as yet incomplete. Also, resistance genes which are known have often not been characterized as to their origin (allopatric, sympatric-coevolved and sympatric-unselected, would be one such system of classification (Harris, 1980a)). This is also true for the majority of resistances in general (Leppik, 1970; Harlan, 1977; Harris, 1980a) so that even when their genetics are described, this deficiency is likely to remain.

3. Perspective: Information presented in Table 1 was primarily obtained by agriculturists whose major goal was obtaining plants which would produce high yields of good quality of a desired resource on a regular basis at an acceptable cost in the presence of the insect. Browning (1974) addresses a similar problem of perspective in pathology. Painter (1968) observes in regard to past successes in breeding for Hessian fly resistance that 'these goals have been reached without exact knowledge of the mechanism of resistance and, in the early stages, without exact information about the genetic factors involved'. Increased attention to the genetics of resistance has been made during the past decade or so on crops like wheat, raspberries, rice, alfalfa, sorghum, etc., and hopefully a better overall understanding will be obtained in the future. The inclusion of both plant and arthropod population geneticists in teams breeding for resistance would improve the perspective from which data have been gathered in the past (see Gallun & Khush (1980) for methodology).

Data in Table 1 represent a very small sample size from as yet incomplete studies carried out with the primary intent of producing better agricultural plants rather than understanding the genetics of resistance. Thus, whatever interpretations are made must be of a speculative nature and tentative pending further studies.

Harlan & Starks (1980) observe that 'Biological theory would suggest that the most likely region in which to find resistance to a given insect would be where the insect is endemic... In actual practice, there may or may not be a correlation'. Presumably, the biological theory referred to implies that sufficient natural selection pressure was exerted on the plant population by the arthropod population through the course of coevolution so that the present situation has resulted in plant individuals and populations which are both biologically and agriculturally fit. Biologically fit? Yes. This is self-evident because of the continued presence of both species in the same habitat. Agriculturally fit? No or maybe. Agricultural value is dependent upon plants or plant parts relating to human resources which need not be related to plant characters important for survival at all. In fact, numerous defense mechanisms in plants involving escape in space and time and biological associations may be extremely important in the wild for the production of fertile offspring, and completely unacceptable in agriculture (Harris, 1980b). Of the 69 resistances (Table 1), 27 rely primarily on allopatric resistance, 1 on sympatric resistance, 16 on possibly sympatric resistance and sources of 25 are unknown. Allopatric resistance has been heavily relied upon in agriculture for the construction of genetic defenses against arthropods and this shows that practice has not been consistent with the 'biological theory' addressed above.

There are additional factors which should be considered if one were to investigate biological foundations of resistance in agriculture. For example, resistance should be most abundant for those arthropods which have exerted the greatest natural selection pressures on the plant through time. Genes for resistance should be dominant, linked and consist of multiple alleles to facilitate and stabilize inheritance. Where multiple selection pressures are consistently experienced by the plant through time, as from an associated arthropod complex, again linkage among genes and multiallelic inheritance would appear to be expected as the progeny of generation after generation of the plant are selectively attacked and their ability to produce fertile offspring is affected by these selection pressures. Furthermore, arthropods which are exposed to these defenses should reflect similar genetic makeups as their own abilities to produce fertile progeny are affected (see Person, 1959, 1966; Person & Mayo, 1974; Groth & Person, 1977, for discussions in plant/pathogen systems, and Lewontin, 1974).

Agriculturally useful resistance has not been associated with arthropods which are or have been observed to exert the greatest selection pressure on the plant population. Genes for resistance to a given species are seldom linked and rarely if ever is resistance to more than one species obtained from a single gene. Multiple alleles for agriculturally useful resistance to arthropods are unknown. Furthermore, arthropod biotypes inherit their virulence recessively. Only in the matter of dominance of plant inheritance of resistance does biological theory agree with agricultural experience. This can also be explained through traditional plant breeding practices and host plant resistance investigations preferentially identifying dominant plant genes rather than recessive ones, thereby biasing these results.

If one postulates that resistance can be an artifact of agriculture and represents a new genotype to which the arthropod has not previously been exposed, then virulence genes will endow individual arthropods possessing them with suddenly increased fitness in this new circumstance. By a rapid presentation of resistance genes, corresponding virulence genes (if they exist at all), could increase in frequency in the arthropod population with almost equal rapidity and result in homozygous recessively virulent biotypes which still possess modifiers evolved for the dominant but now avirulent alleles at these loci. Further, if both the 'avirulent' population and the 'virulent' population are mixed and exposed to a host that is susceptible to both, the 'avirulent' population should exhibit greater fitness and predominate (see MacKenzie, 1980, for further discussion). Reduced fitness of bacterial 'biotypes' resistant to antibiotics (i.e. McVeigh & Hobdy, 1952) and arthropods resistant to pesticides as compared to populations in antibiotic and pesticide-free environments, respectively, indicate this is a general phenomenon associated with man-manipulated selection forces. MacKenzie (1980) observes that 'more research emphasis on differential survival ... is needed'.

This becomes particularly important when one considers taking an indigenous plant and selecting an archetype to be produced in even-aged monocultures in the presence of its coevolved biota for an extended period of time. Henry Ford's 2.5-million-acre rubber plantation failure in its' native Brazil in the 1930's provides a bitter example of the problem and may be contrasted to the more recent multimillion acre endeavor of Daniel Ludwig at Jari, Brazil, where non-indigenous *Gmelina arborea* Van Hics and *Pinus caribaea* Morelet tree plantations are being grown successfully for the time being on an equally vast scale. The latter effort is more typical of traditional agricultural approaches in that centers of production are usually removed from plant centers of origin. This is true for other tree crops as well such as coffee, cacao, apple, banana, coconut, almond, and rubber. Only a few important tree food crops have not followed this pattern and, of course, tree fiber-crops are only recently moving toward exotic plantations.

Considering the mongrel origins of most of our agricultural plants and the often unclear relationships these plants have had with their attendant biota (Harris, 1980b), a better understanding of what does happen in plant domestication to influence susceptibility to arthropods may be obtained from a closer inspection of a plant as cultivation of it proceeds within its original habitat.

The pecan, *Carya illinoensis* (Wang.) K. Koch (Juglandaceae), is indigenous to the southwestern U.S. (Little, 1971), as are the arthropods which commonly attack it (Harris, 1980b, 1980c). The pecan is also the most important horticultural crop native to the U.S. (Brison, 1974), and is presently undergoing domestication so that experimental situations exist varying from wild ecosystems to monocultures of single cultivars grown over hundreds of hectares. Nuts produced from wild trees are highly palatable and present estimates indicate that ca. 85 % of the hectarage in Texas (ca. 300 000) consists of trees nature planted and ca. 75 % of the nuts sold in an average year are produced by such trees. Thus, in this spectrum of wild ecosystem to monoculture, the 'managed native' (wild pecan trees which have been cleared of competing trees and brush, thinned, pruned, fertilized and sprayed with pesticides in varying degrees) presently predominates. Considering that scarcely a century ago, wild trees in the natural ecosystem exclusively reigned, the present status of the pecan appears to represent a dramatic change; however, the next few decades are likely to witness an even more rapid transition toward the managed cultivar in monoculture as 'managed natives' are replaced by deliberately planted monocultures of selected pecan cultivars and other crops such as soybeans.

The walnut caterpillar

Three insects which attack pecan provide some insight into pecan defense mechanisms. The walnut caterpillar *Datana integerrima* G. & R. (Lepidoptera: Notodontidae) is a leaf feeder of plants in the Juglandaceae. It is distributed east of the Rocky Mountains from Canada to Mexico and is univoltine into the latitude of Illinois, bivoltine to Northern Texas and trivoltine farther south. Massive widespread defoliations have been reported caused by walnut caterpillar since the early 1900's (Cutler, 1976; Haseman, 1940; Hixson, 1941). One such infestation in Texas was observed in 1973 and documentation of part of the affected area was made using aerial infrared photography (in cooperation with Bill Hart, USDA-SEA, Citrus Insects Laboratory, Weslaco, TX). Studies at the time throughout the area and follow-up studies using the film indicated no trees (unless sprayed with pesticide) in the Juglandaceae in the area were able to withstand attack and complete defoliation resulted (Harris, 1980c). This was a severe disappointment to the investigators who had hoped to identify sources of walnut caterpillar resistant pecan germ plasm from the affected area. All indications from this study and follow-up investigations indicate that no significant biological differences occur among the thousands of pecans examined which could be used as a source of resistance. We concluded that pecans are, for all practical purposes, universally susceptible to walnut caterpillar and, conversely, walnut caterpillar are universally virulent on pecan. Harris (1980c) has speculated that this situation may have been the result of previous selection; and, if this is true, the pecan gene pool would not contain genes for resistance (agriculturally speaking). In fact, individuals containing 'resistance' genes would be less reproductively fit than the preponderantly susceptible individuals and the culling of resistance would be a continual and ongoing process even today.

The pecan weevil

A second insect which attacks the pecan is the pecan weevil, *Curculio caryae* (Horn) (Coleoptera: Curculionidae). This predominantly biennial obligatory nut feeder infests pecan and hickory in the plant genus *Carya* (Gibson, 1968). Pecan nuts in the water stage of development are destroyed by adult feeding, and subsequent oviposition in more mature nuts results in nut destruction by the larvae. Estimates of damage capacity indicate that one pecan weevil can destroy about 6 nuts by feeding, with the female destroying an additional 22 or so through oviposition and progeny development during one generation (2-3 year life cycle). This attack occurs about 120 days after flowering when the nut has reached full size and begun to fill with energetically expensive oils, fats and proteins. No macrostructure of the pecan tree contains such a large amount of energy in so small a package nor represents a more important element in the effort to continue the species. It appears self-evident, given this expenditure of energy and the biological importance of the nut to the pecan, that the pecan weevil should represent one of the most important biotic selection pressures on the pecan. Consequently, this interaction has been examined in order to identify the nature and extent of pecan defense mechanisms with regard to pecan weevil.

Defense mechanisms are present and appear to consist of escape in time and space exemplified at the nut cluster, the individual plant and the population level. Pecan nuts are susceptible to pecan weevil oviposition based on their phenology. The period from gel stage to shuck split is the window in time when successful oviposition is possible (Harris & Ring, 1979). Nuts attacked during the water stage, just prior to the onset of the gel stage, are shed by the tree and it is questionable if oviposition is ever seriously attempted in such nuts although adult feeding and subsequent nut shedding certainly occurs. All pecan nuts studied have been found to be acceptable for oviposition between gel stage and shuck split and to sustain larval development once oviposition has occurred. Confrontive defense mechanisms to prevent pecan weevil oviposition between gel stage and shuck split have yet to be found and I believe they do not occur in any biologically meaningful capacity, if at all. Given stable supplies of nuts each year, the pecan weevil appears capable of infesting most if not all of them so that few if any nuts escape (see Harris & Ring, 1980; Harris et al., 1981, for more discussion).

However, nut supplies are not stable from year to year. In fact, nut numbers vary widely through time in wild and native populations of trees from no measurable production in some years to production in excess of 100 000 nuts/hectare in others. Furthermore, nut production and barrenness appear to be relatively synchronized from tree to tree over wide areas (100 km^2 or more) so that whole areas tend to produce nuts in a given year followed by 1 or more barren years, etc.

The magnitude of this fluctuation and its effects on the pecan weevil are readily apparent when wild trees are examined for pecan weevil. Typically, such trees have few if any pecans and thus weevil populations are quite low. The few pecan weevils present in the area when a mast year occurs are quite successful in increasing their population since oviposition sites are temporarily non-limiting, but the synchronized low nut production which follows starves the population back to low levels. Effects on the pecan are that production is sufficient in mast years to satiate the nut predator with only a small portion of the nuts produced, while the rest escape. No doubt this defense mechanism also reduces selection pressures from other nut feeders.

The pecan phylloxera

A third insect, the pecan phylloxera *Phylloxera devastatrix* (Perg.) (Homoptera: Phylloxeridae) produces conspicuous galls on pecan petioles, leaflet midribs and occasionally nutlets. These galls are formed by nymphs hatching from overwintering eggs shortly after budbreak when they settle and feed on the young rapidly growing tissue which is then induced to form a gall around them. Galls split open 75 days or so after budbreak and affected leaves begin to dehisce from the tree leaving only a few galled leaves until leaf drop in the fall. Heavily galled trees have less leaf surface and where flowers or flowering shoots are galled, nut production is reduced or eliminated, but high nut reduction due to infestation only occurs in the severest outbreaks. Wild trees rarely can be found to be infested with *P. devastatrix* although close inspection will sometimes reveal a gall now and again. Only twice in the past 8 years of observing several thousands of genetically different pecans have I found a wild tree to be heavily infested and then the infestation was limited to 1 tree. Vegetatively reproduced cultivars, however, are another matter. The cultivars 'Stuart', 'Schley' and 'Success' are reported susceptible by some authors and resistant by others (Table 2). J.W. Stewart (TAEX Area Entomologist, Uvalde, TX) has observed a tree grafted to 'Texas Prolific' and 'Success' in which the 'Texas Prolific' has been heavily galled year after year, yet the 'Success' is virtually gall-free. These kinds of findings suggest a differentiation of the pecan by the phylloxera and a differentiation of the phylloxera by the pecan as has been reported in numerous plant/pathogen interactions, (Wood & Graniti, 1976; Day, 1977) as well as

Table 2. Responses of various pecan cultivars to *Phylloxera devastatrix* from various locations at various times.

Location	Cultivar	Resistant	Susceptible	Reference
Georgia	Stuart		X	Moznette et al. (1940)
	Schley		X	
	Success		X	
Louisiana	Stuart		X	Baker (1935)
	Schley		X	
Louisiana	Caspiana		X	Calcote & Bagent (1974)
	Schley		X	
	Success		X	
	Stuart	X		
Louisiana	Caspiana		X	Boethel, in: Carpenter et al. (1979)
	Stuart	X		
Texas	Texas Prolific		X	Stewart (pers. comm.)
	Success	X		
Texas	San Saba Improved		X	Calcote (pers. comm.) and personal investigations at Big Valley, TX (see text)
	Western Schley	X		
	Eastern Schley	X		
	Caddo	X		
	Cheyenne	X		
	Mahan	X		
	Sioux	X		
	Ideal	X		
	Burkett	X		

wheat/Hessian fly, *Rubus*/raspberry aphids, alfalfa/alfalfa aphids (Maxwell & Jennings, 1980); and conifer/scale (Edmunds & Alstad, 1978) interactions.

DISCUSSION

Summarizing resistance to walnut caterpillar, pecan weevil and *P. devastatrix* in pecan: walnut caterpillar susceptibility appears to occur among all trees at similar levels so that when an outbreak occurs, defoliation is evenly spread across the trees; pecan weevil susceptibility appears to be mitigated primarily by the capacity of individual trees to synchronize extended barren periods with years of high nut production across the tree population so that entire areas are producing nuts or are relatively barren at the same time. Additionally some within tree and between tree escape is possible in bearing years due to the number and manner in which clusters are produced and the distribution of onset of gel stages from tree to tree, respectively, but this is minor compared to the contribution masting makes to escape from pecan weevil attack; *P. devastatrix* resistance appears to be under the control of major genes and subject to biotype development in the insect.

The pecan defense mechanism to each insect discussed appears to be phenotypically and, presumably, genotypically different and only resistance to *P. devastatrix* would fall within the realm of traditional host plant resistance to insects commonly selected for use in agriculture. Furthermore, this classic kind of resistance has already been shown to be vulnerable to what appears to be biotype development, and the likelihood of developing a single cultivar which would remain resistant for 100 years or so seems extremely optimistic.

In a recent review of host plant resistance of pecan to insects and mites, Carpenter et al. (1979) summarize reports from 1917-1979 on 22 insects and note: 'Partial resistance to the pecan weevil ... was demonstrated to result from host evasion. No basis for resistance to other pests had been confirmed'. This indigenous plant population has yet to provide us with agriculturally useful resistance to most endemic arthropods associated with it. However, major production of this commodity within its native range still relies primarily on managed natives and as this 'indigenousness' (Browning, 1974) is lost, greater problems like *P. devastatrix* can be expected to emerge and better understanding of mechanisms of pecan/arthropod interaction will, of necessity, result. Close attention to these harbingers of potential devastation may allow better anticipation of future problems and will certainly contribute to a better understanding of plant/insect/agriculture interaction in general.

ACKNOWLEDGEMENTS

Partial assistance for this research was supported by the Science and Education Administration of the U.S. Dept. of Agric. under Grant No. 7800156 from the Competitive Research Grants Office. Support from the sponsors of this workshop is appreciated. Thanks also to Dr. R.L. Gallun and Dr. Clayton Person for discussions on this manuscript. The author assumes responsibility for views taken.

REFERENCES

Baker, H., 1935. Phylloxera devastatrix Pergande on pecans. J. Econ. Entomol. 28: 681-5.
Brison, F.R., 1974. Pecan culture. Capital printing. Austin, Tex. 294 p.
Browning, J.A., 1974. Relevance of knowledge about natural ecosystems to development of pest management programs for agro-ecosystems. Proc. Am. Phytopath. Soc. 1: 191-9.
Calcote, V.R. &. J.L. Bagent, 1974. Phylloxera control sought. Pecan South. 1: 8-9.
Carpenter, T.L., W.W. Neel & P.A. Hedin, 1979. A review of host plant resistance of pecan, Carya illinoensis, to insecta and acarina. Bull. Entomol. Soc. Amer. 25: 251-7.
Cutler, B.L., 1976. Instar determination and foliage consumption of the walnut caterpillar on pecan in Texas. M.S. Thesis. Texas A & M Univ. College Station.
Day, P.R. (Ed.), 1977. The genetic basis of epidemics in agriculture. Ann. N.Y. Acad. Sci. Vol. 287. 400 p.
Edmunds, G.F., Jr. & D.N. Alstad, 1978. Coevolution in insect herbivores and conifers. Science 199: 941-5.

Gallun, R.L. & G.S. Khush, 1980. Genetic factors affecting expression and stability of resistance. In: Breeding plants resistant to insects. John Wiley and Sons, Inc. p. 63-85.
Gibson, L.P., 1968. The genus Curculio in the United States. Entomol. Soc. Amer. Misc. Pub. 6: 239-85.
Groth, J.V. & C.O. Person, 1977. Genetic interdependence of host and parasite in epidemics. In: P.R. Day (Ed.): The genetic basis of epidemics in agriculture. NYAS 287: 97-106.
Harlan, J.R., 1977. Sources of genetic defense. In: P.R. Day (Ed.): The genetic basis of epidemics in agriculture. NYAS 287: 345-56.
Harlan, J.R. & K.J. Starks, 1980. Germplasm resources and needs. In: F.G. Maxwell & P.R. Jennings (Eds): Breeding plants resistant to insects. John Wiley and Sons, Inc. p. 253-276.
Harris, M.K., 1975. Allopatric resistance: searching for sources of insect resistance for use in agriculture. Environ. Entomol. 4: 73-85.
Harris, M.K. (Ed.), 1980a. Biology and breeding for resistance to arthropods and pathogens of cultivated plants. TAES, MP 1451. 605 p.
Harris, M.K., 1980b. Arthropod-plant interactions related to agriculture, emphasizing host plant resistance. In: M.K. Harris (Ed.): Biology and breeding for resistance to arthropods and pathogens of cultivated plants. TAES, MP 1451. 605 p.
Harris, M.K., 1980c. Searching for host-plant resistance to arthropods in agriculture. EPPO Bull. 10: 349-355.
Harris, M.K. & D.R. Ring, 1979. Biology of pecan weevil from oviposition to larval emergence. Southwestern Entomol. 4: 73-85.
Harris, M.K. & D.R. Ring, 1980. Adult pecan weevil emergence related to soil moisture. J. Econ. Entomol. 73: 339-343.
Harris, M.K., D.R. Ring, L. Aguirre & J.A. Jackman, 1981. Longevity of post-emergent pecan weevil in the laboratory and field. Environ. Entomol. 10: 201-205.
Haseman, L., 1940. The walnut caterpillar. MO. Agri. Exp. Stn. Bull. 418: 1-14.
Hedrick, V.P., 1919. Sturtevant's notes on edible plants. N.Y. Agr. Exp. Stn. State of N.Y. Dept. of Agr. 27th Annual Rept. Geneva, N.Y. 2. 686 p.
Hixson, E., 1941. The walnut datana. Okla. Agri. Exp. Stn. Bull. 246: 1-30.
Johnson, C., 1976. Introduction to natural selection. Univ. Park Press, Baltimore, MD. 213 p.
Leppik, E.E., 1970. Gene centers of plants as sources of disease resistance. Annu. Rev. Phytopath. 8: 324-344.
Lewontin, R.C., 1974. The genetic basis of evolutionary change. Columbia Univ. Press. 346 p.
Little, E.L., 1971. Atlas of United States trees. Vol. 1. Conifers and important hardwoods. USDA For. Serv. Misc. Publ. 1146.
MacKenzie, D.R., 1980. The problem of variable pests. In: F.G. Maxwell & P.R. Jennings (Eds): Breeding plants resistant to insects. John Wiley and Sons Inc. p. 183-213.
McVeigh, I. & C.J. Hobdy, 1952. Development of resistance by Micrococcus pyogenes var. aureus to antibiotics: morphological and physiological changes. Amer. J. Bot. 39: 352-359.
Maxwell, F.G. & P.R. Jennings (Eds), 1980. Breeding plants resistant to insects. John Wiley and Sons, Inc. 683 p.
Metcalf, C.L., W.P. Flint & R.L. Metcalf, 1962. Destructive and useful insects. McGraw-Hill Book Co., New York. 1087 p.
Moznette, G.F., C.B. Nickels, W.C. Pierce, T.L. Bissell, J.B. Demaree, J.R. Cole, H.E. Parson & J.R. Large, 1940. Insects and diseases of the pecan and their control. USDA Farmers Bull. 1829. 70 p.
Painter, R.H., 1968. Insect resistance in crop plants. Univ. of Kansas Press, Lawrence. 520 p.
Person, C., 1959. Gene-for-gene relationships in host: parasite systems. Can. J. Botany 37: 1101-1130.
Person, C., 1966. Genetic polymorphism in parasitic systems. Nature 212: 266-267.
Person, C.O. & G.M.E. Mayo, 1974. Genetic limitations on models of specific interactions between a host and its parasite. Can. J. Bot. 52: 1339-1347.
Robinson, R.A., 1980. The pathosystem concept. In: F.G. Maxwell & P.R. Jennings (Eds): Breeding plants resistant to insects. John Wiley and Sons, Inc.
Russell, G.E., 1978. Plant breeding for pest and disease resistance. Butterworth Inc. Boston, Mass. 485 p.
Wood, R.K.S. & A. Graniti (Eds), 1976. Specificity in plant diseases. Plenum Press, N.Y. 354 p.

On genes for disease resistance in plants

R.R. Nelson

Evan Pugh Professor, Department of Plant Pathology, The Pennsylvania State University, University Park, Pennsylvania 16802, USA

ABSTRACT

Balance between host and parasite may have come about in a process of co-evolution, in which many genes for resistance and virulence respectively accumulated. The distinction between major and minor genes is artificial: whether a gene acts 'major' or 'minor' depends on the genetic background. It is shown that resistance genes that have been overcome by the pathogen can still be useful for the plant and its breeder.

INTRODUCTION

The Organizing Committee for this Workshop has asked me to provide a summary presentation about the current thinking in the area of host-parasite interactions and genetics on the individual plant level. The Committee viewed my presentation as one that would help to set the stage for your later discussions on breeding for balance to certain pests of forest trees. Although I am indeed honored to play an active role in this Workshop, I must confess that I approach my task with considerable trepidation. As a matter of fact, it has become extremely difficult in recent years to speak or write about genes for disease resistance in plants. There is an enormous amount of confusion and disagreement about terminology and concepts. There is a strong undercurrent of dogma generated by many, including myself. We rely on dogma because the dogmatic approach requires few, if any, facts; and we have generated remarkably few facts. We hear so much about weak genes and strong genes, major genes and minor genes, durable genes and defunct genes, race specific genes and race non-specific genes, vertical genes and horizontal genes, and so on. For example, more than 20 terms have been or are being used to characterize a rate-reducing form of disease resistance e.g. horizontal, partial, durable, non-specific, generalized, uniform to mention but a few. It seems to me that all of the terms are essentially, if not absolutely, synonymous. To make matters worse, the list seems to grow rather than diminish. We generate theory after theory, but rarely conduct appropriate or adequate research to test their validity. The ultimate highlight of the present state of confusion, if not conflict, is the saddening fact that we can't even agree to disagree. This rather sobering state of affair will not deter me from attempting to summarize the current thinking on genes for disease resistance in plants. I appreciate that some of my remarks may have no direct relevance to those of you who serve as keepers of the trees. If my chapter whets even a single scientific appetite, my efforts will have been worthwhile.

Plant scientists have found it convenient to arbitrarily characterize disease resistance genes into 2 major categories even though we acknowledge that resistance and susceptibility are merely extremes of a broad continuum of host-parasite interactions. The difference between a hypersensitive reaction and a small necrotic lesion is a relative indication of host response to a parasite. The difference between a small necrotic lesion and a somewhat larger one is just as relative. Different scientists classify the same host response as being a resistant or a susceptible reaction depending upon their personal, if not arbitrary, opinion of what is resistance and what is susceptibility.

At any rate, one category of resistance genes embraces those genes that are said to enable the host to restrict the successful establishment of infection sites and, thus, inhibit the infection process. From the standpoint of population dynamics, these genes are said to be extremely effective against certain races, but totally ineffective against other races. Placed in an epidemiological framework, these race-specific genes are said to affect the extent of disease onset by restricting the amount of effective initial inoculum. Races lacking virulence genes to overcome race-specific resistance genes are disqualified from epidemic encounters. Race-specific genes are considered to be of no value when confronted by pathogen populations with matching virulence genes. From a genetic standpoint, race-specific resistance is usually, but not necessarily, conditioned by a single gene and a major gene at that according to current belief. Playing the word game, resistance conferred by this category of genes has been termed variably as hypersensitivity, race-specific, non-uniform, major gene, vertical, and so on.

The other category of resistance genes is said to enable the host to resist and retard the colonization and reproduction of the parasite following successful infection. From the standpoint of population dynamics, the genes are said to retain some degree of effectiveness regardless of racial exposure. It would be naive, however, to assume or expect that non-specific resistance would be uniformly effective against all races or sub-populations of races in light of the known ability of racial populations to differ in virulence and parasitic fitness.

Epidemiologically, non-specific resistance is rate-reducing in nature;

it affects the rate of disease increase (measured in one way by the apparent infection rate) and, thus, the terminal amount of disease. Reduction in the amount and rate of disease increase is accomplished by host mechanisms that restrict the number of successful infections, extend the latent period (the time required from infection to the subsequent production of inoculum), restrict the amount of tissue that is colonized from a single infection site, and reduce the amount and duration of inoculum production or sporulation.

From a genetic standpoint, non-specific resistance is usually, but not necessarily, conditioned by several genes and minor genes at that according to current acceptance.

My case for the 2 kinds of resistance has been made in a simplistic fashion by design. Naturally, all forms of resistance which allow some level of sporulation are not race-non-specific. Race-specific resistance does occur in which sporulation is not inhibited entirely. In a similar vein, hypersensitive infection sites are known to increase in size and sporulate on more mature plants when the host's resistance mechanism appears to erode. It is further evident that pre-infection factors play a role in some forms of rate-reducing resistance. Thus, rate-reducing resistance is not just a resistance to growth and sporulation; there are many examples in which non-specific resistance can be characterized partially by the presence of fewer lesions.

STABILIZING SELECTION: The saga of weak and strong genes

Some populations of plant parasitic fungi are more fit than other populations. Such a biological principle stands without challenge. Why some populations are less fit than others merits some discussion. It is tempting to de-fuse the topic by simply stating that populations are less fit because they are less fit; that they are less fit because they lack certain vital genes for fitness. Some developments in recent years have placed certain complications in opting for such a convenient dismissal of the subject.

There is a school of thought that believes that nature tends to place some constraints on parasitic populations munching happily towards unbridled virulence. Nature favors the more rational populations who conserve their energies for better endeavors; or so the argument goes. The same school of thought concedes that nature has no control, before the fact, over populations accumulating more virulence genes than are necessary for survival and reproductive success. What nature does, the school contends, is see to it that they are less fit to succeed than their less precocious relatives. Thus, the derivation of the Vanderplankian axiom that populations with unnecessary genes for virulence are less fit to survive.

There are at least 2 theories regarding the ultimate fate of unnecessary genes of any kind. One theory presumes that genes that are not called upon to perform in a vital function will drop out of the population or remain at best in very low frequencies. The other theory, which I have come to support, contends that the issue is not whether a gene is unnecessary, but rather whether its presence in a genome is a detriment to the population carrying it.

The matter of whether or not populations carrying unnecessary genes for virulence are less fit to succeed has become something more than a mere exercise in intellectual jousting. It has become a controversial issue in plant pathology, particularly among those who are concerned with plant disease control by means of disease resistance. The idea that populations with a few (if any) virulence genes are more fit to succeed on hosts with few (if any) resistance genes than are populations with unnecessary virulence genes (in context with the same hosts) has penetrated our thinking on the potential value of gene deployment and on the genetic composition of multilines. Much has been said and written, pro and con, about the hypothesis of stabilizing selection. Some, including myself, have given up all hope of testing the hypothesis, since the built-in escape hatches preclude any valid test. Nonetheless, the idea of stabilizing selection is used by some who reflect on population dynamics and parasitic fitness. It has been theorized that within all parasitic species there are populations of super races that carry virtually every virulence gene the species has ever manufactured. It lies there in nature as a slumbering giant in minute frequency because its gamut of virulence genes has rendered it less fit to succeed on host populations that are no match for it on a gene-for-gene basis. There are those who believe that a substantial portion of a multiline must consist of known susceptible tissue so that stabilizing selection can protect the rest of the multiline from the ravages of a complex race. The proponents of deploying different resistance genes in different but contiguous regions believe that complex races will be stabilized against and would be unable to continue their travels to the next region.

Let us suppose, for the moment, that there are genes that function to induce disease and that there are other genes that condition overall fitness. It is really not an unreasonable tenet; in fact there is some evidence for it. Let us further suppose that at least some of these different kinds of genes are inherited independently of one another. Again, not an unreasonable tenet; in fact there is some evidence for it. If these suppositions are true, as I believe them to be, then one might expect to find populations with few genes for virulence that are fit and other such populations that are unfit to succeed. One might also expect to find populations with many virulence genes, some or many of which may be unnecessary by our criteria, that are fit and other such populations that are unfit to succeed. There is no reason, to my way of thinking, why all of these combinations of virulence and fitness genes should not exist in natural populations at some point in time. If this is true, as I believe it is, then I am forced to reject the equation that simplicity equals fitness and return to the tenet that fit populations are fit because they are fit, pathogenic simplicity aside.

NON-SPECIFIC RESISTANCE AND GENETIC EQUILIBRIUM: The natural way of life

Race-nonspecific resistance, when viewed as a rate-reducing resistance, appears to be conditioned by several to many genes in most instances. Intensity usually is a quantitive trait. A quantitive trait is conditioned by many genes because 1 gene cannot accomplish the task. Each of the genes conditioning a quantitative trait contributes something to a collective venture that none can accomplish alone. Removing a single gene from a many-gene system should not drastically alter the expression of the trait. To what degree the trait may be altered would depend on the relative importance of the gene in the collective venture and the total number of genes controlling the trait. Greater fluctuations in the reaction of a host to different races should occur as more and more genes are removed. As all but 1 gene are removed, the host probably would react quite differently than it had before to the same races. It would react so differently, in fact, as to perhaps simulate a race-specific reaction to the races, i.e., resistant to some and susceptible to others.

A reverse of the sequence of events just described would, as I see it, represent the evolution of genes for disease resistance in plants. Co-epicenters, geographic areas in which both host and parasite have co-evolved, most accurately depict the story of the evolution of genes for virulence and resistance. The story is an incredible Darwinian classic. *Solanum demissum* Lindl. and the late blight fungus in Mexico, *Tripsacum* and teosinte and the northern leaf blight fungus in Mexico, and the old land races of rice and the rice blast fungus vividly depict the classic saga of the co-evolution of host and parasite. The relationship today between each of these host-parasite combinations is not sharply antagonistic. Each of the hosts sustains modest levels of disease, but disease is of no major concern to the host insofar as its ultimate reproductive success and persistent survival are concerned. Both host and parasite had learned that the price of co-existence was preferable to the alternating thrill of victory and the agonies of defeat. A relaxation of selection pressure for both species highlighted their new relationship. Neither was or is today in serious jeopardy of extinction. They had become, in a sense, the 'odd couple' of the biological world.

Philosophical or teleological as these speculations may seem, they would seem to represent the scientific explanation embodied in the co-evolution and co-existence of host and parasite in natural ecological niches. We can return to the beginning of their unique affair to learn how they arrived at their current status of equilibrate co-existence. The initial resistance of a host to a 'new' parasite probably was accomplished by a genetic change at a single locus. Plants with that resistance gene probably reacted to the parasite in a hypersensitive manner; probably like a modern-day cultivar whose resistance is conditioned by a single gene. It is unconceivable to me that they reached equilibrium at the very beginning of their co-evolution. Subsequently, new populations of the parasite evolved and strains pathogenic or virulent to the then resistant host were selected. The host, in turn, generated a new source of effective resistance; and so on. Such must have been the case, since both host and parasite still exist.

The process of co-evolution proceeded, perhaps stepwise in a gene-for-gene manner or perhaps in another way; it really doesn't matter. Host genotypes with fewer genes for resistance and pathogen genotypes with fewer genes for virulence probably were selected against and either dropped out of their respective populations or remained in low frequency. There can be little doubt that the most fit to persist would dominate both populations. At any rate, it is difficult for me to imagine that all genotypes that existed throughout the process of co-evolution remain a part of the current populations of host and parasite.

How long the process of co-evolution has occurred and what stage of co-evolution currently exists are not at all important. By the time the host and parasite had reached a stage of acceptable co-existence, each had accumulated a substantial number of resistance genes or virulence genes. The fact that they now exist in relative harmony indicates that the last resistance gene incorporated into the host genome did not confer a massive or hypersensitive response against the pathogen. Nor did the last virulence gene pose a serious threat to the host. There was and is no immediate selection pressure on either adversary. Each had found a mutual safety in numbers of genes for resistance or virulence. Genes that once functioned separately with only temporary success in the earlier stages of co-evolution now function collectively with a more permanent success.

If the assumption is correct that the long process of co-evolution resulted in the ultimate accumulation of many resistance and virulence genes in the host and parasite, genetic probabilities suggest that subsequent changes in the future course of their co-evolution will be largely subtle ones. At least it seems unlikely that either opponent will enjoy the massive superiority exhibited in the earlier stages of their co-evolution.

SOME THOUGHTS ON GENES FOR DISEASE RESISTANCE

Man has not only found it convenient to characterize disease resistance into 2 major types, but he has also found it convenient to conclude that the 2 types of disease resistance are conditioned by different kinds of genes. The implications, for example, in the terms 'major genes' and 'minor genes' are reasonably obvious. If major gene resistance is the same as race-specific resistance and minor gene resistance is the same as non-specific resistance, the 2 different kinds of resistance must be conditioned by different genes; or so the argument goes.

Several years ago, I proposed the concept that genes for race-specific and non-specific resistance are the *same* genes. The concept since has been developed in greater detail. The 2 major kinds of disease resistance, my argument contended, are not indications of the action of different genes, but rather are expressions of different actions of the same genes in different genetic backgrounds or when faced with different genotypes of the parasite. There are, in fact, no major genes and minor genes. There are only genes for disease resistance. Certainly, different genes may confer varying levels of expression or effectiveness and the expression of resistance genes may be influenced or modified by other resistance genes in different genetic backgrounds; background is the key to gene action.

The implication of the concept is that we recognize and characterize genes as being 'major' when they are separate and race-specific and characterize the same genes as 'minor' when the host response suggests a non-specific or rate-reducing type of resistance.

I again submit that genes function one way when they are separate and another way when they are together. Again, the *Solanum demissum*--potato late blight story has a relevance. Niederhauser stated, 'In Mexico we were forced to concentrate on the multigenic field resistance when it was found that no tuber-bearing *Solanum* species was immune or hypersensitive to the pathogen when exposed in the field'. He also stated, 'Field resistance is characterized by slow-spreading lesions in which sporulation is sparse. The lesions per plant are fewer, and tend to be on older, lower leaves'. Niederhauser apparently was discussing a resistance identical to what is now considered to be a rate-reducing type of resistance.

Man attempted to make use of the impressive resistance demonstrated by *S. demissum* to a multitude of races of *Phytophthora infestans* (Montagne) de Bary in the Toluca Valley of Mexico. Although *S. demissum* displayed a rate-reducing resistance, genes were extracted singly from the species and a race-specific or hypersensitive resistance was the result. Thus, the birth of the R-genes which were considered major genes because man has been scientifically and philosophically schooled to react with major favor to genes eliciting such a dramatic host response. It is difficult for me to comprehend how man can extract a major gene from a host genotype expressing minor gene resistance.

The first R-gene, R_1, soon fell prey to a race of *P. infestans* and man went back to *S. demissum* and extracted another R-gene, R_2. That gene broke down and man went back for another. And so on. One gene could not contain the highly variable late blight fungus. If man had wanted his cultivated potato, *S. tuberosum* L., to behave like *S. demissum* against *P. infestans*, he should have taken all of the genes that protected *S. demissum* and put them back together again. Major genes have been obtained from wild land varieties of rice that exhibit generalized or slow-blasting traits.

One final comment. If a race-specific gene is ineffective against a race, it is assumed not to function at all in resisting that race. Similarly, if 5 race-specific genes are all ineffective against a race, they are all presumed to be functionless against the race. It seems more reasonable to me that 5 singularly ineffective genes can be collectively functional and effective. If each contributes something in the way of resisting the parasite at some point after infection, the net, collective result logically should be a collective resistance to growth and/or reproduction.

SOME THOUGHTS ON 'DEFEATED' RESISTANCE GENES

The periodic arrival of pathogen populations with a virulence gene to overcome single, race-specific resistance genes has sent many such genes to the biological cemetery. They rest there with full honors having done their task gallantly if not indefinitely. It seems implicit that the tendency to bury defeated resistance genes is done with the presumption that they no longer are of any value at all; why else would they be buried.

My colleagues and I have been working with the winter wheat - powdery mildew system to evaluate the merits and demerits of certain genetic strategies to manage plant diseases at some acceptable level. Having access to the near-isogenic lines developed by Briggle in a Chancellor background, each containing a single, different powdery mildew (Pm) resistance gene, we have bred 2 four-gene pyramids to test the theory that the additive effect of several defeated resistance genes would create an equilibrium with a pathogen genotype possessing all the virulence genes needed to match all of the resistance genes in the pyramid. The unbounded, if not biased, faith in the validity of my pyramid theory prompted us to assume that the pyramided resistance genes would create an equilibrate, slow-mildewing relationship with the pathogen. If such were the case, it would be necessary to evaluate each resistance gene and all possible two-gene combinations to ascertain what each gene may be contributing, if anything, to the successful collective venture. On the surface, it may seem naive to evaluate single, defeated genes since it is assumed that they are of no value when once defeated. Nonetheless, 5 near-isogenic lines with different Pm genes were compared to the susceptible recurrent parent, Chancellor, for their reaction to a pathogen isolate possessing all of the virulence genes needed to overcome the 5 Pm resistance genes. Three of the 5 isogenic lines, containing genes Pm3c, Pm4, or a gene identified as Michigan Amber, exhibited a dramatic residual effect expressed as reduced number of lesions and reduced sporulation when compared to Chancellor. These defeated genes were not totally overcome as one might have assumed.

Why did such a discovery remain undetected for so long? Defeated genes are diagnosed as being defeated based on a qualitative reaction. If a susceptible type reaction is observed on a genotype, it is assumed that a genotype is susceptible. The presumed all or nothing syndrome associated with race-specific genes dictates that such genes cannot have an intermediate reaction or some degree of residual effectiveness.

The residual effects of defeated resistance genes provide a beautiful and solid explanation to account for the fact that defeated genes were retained in wild populations, even after their defeat at some early point in the co-evolution of host and parasite. They were retained because they contribute something of value.

THE EFFECTS OF GENE DOSAGE ON THE RESIDUAL EFFECTS OF DEFEATED RESISTANCE GENES

The F_1 of the crosses between each of the Pm isolines and the recurrent parent Chancellor were heterozygous for each of the Pm resistance genes. The availability of that F_1 seed permitted us to assess the effect of gene dosage on the residual effects of certain defeated powdery mildew resistance genes. Three-way comparisons among Chancellor, the homozygous isoline and the heterozygous F_1 lines revealed that the lines heterozygous for Pm genes 3c, 4, or Michigan Amber displayed approximately half of the residuality demonstrated by the isolines homozygous for those genes.

Vanderplank states that race-specific, vertical resistance genes function against epidemic development of plant diseases by reducing the amount of effective initial inoculum (X_o) available for disease onset. Vanderplank further states that vertical resistance genes have no influence on disease increase by races with virulence genes to match them. Mathematical analysis and modeling of plant disease epidemics are conducted with those assumptions of the role of vertical resistance genes. The dramatic effect on disease efficiency and sporulation by certain vertical Pm genes against races with virulence genes to overcome them clearly demonstrates that race-specific resistance also may have an effect on disease increase and can indeed influence the apparent infection rate (r), a trait heretofore attributed solely to genes for horizontal resistance.

Any researcher working with the genetic lines of winter wheat used in this study, but not knowing that the lines were isogenic for so-called 'major' Pm genes, would have concluded quite logically that the lines carrying genes Pm3c, Pm4, or MA possessed some level of rate-reducing or rate-limiting resistance. And yet, rate-reducing resistance, with rare exception, is considered to be under the control of so-called 'minor' genes or polygenes. The present research demonstrates that a gene may perform in a qualitative or a quantitative manner depending upon the genotype of the pathogen that confronts it.

CONCLUDING REMARKS

As keepers of the trees, you have pondered the question of breeding for balance. There is no simple, universal answer to the question. Breeding for balance or equilibrium is essentially a compromise between feast and famine. There is no need to compromising to those parasites which can be succesfully contained by imbalance. No one, for example, would suggest that a workshop be convened to discuss breeding for balance with a parasite for which a single resistance gene has successfully controlled it out of balance for 65 years. My best advice on breeding for balance would be to do so if breeding for imbalance doesn't work. My treatise ends on that profound thought.

Non-specific host-tree processes occurring in bark in response to damage and their role in defense

George S. Puritch & Garry D. Jensen

Environment Canada Pacific Forest Research Centre, 506 W. Burnside Rd., Victoria, B.C., Canada

ABSTRACT

Investigations of host-pathogen interactions and wound healing in conifer bark by the late Dr. D.B. Mullick led to the concept of 2 distinct categories of periderm. Periderms in the exophylactic category function as the normal outer covering of stems and branches, protecting living tissues from the external environment. Those in the necrophylactic category serve as barriers between necrotic and living tissues. The unique association of necrophylactic periderm with successful healing following bark necrosis, regardless of the nature of the external casual agent, indicates that a common host process occurs in response to diverse stimuli. The same process must also occur endogenously at abscission sites and in rhytidome formation. This common, endogenous process called phellogen restoration, is triggered by the loss of a functional phellogen, and is viewed as a major non-specific component of host defense. The non-specific healing process can serve as a basis for unravelling the complexities of host pathogen interactions.

One of the major obstacles to genetic selection of trees resistant or tolerant to pathogens[1] is the meager knowledge of the host response to pathogen attack. Invasion of a pathogen through bark normally triggers a sequence of complex physiological, chemical and anatomical events within the living cells of the tree. These events vary in both time and space and are influenced by the specific pathogen and host, as well as by numerous other factors including metabolic condition of both organisms, environmental conditions at the time of attack, time of year and nutrition. Identification of this myriad of events and clarification of which events are the cause or effect of resistance to the pathogen is a task that is monumental

1. Pathogen is defined as any agent microbe, insect, parasitic plant which causes chronic physiological disorders (pathogenesis) in the host (Treshow, 1970).

but essential for the unambiguous selection of genetically resistant stock.

During the course of research at this laboratory, investigations by the late Dr. D.B. Mullick provided new insights into the non-specific healing reactions of trees and provided a valuable foundation for interpreting host-pathogen interactions. This paper presents an overview of Mullick's work and its contribution to the study of host-pathogen relationships.

Mullick's investigations began with an attempt to chemically categorize *Abies* spp. susceptible and resistant to attack by *Adelges piceae* (Ratz.), the balsam woolly aphid (bwa). This 'aphid' is a tiny, parthenogenic insect that feeds by inserting its stylet into living bark cells and extracting their contents (Balch, 1952). During the course of its feeding, the aphid secretes saliva which contains various compounds including an auxin-like component (Puritch, unpublished data). As a chemical marker the reddish purple pigments that occurred in the phellem of periderm around healed bwa feeding sites were selected (Fig. 1). Utilizing new chromatographic techniques (Mullick, 1969a; 1969b) it was discovered that the pigments were a new class of non-anthocyanic compounds that occurred as mixtures of several distinct pigments (Mullick, 1969c). Further investigation revealed that similar pigments occurred at a variety of other sites in the bark such as dead resin blisters, leaf abscission zones, healed injuries and areas of old pathogen attack (Mullick & Jensen, 1973b). In these sites, pigments always occurred in the phellem of the sequent periderm. Thus, although the pigments were not unique to bwa attack, they did appear to be associated with the healing process of the tree and particularly, periderm. Consequently attention was focussed on the periderms and their role in wound healing.

Literature (Fahn, 1967; Esau, 1977) identifies 2 basic types of 'natural' periderm in gymnosperms: primary, which replaces the epidermis and secondary or sequent, which occurs in deeper bark tissues and replaces the primary. These 2 periderms were considered to differ only in their time of origin. In addition, it was recognized that another periderm, wound or pathological periderm, forms in response to biotic or abiotic injury. In many cases, wound periderms have been associated with specific interactions with pathogens (Hare, 1966; Akai, 1959; Carter, 1962). Rhytidome, the dead bark tissues external to the last formed periderm, (Srivastava, 1964) was suggested to be caused by impermeability of the suberized periderms (Esau, 1977). The periderms were reported to arise by meristematic activity of a layer of bark cortical cells, the phellogen (cork cambium) which produced phellem (cork) towards the outer bark surface and phelloderm towards the vascular cambium (Fig. 2). The periderms were suggested to play only a passive role in defense because they were distant from the site of interaction with the pest and were often absent at the time of pathogen inactivation or delimination (Wood, 1967).

Mullick (1971) found that a distinct difference occurred in the pigmen-

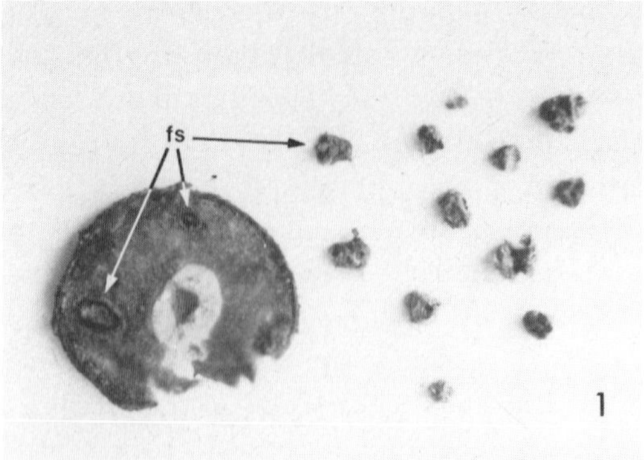
fs
1

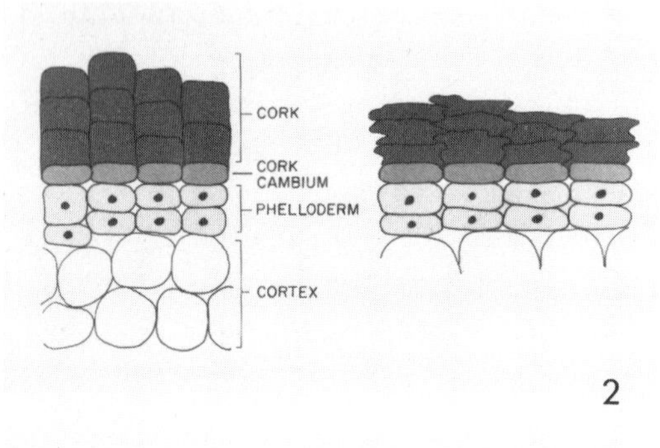
CORK
CORK CAMBIUM
PHELLODERM
CORTEX
2

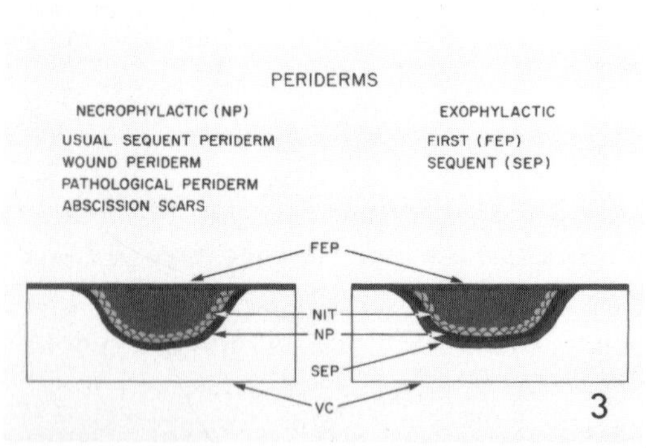
PERIDERMS
NECROPHYLACTIC (NP)
USUAL SEQUENT PERIDERM
WOUND PERIDERM
PATHOLOGICAL PERIDERM
ABSCISSION SCARS
EXOPHYLACTIC
FIRST (FEP)
SEQUENT (SEP)
FEP
NIT
NP
SEP
VC
3

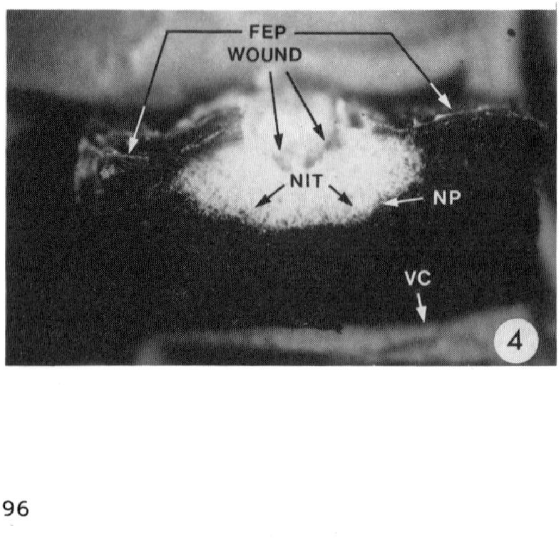
FEP
WOUND
NIT
NP
VC
4

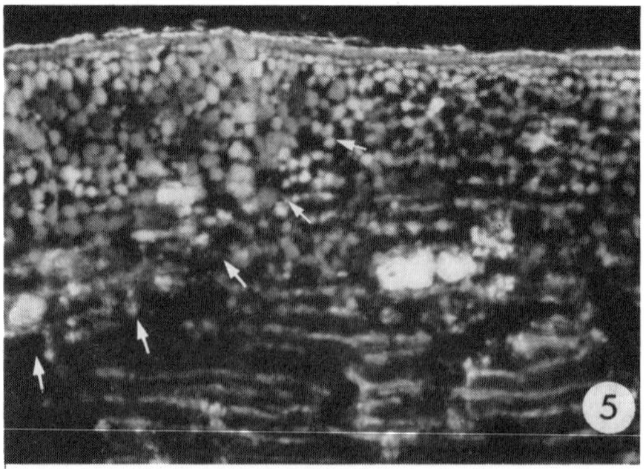
5

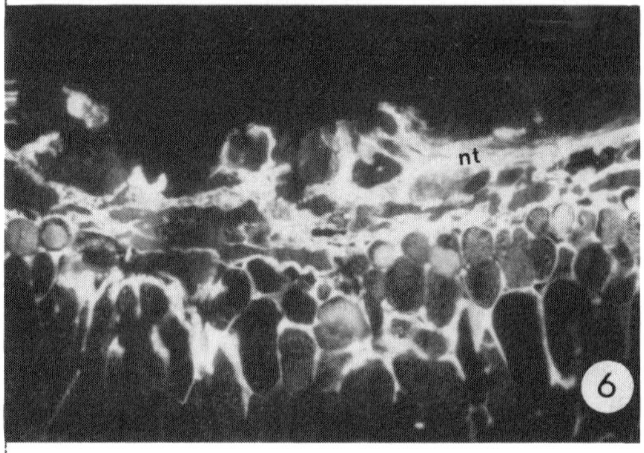
nt
6

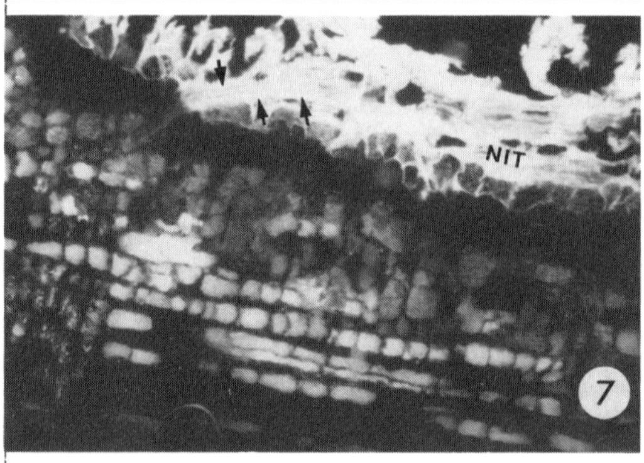
NIT
7

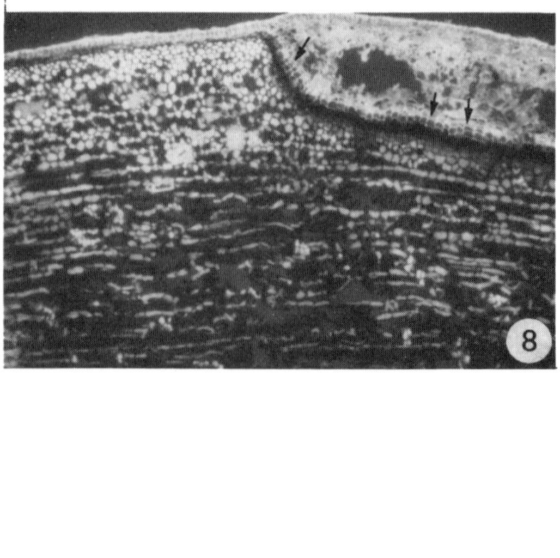
8

Figure 1. Cross section of balsam woolly aphid induced gout in *Abies amabilis*. Dark areas within the swollen bark tissue are the reddish-purple feeding sites (fs). Several of these have been isolated on the right (× 3).

Figure 2. Structure of periderms. The cork cambium (phellogen) produces the cork (phellem) externally towards the outer bark surface and phelloderm towards the vascular cambium. Dead phellem cells become distorted due to growth stresses.

Figure 3. The 2 categories of periderms in woody plants. The necrophylactic category includes usual sequent, wound, pathological and abscission scar periderm, while the exophylactic category includes the first periderm and can occur as a sequent periderm. (FEP = first exophylactic periderm, NIT = non suberized impervious tissue, NP = necrophylactic periderm, SEP = sequent exophylactic periderm, VC = vascular cambium).

Figure 4. Cross section of an area of *Abies grandis* bark 30 days after mechanical wounding. The bark sample was F-F tested (see text) through the cambial surface. The dye failed to penetrate the NIT and necrophylactic periderm (× 6).

Figure 5-8. Cryofixed sections photomicrographed using Carl Zeiss fluorescence exciter barrier filter combination I/53.

Figure 5. Radial section through the boundary (arrows) of a forming rhytidome in *T. heterophylla*. The cortical and outer secondary phloem parenchyma are hypertrophied below a region where some of the phellogen cells of FEP have developed fluorescent walls. Compare to the outer tissues on the right of the figure (× 20).

Figure 6. Radial section through the central region of a 14 day old mechanical injury to *Abies grandis*. The outer necrotic tissue (nt) has collapsed. A zone of hypertrophied cells internally abuts the necrotic tissue. The innermost of the hypertrophied cells have developed strongly fluorescent walls characteristic of NIT. The zone internal to these cells, which has little fluorescence, is the site of NP phellogen formation (× 20).

Figure 7. Radial section of more advanced healing than in Fig. 6. NIT formation has been completed including modifications producing fluorescence in sieve cell walls (arrows). A non-fluorescent phellogen zone is present, producing phellem centrifugally and phelloderm centripetally. The outermost layer of phellem has developed wall fluorescence (× 20).

Figure 8. Late stage of NP formation at rhytidome in *T. heterophylla*. Tiers of maturing phellem with fluorescent walls are clearly seen (arrows). The outermost phellem cells are becoming compressed and their contents are developing the typical reddish purple pigment. The necrotic tissues external to NP have desiccated to the point where intact cryostat sections could not be obtained (× 20).

tation of the primary and sequent periderms in 40 different conifer species. A technique called cryofixation was developed whereby the bark was fixed by freezing, sectioned on a cryostat and observed while frozen under a microscope using fluorescence and other modes of illumination (Mullick, 1971). This cryofixation technique permitted the observation of the pigments in their natural state. Detailed analyses of 3 conifer species showed that the primary periderm was of one type, with a solid brown phellem content whereas the sequent periderm consisted of 2 different types. The most common type of sequent periderm was reddish purple in color with fluid contents in its phellem while the other was identical in pigmentation and characteristics to the primary periderm. On the basis of 15 characteristics revealed primarily by cryofixation, Mullick & Jensen (1973a) proposed that the brown primary periderm and brown sequent periderm were essentially the same and that the reddish-purple sequent periderm comprised another distinct type of periderm. Mullick & Jensen (1973b) subsequently verified this hypothesis and advocated a new terminology and concepts for periderms in woody plants.

The 2 brown periderms constituted a single category of periderms and were termed exophylactic (*exo*, external; *phylacta*, a guard) (Fig. 3). This exophylactic periderm was considered to be the natural 'skin' of the trees providing protection of living tissues from the external environment (Mullick & Jensen,1973b). Two exophylactic periderms were proposed. The first exophylactic periderm replaced the epidermis of the stem and was the first periderm formed, while the sequent exophylactic periderm was associated with exfoliation or sloughing of dead and diseased tissues and developed only rarely in trees with adhering dead bark.

The reddish purple periderm constituted a second category and was termed necrophylactic (*necrus*, a corpse; *phylacta*, a guard) (Fig. 3). Mullick & Jensen (1973b) considered that the prime function of the necrophylactic periderm was the protection of living tissue from the adverse effects associated with death of cells. Whenever phellogen ceased to function, regardless of the reason, necrophylactic periderm would form to maintain the integrity of the outer stem covering. Formation of this periderm was thus *non-specifically* induced, occurring through wounding, disease, needle scars, and physiological disorders all of which damage the phellogen. Damaged and dying cells therefore led to periderm formation and not the opposite as suggested in the literature (Esau, 1977). Terminology in the literature referring to sequent, wound and pathological periderms was also misleading since according to Mullick's results these periderms were all one and the same.

The concepts of exophylactic and necrophylactic periderms were primarily based on detailed studies of 3 *Pinaceae* species. To test the general validity of this concept in relation to woody plants, Soo (1977) observed 15 species of trees including representatives from all conifer families as well as a few families of angiosperms. Without exception, all species assessed contained only the 2 classes of periderms, exophylactic and necrophylactic. Also, as suggested by Mullick & Jensen (1973b), the necrophylactic periderms were always associated with wounds and rhytidome while the exophylactic was associated with the normal first periderm.

Having clarified the types and characteristics of periderms in trees, investigations were initiated on the process of necrophylactic periderm formation in response to mechanical wounding and during rhytidome formation. In the latter case, Mullick (1977) considered rhytidome to be a process of necrophylactic periderm formation at a pathogen-free, 'non-induced' site. Early in these studies it was discovered that prior to periderm formation, an impervious layer of tissue formed around the periphery of the wounded site (Mullick, 1975). This impervious tissue was detected by the failure of a ferric chloride-potassium ferricyanide (F-F) solution to penetrate the impervious zone (Fig. 4).

Histological examination showed the F-F solution stopped at a zone of hypertrophied parenchyma cells which, in cryofixed section, showed a characteristic yellow green wall fluorescence. Sieve elements crossing the zone also showed wall fluorescence. In living bark and in most of the necrotic bark in the injured area, parenchyma and sieve elements showed no intrinsic fluorescence under blue light stimulation. Necrophylactic periderm occurred in tissues internally abutting these hypertrophied cells. Histochemical test for suberin showed the walls of the hypertrophied cells to be non-suberized and the layer was designated non-suberized impervious tissue or NIT.

Examination of necrophylactic periderm at the various sites of occurrence showed that NIT was invariably present externally abutting the phellem. In addition, developmental studies of mechanical injuries showed that NIT preceded formation of phellem and developed specifically from tissues internally adjacent to NIT. This led to the conclusion (Mullick, 1975) that NIT formation was a non-specific process that provided the environment for necrophylactic periderm formation and was a prerequisite for its formation. It was suggested that the process of NIT production was the physiological basis of host response to pathogens in bark.

By examining the process of rhytidome formation in *Tsuga heterophylla* (Raf.) Sarg. and healing following mechanical injury in *T. heterophylla* and *Abies* spp., the step by step sequence of cellular and morphological changes that lead to necrophylactic periderm formation was delineated (Mullick, 1977). Initiation of this process in rhytidome formation appeared to be non-functioning phellogen that was distinguishable by its fluorescent walls (Fig. 5). Following mechanical injuries, initiation was triggered by the physical destruction of phellogen. Existing cortical and secondary phloem cells adjacent to the altered phellogen or injured surface, underwent structural and chemical changes (as indicated by fluorescent modifications) becoming hypertrophied (Fig. 6). These cells then dedifferentiated and some of the cells developed fluorescent reticulum although no fluorescence was observable in the cell walls. This reticulum eventually disappeared and the cells developed varied fluorescence including areas in the cell wall corners. A zone of cells near the internal boundary of the hypertrophied zone showed sporadic fluorescence in their walls (Fig. 7). This zone developed into NIT. NIT formation, although encompassing a diversity of cell types, did not involve any meristematic activity. On the cambial side of NIT, the cells lost all fluorescence, became meristematic and developed into phellogen. The phellogen divided, producing phellem cells towards NIT which developed reddish purple pigments and produced phelloderm cells towards the cambium (Fig. 8).

The process of necrophylactic periderm formation was thus found to be very similar in rhytidome formation and in mechanically induced injuries (Mullick, 1977). In both cases cellular reactions occurred following damage to the phellogen following which NIT developed. NIT, however, did not form at the same rate throughout the year and varied from about 15 days following mechanical injuries made in June and July to over 200 days for injuries made in November (Mullick & Jensen, 1976). Stress conditions also adversely affected this non-specific healing process. Drought stress delayed the formation of NIT in direct proportion to the degree of stress (Puritch & Mullick, 1975). Thus necrophylactic periderm formation was a dynamic process dependent upon the physiological condition of the tree and was impaired under adverse conditions.

Based on these findings, Mullick (1977) presented a new concept of the process of host-pathogen interaction in the stems of trees. Since the process of phellogen restoration was autonomous and non-specific he concluded that it should have a common genetic mechanism in all woody plants (Mullick, 1977). The primary step in establishing the nature of the host response was the requirement for an understanding of the interaction of the pathogen with the first layer of living cells, the phellogen. The reactions might vary from rapid death to no response. If phellogen restoration was triggered, resistance or susceptibility could be determined by how successfully the host completed the formation of necrophylactic periderm while under the influence of the pathogen. For instance, resistance could result from situations where the pathogen was unable to successfully prevent the restorative process or was inactivated prior to or without activating the process. Susceptibility could result from successful interference of the pathogen with the process or failure to trigger it (Mullick, 1977). Once pathogens initiated the process of phellogen restoration they would be interacting not only with chemicals present in the normal bark but also with a host of new chemicals produced during the cellular dedifferentiation.

In this regard, Mullick felt that since phytoalexins were non-specific

(Vanderplank, 1975), they were components of the phellogen restorative process, occurring in the dedifferentiating cells surrounding injury. Following this suggestion, Rahe & Arnold (1975) found that the bean phytoalexin, phaseollin, was located entirely in tissues abutting the injured zone. Mullick (1977) subsequently suggested that phaseollin as well as other non-specific compounds in wound healing, e.g. ethylene, cell-wall degrading enzymes, early RNase, phenolics and terpenoids, resulted from the processes of dedifferentiation leading to NIT and necrophylactic periderm formation rather than non-specific defense factors. He concluded that the response to injury was not one of defense but rather a response to the need for phellogen renewal. Mullick (1977) considered that defense, in the strict sense was only a fortuitous result of the restorative process. It was only through the understanding of this non-specific process that a true understanding of host-pathogen interactions could be obtained and the components of specific interactions isolated and identified.

Mullick's research has provided major contributions to basic bark anatomy, periderm physiology and wound healing. Besides the phellogen restorative process described here he also discovered 2 other non-specific healing processes viz. vascular cambium restoration and sapwood blockage (Mullick, 1977). These findings have given a solid foundation to the investigations of host-pathogen relationships and are a major contribution towards identifying the host component and genetic base of host-pathogen interactions in trees.

REFERENCES

Akai, S., 1959. Histology of defense in plants. In: J.G. Horsfall & A.E. Dimond (Eds): Plant pathology. Academic Press, New York. 674 p.

Balch, R.E., 1952. Studies of the balsam woolly aphid, Adelges piceae (Ratz.) and its effects on balsam fir, Abies balsamea (L.) Mill. Canadian Department of Agriculture Publication 867. 76 p.

Carter, W., 1962. Insects in relation to plant disease. Interscience Publishers, New York. 705 p.

Esau, K., 1977. Anatomy of seed plants. 2nd edition. John Wiley and Sons, Toronto. 550 p.

Fahn, A., 1967. Plant anatomy. Pergamon Press, New York. 534 p.

Hare, R.C., 1966. Physiology of resistance to fungal diseases in plants. Botanical Review 32: 95-137.

Mullick, D.B., 1969a. Thin-layer chromatography of anthocyanidins, I. Journal of Chromatography 39: 291-301.

Mullick, D.B., 1969b. New tests for microscale identification of anthocyanidins on thin-layer chromatograms. Phytochemistry 8: 2003-2008.

Mullick, D.B., 1969c. Reddish purple pigments in the secondary periderm tissues of western North American conifers. Phytochemistry 8: 2205-2211.

Mullick, D.B., 1971. Natural pigment differences distinguish first and sequent periderms through a cryofixation and chemical techniques. Canadian Journal of Botany 49: 1703-1711.

Mullick, D.B., 1975. A new tissue essential to necrophylactic periderm formation in the bark of four conifers. Canadian Journal of Botany 53: 2443-2457.

Mullick, D.B., 1977. The non-specific nature of defense in bark and wood during wounding, insect and pathogen attack. In: F.A. Loewus & V.C. Runeckles (Eds): Recent advances in phytochemistry, Vol. II. Plenum Publishing Co. New York. 527 p.

Mullick, D.B. & G.D. Jensen, 1973a. Cryofixation reveals uniqueness of reddish-purple sequent periderm and equivalence between brown first and brown sequent periderms of three conifers. Canadian Journal of Botany 51: 135-143.

Mullick, D.B. & G.D. Jensen, 1973b. New concepts and terminology of coniferous periderms: necrophylactic and exophylactic periderms. Canadian Journal of Botany 51: 1459-1470.

Mullick, D.B. & G.D. Jensen, 1976. Rates of nonsuberized impervious tissue development after wounding at different times of the year in three conifer species. Canadian Journal of Botany 54: 881-892.

Puritch, G.S. & D.B. Mullick, 1975. Effects of water stress on the rate of non-suberized impervious tissue formation following wounding in Abies grandis. Journal of Experimental Botany 26: 903-910.

Rahe, J.W. & R.M. Arnold, 1975. Injury-related phaseollin accumulation in Phaseolus vulgaris and its implications with regard to specificity of host-parasite interaction. Canadian Journal of Botany 53: 921-928.

Soo, B.V.L., 1977. General occurrence of exophylactic and necrophylactic periderms and non-suberized impervious tissues in woody plants. Ph.D. Thesis, University of British Columbia. 319 p.

Srivastava, L., 1964. Anatomy, chemistry and physiology of bark. International Review of Forestry Research 1: 203-277.

Treshow, M., 1970. Environment and plant response. McGraw Hill, New York. 422 p.

Vanderplank, J.E., 1975. Principles of plant infection. Academic Press, New York. 216 p.

Wood, R.K.S., 1967. Physiological plant pathology. Blackwell Scientific Publications, Oxford. 570 p.

Trees resistant to spread of decay associated with wounds

Alex L. Shigo

Chief Scientist, U.S. Department of Agriculture, Forest Service, Northeastern Forest Experiment Station, Durham, New Hampshire 03824, USA

ABSTRACT

Trees are higly compartmented perennial plants. Normal growth is dependent on shedding, walling off, and regenerating new tissues. Trees cannot move. They have evolved under the constant stress of wounds and a wide variety of organisms that infect wounds. Trees cannot prevent wounds and infection. Trees survive after injury and infection by also shedding, or walling off the infected tissues and then by regenerating similar tissues in a new position. Trees that have the ability to shed or wall off rapidly and effectively, and to regenerate new tissues rapidly, will survive for a long time under the stress of many infections. Walling off processes in the bark and xylem are non-specific to type of injury or organism. Therefore, controlled wounding techniques where the response of the tree can be measured accurately, may be an effective way to determine a tree's genetic capacity to survive after injury and infection by a wide variety of organisms. Preliminary results from wounding experiments on several tree species indicate that the walling off processes are under strong genetic control. Some individuals in a species walled off decayed wood to very small volumes while other individuals still walled off the decayed wood, but to much larger volumes. A large scale pilot project using this information is now in progress to select trees with strong resistance to the spread of decay associated with wounds.

INTRODUCTION

Decay is a major cause of low quality in trees throughout the world. The economic loss is extremely high. Decay is also a major cause of hazard trees. When such trees fall they cause a great amount of damage to power lines, property, and people. With all this damage caused by tree decay, why has so little been done to produce trees resistant to decay?

The literature provides very little information on the subject. The textbook by Wright (1976) on forest genetics does not discuss it. Boyce (1961), in his fine forest pathology text, gives only a few comments on resistance to tree decay. Proceedings from past conferences on forest genetics show that tree decay was not discussed. Several investigators have mentioned decay resistance, but in wood products. It must be kept in mind that root diseases are different in many ways from those initiated by mechanical wounds. For a current review and excellent background information on resistance to root and butt rot in conifers, see von Weissenberg (1980).

The question remains, why so little progress? The answer is that geneticists have not been given an accurate account of the decay process and reliable simple methods to work with tree decay. Geneticists need to know how a tree is set anatomically and biochemically to resist decay, and how the decay process operates. Geneticists need accurate methods for detecting, measuring, and working with tree decay. Given these basics they can progress rapidly with the many fine techniques and basic information already well developed in the field of genetics.

Advances in science are made not on the presentation of new observations, hypotheses, and methods alone, but on the combination or connection of these, whether they are old or new, in such a way that a new perspective emerges. It is with this rationale that this paper is presented: it is hoped that a new perspective will emerge.

BACKGROUND

Trees are perennial plants with well developed systems for shedding parts after they have been used or injured. Trees shed fruits, leaves, twigs, branches, bark, and roots. The openings left by the parts shed are often where microorganisms attack. To defend itself, the tree must have systems to close or to wall off the opening. Timing is extremely important. If the shedding and walling off are not well synchronized, the pathogen will gain the advantage to grow into the tree. This process is seen best in the shedding of branches. Trees also wall off injured and infected tissues in bark (Mullick, 1977; see also the preceding chapter), trunk wood (Shigo, 1979b), and root wood (Tippett & Shigo, 1980). It is conjectured that the same processes that became well developed through evolution for shedding parts, were used for walling off injured and infected tissues that by nature of their position in the tree could not be shed easily. Just as the normal shedding processes are timed to best benefit the tree, there must be times when the walling off processes induced by injury and infection function most rapidly and effectively, and times when they do not.

Trees are constructed as highly compartmented plants. After wounding they wall off the injured and infected wood (Shigo, 1979b). After wounding, many types of microorganisms infect the wound surface and exert a force to grow into the tree, and to digest the wood. The microorganisms may spread and determine the rate of decay, but the tree defines the limits of the decayed wood. To understand the compartmentalization process more easily, we developed a simple model, CODIT, an acronym for Compartmentalization Of Decay In Trees (Shigo & Marx, 1977).

The CODIT model makes it possible to understand the 3-dimensional configuration of the developing column of discolored and decayed wood. The terms in the model - Walls 1, 2, 3 and 4 - must not be confused with the real process of compartmentalization.

After wounding, the cambium begins to form cells that differentiate into a barrier zone, wall 4 of CODIT (Sharon, 1973; Moore, 1978; Mulhern et al., 1979). Once the barrier zone forms, the continued spread of the microorganisms is confined to tissues extant at wounding.

Summaries of work on decay resistance and a description of a non-destructive method using the Shigometer for selection of strongly compartmentalizing trees in a species have been published (Shigo, 1979c; Schmitt et al., 1978).

RESEARCH RESULTS

Anatomical features

Geneticists have shown that many anatomical features are under genetic control (for a review see Hattemer, 1963).

Vessel features were strongly related to the degree of resistance to Dutch elm disease (Elgersma, 1970, and page 143 in this book; McNabb et al., 1970; Sinclair et al., 1979). Vessel features were also strongly associated with wounds in American elm (Shigo et al., 1980). Vessel grouping was the important feature. Trees that had rapid lateral spread of discolored wood had vessels closely grouped. The strongly compartmentalizing trees had large single vessels surrounded by many fibers. A similar situation was reported by Eckstein et al. (1979) for hybrid poplar.

Vessel grouping may be the main feature limiting the lateral spread of infection in wood, but vessels may also be important in limiting vertical spread (Bauch et al., 1980). In red maple (*Acer rubrum* L.) the ability to plug vessels rapidly after wounding appeared to be the main feature limiting vertical spread. For elm, see page 143 in this book.

In the wound studies, the column sizes of discolored wood associated with all the wounds were fairly uniform for each tree, suggesting that it was the tree that mainly regulated the dimensions of the discolored wood. A similar result was reported by Ekman & von Weissenberg (1979) for *Picea abies* (L.) Karst. Little is known about the plugging features in conifers.

In birch (*Betula papyrifera* Marsh., and *B. alleghaniensis* Britt.) lengths of discolored columns associated with drill wounds were correlated with lengths of vessels (Bauch et al., 1980).

Biochemical features

Shortle (1979) gives an excellent discussion of the biochemical mechanisms of compartmentalization. He shows how biochemistry relates to the CODIT model. In other studies not yet published, he has shown that wood from strongly compartmentalizing red maples and hybrid poplars does not support rapid growth of selected wood decay fungi in laboratory experiments. Wood from weakly compartmentalizing trees supported more growth of the fungi.

Compartmentalization in hybrids

The original studies that suggested genetic control of compartmentalization of decayed wood associated with wounds were done on hybrid poplars (Garrett et al., 1976; 1979; Shigo et al., 1977). Santamour (1979) did studies on compartmentalization in red maple (*Acer rubrum* L.), and silver maple (*A. saccharinum* L.) and hybrids of these. His conclusion was that compartmentalization of discolored wood was under genetic control.

Roots

Decayed wood associated with *Heterobasidion annosum* (Fr.) Bref. was compartmentalized in roots of red pine (*Pinus resinosa* Ait.) (Shigo, 1979a). Compartmentalization may help to explain why trees that appear healthy and green can still have so many decayed roots. Trees may remain alive after many infections because they constantly compartmentalize infected tissues to small volumes. Recent results suggest the same pattern of compartmentalization in roots of several tree species infected with *Armillaria mellea* (Vahl. ex Fr.) Quel.

Great care must be taken here with the root infecting fungi that are also associated with decayed wood. They have the ability to act as pathogens first: they kill cambial tissues and a lesion forms. The tree responds by limiting the size of the lesion. In some cases the fungi spread so rapidly in the bark that the root or trunk is killed. But after the spread of the lesion is stopped by the tree, the fungi have the first opportunity to grow rapidly into the wood beneath the lesion.

Bark

It may be that some of the information gained from studies of walling off in wood will be applicable to bark. The work of Mullick (1977) does show that bark forms a protective new periderm after injury. It is a type of walling off. His work will have great application to many bark diseases, and may well fit with *H. annosum* and *A. mellea*. Again, how fast and how effectively the tree walls off infections in the bark determines the size of the lesion.

Wood decay

To this point 2 types of decay-causing fungi have been discussed. First, the fungi that enter wood through wounds. They stay in the wood. This is the type that this paper addresses mainly. Second, the type that start in the bark and then move into the wood. Here the tree must first limit their spread in the bark, and then in the wood. When they grow into the wood, then the information given here is applicable. A third type, not discussed in this paper should be mentioned. These fungi start in the wood after wounding and later grow into the bark. They are the canker rot fungi (Shigo, 1969).

Spread

It may not be the number of infections that injure or kill a tree, but the extent or spread of each infection. A tree that cannot respond rapidly to the spread of infection and hold the infection in place will die. It is extremely important to note that infections per se do not kill a tree. If they did, no tree would live more than a few years, because every tree has small and large wounds and branch stubs that are infected. There is hardly a tree that does not have hundreds of infections at any one time. Trees survive as long as they are able to wall off the spread of infections to small volumes, and then regenerate new healthy tissues.

Heartwood compartmentalization

The heartrot concept states that decay-causing fungi infect dead, non-responsive heartwood exposed by wounds, and that fungi grow through the heartwood until all is digested, leaving a hollow (Boyce, 1961). It was also believed that because the decay fungi infect dead, non-responsive heartwood, that decay was not a disease. And because heartwood was dead, it would be impossible to select or breed for features in dead, non-responsive tissue. These beliefs have held back progress on selecting and breeding trees resistant to decay.

Heartwood may be dead according to our definition, but it is surely not non-responsive. When holes were drilled into healthy heartwood, a column of discolored heartwood formed and was compartmentalized (Shigo & Shortle, 1979). The CODIT model is applicable to heartwood!

This point is very important: investigators must not think it impossible to select and breed for strong compartmentalization of decay in heartwood. The methods described for selection of strongly compartmentalizing trees are also applicable for heartwood.

The heartwood confusion is at the very core of the reason there is no progress in the development of a decay-resistant tree. Rennerfelt (1946) stated that because decay goes on in dead wood of spruce it would be impossible to have physiologically controlled resistance.

Pilot study for selecting decay-resistant trees

The United States Forest Service is now conducting a large pilot study in natural forests in the north-eastern United States to select trees resistant to the spread of decay. The trees in the study, over 200, were those already selected as superior on the basis of form and growth rate. The species are *Acer saccharum* Marsh., *Betula alleghaniensis* Britt., and *B. papyrifera* Marsh. Each tree received 4 large, deep drill wounds at 1.4 m aboveground. After 2 growing seasons the Shigometer method will be used to determine the length of the columns associated with the wounds. The trees with the smallest columns will be selected as superior, not only for growth rate and form, but for their ability to compartmentalize defects associated with wounds. Details on methodology have been published (Shigo et al., 1977).

REFERENCES

Bauch, J., A.L. Shigo & M. Starck, 1980. Wound effects in the xylem of Acer and Betula species. Holzforschung 34: 153-160.

Boyce, J.A., 1961. Forest pathology. McGraw-Hill Book Company, New York/London. 572 p.

Eckstein, D., W. Liese & A.L. Shigo, 1979. Relationship of wood structure to compartmentalization of discolored wood in hybrid poplar. Canadian Journal of Forest Research 9: 205-210.

Ekman, R. & K. von Weissenberg, 1979. Sapwood extractives in Norway spruce inoculated with Fomes annosus. Acta Academiae Aboensis, B. 39, No 7, 8 p.

Elgersma, D.M., 1970. Length and diameter of xylem vessels as factors in resistance of elms to Ceratocystis ulmi. Netherlands Journal of Plant Pathology 76: 179-182.

Garrett, P.W., A.L. Shigo & J. Carter, 1976. Variation in diameter of central columns of discoloration in six hybrid poplar clones. Canadian Journal of Forest Research 6: 475-477.

Garrett, P.W., W.K. Randall, A.L. Shigo & W.C. Shortle, 1979. Inheritance of wound healing in sweetgum (Liquidambar styraciflua L.) and Eastern cottonwood (Populus deltoides Bartr.). United States Forest Service Research Paper NE-443. 4 p.

Hattemer, H.H., 1963. Estimates of heritability published in forest tree breeding research. Proceedings of the World Consultation on Forest Genetics and Tree Improvement. Stockholm, Sweden. Vol 1, session 2a, paper 3. FAO/FORGEN-63. 14 p.

McNabb, H.S., H.M. Heybroek & W.L. MacDonald, 1970. Anatomical factors in resistance to Dutch elm disease. Netherlands Journal of Plant Pathology 69: 1158-1159.

Moore, K.E., 1978. Barrier zone formation in wounded stems of sweetgum. Canadian Journal of Forest Research 8: 389-397.

Mulhern, J., W.C. Shortle & A.L. Shigo, 1979. Barrier zones in red maple: an optical and scanning electron microscope examination. Forest Science 25: 311-316.

Mullick, D.B., 1977. The non-specific nature of defense in bark and wood during wounding, insect and pathogen attack. Recent Advances in Phytochemistry 11: 395-411.

Rennerfelt, E., 1946. Om rotrotan (Polyporus annosus Fr.) i Sverige. Dess utbredning och satt att upptrada. Meddelanden fran Statens Skogsforskningsinstitut 35(8): 1-88.

Santamour, F.S. Jr., 1979. Inheritance of wound compartmentalization in soft maples. Journal of Arboriculture 5: 220-225.

Schmitt, D., P.W. Garrett & A.L. Shigo, 1978. Decay resistant hardwoods? You bet! Northern Logger and Timber Processor, September. p. 220-225.

Sharon, E.M., 1973. Some histological features of Acer saccharum wood formed after wounding. Canadian Journal of Forest Research 3: 83-89.

Shigo, A.L., 1969. How Poria obliqua and Polyporus glomeratus incite cankers. Phytopathology 59: 1164-1165.

Shigo, A.L., 1979a. Compartmentalization of decay associated with Heterobasidion annosum in roots of Pinus resinosa. European Journal of Forest Pathology 9: 341-347.

Shigo, A.L., 1979b. Tree decay: an expanded concept. United States Department of Agriculture Information Bulletin 419. 73 p.

Shigo, A.L., 1979c. Decay resistant trees. North-eastern Forest Tree Improvement Conference Proceedings. School of Forest Resources. Pennsylvania State University. p. 64-69.

Shigo, A.L. & H.G. Marx, 1977. Compartmentalization of decay in trees (CODIT). United States Department of Agriculture Information Bulletin 405. 73 p.

Shigo, A.L. & W.C. Shortle, 1979. Compartmentalization of discolored wood in heartwood of red oak. Phytopathology 69: 710-711.

Shigo, A.L., W.C. Shortle & P.W. Garrett, 1977. Genetic control suggested in compartmentalization of discolored wood associated with tree wounds. Forest Science 23: 179-182.

Shigo, A.L., R. Campana, F. Hyland & J. Anderson, 1980. Anatomy of elms injected to control Dutch elm disease. Journal of Arboriculture 6: 96-100.

Shortle, W.C., 1979. Mechanisms of compartmentalization of decay in living trees. Phytopathology 69: 1147-1151.

Sinclair, W.A., J.P. Zahland & J.B. Melching, 1975. Anatomical marker for resistance of Ulmus americana to Ceratocystis ulmi. Phytopathology 65: 349-352.

Tippett, J.T. & A.L. Shigo, 1980. Barrier zone anatomy in red pine roots invaded by Heterobasidion annosum. Canadian Journal of Forest Research 10: 224-232.

Weissenberg, K. von, 1980. Resistance of Picea abies to Heterobasidion annosum. Proceedings of the Fifth International Conference on Problems of Root and Butt Rot in Conifers. Ed. L. Dimitri. Hessische Forstliche Versuchsanstalt, Hann. Münden. p. 67-74.

Wright, J.W., 1976. Introduction to forest genetics. Academic Press, New York/London. 463 p.

Histopathology of anatomical mechanisms for resistance to fusiform rust in slash pine

F.F. Jewell, Daphne C. Jewell & C.H. Walkinshaw[1]

ABSTRACT

Anatomical expressions of resistance to *Cronartium quercuum* f.sp. *fusiforme*, Southern fusiform rust, in inoculated slash pine seedlings (*Pinus elliottii* var. *elliottii*) were apparently initiated in the primary tissue system of the host. Artificial wounding of slash pine seedlings resulted in host tissue reactions similar to reactions from fusiform rust infection. This in turn indicates that the initial tissue change in slash pine reported due to susceptibility or resistance to fusiform rust is not unique to this organism, but is at first a general host response to injury. As host growth and stimulation by the rust continues, differences in reaction of the rust-susceptible and resistant progenies are probably due to their genetics. The resistant progenies were able to maintain the differentiation of tissues to contain the rust. These tissues apparently were incompatible to growth and expansion of the rust. Susceptible progenies, although initially showing a resistance to rust infection, were unable to perpetuate this reaction through differentiation of incompatible host tissues. Instead, host tissues were produced that were compatible with growth and spread of the rust fungus.

INTRODUCTION

Certain early reports of histological studies of gall-forming pine rusts indicated some type of anatomical reaction of the host to the

1. Respectively, Professor and Research Assistant, School of Forestry, Louisiana Tech. University, Ruston, LA 71272, and Principal Plant Pathologist, Forest Disease Lab., USDA, Gulfport, MS 39503. This work was supported by the McIntire-Stennis Cooperative Forestry Research Program, and in cooperation with the Southern Forest Experiment Station, USDA, Forest Disease Lab., Gulfport, MS.

pathogen, which was concluded to be a form of resistance. Hutchinson (1935), True (1938), and McKenzie (1942) reported such findings working with Scots pine, *Pinus sylvestris* L., infected with a *Peridermium* rust (*Cronartium* or *Endocronartium*). More recently, Jewell & Snow (1972), Jewell & Speirs (1976), and Miller et al. (1976) described resistant reactions in 1- and 2-year-old slash pine (*P. elliottii* var. *elliottii* Engelm.) and loblolly pine (*P. taeda* L.) control-inoculated with the Southern fusiform rust, *Cronartium quercuum* f. sp. *fusiforme* (Burdsall & Snow, 1977). In slash pine, the host areas expressing resistance were termed resistance-zones (Jewell & Speirs, 1976).

Although the recent studies on resistance-reactions in slash and loblolly pine were rather intensive, the initiation of the host tissues associated with the resistance-zones was not determined conclusively. However, Jewell & Snow (1972) and Jewell & Speirs (1976) surmized that the resistant-reactions were initiated in the primary tissues of the slash pine hosts studied.

This paper will summarize a portion of the research from our laboratory concerning the slash pine/fusiform rust host-parasite relationship, the ontogeny of anatomical expressions of host resistance to the rust parasite, and the reaction of the host to artificial stimulus.

METHODS AND MATERIALS

Slash pine tissue for examination was obtained from half-sib progeny of known parentage (resistant) or bulk (susceptible) seed collections. Seedlings 4-8 weeks of age from seed were control-inoculated with known or bulk cultures of fusiform rust by the technique of Snow & Kais (1972). Sample collections were made at various periods following inoculation: 1 and 2 years, including samples from seedlings classed as recovered or galled; and daily for the first 30 days. Uninoculated samples from seedlings of similar ages served as check samples. Six-week-old bulk slash pine seedlings were artificially wounded by the insertion of sterile, stainless steel pins. Thirty days after wounding, samples were collected from the wound-site. Unwounded seedlings of the same seed source and plant age served as checks. Tissue samples were processed as previously reported (Jewell et al., 1962; Jewell & Walker, 1967). Observations were made by light microscopy.

RESULTS

Nine month and older infections

The earlier work on the histopathology of fusiform rust galls on 9-12 month old slash pine revealed in part, the common presence of vertical (wood) parenchyma, termed reaction parenchyma, in gall tissue. This cellular configuration, unusual in pine (Esau, 1953), was of particular interest

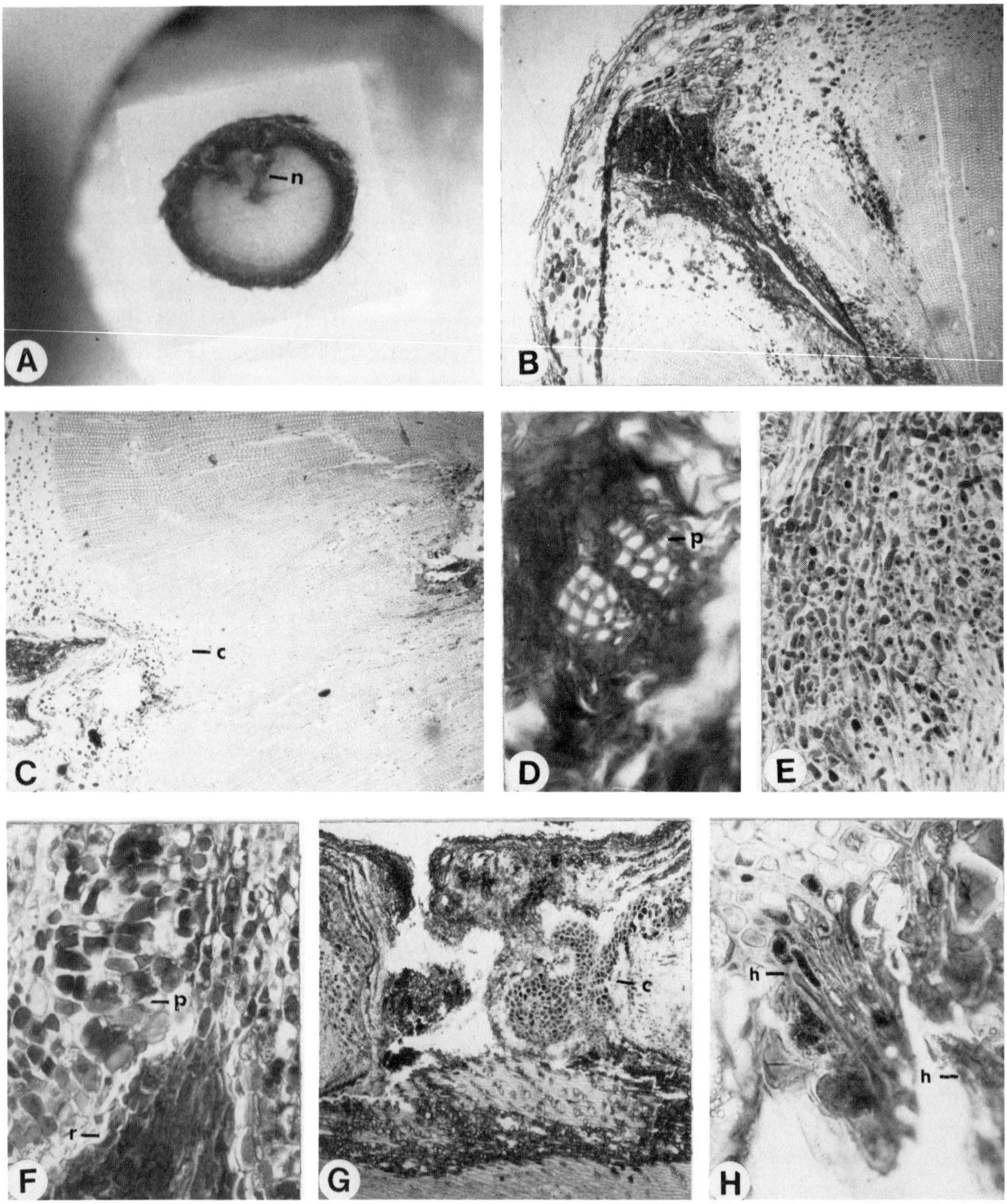

Figure 1.

A. Macro-transverse section of a 1-year-old slash pine with internal necrotic symptoms (n) typical of resistance-zones (6 ×).

B-C. Respectively, micro-transverse view of necrotic resistance-zones in 1- and 2-year rust infections. Note separation of pith and cortical areas by normal host tissue in C and inward bending of host cambium (c) (18 ×).

as it consistently marked the innermost extent of the mycelium of the parasite, and the corresponding host-tissue abnormality. At this time, the origin and importance of these cells were not well understood.

Subsequently, observations on recovered and gall samples from 1- and 2-year infections revealed the presence of reaction parenchyma defining affected tissue areas. However, samples from recovered seedlings contained areas appearing necrotic macroscopically (Fig. 1A) and microscopically similar to but more extensive than the reaction areas reported earlier for slash pine (Jewell & Snow, 1972). In the recovered seedlings, the necrotic-appearing areas, termed resistance-zones, varied in amount of stem area involved. Resistance-zones extended from pith through the epidermis in some samples (Fig. 1B), or were present in the cortex and/or xylem at varying distances from the pith. Where cortical and xylem zones were radially opposite in a sample, the separating host tissues were free of the abnormal cells typical of resistance-zones (Fig. 1C). The latter was consistent with 2-year old infections. The resistance-zones, regardless of sample age, appeared to originate in the outer pith tissues, which comprised portions of the metaxylem of the primary plant body present at inoculation (Fig. 1D).

The resistance-zones were sharply defined from the normal host stem tissues by cell areas considered, in transverse view, similar to reaction parenchyma described for pine-gall rusts (Jewell et al., 1962; Jewell & Walker, 1967). However, longitudinal sections revealed not only vertical parenchyma but also dark-staining, small, nearly isodiametric parenchyma-like cells (Fig. 1E). The cells central to the zones were heavily tanninized and considered non-living, while others, along the borders of the zone, although dark-staining, were considered to have been living at fixation (Fig. 1F). In some instances, meristematic activity of certain of these cells appeared to have resulted in production of additional parenchyma or callus-type cells. These cells apparently were produced to "heal" the area of stem associated with the resistance-zone where tearing of host tissue often resulted due to growth stress (Fig. 1G). The latter types of

D. Remnants of primary needle tissue (p) buried in a resistance-zone pushed by normal growth into the cortex (150 ×).
E. Longitudinal view of multi-type reaction parenchyma cells in xylem (70 ×).
F. Tanninized non-functional cells of resistance-zone (r) and dark-staining viable parenchyma (p) bordering the zone (150 ×).
G. Meristematic cells (callus) (c) differentiated for closing of wound developing due to resistance-zone in the inner cortex and cambial area (30 ×).
H. Tanninized non-functional rust hyphae (h) in resistance-zone (150 ×).

cells were absent in typical gall tissue.

Establishment of fusiform rust had occurred in all inoculated samples studied, but no sign of infection was observed in the check samples. The rust was, in the infected samples from resistant progeny, confined to the resistance-zones and was not found in the normal host tissue surrounding the zones. Hyphae were present in the reaction parenchyma and, based on staining reaction, considered viable. Radial and vertical hyphal extension had occurred in the zones. Haustoria, typically abundant with fusiform rust, were difficult to observe. When obvious, haustoria were considered scant in number related to the amount of hyphae present. Hyphae and haustoria were abundant in tanninized areas of the zones, but degeneration and hypertrophy of the rust was evident and it was considered non-functional in this situation (Fig. 1H). Susceptible progeny had developed the anatomy considered typical for fusiform rust (Jewell et al., 1962).

Cambial cylinder interruption by the resistance-zones was common in the 1-year but not in the 2-year infections (Fig. 1B,C). The latter consistenly had an inward curvature of apparently normal host cambium opposite xylem and/or cortical resistance-zones (Fig. 1C). This cambial abnormality was considered indicative of previous injury or interruption of typical activity of this tissue. Also indicated was that the resistant seedlings, in time, would bury the xylem-zone in normal stem tissue and slough-off the cortical zones and continue the normal growth process. It was obvious that there was a degeneration or regression of the resistance-zones with an increase in host age and growth, at least from 1 year to 2 years of age, as observed.

30-day infection

Susceptible and resistant plants exhibited differences in visible rust-infection symptom appearance. The resistant progeny exhibited symptoms as purple spots on the host hypocotyl at the primary needle bases. By 30 days, the hypocotyl was split at the purple area, but little or no swelling was evident. The susceptible plants exhibited symptoms after inoculation primarily as spots on the primary needles. At 30 days, these plants exhibited noticeable swelling but no splitting of the hypocotyl.

Microscopically, the most obvious reaction to rust-infection occurred in the resistant seedlings. This was evidenced by localized areas of heavily tanninized cortical cells radially opposite abnormal cellular configuration extending from the pith. In the affected cortical areas, the cells were conspicuously rounded with a loss of intercellular space (Fig. 2A). Host tissues adjacent to these abnormalities were considered normal for pine and no rust hyphae or haustoria were observed. Functional and non-functional hyphae were present in the affected tissue areas, but haustoria were rare.

The susceptible plants had limited tanninization in affected cortex, phloem, or xylem. Functional hyphae were present in the parenchyma tissues of the host. Typical rust haustoria were abundant and well developed. The cellular characteristics developed appeared characteristic for typical fusiform rust galls (Jewell et al., 1962).

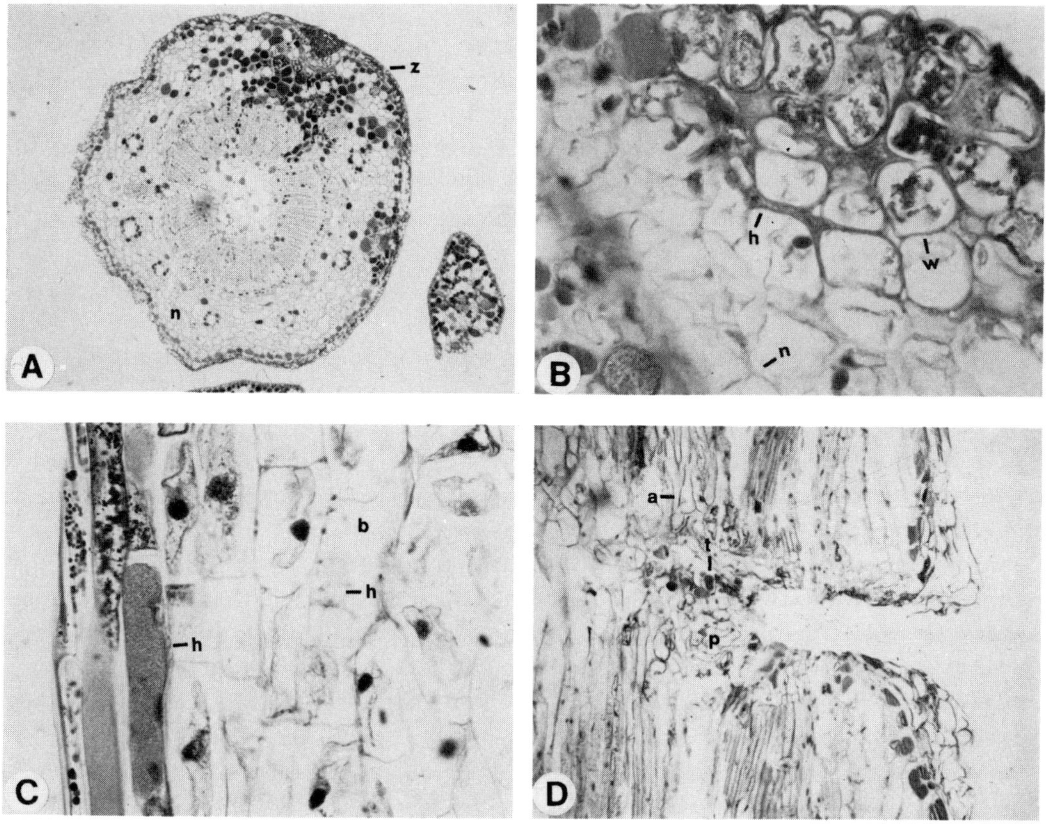

Figure 2.

A. Transverse view of resistance-zone (z) developing 30 days after infection, in contrast to the normal tissue (n) (18 ×).

B. Resistance-zone development 9 days following infection. Note hyphae (h) and cell wall thickening (w) in zone area contrasted to unaffected tissue (n) (70 ×).

C. Longitudinal view of resistance-zone 14 days following infection. Intra- and intercellular hyphae present (h). Base meristem (b) cells are hypertrophied (150 ×).

D. Longitudinal view of a 30-day-old wound. Tanninization (t) of cells present with proliferation and misalignment of parenchyma (p) in reaction to wound. Note atypical development of cells (a) removed from the wound (30 ×).

Infection prior to 30 days

The host reactions present 30 days after inoculation appeared, due to their location in the stem, to have been initiated in primary tissues of the stem and adjoining primary needle traces early in the infection process. Although macroscopic symptom appearance varied between resistant and susceptible plants, initial microscopic host reactions were similar in both progeny types and occurred earlier in the infection process. The first host response was observed 8-9 days after inoculation immediately above the primary needle/stem junction or in the primary needle traces. In the latter tissue, much of the reaction was in the base meristems and their cellular derivatives (Speirs & Jewell, 1976) associated with the primary needle traces. The affected zones were most obvious in the outer primary cortex. This was evidenced by host cell tanninization, cell wall thickening, cell content dissolution, and the presence of inter- and intracellular rust hyphae (Fig. 2B), the latter being atypical for fusiform rust. Haustoria were rare.

With an increase in time following inoculation, differences in the reaction to infection of susceptible and resistant plants became evident. At the time of external symptom appearence on the susceptible plants well developed inter- and intracellular rust hyphae were present in and among the base meristem cells, procambium, and adjacent parenchyma cells. Small, newly developed haustoria were often observed. Cells of the affected areas exhibited limited tanninization, as well as hypertrophy in the primary cortex and the pith. Conversely, the resistant plants exhibited tanninized cells in the affected areas, conspicuous cell hypertrophy in the outer primary cortex, and cell wall thickening in these areas. Inter- and intracellular rust hyphae were present, but were usually larger than normal and frequently distorted, indicating the hyphae were under stress (Fig. 2C). Haustoria, normal or atypical were rare. Of considerable interest was the appearance in the affected areas of abnormal host cells in advance of or somewhat removed from direct contact with the rust hyphae and/or haustoria. Frequently, as many as 3 cell layers separated the rust from the host cell abnormality.

Artificial wounding

Thirty days after wounding, all samples exhibited a limited brown discoloration around and along the path of the needle, and a slight stem swelling. This is somewhat similar to typical symptoms of fusiform rust infection on slash pine. In addition, macroscopic reactions often appeared to be in conjunction with primary needle bases.

Microscopically, cell destruction due to the actual act of wounding extended from pith to epidermis. However, the effect on adjacent host tissue was what was of interest. Similar to the host response in rust-resistant progeny, cell tanninization with limited necrosis was frequent in the pith, xylem, phloem-cortex, and primary needle trace elements, and needle bases of the wounded tissue. Little or no necrosis was observed in the secondary vascular elements, although tanninization, misalignment, and abnormal pitting were observed (Fig. 2D). Normal cell deposition delimited the wound areas of host tissue. However, atypical host cells were present in areas removed from the wound-site, indicating the host effect was not limited to the actual wound-site.

Atypical parenchymatous tissues appeared to have developed near the affected areas. These atypical cells, which differentiated in the vascular rays, resin ducts, and primary cambium, possessed contents, but were tracheal-like in shape. In general, these atypical parenchyma resembled the reaction parenchyma reported in fusiform rust galls (Jewell et al., 1962). In addition, peridermal-like cells were observed separating the wound areas from the normal.

In the wound area, light-staining, callus-like parenchyma developed between the normal and wound tissue. These cells appeared regenerative in nature indicating a meristematic potential for closing the wound area (Fig. 2D). This cellular deposition was similar to but less extensive than the callus-like cells associated with resistance-zones in rust-resistant slash pine.

DISCUSSION

Evidence indicates that anatomical expressions of resistance to fusiform rust, characterized as resistance-zones, were consistently present in certain progeny of rust-resistant slash pine parents. The resistance-zones were quite uniform in the types of abnormal host-stem tissue produced in the zone. There was no evidence of identifiable type of reaction to infection, although some variation occurred due to age differences.

The initiation of the resistance-zones occurred very early in the hostparasite relationship. Observations indicate that the resistance-zones, quite evident 30 days after inoculation, were initiated in the primary tissues of the seedlings studied and were a direct result of host reaction to the fusiform rust fungus used for inoculation. The resistant reaction resulted in the primary meristems (procambium, base meristems, etc.) (Speirs & Jewell, 1976) developing a limited but self-perpetuating abnormal parenchyma-like area of tissue in the host that contained the rust but prevented its spread or concurrent growth with the host. As the host matured and produced secondary tissue systems, the zones with age regressed. Eventually, the normal tissue production of the host overgrew the zone resulting in burying portions of the zone deep in the xylem and sloughing-off the outer portions. This would result in development of a seedling, recovered from rust infection, with the possibility of continued normal growth.

In contrast, susceptible progeny, although initially showing a reaction to infection similar to the resistant progeny, exhibited no restraint on the rust fungus and by 30 days after inoculation a typical gall anatomy was developing.

Artificial wounding of slash pine resulted in host reactions similar in nature to those initially expressed in rust-infected slash pine. The formation of periderm-like cells limiting the wound areas indicates the pin-wounded slash pine attempted to isolate the injury or stimulus in degenerating tissue. The pin-wounding appeared to stimulate parenchymatous tissue systems to become meristematic and differentiate callus-like cells to close the wound area. In time, normal host growth would continue until all signs of injury were hidden deep in the stem or sloughed off by radial growth of the host.

Based on similarities between pin-wounded and fusiform rust-infected slash pine, it appears that the limited initial host response to non- or pathologic injury is basically the same.

REFERENCES

Burdsall, Jr., H.H. & G.A. Snow, 1977. Taxonomy of Cronartium quercuum and C. fusiforme. Mycologia 69: 503-508.

Esau, K., 1953. Plant anatomy. John Wiley and Sons, New York. 735 p.

Hutchinson, W.G., 1935. Resistance of Pinus sylvestris to a gall-forming Peridermium. Phytopathology 25: 819-843.

Jewell, F.F. & G.A. Snow, 1972. Anatomical resistance to gall-rust infection in slash pine. Plant Disease Reporter 56: 531-534.

Jewell, F.F. & D.C. Speirs, 1976. Histopathology of one- and two-year old resisted infections by Cronartium fusiforme in slash pine. Phytopathology 66: 741-748.

Jewell, F.F., R.P. True & S.L. Mallett, 1962. Histology of Cronartium fusiforme in slash pine seedlings. Phytopathology 52: 850-858.

Jewell, F.F. & N.M. Walker, 1967. Histology of Cronartium quercuum galls on shortleaf pine. Phytopathology 57: 545-550.

McKenzie, M.A., 1942. Experimental autoecism and other biological studies of a gall-forming Peridermium on northern hard pines. Phytopathology 32: 785-798.

Miller, T., E.B. Cowling, H.R. Powers, Jr. & T.E. Blalock, 1976. Types of resistance and compatibility in slash pine seedlings infected by Cronartium fusiforme. Phytopathology 66: 1229-1235.

Snow, G.A. & A.G. Kais, 1972. Technique for inoculating pine seedlings with Cronartium fusiforme. In: R.T. Bingham, R.J. Hoff & G.I. McDonald (Eds): Biology of rust resistance in forest trees. Proceedings of the NATO-IUFRO advanced study institute. USDA Miscellaneous Publication 1221. p. 325-326.

Speirs, D.C. & F.F. Jewell, 1976. Observations on the anatomy of the hypocotyl and epicotyl of slash pine seedlings. Bulletin No. 9, School of Forestry, Louisiana Tech University, Ruston, LA. 29 p.

True, R.P., 1938. Gall development on Pinus sylvestris attacked by the Woodgate Peridermium and morphology of the parasite. Phytopathology 28: 24-49.

Mechanisms and inheritance of rust resistance in conifers

Bohun B. Kinloch

Geneticist, Pacific Southwest Forest and Range Experiment Station, Forest Service, U.S. Department of Agriculture, Box 245, Berkeley, California 94701, USA

ABSTRACT

Relatively few of the many rust diseases of conifers cause serious damage to their host populations. Exceptions include blister rust of white pines, a disease introduced to North America, and fusiform rust, a native. Both have become epidemic but striking differences in the pattern of genetic variation in resistance between hosts of these diseases are evident. Some common mechanisms of resistance are shared by hosts of these and related pine stem rusts, including tolerance, hypersensitivity, and various expressions of incompatibility. Both oligogenic and polygenic modes of inheritance are indicated, but few specific genes have been identified. More rapid progress in gene identification might be made by shifting the focus of genetic analysis from hosts to pathogens, which can be genetically manipulated with much lower investment in time, space, and cost. Because of the complementarity that exists in gene-for-gene relationships, genes for avirulence or virulence infection type identified in pathogens can be used to deduce their complements for resistance and susceptibility in hosts.

INTRODUCTION

Of approximately 87 known rust diseases on North American conifers, less than a quarter cause more than moderate or localized damage to their host populations, and only about 6 are recognized as being actually or potentially epidemic (Hepting, 1971; Ziller, 1974). Most of these parasites seem to have evolved ways of coexisting with their hosts that avoid significant biological damage, even when the apparent amount of disease is enormous. Hepting (1971, p. 365) vividly describes a case of *Coleosporium* needle rust on loblolly pine: 'Attack on planted stock is sometimes so heavy that one cannot walk through an infected plantation without choking on the aeciospore load in the air, yet growth effects have usually been temporary and light.' A more graphic example of host tolerance is difficult to imagine, yet it seems likely that many conifers coexist with their more innocuous rust parasites through this poorly understood adaptation, rather than through more obvious and dramatic mechanisms of resistance. In any case, there has been little economic motivation for studying the specific coadaptive mechanisms of other than seriously destructive pathosystems of commercial importance. This is unfortunate, since problems in developing genetic stability could be more effectively approached through better understanding of parasite populations that have evolved such benign relationships.

In spite of the promise of spectacular gains in resistance over more than 3 decades of selection and progeny testing (Bingham et al., 1973; Zobel, page 417 of this book), our knowledge of genetic interactions even of very damaging pathosystems remains meager. Few specific genes for resistance have been identified, and some of these only tentatively. Documented mechanisms of resistance - and their inheritance, when known - are listed in Table 1 for several pine rusts. Most of the data pertain to 2 of the most serious and intensively studied of these: blister rust of white pines, caused by *Cronartium ribicola* Fisch., and fusiform rust of southern pines, caused by *C. quercuum* f. sp. *fusiforme* (Cumm.) Burds. et Snow. These 2 rusts, 1 exotic and the other native in North America, offer intriguing contrasts in genetic interactions with their hosts.

BLISTER RUST - WHITE PINES

In sugar pine (*Pinus lambertiana* Dougl.), a hypersensitive reaction to invading mycelium in needle tissues is conditioned by a single dominant gene (Kinloch & Littlefield, 1977). Seedlings carrying this gene develop small, brown, necrotic flecks, which are macroscopically distinct from the large, lustrous, yellow or red lesions typical of susceptible genotypes. The gene is expressed in trees of all ages and in secondary, primary, and even cotyledon leaves (Kinloch & Comstock, 1980). Histologically, the hypersensitive reaction is characterized by rapid degeneration of affected cells into granular cytoplasm devoid of organelles and filled with dense tannin deposits, and is essentially similar to hypersensitive reactions described in other rust pathosystems. Yet, the development of mycelium in tissues of resistant genotypes varies considerably, from no visible traces to extensive colonies that occasionally penetrate into the stele, and, less frequently, grow down through phloem tissue into the bark. When this occurs, another hypersensitive reaction is triggered in bark tissue, further development of the pathogen is arrested, and active healing of the lesion by periderm formation ensues (Kinloch & Littlefield, 1977). The occasional and limited expression of pathogenicity in needle tissues might be due to variation in aggressiveness or virulence in the pathogen (Kinloch &

Comstock, 1980). Recently, we found a distinct virulent race causing typical yellow needle lesions and subsequent bark blisters on seedlings known to carry the dominant resistance gene (Kinloch & Comstock, 1981).

Similar hypersensitive reactions occur in needles of *P. armandii* Franch., a presumed native host of the pathogen (Bingham, 1977). The mode of inheritance could not be determined, because the seedlot studied was from a bulk collection (Hoff & McDonald, 1975), but the involvement of major genes seems likely. Hypersensitive and yellow lesions were mixed on 3/4 of the seedlings, while only hypersensitive lesions were observed on the remainder. This suggested the possibility not only of differential races of the pathogen, but also of more than 1 major gene in the host.

Needle infection in white pines is only the first stage in disease development, and is not very harmful by itself. McDonald & Hoff (1970) showed that premature shedding of infected needles in certain western white pine (*P. monticola* Dougl.) families was an effective mechanism in reducing the probability of the fungus establishing in bark tissue. Another expression of resistance in the 'needle spots only' syndrome was the development of a necrotic area of host and fungal cells near the base of the needle fascicle (Hoff & McDonald, 1971). Mycelium developed normally in mesophyll and vascular tissues of the needle and was arrested only after contact with stem tissues at or immediately below the small, dormant meristem at the base of the needle fascicle. Segregation ratios in self and full-sib families led McDonald & Hoff (1971) to hypothesize that each of these mechanisms is controlled by independently inherited recessive genes. However, consistent deficiencies in numbers of resistant seedlings in most families, and discrepancies in performance between selfed and outcrossed families of the same parents, leave this hypothesis tentative.

Two additional genes for resistance to specific races of *C. ribicola* were proposed by McDonald & Hoff (1975). Identification of these genes proceeded from the observation that 2 distinct foliar lesion types occur in western white pine following infection. Some seedlings developed only yellow spots, some only red, while approximately a quarter developed both types. Evidence for a gene-for-gene relationship was based on the additive property the color types exhibited: the average number of lesions per seedling on seedlings with both color types nearly equaled the combined average of the single types. Although the relative frequencies of the yellow and red phenotypic classes were highly variable, the additive property was consistent in 2 experiments in different years. This led McDonald & Hoff (1975) to conclude that one host gene prevented infection by the 'red race' and another prevented infection by the 'yellow race', but their hypothesis was not tested by segregation ratios of seedlings in the full-sib families used. In a different approach, McDonald (1978) analyzed segregation ratios of 245 monoaeciospore cultures derived from seedlings bearing yellow spots only, red spots only, and both spot types. Under the hypothesized gene-for-gene relationship, cultures from yellow-only and red-only spotted seedlings should have 'bred true' when susceptible test seedlings were inoculated with basidiospores resulting from these cultures, while cultures from seedlings with both spot types should have segregated into 1:2:1 ratios of yellow-only: mixed yellow and red: red-only classes. Although the frequency of spot types induced by basidiospore progeny was strongly associated with the type of the parent culture, the putative homocaryon cultures did not breed true, and there was a large deficiency of mixed spot types in the putative heterocaryon cultures. McDonald (1978) proposed that the lack of fit to expected ratios observed might have been due to irregularities in fertilization of the pycnia or thalli of the original sources of parent cultures. Thus, the genetic control of lesion types in this host-pathogen interaction remains unresolved.

Other foliar mechanisms of resistance have been reported but their inheritance is even less certain. Reduced numbers of needle lesions among different progenies have been noted in western white pine (Bingham et al., 1960), eastern white pine (*P. strobus* L.; Patton, 1967), and sugar pine (Kinloch, unpublished data), and suggested to be under polygenic control (Hoff & McDonald, 1980). Patton (1967) described a mechanism in secondary needles of grafted ramets from a resistant eastern white pine selection that completely inhibited the formation of substomatal vesicles of the fungus. This mechanism was transmitted to seedling progenies in different degrees.

Bark reactions are the most common mechanisms of resistance in conifers not only to stem rusts, but to injurious agents in general. They are expressed as necrosis and collapse of tissue in response to trauma, followed by formation of periderm tissues that effectively isolate the lesion (Hutchinson, 1935; True, 1938; Struckmeyer & Riker, 1951). Mullick (1977) and Puritch & Jensen (page 94 in this book) have shown that they are nonspecifically induced defense reactions to a variety of traumas incited by mechanical injury or insect or pathogen attack. This implies, then, that normal pathogenesis involves specific inhibition of the regulatory processes (and genes) that trigger these reactions.

Bark reactions have been described in several pine-rust systems (Table 1). Their salient features are their common mode of action and expression across different host-pathogen combinations, and the wide range in the degree of incompatibility expressed. They range continuously in size from microscopic (Hoff & McDonald, 1971) or visible necrotic flecks (Hutchinson, 1935; True, 1938), to sunken cankers several centimeters square. Bingham et al. (1960) discriminated between relatively small ('hypersensitive') and large ('corked-out') bark lesions, but there is no anatomical or physiological basis for this distinction. Often, incompatibility is only partial, as when mycelium 'escapes' the collapsing lesion and invades periderm and healthy tissue beyond, which then collapses again (Hutchinson, 1935; True,

Table 1. Mechanisms and inheritance of resistance in pines to pine stem rusts, caused by *Cronartium* spp. and *Endocronartium harknessii*.

Mechanism	Host-pathogen	Hypothesized inheritance	Authority
Foliage			
Hypersensitivity	*lambertiana - ribicola*	dominant	Kinloch & Littlefield, 1977
	armandii - ribicola		Hoff & McDonald, 1975
Premature needle shed	*monticola - ribicola*	recessive	Hoff & McDonald, 1971
Exclusion of differential races	*monticola - ribicola*		McDonald & Hoff, 1975; McDonald, 1978
Reduced receptivity to infection	*monticola - ribicola*	polygenic	Hoff & McDonald, 1980
Inhibition of fungal infection structures (vesicles)	*strobus - ribicola*		Patton, 1967
Bark			
Incompatibility (variable expression)	*monticola - ribicola*		Bingham et al., 1960
	lambertiana - ribicola		Kinloch & Littlefield, 1977
	strobus - ribicola		Struckmeyer & Riker, 1951
	armandii - ribicola		Hoff & McDonald, 1972
	elliottii - fusiforme		Miller et al., 1976; Jewell & Speirs, 1976
	taeda - fusiforme		Lundquist, 1979
	sylvestris - harknessii		Hutchinson, 1935; True, 1938
	ponderosa - harknessii		Quick, 1966
	pinea - flaccidum		Mittempergher & Raddi, 1977
Hypersensitivity (as a special case of incompatibility)	*monticola - ribicola*	recessive	McDonald & Hoff, 1971; Hoff & McDonald, 1971
	lambertiana - ribicola	dominant	Kinloch & Littlefield, 1977
	elliottii - fusiforme	oligogenic	Miller et al., 1976; this paper
	taeda - fusiforme	oligogenic	Lundquist, 1979; this paper
Slow fungal growth	*monticola - ribicola*	polygenic	Bingham et al., 1971
Site undetermined			
Reduced infection rate (slow rusting)	*lambertiana - ribicola*	polygenic	Kinloch & Byler, 1981
	elliottii - fusiforme		Griggs & Dinus, 1977
Ontogenetic resistance	*strobus - ribicola*		Patton, 1961, 1967
	monticola - ribicola		Bingham, 1966
	lambertiana - ribicola		Kinloch & Byler, 1981
Differential interactions (host genotype-pathogen 'race')	*taeda - fusiforme*	oligogenic	Powers et al., 1977; this paper
	elliottii - fusiforme	oligogenic	Snow et al., 1975; this paper
	pinea - flaccidum	oligogenic	Mittempergher & Raddi, 1977; this paper
	pinaster - flaccidum	oligogenic	Mittempergher & Raddi, 1977; this paper
	nigra - flaccidum	oligogenic	Mittempergher & Raddi, 1977; this paper

1938; Hoff & McDonald, 1972). In sugar pines lacking the major gene for hypersensitivity described above, we have observed a variety of incompatible and compatible reactions on the same tree. The frequency of bark reactions, as well as their size, seems to vary among different families. These observations suggest additional complementary genetic variability in host and pathogen populations, but proof is lacking. The only other tentatively identified gene for bark reaction resistance is the 'short shoot fungicidal reaction' described in western white pine by Hoff & McDonald (1971). Paired recessives confer hypersensitivity in bark but - in contrast to sugar pine - not needle tissues.

A single phenotypic expression that simplifies and integrates complexly inherited - and perhaps unrecognized - mechanisms of resistance is 'slow rusting', which can be measured by the apparent infection rate per unit of time (Vanderplank, 1963). This parameter (r) is a more sensitive indicator of resistance than cumulative infection at any single point. A more than threefold range in r values among sugar pine families was demonstrated, compared with differences in cumulative infection of less than 20 % (Kinloch & Byler, 1981). Average r values among families from slow × slow rusting parents were only half as large as those from slow × fast rusting crosses, and only a quarter of fast × fast crosses. The pattern strongly suggests additive inheritance, with the potential for rapid gains in resistance in early breeding generations.

Adult plant resistance is well known in cereal rust pathosystems, but the magnitude of variability in this kind of age-related or ontogenetic resistance has not been fully appreciated in tree rusts. Patton (1961) showed that grafted ramets of different eastern white pines decreased in susceptibility to blister rust as ortet age increased. Because his choice of ortets was essentially random, the increase of resistance with age was seen as a general population characteristic. In sugar pine, however, an astonishing degree of variability in susceptibility was found among grafted clones from sexually mature ortets of roughly the same age in a common garden (Kinloch & Byler, 1981). Different clones varied from 0 % to 100 % infected, and from 0 to 50 cankers per ramet (Kinloch, unpublished). Since all of the clones in question were from ortets highly selected for resistance, the true range of susceptibility is yet to be determined. Nevertheless, the amount of infection among clones was far less than that of their seedling offspring. Thus, although resistance increases with age generally, the degree to which it is expressed varies greatly among genotypes. The specific mechanisms and inheritance of ontogenetic resistance are not known either, but do operate in both needle and bark tissue (Patton, 1961, 1967). Although one of the strongest and perhaps most variable kinds of resistance, ontogenetic resistance unfortunately may be the least useful, since it is not fully expressed until after a tree crop has passed its most vulnerable growth stage. In programs based on phenotypic selection in wild stands, it can also be deceptive, since it often masks juvenile susceptibility. It nevertheless has value in stabilizing resistance in later stages of a rotation and reducing levels of inoculum, especially of races of rust virulent to resistance genes that are effective in more juvenile stages. In at least 1 pine-stem rust system *(P. sylvestris - Peridermium pini)*, the ontogenetic relationship is reversed and susceptibility increases with age (Van der Kamp, 1968).

Slow rusting and ontogenetic resistance appear to be inherited independently of each other and of major gene resistance, since parent-offspring relationships for these traits form no discernible pattern (Kinloch & Byler, 1981).

FUSIFORM RUST - SOUTHERN PINES

Wide variation in resistance to fusiform rust of southern pines is amply documented (see Zobel, page 417 of this book). Although the mechanisms and expression of resistance have been studied intensively, no specific genes have yet been identified. Miller et al. (1976) found that most resistance of young slash pine seedlings was expressed in bark tissues as hypersensitivity. Both the size and rate of lesion development varied considerably, from early necrosis of only a few cell layers in the cortex to incipient formation of galls which subsequently aborted in a mass of necrotic tissue surrounding the 'reaction zone'. Occasionally, the fungus escaped the reaction zone and resumed active parasitism. These resistant reactions were observed on both relatively resistant and susceptible wind-pollinated families, but were more frequent on the former. Another family-related expression of resistance was complete absence of incipient symptoms and any evidence of tissue penetration or infection. Snow et al. (1975, and page 243 of this book) measured significant variation among different families in the shape and growth rate of active galls. Thus, a complete spectrum of both qualitative and quantitative host responses is apparent in this pathosystem.

Equally dramatic variation in pathogenicity of inoculum derived from different single gall isolates and even monoaeciospore isolates from a single gall has been described (Powers et al., 1977; Powers, 1980; Snow et al., 1975). In most cases studied, family × isolate variance was highly significant, with frequent rank changes among either families or isolates. Differential interactions of this kind are characteristic of complementary gene-for-gene relationships, which almost certainly exist in this pathosystem. Similar relationships apparently exist between certain Mediterranean pines and *C. flaccidum* (Mittempergher & Raddi, 1977).

Patterns of variation in resistance in hosts of fusiform rust and white pine blister rust are fundamentally different, and offer useful paradigms of the potential genetic resources available in host populations for coping with native and exotic pathogens. The pathogens are closely related taxonomically, and relatively minor differences distinguish the diseases in their life cycle, etiology, and epidemiology. But *C. q.* f. sp. *fusiforme* coevolved with its southern pine hosts whereas *C. ribicola* was recently introduced (ca. 1900) from Asia via Europe. In Asia, the latter is endemic on white pines related by common progenitors to North American species, but the Asiatic and American species have long been geographically separated (Bingham, 1977). In the absence of coadaptive selection pressures, the North American white pines may differ from either their Asiatic or southern pine relatives in the kinds of resistance genes (major effect, minor effect), their frequencies, and even numbers of loci. This is strongly suggested by the relative lack of variability in resistance to blister rust in highly selected families of western white pine compared with the high levels of resistance found in an arbitrarily chosen bulk seed lot of Armand pine, which is native to Asia (Hoff & McDonald, 1975). Even more striking are differences in patterns of infection between southern pines and North American white pines under epidemic conditions for their respective rusts. An example is given in Fig. 1, but it understates the differences because the sugar pine parent population was intensively selected for freedom from rust in several heavily infected stands, whereas the loblolly pine parents were randomly chosen from a single wild stand. Even so, much more variability in resistance, distributed nearly continuously, is evident in the loblolly pine families. The relatively small degree of resistance in the sugar pine families is largely due to a single dominant gene at low frequency in the parent population, which is apparent in the trimodal distribution of offspring. The situation with sugar pine and other North American white pines is in some ways analogous to the 'vertifolia' effect in the potato-late-blight pathosystem described by Vanderplank (1963), wherein horizontal resistance conditioned by polygenes with individual minor effects has been lost in commercial cultivars through selection primarily for major genes. Loss of polygenes reduces the apparent complexity of host-parasite interactions, further highlighting the effects of major genes. Thus, it is perhaps not surprising that major genes have been more clearly perceived in North American hosts of blister rust, which are relatively depauperate in variability, than in native hosts of fusiform rust, where they may be obscured in an apparently large background of quantitative variability.

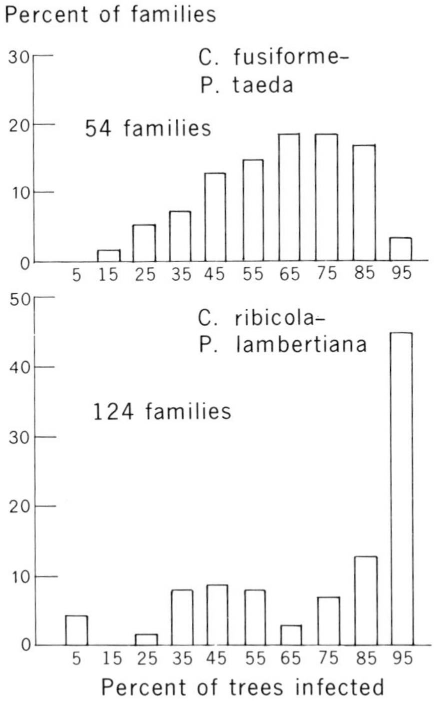

Figure 1. Distribution of full-sib family means in infection of loblolly pine by fusiform rust and of sugar pine by blister rust, after 4 years in disease gardens in Georgia and California, respectively. On each site alternate hosts were interplanted among test families and infection occurred naturally. (Data on loblolly pine from Kinloch & Stonecypher, 1969; on sugar pine from Kinloch, unpublished.)

CONCLUSION

Since diversity of resistance genes is the most effective means of stabilizing pathogen populations (Browning et al., page 371 of this book) maximum protection of host populations will be possible when all genes for resistance/virulence and their frequencies are identified. The main focus of current tree improvement programs has been on progeny testing for relative levels of resistance. While effective for this purpose, the use of open-pollinated families and heterogeneous sources of inoculum is unlikely to result in the more precise identification of resistance genes needed.

Because tree breeding by controlled pollination is so expensive and time-consuming, more rapid progress might be made by shifting emphasis from host to pathogen. In the *Cronartium* rusts, for example, the spore stage infective on pine is haploid, so that the phenotype of the lesion (low or high infection type) also denotes the genotype of the spore that induced it for avirulence or virulence. Thus, an observed 1:1 segregation ratio of lesion types induced by basidiospores derived from monoaeciospore or monourediospore clones would identify a gene locus in the pathogen, and - by deduction from the gene-for-gene theory - a corresponding locus for resistance/susceptibility in the host. The eventual goal would be to assign resistance/virulence formulae to parent trees selected for breeding programs, so that the genes identified could be deployed to maximum effect in commercial crops through appropriate matings. At the same time, important insights would be gained into the genetic structure of host-parasite interactions in wild populations.

REFERENCES

Bingham, R.T., 1966. Breeding blister rust resistant western white pine. III. Comparative performance of clonal and seedling lines from rust free selections. Silv. Genet. 15: 160-164.

Bingham, R.T., 1977. Breeding white pines resistant to the white pine blister rust disease. In: R. Bogart (Ed.): Genetics lectures, Vol. 5, Oregon State Univ. Press, Corvallis. p. 33-44.

Bingham, R.T., R.J. Hoff & G.I. McDonald, 1971. Disease resistance in forest trees. Ann. Rev. Phytopath. 9: 433-452.

Bingham, R.T., R.J. Hoff & G.I. McDonald, 1973. Breeding blister rust resistant western white pine. VI. First results from field testing of resistant planting stock. USDA For. Serv., Res. Note INT-179. Ogden, UT. 12 p.

Bingham, R.T., A.E. Squillace & J.W. Wright, 1960. Breeding blister rust resistant western white pine. II. First results of progeny tests including preliminary estimates of heritability and rate of improvement. Silv. Genet. 9: 33-41.

Griggs, M.M. & R.J. Dinus, 1977. Patterns of fusiform rust increase their implications for selection and breeding. Proc. South. For. Tree Improv. Conf. 14: 43-52.

Hepting, G.H., 1971. Diseases of forest and shade trees of the United States. USDA For. Serv. Handbook 386. Washington, D.C. 658 p.

Hoff, R.J. & G.I. McDonald, 1971. Resistance to Cronartium ribicola in Pinus monticola: short shoot fungicidal reaction. Can. J. Bot. 49: 1235-1239.

Hoff, R.J. & G.I. McDonald, 1972. Resistance in Pinus armandii to Cronartium ribicola. Can. J. Forest Res. 2: 303-307.

Hoff, R.J. & G.I. McDonald, 1975. Hypersensitive reaction in Pinus armandii caused by Cronartium ribicola. Can. J. Forest Res. 5: 146-148.

Hoff, R.J. & G.I. McDonald, 1980. Resistance to Cronartium ribicola in Pinus monticola: reduced needle spot frequency. Can. J. Bot. 58: 574-577.

Hutchinson, W.G., 1935. Resistance of Pinus sylvestris to a gall-forming Peridermium. Phytopathology 25: 819-843.

Jewell, F.F. & D.C. Speirs, 1976. Histopathology of one- and two-year-old resisted infections by Cronartium fusiforme in slash pine. Phytopathology 66: 741-748.

Kinloch, B.B. & J.W. Byler, 1981. Relative effectiveness and stability of different resistance mechanisms to white pine blister rust in sugar pine. Phytopathology 71: 386-391.

Kinloch, B.B. & M. Comstock, 1980. Cotyledon test for major gene resistance to white pine blister rust in sugar pine. Can. J. Bot. 58: 1912-1914.

Kinloch, B.B. & M. Comstock, 1981. Race of Cronartium ribicola virulent to major gene resistance in sugar pine. Plant Disease 65: 604-605

Kinloch, B.B., Jr. & J.L. Littlefield, 1977. White pine blister rust: hypersensitive resistance in sugar pine. Can. J. Bot. 55: 1148-1155.

Kinloch, B.B., Jr. & R.W. Stonecypher, 1969. Genetic variation in susceptibility to fusiform rust in seedlings from a wild population of loblolly pine. Phytopathology 59: 1246-1255.

Lundquist, J.E., 1979. The pre-gall symptoms of fusiform rust. Unpub. Ph.D. thesis. Univ. Georgia, Athens. 115 p.

McDonald, G.I., 1978. Segregation of "red" and "yellow" needle lesion types among monoaeciospore lines of Cronartium ribicola. Can. J. Genet. Cytol. 20: 313-324.

McDonald, G.I. & R.J. Hoff, 1970. Resistance to Cronartium ribicola in Pinus monticola:

early shedding of infected needles. USDA For. Serv. Res. Note INT-124.
McDonald, G.I. & R.J. Hoff, 1971. Resistance to Cronartium ribicola in Pinus monticola: genetic control of needle-spots-only resistance factors. Can. J. Forest Res. 1: 197-202.
McDonald, G.I. & R.J. Hoff, 1975. Resistance to Cronartium ribicola in Pinus monticola: an analysis of needle spot types and frequencies. Can. J. Bot. 53: 2497-2505.
Miller, T., E.B. Cowling, H.R. Powers, Jr. & T.E. Blalock, 1976. Types of resistance and compatibility in slash pine seedlings infected by Cronartium fusiforme. Phytopathology 66: 1229-1235.
Mittempergher, L. & P. Raddi, 1977. Variation of diverse sources of Cronartium flaccidum. Eur. J. For. Path. 7: 93-98.
Mullick, D.B., 1977. The non-specific nature of disease in bark and wood during wounding, insect and pathogen attack. Rec. Adv. in Phytochem. 11: 395-440.
Patton, R.F., 1961. The effect of age upon susceptibility of eastern white pine to infection by Cronartium ribicola. Phytopathology 51: 429-434.
Patton, R.F., 1967. Factors in white pine blister rust resistance. Proc. 14th IUFRO Congress, Sept. 4-9, 1967. Munich, Germany. Part 3, Sec. 22 and Intersectional Working Group 22/24. p. 876-890.
Powers, H.R., Jr., 1980. Pathogenic variation among single aeciospore isolates of Cronartium quercuum f. sp. fusiforme. For. Sci. 26: 280-282.
Powers, H.R., Jr., F.R. Matthews & L.D. Dwinell, 1977. Evaluation of pathogenic variability of Cronartium fusiforme in loblolly pine in the southern USA. Phytopathology 67: 1403-1407.
Quick, C.R., 1966. Experimental inoculation of ponderosa pine with western gall rust. Plant Dis. Rep. 50: 550-552.
Snow, G.A., R.I. Dinus & A.G. Ross, 1975. Variation in pathogenicity of diverse sources of Cronartium fusiforme on selected slash pine families. Phytopathology 65: 170-175.
Struckmeyer, B.E. & A.J. Riker, 1951. Wound periderm formation in white pine trees resistant to blister rust. Phytopathology 41: 276-281.
True, R.P., 1938. Gall development on Pinus sylvestris attacked by the Woodgate Peridermium and morphology of the parasite. Phytopathology 28: 24-49.
Van der Kamp, B.J., 1968. Peridermium pini (Pers.) Lev. and the resin-top disease of Scots pine. I. A review of the literature. Forestry 41: 189-198.
Vanderplank, J.E., 1963. Plant diseases: Epidemics and control. Academic Press, New York. 349 p.
Ziller, W.G., 1974. The tree rusts of western Canada. Can. For. Serv. Publ. 1329, Dep. Environ., Victoria, B.C.

Resistance mechanisms in interactions between poplars and rust[1]

R. Siwecki[2], A. Werner[2], Z. Krzan[3] & F. Młodzianowski[4]

ABSTRACT

Predisposing factors on poplar leaves to infection by *Melampsora larici-populina* and resistance mechanisms involved in the penetration and colonization of poplar leaf tissues by the uredinial stage have been discovered. Disease incidence was estimated during 4 consecutive years, and the degree of resistance of various poplar clones growing in a stool bed was determined. Studies on germination of urediniospores and growth of germ tubes on poplar leaves showed that these processes are determinants of varietal resistance of hosts. Histological investigations of the penetration and pre-penetration growth of the pathogen revealed that in the incompatible association the host cells actively react against the pathogen. As an attempt to identify different resistance mechanisms against poplar rust a model is presented of the studies of host-pathogen interactions which are in progress or are contemplated in the near future.

INTRODUCTION

Poplar plantations, regions of natural occurrence of the more important poplar species, and rich collections of poplar cultivars maintained by various research institutions are presently under constant biological study aimed at the determination and comparative analysis of variability in resistance or susceptibility of poplars to infection by the common poplar rust fungus, *Melampsora larici-populina* Kleb.

1. This work has been supported partially by grant II/15 from the Polish Academy of Sciences and partially by grants FG-Po-298 and P-109 from the U.S. Department of Agriculture under PL-480.
2. Polish Academy of Sciences, Institute of Dendrology, 62-035 Kórnik, Poland.
3. Tatra National Park, 34-500 Zakopane, Poland.
4. A. Mickiewicz University, Institute of Biology, 61-713 Poznań, Poland.

The interdisciplinary studies involving phytopathology, genetics, physiology, and biochemistry create a good base for the selection of poplar clones for resistance in order to recommend them for cultivation and further breeding. At the same time results of studies on the interaction between both resistant and susceptible hosts and the pathogen may constitute a basis for the identification of some biological resistance mechanisms.

Various studies conducted by a group of scientists at the Institute of Dendrology, Polish Academy of Sciences in Kórnik permitted a detailed description of the interactions between poplars and rust (Siwecki, 1978). Pre-infection factors existing on the poplar leaf and resistance mechanisms involved in the penetration and colonization of poplar leaf tissues by *M. larici-populina* were discovered. Also permitted was evaluation of the genetic resistance of poplars to infection by the rust (Siwecki, 1981).

LIFE CYCLE OF THE PATHOGEN AND ESTIMATION OF DISEASE

The life cycle of *M. larici-populina* on larch (*Larix decidua*) and poplars from sections *Tacamahaca* and *Aigeiros* and intra- or inter-sectional hybrids has been described with particular consideration given to the most pathogenic stage of the parasite to poplars, the uredinium. Characters of spore size, and particularly those of urediniospores, cannot constitute a basis for the identification of rust species on various poplars. There are probably numerous physiological races of the pathogen, conditioned by the response and variability of the hosts, namely the numerous poplar cultivars grown around the world (Krzan, 1981a).

Optimal germination of urediniospores and development of germ tubes on the lower surface of poplar leaves takes place at 100 % relative humidity and an air temperature of +16 °C (Krzan, 1980). In such conditions urediniospore germ tubes are able to infect the mesophyll of poplars within 2-4 hours. Successful infections, however, are dependent on numerous mechanisms that hinder or hasten this process.

The studies conducted on ultrastructure permitted us to describe in greater detail some of the stages of the pathogen and follow the ultrastructural changes that occur in poplar clones differing in resistance (Młodzianowski & Siwecki, 1976, 1978; Młodzianowski et al., 1978).

During 1975-1978 rust incidence was estimated and the degree of resistance to rust was determined for 268 poplar clones growing in a stool bed, each clone represented by 5 stumps (Krzan, 1981b). Observations included data on the occurrence of uredinial and telial stages on poplar leaves. The following six-point scale of infection was used:

1. All leaves on the stump completely free of rust sori - Immune (I).
2. Infrequent sori of uredinial or telial stages on one or only a few leaves per stump - Very Resistant (VR).
3. More numerous sori on many leaves. No necrosis - Resistant (R).

4. Sori common on all or almost all leaves. Localized necrotic spots - Medium Resistant (MR).
5. Sori common on all leaves, numerous necrotic spots or patches, slight defoliation - Susceptible (S).
6. Sori common, almost the entire leaf blade necrotic, severe defoliation - Very Susceptible (VS).

Observations were conducted for 4 consecutive years at several times during the year when the uredinial and telial stages were occurring. From these observations for each clone a mean infection index (MII) was calculated, which characterized the degree of resistance and susceptibility to natural rust infection.

Table 1 (from data of Krzan, 1981a, b) shows that there was great variation in the different years in estimates of the degree of rust resistance of the clones. During these 4 years it was established that the final extent of rust development was determined by the pattern of preceding weather conditions. For example the influence of some climatic factors, especially relative humidity and air temperature during June and July, was also variable over this period (compare Table 1 with Fig. 1).

In his doctoral investigations Krzan (1980, 1981a, b) studied resistance mechanisms and confirmed the existence of ecological and genetic factors which influence resistance or susceptibility of poplar clones to rust and which make possible selection for this trait.

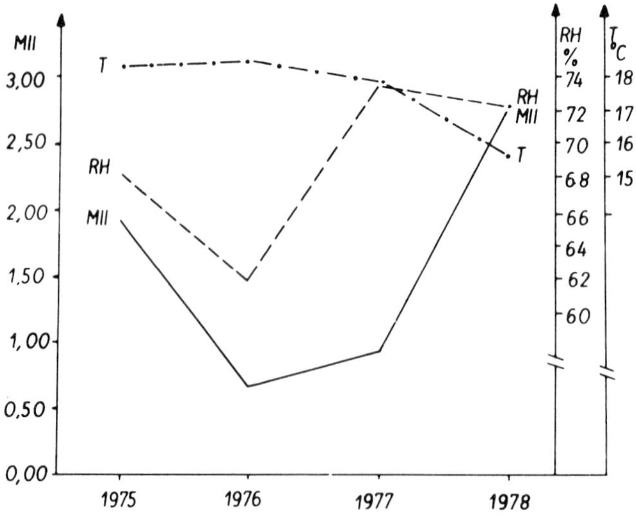

Figure 1. The relationship between severity of rust infection on poplar leaves (MII), relative air humidity (RH), and air temperature (T) in June and July during 1975-1978 (From Krzan, 1980).

Table 1. Number and percentage of poplar clones included in the resistant and susceptible groups characterized by the Mean Infection Index (MII) calculated for 4 years of observations.

Resistant and susceptible groups	1975			1976			1977			1978		
	MII* range	No	%	MII range	No	%	MII range	No	%	MII range	No	%
Immune clones (I)	0.00	8	3.0	0.00	33	12.3	0.00	29	10.8	0.00	3	1.1
Very resistant clones (VR)	0.01-0.84	23	8.6	0.01-0.52	102	38.1	0.01-0.67	92	34.3	0.01-0.87	10	3.7
Resistant clones (R)	0.85-1.69	99	36.9	0.53-1.05	95	35.4	0.68-1.35	82	30.6	0.88-1.75	25	9.3
Medium resistant clones (MR)	1.70-2.54	87	32.5	1.06-1.58	13	4.9	1.36-2.03	50	18.7	1.76-2.63	84	31.3
Susceptible clones (S)	2.55-3.39	33	12.3	1.59-2.11	19	7.1	2.04-2.71	9	3.4	2.64-3.51	95	35.4
Very susceptible clones (VS)	3.40-4.24	18	6.7	2.12-2.64	6	2.2	2.72-3.39	6	2.2	3.52-4.39	51	19.0
Sum		268	100		268	100		268	100		268	100

* Range of Mean Infection Index was changed for 4 consecutive years.

Infection of poplar leaves by rust aeciospores and urediniospores in specific microclimatic conditions existing on the leaf surface is determined by the structure of the stomata, structure of the leaf surfaces, and stomatal diffusion resistance and stomatal movement. The course of the transpiration process in several resistant or susceptible poplar clones was studied by Siwecki & Przybył (1981). Comparisons were made of detached and undetached non-infected poplar leaves under different experimental conditions. The poplar clones were characterized by variable frequency and size of stomata on the adaxial and abaxial leaf surfaces (Table 2).

The rate of transpiration of detached leaves measured in a wind tunnel and calculated on the basis of leaf dry weight was decidedly lower for the resistant poplar clones than for susceptible ones (Table 3). This is particularly emphasized by the rate of transpiration in the time intervals 0-15 min and 15-30 min.

Results of measuring transpiration of undetached leaves in the porometer are presented on Fig. 2. The most interesting result is that in the morning hours the 2 susceptible clones *P.* × *petrowskyana* and *P. tacamahaca* demonstrated a more rapid water loss in both wind tunnel and porometer experiments than in the afternoon or evening.

This phenomenon suggests a longer period of stomatal opening in the morning and an altered cycle of water management in these clones compared to that in some resistant clones, for example *P. deltoides* B-60. The entire

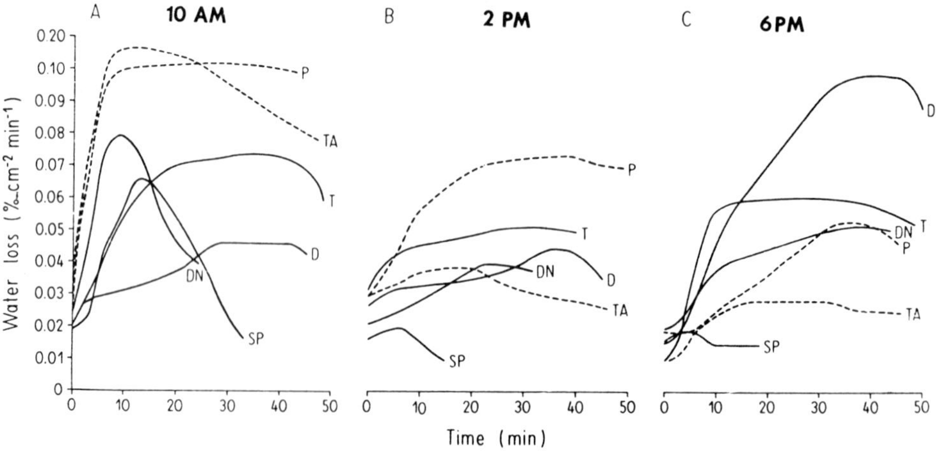

Figure 2. Water loss by undetached leaves of resistant (———) and susceptible (-----) poplar clones at 10 am (A), 2 pm (B) and 6 pm (C). *P. deltoides* × *P. nigra* 490-1 (DN), *P.* 'Serotina de Poitou' (SP), *P. deltoides* B-60 (D), *P. trichocarpa* 404 (T), *P.* × *petrowskyana* (P) and *P. tacamahaca* (TA) (From Siwecki & Przybył, 1981).

Table 2. Mean stomatal frequency and mean size of stomata on adaxial and abaxial leaf surfaces (From Siwecki & Przybł, 1981).

Clones	Stomatal frequency (cm^{-2}) Adaxial / Abaxial	Mean length and width of stomata (μm) Adaxial / Abaxial	Mean Infection Index (MII)
Resistant clones			
P. deltoides B-60 (11892)	15 271 ± 361 / 22 589 ± 437	27.30 × 16.94 / 28.07 × 17.50	0.00
P. deltoides ssp. *angulata* (11621)	13 664 ± 265 / 27 263 ± 514	29 47 × 18.48 / 28.91 × 18.27	0.00
P. deltoides × *P. nigra* 491-1 (12299)	10 415 ± 215 / 30 088 ± 572	28.48 × 16.12 / 25.94 × 15.50	0.11
P. trichocarpa 1200 (404)	5 997 ± 112 / 10 764 ± 220	32.24 × 25.17 / 33.56 × 17.55	0.96
P. 'Serotina de Poitou' (11309)	11 046 ± 342 / 14 146 ± 341	27.37 × 16.17 / 32.55 × 20.09	1.05
P. deltoides × *P. nigra* 490-1 (11888)	10 513 ± 217 / 14 883 ± 327	25.40 × 14.79 / 26.50 × 16.23	1.17
Susceptible clones			
P. nigra PW 3	8 552 ± 308 / 21 444 ± 427	26.39 × 15.19 / 29.61 × 18.54	1.38
P. × *petrowskyana* (5163)	7 137 ± 252 / 20 673 ± 596	27.85 × 16.73 / 27.09 × 16.03	2.12
P. trichocarpa cl. 30	6 237 ± 309 / 19 033 ± 373	32.05 × 19.52 / 37.02 × 23.10	2.13
P. tacamahaca (3747)	6 880 ± 260 / 16 718 ± 347	26.60 × 15.47 / 27.58 × 16.16	3.03

Table 3. Mean transpiration rates of detached leaves of rust-resistant and rust-susceptible poplar clones in the wind tunnel as calculated on the basis of leaf area and dry weight in the morning and afternoon (From Siwecki & Przybył, 1981).

Clones	Transpiration			
	$mg \cdot cm^{-2}$ per 15 min		$mg \cdot mg^{-1}$ dry wt per 15 min	
	a.m.	p.m.	a.m.	p.m.
Resistant clones				
P. deltoides B-60 (11892)	0.1608	0.1117	0.0566	0.0398
P. deltoides ssp. *angulata* (11621)	0.2150	0.1622	0.0745	0.0574
P. deltoides × *P. nigra* 491-1 (12299)	0.2336	0.1013	0.0851	0.0400
P. 'Serotina de Poitou' (11309)	0.1599	0.0834	0.0573	0.0301
Mean	0.1923	0.1146	0.0684	0.0418
Susceptible clones				
P. nigra PW 3	0.2279	0.1462	0.0742	0.0495
P. × *petrowskyana* (5163)	0.1835	0.0941	0.0765	0.0377
P. tacamahaca (3747)	0.2667	0.1539	0.0893	0.0534
P. trichocarpa (11651)	0.3469	0.2269	0.1118	0.0799
Mean	0.2562	0.1552	0.0879	0.0551

aspect of leaf water relations and water management in the 2 susceptible poplar clones favors infection by germ tubes of rust urediniospores. Also, the water relations of these poplars govern pre-infection mechanisms that affect and facilitate the next steps in compatible host-pathogen relationships.

Werner (1981a), according to his studies on 59 poplar clones, claimed that the number of stomata per unit of leaf surface was not correlated with frequency of infection sites on leaves affected by the rust. However, the frequency of stomata on a leaf surface could have some influence on the number of contacts between hyphae and stomata. In his opinion the course of the pre-penetration growth of the pathogen is dependent on several pre-infection factors which limit markedly the intensity of infection. The dimensions of the stomata, sculpturing of the leaf surface, anatomical struc-

Table 4. Mean dimensions of the surface of areoles of resistant and susceptible poplar clones (From Werner, 1981a).

Clones	Surface of areoles (mm^2)
Resistant clones	
P. deltoides B-60 (11892)	0.138
P. deltoides ssp. *angulata* (11621)	0.114
P. maximowiczii (4544)	1.116
P. 'Serotina de Poitou' (11309)	0.089
Susceptible clones	
P. nigra PW 3	0.231
P. × *petrowskyana* (5163)	1.189
P. trichocarpa (11651)	1.277
P. balsamifera (8348)	1.975

ture of the guard cells and subsidiary cells all are factors determining the behavior of the germ tubes on the leaves and appressorial formation over the stomata.

Werner (1981a) indicated that the structure of the vascular bundles and the dimensions of the areoles can determine the size of the rust colonies developing on poplar leaves (Table 4).

In compatible associations the mesophyll cells inhibit the development of the fungus. In compatible associations, however, a stimulation of fungal growth may result. On the basis of an analysis of compactness of poplar leaf mesophyll we can conclude that the development of the pathogen in the tissue of the host is determined by a reaction of the tissue that is the sum of the reactions of all the individual cells. Inhibition of growth of the pathogen was stronger in resistant poplars with a compact mesophyll (e.g. *P. deltoides* B-60) than in those with a less compact mesophyll (*P. maximowiczii*). On the other hand, in susceptible poplars with a compact mesophyll the effect of pathogen growth stimulation was stronger than in poplars with a less compact mesophyll. Therefore, a greater compactness of the mesophyll increases resistance in resistant poplars and increases susceptibility in susceptible ones.

Werner (1981a, b) showed that several important resistance mechanisms of poplars to rust become apparent during the penetration and post-penetration growth of the pathogen. Some of those mechanisms were presented in the paper by Siwecki & Werner (1980).

THE EPIPHYTIC PHASE OF PATHOGEN DEVELOPMENT

The germination of urediniospores of *M. larici-populina* and the behavior of germ tubes are different on resistant and susceptible poplars. In the resistant poplar clones (*P. maximowiczii*, *P. deltoides* B-60) the processes of urediniospore germination and germ tube development were distinctly inhibited during the first hours after inoculation. Some chemical factors present on the surface of the poplar leaves were responsible for this inhibition (Krawiarz & Werner, 1978).

The structure of the stomatal surface and certain anatomical features of the guard and subsidiary cells constitute a system of factors that condition the appropriate 'recognition' of the stomata by the hyphae. Such features as: substantial cuticular thickenings present on the external walls of the subsidiary cells, depressions between the guard cells and subsidiary cells, as well as the small diameter of stomatal pores and the large external ridges, all hinder the free growth of hyphae over the stomata and their penetration. The physical characteristics conditioning recognition of the stomata by the hyphae induce the formation of appressoria over the stomata. In susceptible poplars that were characterized by large stomata, small external ridges and weakly marked sculpturing of the leaf surface (e.g. in *P. trichocarpa*), the germ tubes of urediniospores accomplished penetration of the stomata without forming appressoria over them. On other poplars, both susceptible and resistant ones, the germ tubes formed appressoria over the stomata. Their number, and as a consequence the number of penetration sites, was decidedly greater on susceptible poplars. On the other hand, the smaller number of penetration sites in the resistant poplars was a consequence of the poorer 'recognition' of poplar stomata by germ tubes. On the leaves of these poplars frequently growth of the hyphae near stomata or over them without appressoria formation was observed. This phenomenon can be explained by the fact that the small stomata of the resistant poplars could not induce appressoria formation and therefore in the majority of cases remain 'unrecognized'. When they were 'recognized', the hyphae always produced appressoria over the stomata.

PENETRATION OF THE PATHOGEN INTO THE LEAF

With poplars resistant to rust distinct reactions of stomata to the presence of the pathogen in the stomatal pore were observed. These reactions were made visible by the strong fluorescence of the guard cells and spongy parenchyma cells, after being labelled with a fluorescent brightener (Rohringer et al., 1977). This fluorescence indicated the formation of compounds of phenolic character in that region. The considerable speed of this reaction allows us to classify it as a typical hypersensitive reaction.

This confirmed the formation of phenolic compounds. Here, the appearance of phenolic compounds was associated with the dying of pathogen hyphae and host cells in the vicinity of the stomata. In spite of differences in opinion concerning the role that hypersensitive reactions play in the active defense of a plant against a pathogen, the rapidity of this reaction in resistant poplars (*P. deltoides* B-60) always resulted in a complete inhibition of the pathogen within the substomatal cavity. The same reactions also took place, but with some delay, in the other poplars. However, in these cases they never led to the complete cessation of growth of the pathogen in this early stage of its development. Therefore, we assume that 'overcoming' of the defense reaction of the host by the pathogen within the substomatal cavity represents the key factor in the establishment of a compatible relation between the host and the pathogen and determines the subsequent progress of infection.

COLONIZATION OF HOST TISSUES

In *P. deltoides* B-60, and to a lesser extent in the other 2 resistant poplars (*P. deltoides ssp. angulata* and *P. maximowiczii*), after the infection hyphae entered the mesophyll the growth of the pathogen was stopped at the first contact of the infection hypha with the mesophyll cell, in the phase of haustorium mother cell formation, during the haustorium formation, or just after that. The cessation of growth of the pathogen was always associated with the death of the fungus and of the adjacent host cells. These were typical necrotic reactions, and when they appeared in large areas of the mesophyll, they were visible as necrotic spots at the point of infection of the leaf.

In Table 5 host-pathogen interactions during the infection process of poplar leaves by the rust are presented descriptively.

NECESSITY FOR FUTURE STUDIES ON RESISTANCE MECHANISMS

Some proposed trends in biological research in order to explain the resistance mechanisms in the confrontation between poplars and rust are presented in Fig. 3, according to the pattern given in the paper of Kaufmann & Wellendorf (1979). These kinds of studies should be performed in the field or laboratory, or both, on poplar clones, hybrid families, and provenances. Possible explanations of the variability in resistance of rust-infected poplars should arise from such studies.

Clones selected for resistance or susceptibility should be the basic material for investigations of the resistance mechanisms. Heritability (h^2) of the features characteristic of resistance can be estimated from observations of the degree of infection of parent trees and their hybrid progeny (Siwecki, 1981).

Table 5. Host-pathogen relationships during the process of infection of poplar leaves by the rust *M. larici-populina*.

Poplars	Percentage of appressoria formation over stomata*	Stage of cessation of fungal growth following distinct reactions of host cells observed as an increase of fluorescence, wall 'darkening' or thickening associated with plant necrosis				Growth of the pathogen in the mesophyll	Regeneration ability of the pathogen
		during penetration or formation of substomatal vesicles	during growth of infection hyphae	at or during formation of haustorial mother cells	during or after formation of haustorium		
Resistant clones							
P. deltoides B-60 (11892)	48	more frequent	frequent	frequent	more frequent	poor	none
P. deltoides ssp. *angulata* (11621)	54	frequent	frequent	less frequent	less frequent	poor	poor
P. maximowiczii (4544)	50	frequent	rare	less frequent	less frequent	poor	poor
P. 'Serotina de Poitou' (11309)	62	rare	rare	rare	rare	intermediate	good
Susceptible clones							
P. nigra PW 3	81	rare	rare	very rare	very rare	good	very good
P. × *petrowskyana* (5163)	88	sporadic	none	none	none	good	very good
P. trichocarpa (11651)	79	none	none	none	none	good	very good
P. balsamifera (8348)	100	none	none	none	none	good	very good

* Calculated in relation to the number of appressoria over stomata on the leaves of the very susceptible *P. balsamifera* clone which has been accepted as 100 %.

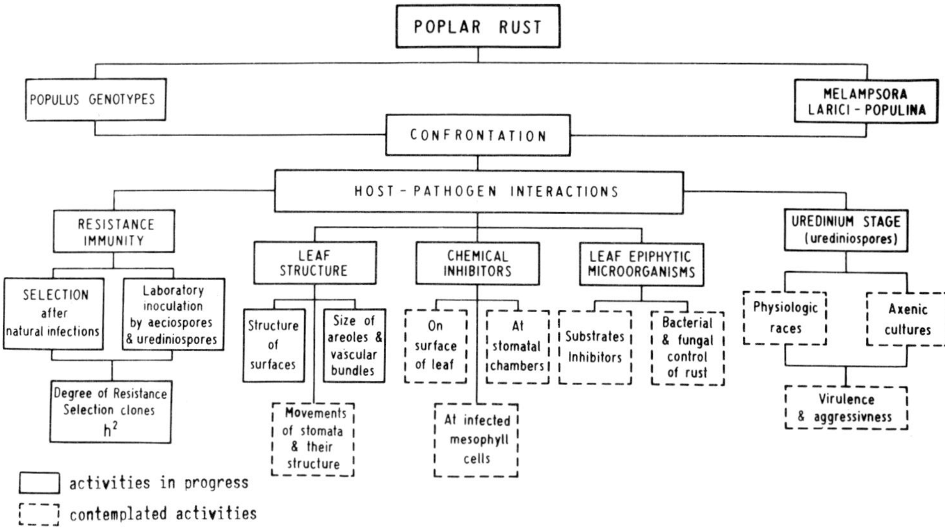

Figure 3. Some proposed trends in biological research in order to explain the resistance mechanisms in the confrontation between poplars and rust.

In such further studies 3 directions should be stressed: (1) leaf structure and its influence on the infection process by pathogen, (2) biochemical inhibitors manifesting during infection process, and (3) the role of epiphytic microorganisms during the infection process.

As shown in Fig. 3, and by the results discussed in this paper, only a few mechanisms of resistance active during the confrontation between poplars and rust have been investigated. The other possible influences should be checked intensively in the near future.

Our knowledge of the pathogen is still insufficient for the study of resistance from the standpoint of establishment of physiological races of the pathogen and their growth in axenic culture. Also, factors influencing the virulence and aggressiveness of the pathogen are still unknown. There is also a lack of information on other than the uredinial stage of development of the pathogen (pycnium, aecium, telium and basidium). Those stages have a direct or indirect influence on the complex relationship between poplars and rust.

REFERENCES

Kaufmann, U. & H. Wellendorf, 1979. A study of mechanisms in Norway spruce having inhibitory effects on the growth of Fomes annosus. Proceedings of the Conference on Biochemical Genetics of Forest Trees Umeå, Sweden 1978. p. 50-59.

Krawiarz, K. & A. Werner, 1978. Studies on the effect of surface chemistry of poplar leaves on urediniospore germination and growth of hyphae in development of Melampsora larici-populina Kleb. In: R. Siwecki: The mechanisms of poplar leaf resistance to fungal infection. Final Report, 1978. Mimeographed. 112 p.

Krzan, Z., 1980. Effect of climatic factors on development of the disease caused by Melampsora larici-populina. Folia Forestalia 422: 10-13.
Krzan, Z., 1981a. Cykl rozwojowy i morfologia grzyba Melampsora larici-populina Kleb. - sprawcy rdzy topoli. Arboretum Kórnickie 26 (in press).
Krzan, Z., 1981b. Poplar resistance to infection by the fungus Melampsora larici-populina in field conditions. Arboretum Kórnickie 26 (in press).
Młodzianowski, F. & R. Siwecki, 1976. Ultrastructure of poplar leaves naturally infected by rust Melampsora larici-populina. Arboretum Kórnickie 21: 375-400.
Młodzianowski, F. & R. Siwecki, 1978. Morphology and ultrastructure of urediospores of Melampsora larici-populina. European Journal of Forest Pathology 8: 43-48.
Młodzianowski, F., A. Werner & R. Siwecki, 1978. Germination of Melampsora larici-populina uredospores on poplar leaves. European Journal of Forest Pathology 8: 119-125.
Rohringer, R., W.K. Kim, D.J. Samborski & N.K. Howes, 1977. Calcofluor: an optical brightener for fluorescence microscopy of fungal plant parasites in leaves. Phytopathology 67: 808-810.
Siwecki, R., 1978. The mechanism of poplar leaf resistance to fungal infection. Final Report, July 1, 1973 - October 31, 1978. Institute of Dendrology, Kórnik (Poland). Mimeographed. 112 p.
Siwecki, R., 1981. Characterization of poplar resistance to more important diseases. In: R. Siwecki & M. Ridé (Eds): Resistance mechanisms in poplar diseases. Proceedings joint symposium FAO/IUFRO, Kórnik 1980. p. 41-49.
Siwecki, R. & K. Przybył, 1981. Water relations in the leaves of poplar clones resistant and susceptible to Melampsora larici-populina. European Journal of Forest Pathology 11: 348-357.
Siwecki, R. & A. Werner, 1980. Resistance mechanism involved in the penetration and colonization poplar leaf tissues by Melampsora rust. Phytopathologia Mediterranea 19: 27-29.
Werner, A., 1981a. Histologiczne badania zależności gospodarz-patogen w przebiegu choroby liści topoli wywołanej przez Melampsora larici-populina Kleb. Arboretum Kórnickie 26 (in press).
Werner, A., 1981b. Rozmieszczenie metabolitów w mezofilu topoli porażonych rdzą Melampsora larici-populina Kleb. Arboretum Kórnickie 26 (in press).

Resistance mechanisms of elms to Dutch elm disease

D.M. Elgersma

Phytopathological Laboratory 'Willie Commelin Scholten', Baarn, the Netherlands

ABSTRACT

Processes provoking symptoms in elms after inoculation with the Dutch elm disease fungus are briefly discussed. A review on research concerning mechanisms of resistance in elms to *Ophiostoma ulmi* is given. Spread of the fungus throughout the tree is hampered in resistant trees compared to susceptible ones. Anatomical factors, such as vessel length and diameter and size of vessel groups, might reduce the speed of transport of conidia in resistant plants. Tylosis formation is faster in resistant trees than in susceptible ones, another factor which might limit the spread of the fungus. The meaning of phytoalexins in disease resistance and the role of phytotoxins in pathogenesis are discussed.

INTRODUCTION

Dutch elm disease has become one of the most severe tree diseases in North America and Europe. Recently an excellent review on Dutch elm disease has been compiled by Sinclair & Campana (1978). This compilation includes a paper by MacHardy (1978) on mechanisms of resistance, and therefore there is no need for a thorough and detailed review of the literature now. The purpose of this paper is to discuss further research on physiological, biochemical or morphological aspects of the expression of genes for resistance to Dutch elm disease in different species or clones of elms. A study of resistance mechanisms can be fruitful only if we know how the normal processes of pathogenesis elapse. We may expect that products of genes for resistance block or counteract some, or all, of those processes that otherwise ultimately lead to symptom expression and death of the host.

Dutch elm disease is caused by the Ascomycete *Ophiostoma* (*Ceratocystis*) *ulmi* (Buisman) Nannf., which after introduction into the xylem by bark beetles, grows and produces conidia in the vessels. Conidia are transported with the transpiration stream throughout the tree. During its growth in the vessels the fungus causes changes in pH of the xylem fluid, and produces metabolites such as cell-wall degrading enzymes, growth substances and toxins. All these factors in concert cause wilting and eventually, after 1 or 2 years, death of the whole tree. It is up to the plant pathologist to determine which metabolites or physiological processes are the crucial ones for disease expression. Processes caused by host-parasite interactions in the vessel lumina are so complex and intermingled that a study of each factor separately seems impossible.

A cross-section of an infected elm twig reveals a dark discoloration of the new annual ring, caused by oxydation and polymerization of phenolic compounds in the xylem sap, xylem parenchyma and ray cells (Gagnon, 1967b). Gums and tyloses are formed in the lumina of the xylem vessels, clearly impeding the sap flow (Gagnon, 1967a). Cell-wall degrading enzymes may release partially degraded wall polymers into the xylem fluid. These polymers as well as high-molecular-weight toxins as cerato-ulmin (Takai, 1974) and glycopeptides (Rebel, 1969), isolated from shake culture, may interfere with water transport, probably by physical plugging of the pit membranes (Van Alfen & Turner, 1975).

Apart from the above-described parasitic stage we know also a saprophytic stage of the Dutch elm disease fungus. The pathogen grows and produces coremia in breeding galleries made by elm bark beetles in the bark. Fungus spores adhere to emerging beetles, and during feeding by beetles in crotches of twigs of healthy elms, conidia may be introduced into the xylem vessels and infection is established. Dying or weakened elms are attractive for beetles as brood trees, thus amplifying the epidemic.

Just after the first World War the Dutch elm disease appeared in Europe and swept through the Old World (Gibbs, 1978). In the late 1920's the fungus was introduced into the United States by the import of veneer wood from France.

At the 'Willie Commelin Scholten' Laboratory at Baarn research started on the cause of the disease and the disease cycle was unraveled (Buisman, 1928; Fransen & Buisman, 1935). Since all native elm species in Holland were susceptible, a breeding program was started. Elms were collected from all over the world and selected for resistance to the disease, a program resulting in the release of several resistant elm clones (Heybroek, 1957, 1976). It became clear that great care should be taken to avoid non-heritable variations in symptom-expression, since trees are less susceptible because of slow growth or after transplanting (Heybroek, 1957; Went, 1954). It also became obvious that resistance was polygenic (Heybroek, 1970), and no substantial evidence has been found for a host-parasite differential interaction (Gibbs et al., 1975).

In the mid 1960's research on mechanisms of resistance to Dutch elm disease was started at the 'Willie Commelin Scholten' Laboratory at Baarn. By then good resistant clones were available and a propagation method had been developed (Tchernoff, 1963) that utilized shoots cut from callused roots. For physiological experiments genetically uniform plant material is a necessity. Because probably many genes are involved in the mechanisms of resistance and no race-specific effect has been discovered, we may expect to find effects of a quantitative nature governed by genes that have a normal function in healthy plants (Vanderplank, 1978).

COMPOSITION OF XYLEM SAP

In studying resistance mechanisms we first studied the composition of the xylem sap in susceptible and resistant clones. Since the fungus must grow in xylem fluid, the nutrient level of such fluid might have an impact

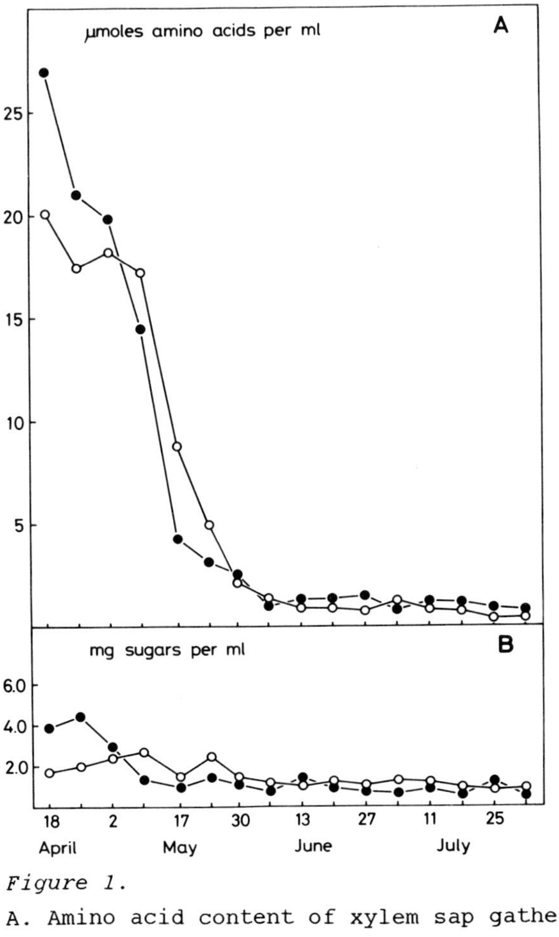

Figure 1.
A. Amino acid content of xylem sap gathered from ten 1-year-old callus cuttings grown in pots outdoors. o Amino acid content of susceptible elms.
● Amino acid content of resistant elms.
B. Total sugar content of the same xylem sap described in A. o Sugar content of susceptible elms. ● Sugar content of resistant elms.

on the growth of the pathogen. Amide and amino acid contents of the xylem sap of susceptible and resistant elms (Fig. 1A) were high from the beginning of April until the beginning of May, decreased rapidly in the middle of May, and then remained rather constant from the beginning of June until August (Elgersma, 1969). During the period of susceptibility, which in the Netherlands is from about the middle of May until the middle of July, no important changes occurred in amino acid or sugar content. In *Ulmus americana* L., however, Singh & Smalley (1969b) found a high amide and a low proline concentration in the xylem sap during the period when the trees were susceptible. They found a positive correlation between proline content and resistance in various elm species (Singh & Smalley, 1969a). We also found a higher proline content in resistant clones than in susceptible ones. By growing elms on a nutrient solution with NH_4^+ instead of NO_3^-, however, we were able to increase the proline content of the sap of a sus-

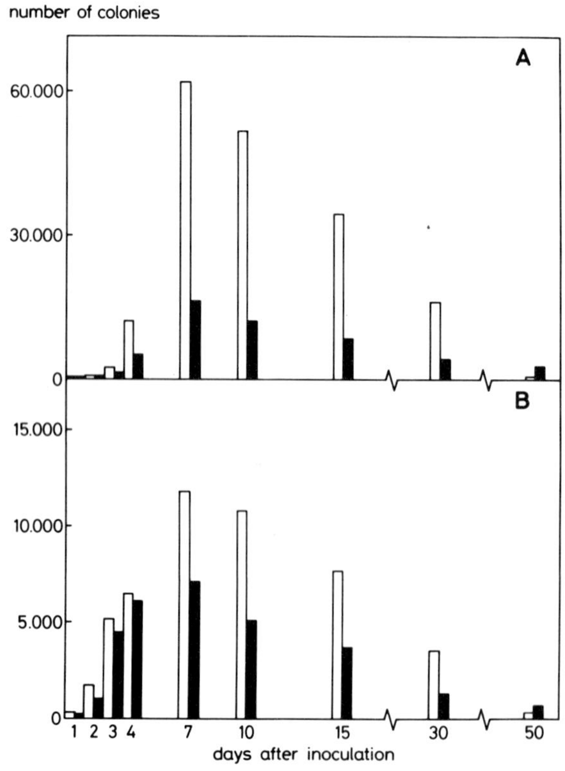

Figure 2. Average number of colonies grown from stem pieces that were cut from susceptible (white) and resistant (black) elms and homogenized on various days after inoculation.
A. stem pieces, 25-30 cm above the site of inoculation.
B. stem pieces, 1-4 cm above the site of inoculation.

ceptible clone to several times higher than that of resistant elms grown on NO_3^- nutrient solution without obtaining resistance (Elgersma, 1969). The relation of proline and resistance is not well understood. Singh & Smalley (1969a) suggest that a high content of proline reflects a tolerance to waterstress.

Data obtained by Singh & Smalley (1969a) and Elgersma (1969) gave no evidence that growth of the fungus in resistant clones or species of elms compared to that in susceptible ones would be limited by shortage of available amino acids or sugars. Growth of the fungus in 2-year-old callus cuttings of resistant and susceptible clones was compared by culturing the pathogen from homogenates of stem pieces selected at various heights above the site of inoculation (Fig. 2A). Colony counts showed that fungal material increased in resistant and in susceptible trees in the lower stem pieces at about the same rate during the first 4 days after inoculation. At 7 days after inoculation the number of propagules was higher in susceptible than in resistant plants, because vessels not directly infected at the time of inoculation became infected in the susceptible plants but not in resistant plants. The number of propagules isolated from stem pieces at higher parts of resistant plants was much less than from the corresponding parts of susceptible elms (Fig. 2B). The conclusion is that the pathogen can grow in xylem vessels of resistant plants, but spread of the fungus through the resistant plant is limited. This limited spread keeps the pathogen localized and also its effect on the trees. Water conduction is interrupted only partially and this reduction has no or only very slight effect on the tree as a whole.

ANATOMICAL DIFFERENCES

The question now is what factors are responsible for limiting the spread of the fungus. It is generally believed that spread of the fungus throughout the tree is accomplished by movement of the conidia in the transpiration stream (Banfield, 1968) and therefore differences in anatomical structure of the vascular system might have an impact on transport of the conidia.

Comparisons of vessel length and vessel diameter of susceptible and resistant clones revealed that the average diameter of the xylem vessels from a resistant clone was lower than that of vessels in a susceptible clone (Elgersma, 1970). Small vessels are a disadvantage in the spread of the fungus because transport can be blocked easier by gums and tyloses. This phenomenon also has been observed in American elms (Sinclair et al., 1975). Another difference was that the effective length of the vessels appeared to be shorter in the resistant clone than in the susceptible one. This resulted in less spread of conidia in the resistant tree, because in short vessels the pathogen has to penetrate more end walls to spread over the same distance than in long vessels. Conductivity for water and air through stem pieces was lower in resistant than in susceptible trees, which is in agreement with the observed differences in anatomical structure of the vessels. Also, the decreased susceptibility of trees when late wood is formed may be due to the small diameter of these vessels. McNabb et al. (1970) found that an increase in disease susceptibility was correlated with an increase in average size of vessel groups. The average size of vessel groups is defined as the product of the average vessel diameter and the average number of contiguous vessels. All these observations indicate that anatomical factors are important in resistance to Dutch elm disease.

TYLOSIS FORMATION

As already mentioned tylosis formation can block the sap flow and is considered of paramount importance in symptom expression. On the other hand tyloses will also hinder or arrest movement of conidia, a host reaction that might be rather effective in small sized vessels. Tylosis formation has been described as a possible mechanism of resistance to *Fusarium* wilt in banana plants. In the incompatible host-parasite combination rapid formation of tyloses seals off the vessels in advance of the pathogen, thus localizing the infection. In the compatible host-parasite combination, however, tylosis formation is delayed, and the pathogen can spread through the whole plant (Beckman & Halmos, 1962; Beckman et al., 1962).

A significant difference in tylosis formation appeared between a susceptible and a resistant elm clone after inoculation with the pathogen (Fig. 3). Tyloses in susceptible plants develop slower than in resistant plants, and therefore they supposedly were not effective in sealing off the vessels in advance of the infection (Elgersma, 1973).

After discovery in England of a highly aggressive strain of *O. ulmi* (Gibbs et al., 1972) we repeated the experiments and included this aggressive strain as well as the vascular parasite *Fusarium oxysporum* f. sp. *lycopersici*, a non-pathogen to elms (Elgersma & Miller, 1976). No significant difference in tylosis formation was found within clone Belgica, a very susceptible clone, after inoculation with the non-aggressive strain of *O. ulmi*, the aggressive strain, or *F. oxysporum* f. sp. *lycopersici*. Similar results were obtained in clone 390, a clone resistant to the non-aggressive strain of *O. ulmi*. Tylosis formation in clone Belgica as well as in clone 390 is not inhibited in individuals of a susceptible reaction type, but seems to be genetically determined by the elm clone alone and therefore we consider tylosis formation a non-specific response of elms to injury or microbial infestation. Since tylosis formation in clone 390 does not appear to be effective in arresting the transport of the aggressive strain of *O. ulmi* (Elgersma & Heybroek, 1979), we must assume that bypassing of obstacles such as tyloses is probable. This has been suggested previously by

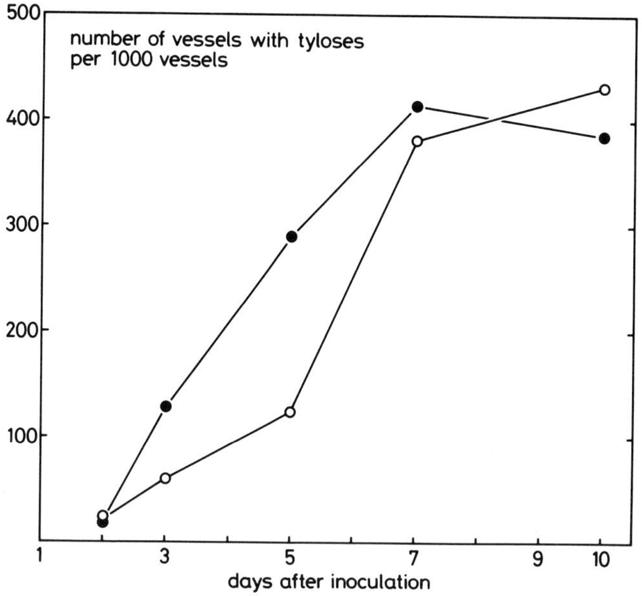

Figure 3. Number of vessels showing tyloses per 1000 vessels in cross-section. o susceptible, • resistant clone.

Miller & Elgersma (1976) as being due to more rapid growth and penetration of vessel pits by the aggressive strain.

Production of cell-wall degrading enzymes by the pathogen might therefore be of importance. A study of cellulase and endopolygalacturonase production by aggressive and non-aggressive strains showed no difference between them (Elgersma, 1976). Research on production of hemicellulase by these strains is in progress.

PHYTOALEXINS AND TOXINS

Accumulation of fungitoxic compounds or phytoalexins after infection with micro-organisms is a widespread phenomenon in angiosperms. It appeared also that infected elms accumulate fungitoxic compounds (Overeem & Elgersma, 1970). The naphthoquinones mansonone E and F were isolated from susceptible as well as resistant clones. The Dutch elm disease fungus, however, can tolerate a relatively high concentration of these compounds.

Results obtained in later studies showed that infected resistant clones accumulated no more of these compounds than susceptible plants (Elgersma & Overeem, 1971). The induction of synthesis of mansonones appears to be non-specific. Accumulation also occurs after chemical injury. Probably we may consider the production of these fungitoxic compounds as a weak resistance mechanism. Such compounds may slow down the infection process and the killing of the tree or inhibit other microbial infestation.

O. ulmi produces phytotoxic compounds in shake culture (Zentmeyer, 1942). Rebel (1969) isolated a toxic glycopeptide fraction, later purified by Strobel et al. (1978), that caused wilting of elm sprouts of susceptible as well as resistant trees. So it appeared to be non-specific. Van Alfen & Turner (1975) showed that even small amounts of this toxin could interfere with water conductivity in the xylem vessels. Probably the high molecular compounds plug the pit membranes of the vessels, thus hampering water transport. Recently we were able to detect this glycopeptide in infected trees by means of an advanced immunological technique (ELISA) (Scheffer & Elgersma, 1981). Important questions, such as can the toxin be neutralized by the resistant trees or does the aggressive strain produce more toxin in the tree than the non-aggressive one, still have to be answered.

CONCLUSIONS

Much evidence for the existence of mechanisms of resistance, as described here, has been presented, but no absolute proof has been given because genetic crossings to determine if correlated phenomena segregate in absolute association have not been performed (Ellingboe, 1976). The polygenic and probably quantitative nature of resistance will make such experiments almost impossible.

Elm breeding programs might use the accumulated information about the mechanisms of resistance, however. In the choice of alternative parent trees the nature of resistance should be taken into consideration. Crossing trees with different sets of genes for resistance would be advisable instead of crossing parents with almost identical ones.

In the future more mechanisms of resistance probably will be detected especially in the Asiatic elm species, which have not been studied very intensively till now.

REFERENCES

Banfield, W.M., 1968. Dutch elm disease recurrence and recovery in American elms. Phytopathologische Zeitschrift 62: 21-60.

Beckman, C.H., 1971. The plasticizing of plant cell walls and tylose formation - a model. Physiological Plant Pathology 1: 1-10.

Beckman, C.H. & S. Halmos, 1962. Relation of vascular occluding reactions in banana roots to pathogenicity of root invading fungi. Phytopathology 52: 893-897.

Beckman, C.H., S. Halmos & M.E. Mace, 1962. The interactions of host, pathogen and soil temperature in relation to susceptibility to Fusarium wilt of bananas. Phytopathology 52: 134-140.

Buisman, C.J., 1928. De oorzaak van de iepenziekte. Tijdschrift der Koninklijke Nederlandse Heidemaatschappij 40: 338-345.

Elgersma, D.M., 1969. Resistance mechanisms of elms to Ceratocystis ulmi. Mededelingen Phytopthologisch Laboratorium Willie Commelin Scholten, Baarn, No. 77. 84 p.

Elgersma, D.M., 1970. Length and diameter of xylem vessels as factors in resistance of elms to Ceratocystis ulmi. Netherlands Journal of Plant Pathology 76: 179-182.

Elgersma, D.M., 1973. Tylose formation in elms after inoculaton with Ceratocystis ulmi,

a possible resistance mechanism. Netherlands Journal of Plant Pathology 79: 218-220.
Elgersma, D.M., 1976. Production of pectic and cellulolytic enzymes by aggressive and non-aggressive strains of Ophiostoma ulmi. Netherlands Journal of Plant Pathology 82: 161-172.
Elgersma, D.M. & H.M. Heybroek, 1979. Spread and survival of an aggressive and a non-aggressive strain of Ophiostoma ulmi in elms. Netherlands Journal of Plant Pathology 85: 235-240.
Elgersma, D.M. & H.J. Miller, 1977. Tylose formation in elms after inoculation with an aggressive or a non-aggressive strain of Ophiostoma ulmi or with a non-pathogen to elms. Netherlands Journal of Plant Pathology 83: 241-243.
Elgersma, D.M. & J.C. Overeem, 1971. The relation of mansonones to resistance against Dutch elm disease and their accumulation, as induced by several agents. Netherlands Journal of Plant Pathology 77: 168-174.
Ellingboe, A.H., 1976. Genetics of host-parasite interactions. In: R. Heitefuss & P.H. Williams (Eds): Encyclopedia of plant physiology. Springer-Verlag, Berlin/ Heidelberg/New York. p. 761-778.
Fransen, J.J. & C.H. Buisman, 1935. Infectieproeven op verschillende iepensoorten met behulp van iepenspintkevers. Tijdschrift over Plantenziekten 41: 221-239.
Gagnon, C., 1967a. Histochemical studies on the alteration of lignin and pectic substances in white elm infected by Ceratocystis ulmi. Canadian Journal of Botany 45: 1619-1623.
Gagnon, C., 1967b. Polyphenols and discoloration in the elm disease investigated by histochemical techniques. Canadian Journal of Botany 45: 2119-2124.
Gibbs, J.N., 1978. Intercontinental epidemiology of Dutch elm disease. Annual Review of Phytopathology 16: 287-307.
Gibbs, J.N., H.M. Heybroek & F.W. Holmes, 1972. Aggressive strain of Ceratocystis ulmi in Britain. Nature, London 236: 121-122.
Gibbs, J.N., C.M. Brasier, H.S. McNabb Jr. & H.M. Heybroek, 1975. Further studies on pathogenicity in Ceratocystis ulmi. European Journal of Forest Pathology 5: 161-174.
Heybroek, H.M., 1957. Elm breeding in the Netherlands. Silvae Genetica 6: 112-117.
Heybroek, H.M., 1970. Three aspects of breeding trees for disease resistance. 2nd World Consultation on Forest Tree Breeding Proceedings Vol. I. FAO, Rome. p. 519-535.
Heybroek, H.M., 1976. Drie nieuwe iepeklonen. Nederlands Bosbouw Tijdschrift 48: 117-123.
MacHardy, W.E., 1978. Mechanisms of resistance. In: W.A. Sinclair & R.J. Campana (Eds): Dutch elm disease. Perspectives after 60 years. Cornell University. Agricultural Experiment Station. Search (Agr.) Vol. 8: 25-26.
McNabb, H.S., Jr., H.M. Heybroek & W.L. MacDonald, 1970. Anatomical factors in resistance to Dutch elm disease. Netherlands Journal of Plant Pathology 76: 196-204.
Miller, H.J. & D.M. Elgersma, 1976. The growth of aggressive and non-aggressive strains of Ophiostoma ulmi in susceptible and resistant elms, a scanning electron microscopical study. Netherlands Journal of Plant Pathology 82: 51-65.
Overeem, J.C. & D.M. Elgersma, 1970. Accumulation of mansonones E and F in Ulmus hollandica infected with Ceratocystis ulmi. Phytochemistry 9: 1949-1952.
Rebel, H., 1969. Phytotoxins of Ceratocystis ulmi. Thesis. State University of Utrecht, the Netherlands.
Scheffer, R.J. & D.M. Elgersma, 1981. Detection of a phytotoxic glycopeptide produced by Ophiostoma ulmi in elm by Enzyme-Linked Immunospecific Assay (ELISA). Physiological Plant Pathology 18: 27-32.
Sinclair, W.A. & R.J. Campana (Eds), 1978. Dutch elm disease. Perspectives after 60 years. Cornell University. Agricultural Experiment Station. Search (Agr.) Vol. 8. 52 p.
Sinclair, W.A., J.P. Zahand & J.B. Melching, 1975. Localization of infection in American elms resistant to Ceratocystis ulmi. Phytopathology 65: 129-133.
Singh, D. & E.B. Smalley, 1969a. Nitrogenous and carbohydrate compounds in the xylem sap of Ulmaceae species varying in resistance to Dutch elm disease. Canadian Journal of Botany 47: 335-339.
Singh, D. & E.B. Smalley, 1969b. Nitrogenous compounds in the xylem sap of Ulmus americana; seasonal variations in relation to Dutch elm disease susceptibility. Forest Science 15: 299-304.
Strobel, G., N. van Alfen, K.D. Hapner, M. McNeil & P. Albersheim, 1978. Some phytotoxic glycopeptides from Ceratocystis ulmi, the Dutch elm disease pathogen. Biochemica et Biophysica Acta 538: 60-75.
Takai, S., 1974. Pathogenicity and cerato-ulmin production in Ceratocystis ulmi. Nature, London 252: 124-126.
Tchernoff, V., 1963. Vegetative propagation of elms by means of shoots cut from callused

roots. Acta Botanica Neerlandica 12: 40-50.
Van Alfen, N. & N.C. Turner, 1975. Influence of Ceratocystis ulmi toxin on water relations of elm (Ulmus americana). Plant Physiology 55: 312-316.
Vanderplank, J.E., 1978. Genetic and molecular basis of plant pathogenesis. (Advanced series in agricultural sciences 6.) Springer-Verlag Berlin/Heidelberg/New York. 167 p.
Went, J.C., 1954. The Dutch elm disease. Tijdschrift over Plantenziekten 60: 109-127.
Zentmeyer, G.A., 1942. Toxin formation by Ceratocystis ulmi. Science 95: 512-513.

Pine wilt and pine wood nematode: histopathological aspects of disease development

Yasuharu Mamiya

Forestry and Forest Products Research Institute, Tsukuba Norin Kenkyu Danchi-nai, Ibaraki, Japan

ABSTRACT

Pine wilt involving the pine wood nematode, *Bursaphelenchus lignicolus*, and the pine sawyer, *Monochamus alternatus*, is the most serious pest of forest trees in Japan. In 1979 it caused the death of 8 million pine trees (*Pinus densiflora* and *P. thunbergii*). The chronological development of pine wilt coincides with the life cycle of the vector, pine sawyer. Nematode inoculated pine seedlings were examined histopathologically to reveal pathological responses of pine to nematode infection especially during early stages of pathogenesis. Death of ray and axial parenchyma cells occurred as the first visible pathological response. This cellular response developed in relation to time after infection occurred. Cell death was observed in wood tissues as early as 24 hours after inoculation. When an inoculated seedling ceased to exude oleoresin at the 6th to 9th day after inoculation, death of axial parenchyma cells had prevailed markedly in its wood.

INTRODUCTION

The primary objectives of this presentation are to provide a summary of pine wilt disease in Japan and to point out the histopathological aspects of disease development emphasizing those during early stages of pathogenesis. The study on histopathological changes of pine tissues induced from pathological responses of pine to nematode infection will provide a basic step to reveal resistance mechanisms to pine wilt.

HISTORICAL REVIEW OF PINE WILT

Pine wilt first occurred in Japan early in this century. Yano (1913) reported the death of pine trees in Nagasaki of Kyushu as epidemic. The damage had spread since 1905 and the causal agent had not been determined at that time. According to his report, however, disease symptoms of pine trees corresponded well to symptoms currently induced by the pine wood nematode. Since then this dreadful disease has spread in epidemic proportion throughout central to southwestern Japan, and now the disease covers the northern region of Japan. In 1978 it caused the loss of 2 million m^3 of pine wood, and death of 8 million pine trees (*Pinus densiflora* Sieb. & Zucc., and *P. thunbergii* Parl.). This volume represented almost 1 % of the growing stock of pine in Japan. The pine wood nematode is the most serious pest of forest trees. The Japanese Government appropriated US $ 35 million in fiscal year 1980 to address the problem.

Numerous investigations had been conducted mainly on insects associated with deteriorated pine trees as suspected causal agents of tree mortality. Kiyohara & Tokushige (1971) first suggested that *Bursaphelenchus lignicolus* Mamiya & Kiyohara, the pine wood nematode, was the causal organism of the pine wilt disease. They found this nematode a common inhabitant in wood of dead pine trees (Tokushige & Kiyohara, 1969), and demonstrated the drastic effect of this nematode on pine trees as the result of inoculation tests. Since then, many investigators have presented experimental evidence which showed a strong correlation between nematode inoculation and death of pine trees (Mamiya, 1972). Among many bark and wood borers examined for their association with *B. lignicolus*, *Monochamus alternatus* Hops, the pine sawyer, was found to be the most important vector of the nematode (Mamiya & Enda, 1972; Morimoto & Iwasaki, 1972).

DISEASE CYCLE

Affected trees are conspicuous because of quick death after the symptoms appear. Pine trees observed to be healthy in early summer die in late summer, showing yellowish foliage. The very same symptoms occurring in naturally infected trees are produced by introducing nematode suspensions in healthy trees. Death of trees follows in 40 to 60 days after inoculation when experiments are conducted in summer.

The chronological development of pine wilt under natural conditions coincides with the life cycle of *M. alternatus* (Mamiya, 1972). Newly emerged adult pine sawyers from dead pine trees carry dauerlarvae of *B. lignicolus* in their tracheae. During maturation feeding of adult beetles on fresh branches of healthy pine trees, nematodes leave the beetle body and enter pine tissues through injuries. Nematode infection occurs from early June to late July, the period of maturation feeding of *M. alternatus*. The first symptom of disease is the failure of a tree to exude oleoresin from wounds. Adult beetles oviposit successfully in diseased trees which still appear healthy at that time, during July to late August. The rapid development of disease results in death of the tree in late August. Huge populations of nematodes build throughout the diseased tree. In wood, nematodes inhabit mostly axial and radial resin canals where they feed on epithelial cells, most likely the source of their food (Tamura & Mamiya, 1979). Hatched larvae of *M. alternatus* bore under bark and then through wood tissues. Larvae overwinter in the tunnels and pupate the following spring During February to May nematodes, the dispersal third stage larvae, gather around the pupal chambers of beetles. In May aggregated nematodes molt to the dauerlarvae. Pupation of beetles begins at mid-May and adult beetles emerge from pupae during late May to July. Just after emerging, adult beetles become contaminated with dauerlarvae.

Closely coordinated life cycles of the pine wood nematode and the pine sawyer cause pine wilt.

HISTOPATHOLOGICAL RESPONSE OF PINE TISSUES TO NEMATODE INFECTION

Histopathological studies have been carried out to demonstrate the nature of host responses in relation to feeding and reproduction of nematodes in wood, especially in early stages of pathogenesis.

Materials and methods

Three-year-old pine seedlings (*Pinus densiflora* and *P. thunbergii*) were inoculated with *B. lignicolus* cultured on *Botrytis cinerea* Pers. Inoculation was made by injecting a water suspension of nematodes, 5 000 nematodes per seedling, into a rubber tube which was attached to cover the end of a branch cut at 5-7 cm distant from the stem. Seedlings were kept at 27-30 °C. Several inoculated and uninoculated seedlings were collected 24 hours after inoculation and then at intervals of 3 days. The nematode population of each seedling was estimated by extracting nematodes from the whole seedling without leaves using the Baermann funnel technique. For a histopathological study wood cylinders 1 cm in length were collected from each inoculated branch, the stem just below the inoculated branch, the basal part of the stem, and a middle part of the shoot, and were fixed in FPA. Fixed specimens were dehydrated and embedded in 10 % celloidine. Celloidine sections, 12 μm thick, were stained with safranin and fast green.

Results

Populations of *B. lignicolus* in wood of collected seedlings are shown in Table 1. Despite the rapid spread of nematodes from the inoculation site, almost all nematodes extracted from a seedling were found in the inoculated branch until the 9th day after inoculation. Results of observations at intervals after inoculation were as follows.

24 hours: At the cut end of inoculated branches nematodes entered most axial resin canals whose epithelial cells were destroyed. Eggs were also observed in those axial resin canals. Beyond the inoculated branch a few nematodes were found in resin canals in the cortex of the shoot and a part

Table 1. Disease development and nematode population of 3 seedlings at each sampling date.

	24 h			3rd day		
Oleoresin exudation	+[a]	+	+	+	+	+
No. of nematodes per seedling	2866 (406)[b]	1297 (237)	1721 (321)	1638 (18)	908 (8)	756 (346)
	6th day			**9th day**		
Oleoresin exudation	-	+	+	-	-	-
No. of nematodes per seedling	1343 (303)	850 (530)	1691 (291)	1582 (282)	5717 (4157)	1122 (322)
	12th day			**15th day**		
Oleoresin exudation	-	-	-	-	-	-
No. of nematodes per seedling	5800	12750	1805	68980	104250	20770

a. +: normal exudation, -: no exudation

b. Number of nematodes except those from inoculated branch

of the stem. Cell death signaled by a granulation of the cytoplasm, deformed nucleus which was stained red with safranin, and browned cell contents, was observed in axial parenchyma cells surrounding epithelial cells of an axial resin canal (Fig. 1) and in ray parenchyma cells at the basal part of the stem. At that time, however, it occurred separately only in a few cells.

3rd day: Destruction of epithelial cells of axial and radial resin canals in the inoculated branch became more advanced. Death of ray cells was commonly observed but sporadically distributed all over the anatomical cross sections or radial sections of the inoculated branch. Only a few nematodes were located in resin canals of the stem. Throughout the stem, death of axial parenchyma cells and ray parenchyma cells occurred sporadically, while no effect was observed on epithelial cells of resin canals which were associated with such dead axial parenchyma cells.

6th day: Almost all resin canals of the inoculated branch were destroyed. One of the collected seedlings showed marked reduction of oleoresin exudation from a cross section of the stem cut at the basal part. In this seedling death of axial parenchyma cells was noticeable, especially at the basal part of the stem (Fig. 2), while no damage of epithelial cells of resin canals was observed. Pathological changes of parenchyma cells did not occur in wood tissues of the shoot.

9th day: All 3 seedlings collected at this time ceased to exude oleoresin on the surface of cross sections cut at the basal part of stems. A nematode population exceeding inoculum level (5 000 per seedling) was found in 1 of these seedlings while in the other 2 seedlings there was no population growth. Dead axial parenchyma cells prevailed throughout the stem of the latter 2 seedlings. Most axial resin canals whose epithelial cells were not affected at all had such dead axial parenchyma cells, at least one of them. Death of ray cells also became more noticeable than at the 6th day (Fig. 3A). A few nematodes were observed in stems, but no destruction of epithelial cells. Tylosoids of epithelial cells were seen in a few axial resin canals. Discoloration appeared sporadically along the cambial zone. Nematode distribution at a low level occurred in resin canals of the shoot, but no sign of cell death was detected there. In the stem of the seedling having more nematodes throughout wood tissues than the inoculum level, a group of axial resin canals whose epithelial cells were completely destroyed by nematodes occupied nearly 1/4 of the cross section of the stem. Axial parenchyma cells of those resin canals were also destroyed. Nematodes were observed commonly in such resin canals (Fig. 3B). Numbers of dead axial parenchyma cells and ray parenchyma cells increased markedly and they were widely distributed throughout the xylem. Discoloration of cambium and phloem cells became advanced and cracks were seen along the cambial zone. Many eggs found in resin canals indicated that reproduction of nematodes progressed rapidly. Smaller larvae, mostly second stage larvae, were observed passing through window-like pits of ray cells and locating in tracheids.

12th day: The extent of destruction of resin canals and parenchyma cell death in wood of those 2 seedlings having nematode populations exceeding 5 000 were similar to effects on seedlings of the 9th day in which nematode populations exceeded the inoculum level. However, it was clear that damages became more advanced as nematode populations became larger. In addition, discoloration of cambium and cavity formation along the cambial zone occurred more frequently than before. There were little visible pathological responses of parenchyma cells in the shoots of these seedlings. Only a very few nematodes were found in resin canals of the cortex, but not in those of the xylem. The other seedling which showed the symptom of ceasing to exude oleoresin had a smaller population of nematodes in its wood than the inoculum level. Just as had been observed at the 6th or 9th day, death of axial parenchyma cells and ray parenchyma cells prevailed throughout the stem while only a few resin canals were affected and a very few nematodes were observed in such resin canals.

15th day: Rapid growth of nematode populations in inoculated seedlings was found at the 15th day after inoculation. It reached more than 10 times

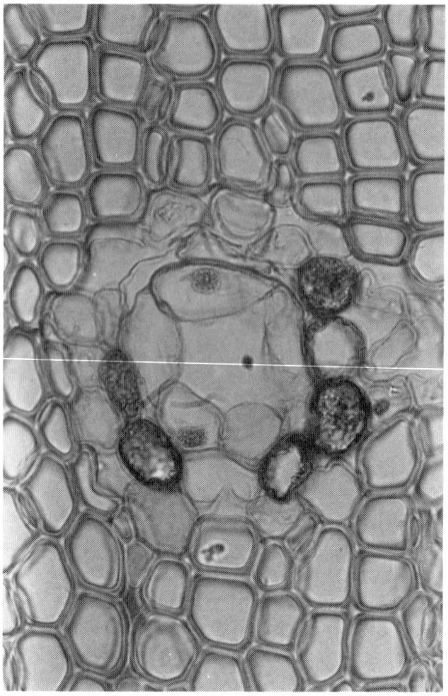

Figure 1. 24 hours after inoculation. Cell death of axial parenchyma cells at the basal part of stem. Cross section.

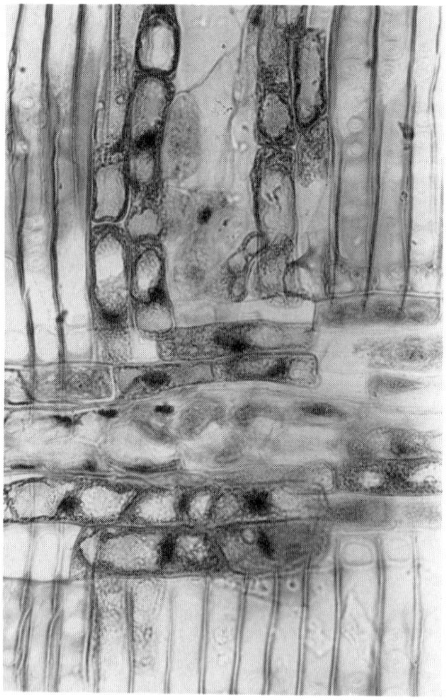

Figure 2. 6 days after inoculation. Cell death of ray and axial parenchyma cells at the basal part of stem. Radial section.

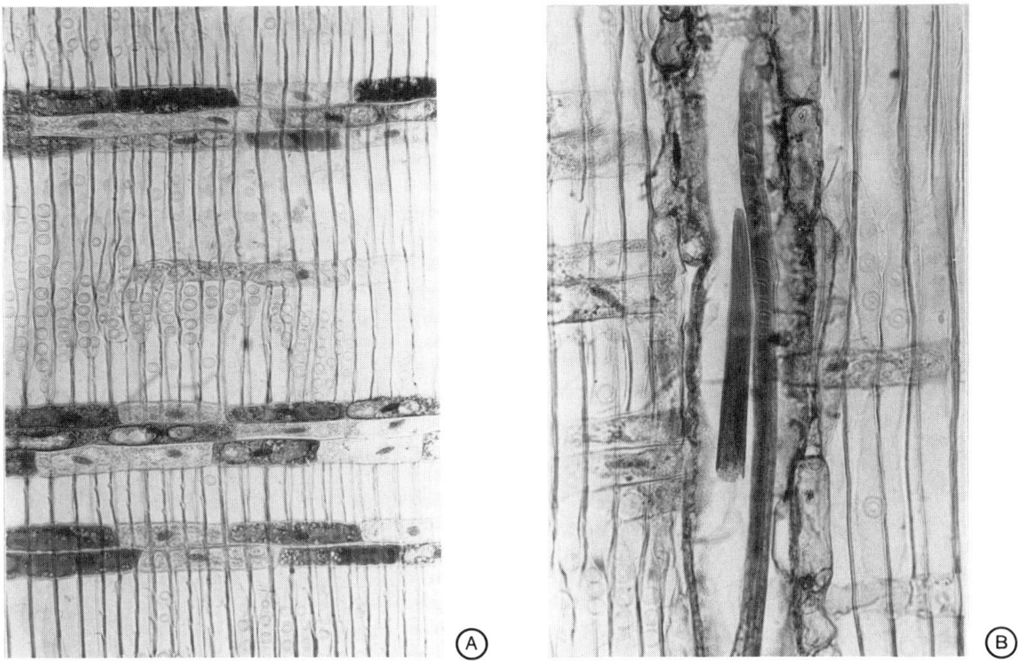

Figure 3. 9 days after inoculation. Cell death of ray and axial parenchyma cells at the basal part of stem. A. Radial section. B. Radial section. Nematodes in an axial resin canal.

the inoculum level. In almost all resin canals, epithelial cells and axial parenchyma cells were destroyed throughout the shoot and the stem. All parenchyma cells of the cortex, phloem, cambium, and rays seemed to be dead judging by total shrinkage, discoloration and granulation of cytoplasm, and disappearing or red stained nuclei. Many cavities that resulted from nematode attacks were located along the cambial zone. There were many nematodes and eggs in resin canals and such cavities.

Discussion

Death of ray and axial parenchyma cells occurred as the first visible pathological response of pine to nematode infection during early stages of pathogenesis. Anatomical observations demonstrated that this cellular response developed in relation to time after infection. Thus, it might be concluded that a steady increase in numbers and extent of dead parenchyma cells reflected the disease development of seedlings. Cell death occurred earlier than reduction or cessation of oleoresin exudation. When an inoculated seedling ceased to exude oleoresin, dead axial parenchyma cells surrounding epithelial cells of axial resin canals prevailed markedly in its wood. This provides an interesting subject regarding the supposed relationship between cell death and reduction of oleoresin exudation. In addition, at this time no death of epithelial cells similar to that of axial parenchyma cells and ray parenchyma cells was observed.

More developed pathological responses were commonly shown in the basal part of the stem than in the upper part, just below the inoculated branch, during early stages of pathogenesis. This indicates that disease development began at the lower part of a seedling, except in the inoculated branch. At the inoculation site, the cross section of a branch, nematodes directly invaded resin canals just after inoculation and then gradually invaded areas extending deep inside the branch. In the inoculated branch, parenchyma cell death and destruction of epithelial cells directly caused by nematode attacks preceeded those of any other parts throughout a seedling.

Cell death occurring prior to nematode population growth and distribution through wood tissues indicates that pathological reactions of pine tissues to some chemicals might be involved. Oku et al. (1979, 1980) reported that some metabolic substances of the pine wood nematode or bacteria associated with the pine wood nematode had toxic effects on a pine tree.

Cell death occurring at initial stages of pathogenesis was detected as one of the most remarkable pathological responses of living cells to nematode infection through this study. Histopathological study of cell death will provide a useful measure to investigate differences of resistance to the pine wood nematode within or among species of pine.

REFERENCES

Kiyohara, T. & Y. Tokushige, 1971. Inoculation experiments of a nematode, Bursaphelenchus sp., onto pine trees. Journal of Japanese Forestry Society 53: 210-218. (In Japanese with English summary).

Mamiya, Y., 1972. Pine wood nematode, Bursaphelenchus lignicolus Mamiya and Kiyohara, as a causal agent of pine wilting disease. Review of Plant Protection Research 5: 46-60.

Mamiya, Y. & N. Enda, 1972. Transmission of Bursaphelenchus lignicolus (Nematoda: Aphelenchoididae) by Monochamus alternatus (Coleoptera: Cerambycidae). Nematologica 18: 159-162.

Morimoto, K. & A. Iwasaki, 1972. Role of Monochamus alternatus (Coleoptera: Cerambycidae) as a vector of Bursaphelenchus lignicolus (Nematoda: Aphelenchoididae). Journal of Japanese Forestry Society 54: 177-183. (In Japanese with English summary).

Oku, H., T. Shiraishi & S. Kurozumi, 1979. Participation of toxin in wilting of Japanese pines caused by a nematode. Naturwissenschaften 66: 210.

Oku, H., T. Shiraishi, S. Ouchi, S. Kurozumi & H. Ohta, 1980. Pine wilt toxin, the metabolite of a bacterium associated with a nematode. Naturwissenschaften 67: 198-199.

Tamura, H. & Y. Mamiya, 1979. Reproduction of Bursaphelenchus lignicolus on pine callus tissues. Nematologica 25: 149-152.

Tokushige, Y. & T. Kiyohara, 1969. Bursaphelenchus sp. in the wood of dead pine trees. Journal of Japanese Forestry Society 51: 193-195. (In Japanese).

Yano, S., 1913. Reports of investigation on the death of pine trees. Sanrin Koho 4 suppl.: 1-22 (In Japanese).

Indirect selection for pest resistance using terpenoid compounds

Michael R. Bridgen & James W. Hanover[1]

ABSTRACT

One problem in developing genetic resistance in trees to insect or disease pests is that of tree longevity. It may be overcome with the use of an indirect selection program. Any secondary trait used in indirect selection must be highly correlated with the primary trait, but it does not need to be causally related to resistance.

The identification of suitable marker traits is difficult and expensive and may be the one factor most limiting the use of indirect selection. In oleoresin-producing conifers, the monoterpenes and resin acids may be useful groups of biochemical markers for such a program. They are under genetic control and in some species are correlated with resistance to insect or disease pests.

The resin acid composition of Scotch pine oleoresin may be involved in mechanisms of resistance to some insect pests. Parallel seasonal variation and larval feeding behavior and the strong negative correlations between geographic variations in resin acids and insect susceptibility show that insect resistance is highest when oleoresin composition is highest. These compounds are strongly correlated with resistance and are relatively easy and inexpensive to measure. They may be very useful markers in indirect selection programs for pest resistance in pines.

The forest industry in the United States is placing increasing emphasis on the importance of basic research in forest regeneration. Under pressures from an inflationary economy, escalating fuel costs, government and environmental restrictions, companies have now recognized the desirability of

1. The authors are respectively: Research Scientist, Westvaco Corporation, Rupert, West Virginia, USA, and Professor, Michigan State University, East Lansing, Michigan, USA.

having the most superior seedlings available to replace stands after harvesting. Although improved seedlings are expensive, high site-preparation, planting, and site maintenance costs overshadow the initial investment of improved planting stock. Not only should planted seedlings grow quickly and possess desirable form and wood quality characteristics, they should also be able to withstand various insect and disease pests which may attack them at any time during their life, in order that the costs of the seedlings may be completely recovered in future growth.

Although research in tree improvement and genetics has produced rapidly growing genotypes, little work has been done to develop insect and disease resistance in the same material. There are several reasons for this (Hanover, 1980): forest genetics is a relatively new field and there are few genetic researchers; the simplicity of natural regeneration makes it's wide use more attractive than planting; and the lack of basic knowledge of host-pest physiology and the long periods of time between generations in trees discourage researchers from working in the field. Concurrent with the long time intervals between generations and the longevity of trees is the possibility that trees may not exhibit resistance to insect or disease pests at an early stage as they may not be attacked by these pests for many years. This makes traditional selection schemes very difficult and their results of limited value.

The use of indirect selection may be one method of overcoming the problem of tree longevity. Indirect selection is the early identification of superior individuals or populations for a trait other than the one for which improvement is sought. The trait being used for selection, the secondary trait, must be highly correlated with the desired or primary trait, whether or not it is causally related. Indirect selection is especially useful if greater or more rapid success can be had by selecting for a correlated trait instead of a primary character.

The theoretical background for indirect selection has been outlined by Falconer (1960). Weissenberg (1976) has previously described the theory, advantages, and use of indirect selection in breeding for various traits in forest tree species.

Direct selection has a simple mathematical basis. With direct selection, a number of individuals from a population are chosen to be the parental stock for future generations. The intensity of selection is defined as the ratio of the mean phenotypic value of the individuals chosen, S, to the phenotypic standard deviation, σ_p.

$$i = S/\sigma_p$$

Heritability, h^2, is defined as the ratio of the mean phenotypic value of the progeny of selected parents, R, to the mean phenotypic value of the parents. Heritability is also the ratio of additive genetic variance to total phenotypic variance of the population.

$$h^2 = R/S = \sigma_a^2/\sigma_p^2$$

The expected gain or change from the mean of the original population, R, is the product of the intensity of selection times the square root of the heritability of the trait times the square root of the additive genetic variance of the selected trait.

$$R = ih\sigma_a$$

By using indirect selection, a secondary trait is chosen which is highly correlated with the primary trait. The gain for the primary trait, then, is the product of the intensity of selection for the secondary trait, the additive genetic correlation coefficient, and the square root of the heritability of the secondary trait times the additive genetic variance of the primary trait.

$$G_x = i_y h_y r_a \sigma_{ax}$$

The relative efficiency of using indirect instead of direct selection is shown by a ratio of the expected gain of indirect selection to that expected with the direct method. This ratio is the correlation coefficient times the selection intensity and square root of heritability of the secondary trait divided by the selection intensity and square root of heritability of the primary trait.

$$G_x/R_x = r_a\ (i_y h_y)/(i_x h_x)$$

Indirect selection can be a superior method to direct selection if:
- a high genetic correlation exists between the 2 traits;
- the secondary trait has a higher heritability than the desired character;
- a higher selection intensity can be applied on the secondary trait than on the primary trait.

The terpenes may be an ideal group of biochemicals to use as markers in an indirect selection program, especially for pines, spruces and other resin producing conifers. Monoterpenes and diterpenoid resin acids comprise the greatest portion of total oleoresin. The monoterpenes are under strong genetical control (Baradat et al., 1975; Hanover, 1966a; Squillace, 1971) and show minimum influence by environmental variation (Hanover, 1966b). Genetic variation in monoterpenes between seed sources has been demonstrated for many pine and spruce species (Arbez et al., 1974; Bridgen et al., 1979; Gansel & Squillace, 1976; Tobolski & Hanover, 1971; Wilkinson & Hanover, 1972). The genetics of resin acids have not been studied very widely. There appears to be no genetic variation in the resin acids of the oleoresin of white pine (Bridgen et al., 1979) but seedlot and regional variation has been found in Scotch pine (Bridgen, 1979). Further studies should identify genetic variation and control of resin acid in other species.

There are 4 characteristics of terpenes that make these compounds useful in indirect selection for pest resistance:

1. Several terpenes have been identified as attractants or repellents to insects and so are directly involved in resistance mechanisms (All & Anderson, 1974; Bordasch & Berryman, 1977; Hanover, 1975). Similarly, they are probably involved in resistance to some fungal diseases (Bailey et al., 1975; Cobb et al., 1968).
2. Oleoresin may be sampled from tree seedlings while many insect and disease pests may not be observed until the tree is many years old. Although there is some variation in oleoresin caused by the age of the tree, correlations of juvenile traits with resistance patterns should be equally useful and valid as correlations traits measured at the same time as resistance.
3. Measuring errors can be minimized and large numbers of trees can be analyzed using the terpenes. The quantitative accuracy of chemical measurements using gas chromatography greatly overshadows the qualitative, often ocular, measurements which are usually applied toward insect and disease infection rates. Many samples can easily be handled as little storage space is necessary for the small quantities of oleoresin used and automatic sampling devices allow a continual analysis, even when lab personnel are not available.
4. Finally, many terpenes are simultaneously analyzed. A typical pine or spruce analysis will yield 4-15 or more monoterpenes and 16 or more resin acids. As each compound is presumably under independent genetic control and insects and disease react differently to various chemicals, the probability of finding a correlated trait is much higher than if only 1 or a few traits or chemicals were being studied.

An example of the use of terpenes as markers for resistance to a disease on forest trees is that of fusiform rust (*Cronartium fusiforme* Hedgc. & Hunt ex Cumm.) in loblolly (*Pinus taeda* L.) and slash (*Pinus elliottii* Engelm.) pines. Rockwood (1973) studied the phenotypic and genetic correlations between fusiform rust infestation and the chemical composition of both stem xylem and branch cortex oleoresin of loblolly pine. He collected oleoresin from about 350 trees representing 36 full-sib families. The trees were 5 years old. None of the 3 primary monoterpenes of stem xylem, α-pinene, β-pinene, or myrcene, showed any relationship with resistance, but branch β-phellandrene was phenotypically correlated with susceptibility. As β-phellandrene composition increased, percent infection also increased. Rust infection is directly in contact with the branch cortex tissues, and it is reasonable to believe that cortex oleoresin would be more involved in any resistance mechanism than stem xylem oleoresin.

To expand his studies on the role of monoterpenes and rust resistance, Rockwood (1974) examined another pine species which also shows variation in resistance to the rust, slash pine. He collected branch oleoresin in about 300 trees from clones, wind pollinated families, full-sib families, and existing rust tests in Florida, Mississippi and Georgia. There are only 4 major monoterpenes in slash pine, α-pinene, β-pinene, myrcene and β-phellandrene. As in loblolly pine, β-phellandrene was the only monoterpene showing a relationship with fusiform rust, but in this case, resistance was associated with high levels of the terpene rather than low. The monoterpene may not be the causal agent of resistance or susceptibility, but rather be genetically linked to the controlling traits. Therefore, it may be successfully used in an indirect selection program.

Several years ago we began to study the physiology and genetics of the oleoresin system of Scotch pine (*Pinus sylvestris* L.). To test the hypothesis that diterpenes are useful indicators of insect resistance, comparisons were made between data collected in the physiology and genetic studies with the known patterns of resistance to several insect pests common on Scotch pine. All of the insect resistance information had been previously collected from provenance tests within Michigan (Steiner, 1974; Wright & Wilson, 1972; Wright et al., 1967, 1975, 1976). Results from some of our observations indicate that a relationship may exist between the diterpenoid resin acids and insect resistance in this species.

Beginning 1 April 1978, oleoresin was collected from 14 Scotch pine trees every 15 days throughout the growing season. The trees were 18 years old, originating in Sweden of the variety *septentrionalis*. Oleoresin was collected in capillary tubes which were tapped in 2-year-old tissue and the resin acids were analyzed using gas-liquid chromatography (Bridgen et al., 1979).

Twelve resin acids or pairs of unseparable resin acids were measured. Four major resin acids, sandaracopimarate, levopimarate + palustrate, strobate + dehydroabietate and neoabietate, 2 minor acids and total resin acids showed significant seasonal variation. The combined peak of strobic + dehydroabietic acids showed an increase in concentrations between 15 May and 15 June, after which it decreased to levels close to those prior to 15 May. All of the other resin acids, and total acids, had a noticeable decrease during this same time period, extending to 1 July. This drop in total resin acid concentrations coincides with the time of most rapid shoot elongation. It also parallels the developmental physiology of several insects which attack Scotch pine. The larvae of the European pine sawfly (*Neodiprion sertifer* Geoff.) feed on year-old foliage between early May and mid-June. Also during this period, larvae of the eastern pineshoot borer (*Eucosma gloriola* Heinrich) bore into lateral and terminal shoots.

Table 1. Simple correlations of resin acids with susceptibility patterns of several Scotch pine insect pests.

	European pine sawfly	Pine shoot borer	Zimmerman pine moth	Pine root collar weevil
Unknown 28	0.23	0.07	0.12	0.23
Unknown 30	-0.06	-0.05	-0.26	-0.15
Unknown 31	-0.23	0.14	-0.16	-0.21
Pimarate	-0.46**	-0.27	-0.37**	-0.34*
Sandaracopimarate	-0.29	-0.33*	-0.23	-0.21
Levopimarate + palustrate	-0.44**	-0.21	-0.36*	-0.36*
Isopimarate	-0.43**	-0.51**	-0.30*	-0.19
Unknown 36	-0.28	-0.32*	-0.20	-0.16
Strobate + dehydroabietate	-0.15	-0.14	-0.20	-0.14
Unknown 38	-0.01	0.08	-0.14	-0.13
Abietate	-0.42**	-0.40*	-0.35*	-0.31*
Neoabietate	-0.04	0.11	-0.15	-0.12
Unknown 41	-0.04	0.11	-0.15	-0.12
Total resin acids	-0.46**	-0.27	-0.37**	-0.36*

*, ** Significant at $p \leq 0.05$ and $p \leq 0.01$ respectively.

The timing of these events corresponds very closely to the decrease of total resin acids in the oleoresin of branches. It may be only coincidental that the developmental physiology of the insect parallels the change of chemical composition in the oleoresin, but it is also possible that the insect has adapted its developmental stages genetically to feed at a time when undesirable compounds in the host are at a low concentration. A similar situation has been described with the winter moth (*Operophtera brumata* L.). The tannin concentration of oak leaves may have caused evolution of the winter moth so that its feeding ceases before tannins increase in concentration in the leaves (Feeny, 1970).

Along with seasonal variation, variation between seedlots of Scotch pine exists for the resin acids. In the latter part of June 1978, oleoresin was collected from the boles of 300 trees representing 50 seedlots and 18 varieties of Scotch pine. Total resin acids ranged in concentration from 8.5 % to 54 % of oleoresin. The Scandinavian varieties had the highest concentrations of total resin acids while the central European varieties had the lowest concentrations of pimaric, abietic, neoabietic and total resin acids. Scandinavian varieties also have the highest resistance to European pine sawfly (Wright et al., 1967) and the eastern pineshoot borer (Steiner,

1974). Correlation analysis between insect resistance and resin acids for all seedlots show a significant negative relationship between these traits (Table 1).

REFERENCES

All, J.N. & R.F. Anderson, 1974. The influence of various odors in host selection of pioneer beetles of Ips grandicollis. Journal of the Georgia Entomological Society 9: 223-228.

Arbez, M., C. Bernard-Dagan & C. Fillon, 1974. Variabilité intraspécifique des monoterpènes de Pinus nigra Arn.: des premiers résultats. Annales des Sciences Forestières 31: 57-70.

Bailey, J.A., G.A. Carter, R.S. Burden & R.L. Wain, 1975. Control of rust diseases by diterpenes of Nicotiana glutinosa. Nature 255: 328-329.

Baradat, Ph., C. Bernard-Dagan, G. Pauly & C. Zimmerman Fillon, 1975. Les terpènes du pin maritime. Aspects biologiques et génétiques. III. Hérédité de la teneur en myrcène. Annales des Sciences Forestières 32: 29-54.

Bordasch, R.P. & A.A. Berryman, 1977. Host resistance to the fir engraver beetle Scolytus ventralis (Coleoptera: Scolytidae). 2. Repellancy of Abies grandis resins and some monoterpenes. Canadian Entomologist 102: 95-100.

Bridgen, M.R., 1979. Genetic variation of Scotch pine oleoresin physiology. Ph.D. Dissertation Michigan State University, East Lansing, Michigan. 78 p.

Bridgen, M.R., J.W. Hanover & R.C. Wilkinson, 1979. Oleoresin characteristics of eastern white pine seed sources and relationship to weevil resistance. Forest Science 25: 175-183.

Cobb, F.W., Jr., M. Krstic, E. Zavarin & H.W. Barber, Jr., 1968. Inhibitory effects of volatile oleoresin components on Fomes annosus and four Ceratocystis species. Phytopathology 58: 1327-1337.

Falconer, D.S., 1960. Introduction to quantitative genetics. The Ronald Press Company, New York. 365 p.

Feeny, P.P., 1970. Seasonal changes in oak leaf tannins and nutrients as a cause of spring feeding by winter moth caterpillars. Ecology 51: 565-581.

Gansel, C.R. & A.E. Squillace, 1976. Geographic variation of monoterpene in cortical oleoresin of slash pine. Silvae Genetica 25: 150-154.

Hanover, J.W., 1966a. Genetics of terpenes. I. Gene control of monoterpene levels in Pinus monticola Dougl. Heredity 21: 73-84.

Hanover, J.W., 1966b. Environmental variation in the monoterpenes of Pinus monticola Dougl. Phytochemistry 5: 713-717.

Hanover, J.W., 1975. Physiology of tree resistance to insects. Annual Review of Entomology 20: 75-95.

Hanover, J.W., 1980. Breeding forest trees resistant to insects. In: F.G. Maxwell & P.R. Jennings (Eds): Breeding plants resistant to insects. Wiley & Sons, Inc. p. 487-511.

Rockwood, D.L., 1973. Monoterpene-fusiform rust relationships in loblolly pine. Phytopathology 63: 551-553.

Rockwood, D.L., 1974. Cortical monoterpene and fusiform rust resistance relationships in slash pine. Phytopathology 64: 976-979.

Squillace, A.E., 1971. Inheritance of monoterpene composition in cortical oleoresin of slash pine. Forest Science 17: 381-387.

Steiner, K., 1974. Genetic differences in resistance of Scotch pine to eastern pineshoot borer. Great Lakes Entomologist 7: 103-107.

Tobolski, J.J. & J.W. Hanover, 1971. Genetic variation in the monoterpenes of Scotch pine. Forest Science 17: 293-299.

Weissenberg, K. von, 1976. Indirect selection for improvement of desired traits. In: J.P. Miksche (Ed.): Modern methods in forest genetics. Springer-Verlag. 288 p.

Wilkinson, R.C. & J.W. Hanover, 1972. Geographical variation in the monoterpene composition of red spruce. Phytochemistry 11: 2007-2010.

Wright, J.W., W.A. Lemmien, J.N. Bright, M.W. Day & R.L. Sajdak, 1976. Scotch pine varieties for Christmas tree and forest planting in Michigan. Michigan State University Agricultural Experiment Station Research Report 293. 16 p.

Wright, J.W. & L.F. Wilson, 1972. Genetic differences in Scotch pine resistance to pine root collar weevil. Michigan State University Agricultural Experiment Station

Research Report 159. 6 p.
Wright, J.W., L.F. Wilson & J.M. Bright, 1975. Genetic variation in resistance of Scotch pine to Zimmerman pine moth. Great Lakes Entomologist 8: 231-236.
Wright, J.W., L.F. Wilson & W.K. Randall, 1967. Differences among Scotch pine varieties in susceptibility to European pine sawfly. Forest Science 13: 175-181.

Monoterpenes and resistance of conifers to fungi

H.J. Schuck

Lehrstuhl für Forstbotanik, Universität München, D-8000 München 40, Amalienstr. 52, BRD

ABSTRACT

Monoterpenes usually inhibit the development of pathogenic fungi. This effect varies over a wide range. In a few cases the composition of the terpene fraction could be correlated with the resistance of a tree against pathogens. After being wounded or infected a coniferous tree can exhibit a change in the composition of the terpene fraction. This reaction perhaps determines at least the actual resistance against pathogens.

Monoterpenes are widely distributed in the plant kingdom. Not only the conifers can synthesize them, but other systematically unrelated families such as the *Lauraceae*, *Poaceae*, *Liliaceae*, *Rutaceae*, *Compositae* and *Lamiaceae* contain monoterpenes, too (Fretz, 1978). Compared with these plant families the conifers produce the greatest amounts of terpenes, which are constituents of resin that can make up to a maximum of 25 % (in *Pinus elliottii* Engelm.) of the dry weight of heartwood (Parham, 1976). In *Pinus sylvestris* L. the heartwood contains 8 % resin consisting of 25 % monoterpenes. Certainly these values cannot be generalized for all conifers, and furthermore great variation occurs among individuals.

It is characteristic in conifers that the resin is located in a special resin duct system, usually existing in all parts of a tree. Only some genera, like *Abies*, normally lack resin ducts, but as a reaction to injuries they can develop such systems, called traumatic resin ducts. Other conifers react in the same way upon wounding.

The development of traumatic resin ducts shows that the resin and the monoterpenes play a part in the defense mechanisms against pathogenic organisms. Surely one of the main functions of resins is closing wounds quickly. They establish mechanical and chemical barriers to prevent infections by fungi (Rishbeth, 1972) and insects (Smith, 1977). The mechanical barrier depends on the resin acids, which are main constituents of the resin, whereas the monoterpenes of the resin act as toxic agents. Generally these barriers are built up by resin that comes from ducts in the bark.

The wood can also form a barrier zone: injuries induce the development of new resin ducts in the wood by cambial cells of the tree, and so more resin can be mobilized to prevent an infection. In older xylem no new ducts can be differentiated.

This traumatic reaction signals the significance of resins in host-parasite interactions. It is not surprising that for several years the effects of monoterpenes on pathogenic fungi have been investigated (Table 1). Here only the experiments with monoterpene hydrocarbons are catalogued; other derivatives, in literature often named as monoterpenes, are not considered. Mostly the influence on mycelial growth of fungi was tested; but in some cases also spore germination, dry matter production, and the peroxidase activity of fungi growing in a terpene atmosphere were studied.

In most cases (Table 1) the mycelial growth was inhibited by monoterpenes. Stimulating effects were rarely observed (Fries, 1973; Väisälä, 1974). Most of the fungi tested were wood-decomposing or wood-inhabiting organisms. The fungi growing in conifer wood seem to be more tolerant against high levels of terpenes than the species growing in wood of deciduous trees (Hintikka, 1970). Fries (1973) recorded that, although linear mycelial growth was reduced, the fungus could increase production of its dry matter when exposed to a terpene atmosphere.

The effects were always correlated with the concentration of the applied terpenes, and the terpenes themselves were also different in their effects. Differences may also exist in the effects of a monoterpene on different pathogens. For example Δ-3-carene seems to be relatively harmless against *Fomes annosus* (Fr.) Cooke (Schuck, 1977; Cobb et al., 1968), but very aggressive against *Diplodia pinea* (Desm.) Kickx (Chou & Zabkiewicz, 1976). Finally, different effects were also related to the test methods. Fig. 1 shows the influence of α-pinene on the germination rate, the linear mycelial growth, and the wood-decomposing activity of *Fomes annosus*. In all 3 experiments the inhibition increased with the concentration of the terpene. The germination rate was not reduced as fast as the mycelial growth. The wood-decomposing activity, measured as weight loss, cannot be compared with the other test-methods, because greater quantities of terpene were applied. This test took 3 1/2 months. About every 10 days the quantities mentioned in the figure were added to the cultures of *Fomes annosus*.

Because of the great variation in their effects, it is not possible to postulate a gradation in the aggression of terpenes. In addition, the results reported in the literature differ sometimes, even though the same fungus was examined. Hintikka(1970) found that Δ-3-carene and limonene reduced the mycelial growth of *Fomes annosus* very strongly whereas α-pinene had only a slight reducing effect. In contrast, in my investigations (Schuck, 1977) carene was more harmless and α-pinene the most aggressive terpene. This difference cannot be explained with certainty. Perhaps it depends on differences in method or on the purity of terpenes applied by the 2 investi-

Table 1. Investigations on the effect of monoterpenes against fungi.

Fungus	α-pinene	β-pinene	camphene	Δ-3-carene	β-phellandr.	limonene	myrcene	cymene	terpinolene	Test-method	Effect: inhibition	Effect: stimulation	Reference
Fomes annosus, 4 *Ceratocystis* spp.	X	X	-	X	X	X	X	-	-	mycelial growth	X		Cobb et al.(1968)
38 diff.wood-inhabit. fungi	X	X	X	X	-	X	-	-	-	mycelial growth	X		Hintikka (1970)
Trichoderma viride, *Lenzites saepiaria*, *Schizophyllum comm.*, *Ceratocystis minor*	X	X	X	-	-	X	X	X	X	mycelial growth	X		de Groot (1972)
Peniophora candida, *Polyporus brumalis*	X	X	-	X	-	X	-	-	-	mycelial growth	X	X (lim. *P.cand.*)	Fries (1973)
										dry-matter production		X (car. *P.brum.*)	
12 diff.wood-decomp. fungi	-	-	-	-	-	-	X	-	-	mycelial growth	X	X[1]	
Ceratocystis ulmi	-	-	-	-	-	-	X	-	-	coremia formation		X	Hubbes (1975)
										mycelial growth	X		
Diplodia pinea	X	X	-	X	X	X	-	-	X	spore germination	X		Chou & Zabkiewicz (1976)
										germ-tube growth	X		
Fomes annosus	X	X	-	X	-	X	-	-	-	mycelial growth	X		Schuck (1977)
Fomes annosus, *Phellinus pini*	X	-	-	-	-	X	X	-	-	mycelial growth	(X)		Väisälä (1978)
										peroxidase activity		X	
Fomes annosus, *Stereum sanguinulentum*	X	X	-	X	-	X	X	-	-	dry-matter production	X	(X)[2]	Flodin & Fries (1978)
Fomes annosus	X	X	-	X	-	-	-	-	-	phenoloxidase activity		X	Flodin (1979)

1. 10 ppm myrcene: *Phellinus pini*, *Lentinus lepideus*, *Fomitopsis pinicola*.

2. Limonene in lower concentrations.

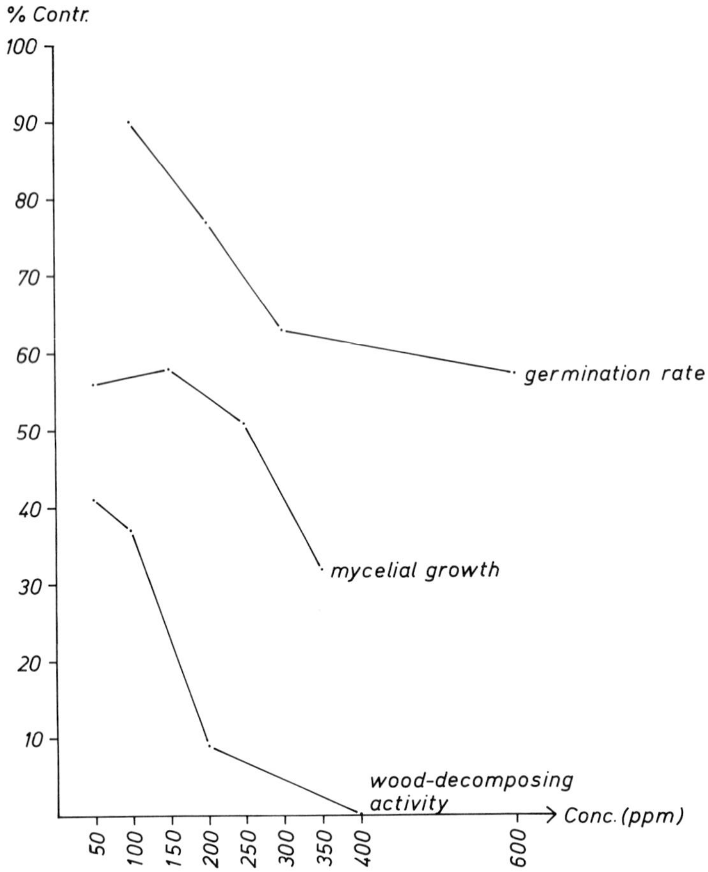

Figure 1. Inhibitory effects (as a percentage of the control) of α-pinene on the development of *Fomes annosus* in 3 different tests.

gators. Nevertheless, all the papers describe the inhibitory effect of terpenes on the vegetative development of fungi. Therefore attempts have been made to distinguish between trees that are resistant and susceptible to a particular fungus according to the composition of their terpene fractions. Such investigations are very difficult, because the composition is highly variable. It is influenced by season (von Rudloff, 1975; Powell & Adams, 1973; Hrutfiord et al., 1974), water balance (Hodges & Lorio, 1975; Gilmore 1977) and even diurnal variations (Adams, 1979). Very great inter- and intra-individual variations complicate such investigations (Schuck & Schütt, 1975; von Rudloff & Hunt, 1977). Nevertheless, recent results show that terpene composition can be related to susceptibility to pathogenic fungi in some cases. Rockwood (1974) demonstrated a correlation between the percentage of β-phellandrene in the resin of *Pinus elliottii* and resistance against *Cronartium fusiforme* Hedgc. & Hunt. Chou & Zabkiewicz (1976) found higher percentages of Δ-3-carene in trees resistant to *Diplodia pinea* than in susceptible ones. There are no differences in the terpene fractions of several western white pines (*Pinus monticola* Dougl.) resistant against *Cronartium ribicola* J.C. Fisher (Hunt & von Rudloff, 1977).

All this information could be useful in breeding for resistance, but only if the composition of the terpene fraction is under genetic control. Esteban et al.(1976) and Hanover (1966) postulated genetic control systems for the heredity of monoterpenes.

Many of the wood-decomposing fungi are not able to penetrate the intact

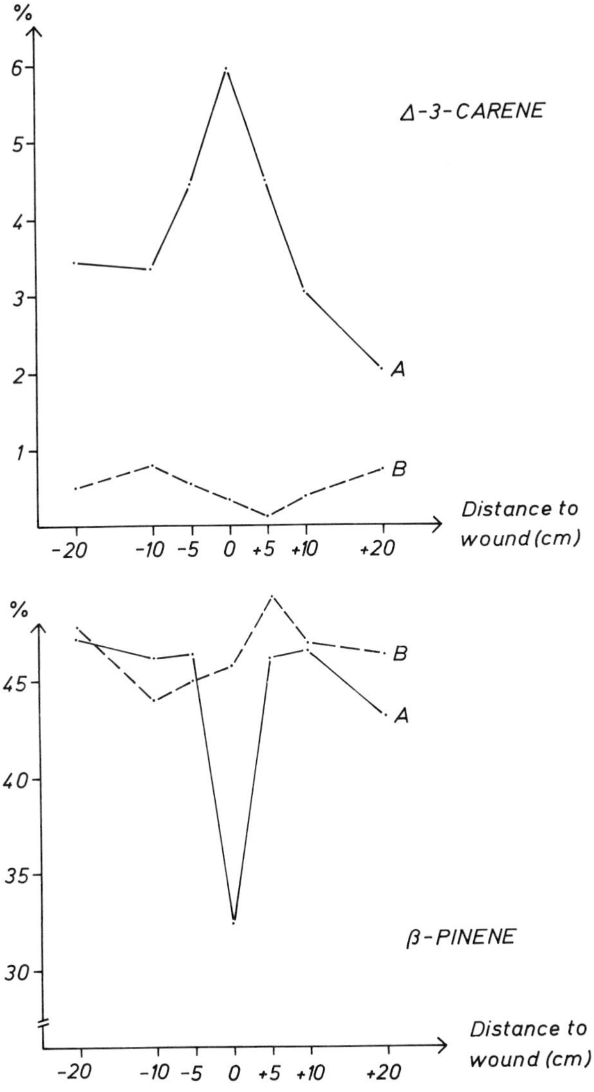

Figure 2. Effect of wounding on composition of the monoterpene fraction of *Picea abies* sapwood. A: wounded; B: control.

bark. They infect a tree only by wounds, where not only the amount of terpenes is increased, but also the composition of the terpene fraction has been changed. These alterations are of great importance for the defense against pathogens, because they at least determine the actual resistance to fungal attack. The significance of increasing amounts of resin in wounded parts of a tree has been recognized by some authors. Cobb et al.(1968) postulated that the accumulation of terpenes can be a factor in resistance to pathogenic fungi, and Rishbeth (1951) supposed that resin production in roots can inhibit *Fomes annosus* infections. So the resistance against pathogens which infect through wounds could be correlated with the ability of trees to mobilize resin (Gibbs, 1968) in the wounded region. This traumatic resin has a different chemical composition. Russel & Berryman (1976) found a high myrcene and Δ-3-carene content in fungus-infected wounds of *Abies grandis* Lindl.

Recently such changes in the terpene fraction of *Picea abies* (L.) Karst. were found. Two remarkable chemical alterations in the terpene fraction of sapwood were demonstrated 2 months after artificially wounding with a borer (Fig. 2). Line A indicates the percentage of a distinct terpene in wounded wood (at position 0) and in wood located 5, 10, or 20 cm below (-) or above (+) the wound. For control (B) we took intact wood samples from comparable positions not influenced by the wound. The β-pinene content decreased in wounded parts, and this effect was strongly limited to the wounded region whereas the Δ-3-carene content increased and was not strongly limited to wounds.

In the samples taken at distances up to 10 cm below and above the wound, the altered resin could be found. But the effect decreased with growing distance from the wound. Not only sapwood but also the heartwood can react by synthesizing traumatic terpenes.

These results cannot be generalized because they were not universal. Not all of our tested trees showed such traumatic reactions, or the reactions were not similar. This is a new aspect of the resistance mechanism due to monoterpenes. It is necessary to investigate whether these different reactions in trees of the same species are correlated with their resistance against fungal attack. It is possible that such changes take place also in infected but not injured wood. No investigations of this phenomenon have been started at our laboratory.

REFERENCES

Adams, R.P., 1979. Diurnal variation in the terpenoids of Juniperus scopulorum (Cupressaceae) - Summer versus winter. American Journal of Botany 66: 986-988.

Chou, C.K.S. & J.A. Zabkiewicz, 1976. Toxicity of monoterpenes from Pinus radiata cortical oleoresin to Diplodia pinea spores. European Journal of Forest Pathology 6: 354-359.

Cobb, F.W., M. Krstic, E. Zavarin & H.W. Barber, 1968. Inhibitory effects of volatile oleoresin components on Fomes annosus and four Ceratocystis species. Phytopathology 58: 1327-1335.

Esteban, I., F. Bergmann, H.R. Gregorius & O. Huhtinen, 1976. Composition and genetics of monoterpenes from cortical oleoresin of Norway Spruce and their significance for clone identification. Silvae Genetica 25: 59-66.
Flodin, K., 1979. Effects of monoterpene of Fomes annosus (Fr.) Cooke and its phenol oxidase activity. European Journal of Forest Pathology 9: 1-6.
Flodin, K. & N. Fries, 1978. Studies on volatile compounds from Pinus silvestris and their effect on wood-decomposing fungi. II. Effects of some volatile compounds on fungal growth. European Journal of Forest Pathology 8: 300-310.
Fretz, T.A., 1978. Chemical identification of woody ornamental plants. Ohio Report 63: 23-25.
Fries, N., 1973. The growth-promoting activity of terpenoids on wood-decomposing fungi. European Journal of Forest Pathology 3: 169-180.
Gibbs, J.N., 1968. Resin and the resistance of conifers to Fomes annosus. Annals of Botany 32: 649-665.
Gilmore, A.R., 1977. Effects of soil moisture stress on monoterpenes in Loblolly Pine. Journal of Chemical Ecology 3: 667-676.
Groot, R.C. de, 1972. Growth of wood-inhabiting fungi in saturated atmospheres of monoterpenoids. Mycologia 64: 863-870.
Hanover, J.W., 1966. Genetics in terpenes. I. Gene control of monoterpene levels in Pinus monticola Dougl. Heredity 21: 73-84.
Hintikka, V., 1970. Selective effects of terpenes on wood-decomposing Hymenomycetes. Karstenia 11: 28-32.
Hodges, J.D. & P.L. Lorio, 1975. Moisture stress and composition of xylem oleoresin in loblolly pine. Forest Science 22: 283-290.
Hrutfiord, B.F., S.M. Hopley & R.J. Gara, 1974. Monoterpenes in Sitka spruce: within tree and seasonal variation. Phytochemistry 13: 2167-2170.
Hubbes, M., 1975. Terpenes and unsaturated fatty acids trigger coremia formation by Ceratocystis ulmi. European Journal of Forest Pathology 5: 129-137.
Hunt, R.S. & E. v. Rudloff, 1977. Leaf-oil-terpene variation in western white pine populations of the Pacific Northwest. Forest Science 23: 507-516.
Parham, M.R., 1976. Stimulation of oleoresin yield in conifers. Outlook on Agriculture 9: 76-81.
Powell, R.A. & R.P. Adams, 1973. Seasonal variation in the volatile terpenoids of Juniperus scopulorum (Cupressaceae). American Journal of Botany 60: 1041-1050.
Rishbeth, J., 1951. Observations on the biology of Fomes annosus with particular reference to East Anglian pine plantations. III. Natural and experimental infections of pines, and some factors affecting severity of the disease. Annals of Botany (N.S.) 15: 221-246.
Rishbeth, J., 1972. Resistance to fungal pathogens of tree roots. Proceedings of the Royal Society, Londen B 181: 333-351.
Rockwood, D.L., 1974. Cortical monoterpene and fusiform rust resistance relationships in slash pine. Phytopathology 64: 976-979.
Rudloff, E. von, 1975. Seasonal variation in the terpenes of the foliage of black spruce. Phytochemistry 14: 1695-1699.
Rudloff, E. von & R.S. Hunt, 1977. Chemosystematic studies in the genus Abies. III. Leaf and twig oil analysis of amabilis fir. Canadian Journal of Botany 55: 3087-3092.
Russel, C.E. & A.A. Berryman, 1976. Host resistance to the fir engraver beetle. 1. Monoterpene composition of Abies grandis pitch blisters and fungus-infected wounds. Canadian Journal of Botany 54: 14-18.
Schuck, H.J., 1977. Die Wirkung von Monoterpenen auf das Mycelwachstum von Fomes annosus (Fr.) Cke. European Journal of Forest Pathology 7: 374-384.
Schuck, H.J. & P. Schütt, 1975. Die Verteilung flüchtiger Terpene im Fichtenstamm und ihre Bedeutung für die Ausbreitungsresistenz gegenüber Fomes annosus. Biochemie und Physiologie der Pflanzen 167: 65-77.
Smith, R.H., 1977. Monoterpenes of ponderosa pine xylem resin in western United States. US. Forest Service. Technical Bulletin 1532.
Väisälä, L., 1974. Effects of terpene compounds on the growth of wood-decomposing fungi. Annales Botanici Fennici 11: 275-278.
Väisälä, L., 1978. Effects of terpene compounds on the growth and peroxidase activity of Phellinus pini. Annales Botanici Fennici 15: 131-137.

Buffering of pH in plant organs and resistance against fungi

F. Scholz & B.R. Stephan

Bundesforschungsanstalt für Forst- und Holzwirtschaft Hamburg-Reinbek, Institut für Forstgenetik und Forstpflanzenzüchtung Schmalenbeck, Siekerlandstrasse 2, D-2070 Grosshansdorf, BRD

ABSTRACT

The present knowledge of relations between resistance to fungal parasites and buffering capacity in vulnerable plant organs is reviewed. Results are discussed with respect to their importance for our understanding interactions in host-parasite systems.

INTRODUCTION

Resistance or susceptibility is the result of an interaction between host and parasite. Physiological and biochemical processes are involved in this interaction. To a great extent they are influenced by hydrogen ion concentration and its regulation. In the following paper pH-relations will be discussed with regard to their possible importance for resistance of trees against fungal parasites.

IMPORTANCE OF PH FOR REGULATION OF METABOLIC PROCESSES

When after infection host plant and pathogen are in contact the interaction between metabolic mechanisms of virulence and resistance determines the course of pathogenesis. Not only the processes themselves but also the medium in which they interact are important. This medium is characterized, for example, by its pH.

The pH is very important for the optimal functioning of metabolic processes. Deviations from optimal hydrogen ion concentration influence biomembranes, affecting their structure and function so that transport processes between cells and within cells are disturbed. Additionally the biochemical reactions themselves can be influenced, as well as the activity of enzymes involved. An example (Fig. 1) is the equilibrium between 3 molecular states of a hypothetical enzyme in relation to pH and enzyme activity. Thus regulation of intracellular pH is a fundamental biological process (Raven & Smith, 1973).

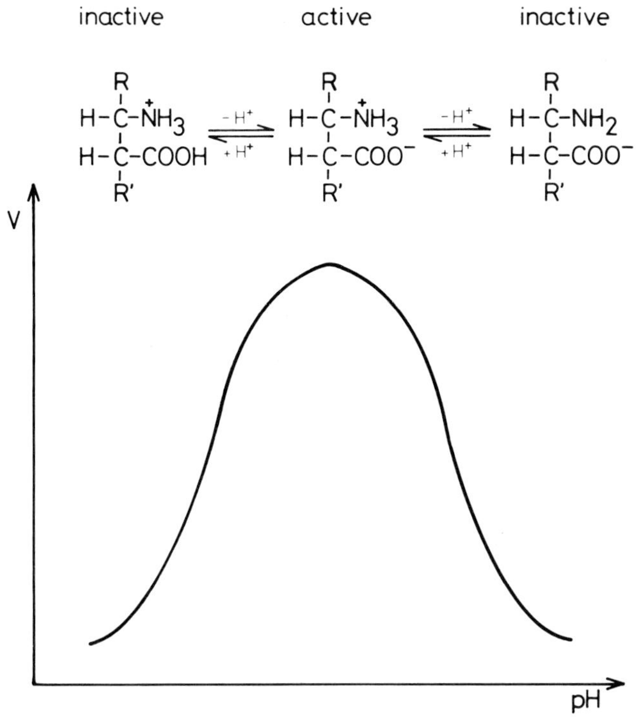

Figure 1. Equilibrium between 3 molecular states of a hypothetical enzyme in relation to pH and velocity of reaction (v) (from Scholz & Knabe, 1976).

Within each organism, varying with the respective organ and the involved cells, an optimal pH is maintained by regulation. The regulation is due to internal metabolic processes consuming or producing H^+ of OH^-, and pH is stable in a centain range against external influences (Smith & Raven, 1976). The objective of the parasitic fungus is to metabolize substrates in the internal milieu of the host, which sumultaneously is the external milieu for the fungus. This means that for the fungus also the hydrogen-ion concentration in the host plant cells is of importance.

In many host-pathogen systems the external optimum pH for fungal substrate metabolism is higher than the pH in the internal milieu of the host plant. Therefore 2 different regulation mechanisms interact with each other. This interaction is illustrated by a cybernetic model (Fig. 2) in which the relationship is shown between the intracellular pH-milieu of the host and the extracellular pH-milieu of the parasite under the assumption that the intracellular pH of the host is suboptimal for the parasite.

Measurements of pH interactions in living plant cells are difficult. Hence, for characterizing pH relations, titration of aqueous plant extracts can be carried out with H^+ or OH^-. The resistance against pH change during titrant addition can be considered as a measure of the stability of the pH,

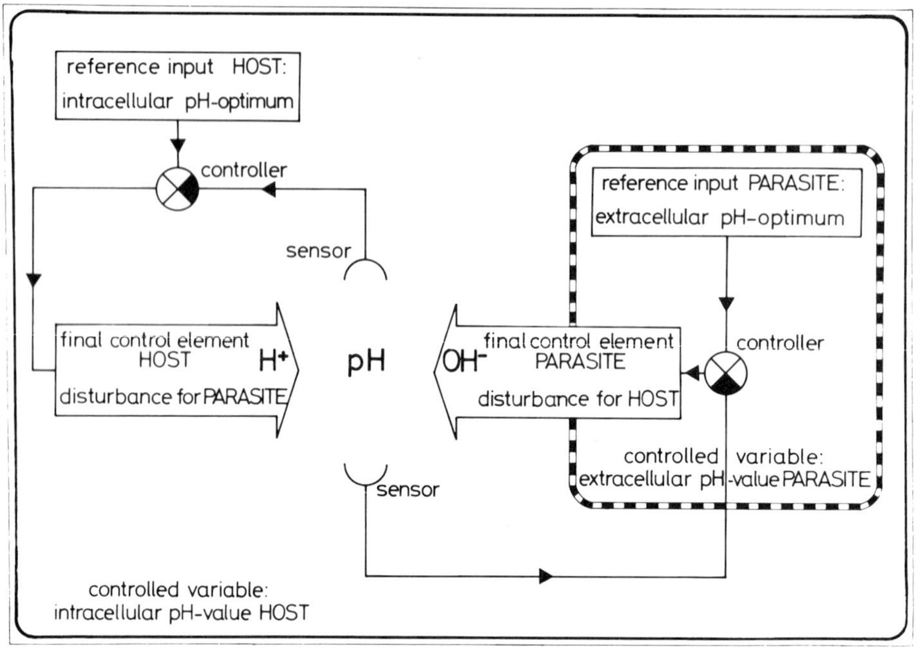

Figure 2. A cybernetic model of the interaction of host and parasite with respect to the intracellular pH-milieu of the host and the extracellular pH-milieu of the parasite when the pH of the host is lower than the optimum pH for the parasite (after Scholz & Stephan, 1975).

expressed as buffering capacity. The hypothesis is that the higher the buffering capacity, the more resistant is the host plant.

These interactions were tested first in the host-parasite system *Pinus sylvestris* L./*Lophodermium pinastri* (Schrad. ex Hook.) Chev. (Scholz & Stephan, 1974).

THE HOST-PARASITE SYSTEM PINUS SYLVESTRIS/LOPHODERMIUM PINASTRI

The most interesting period for studying this host-parasite interaction is soon after infection in autumn and during the latent phase in winter, which is decisive for pathogenesis.

With reference to the cybernetic model (Fig. 2), 2 subjects may be investigated: the contribution of the pH-regulation by the host plant to its resistance, and the contribution of pH-regulation by the fungus to its pathogenicity. First we can test whether the fungus can regulate pH to the level for its optimum growth, which is assumed to be approximately pH 5.0 to pH 5.5, depending on the fungus strain. In trials with a *L. pinastri* strain on agar media containing various pine needle extracts we found that the fungus can shift the pH toward its external optimum pH (Fig. 3).

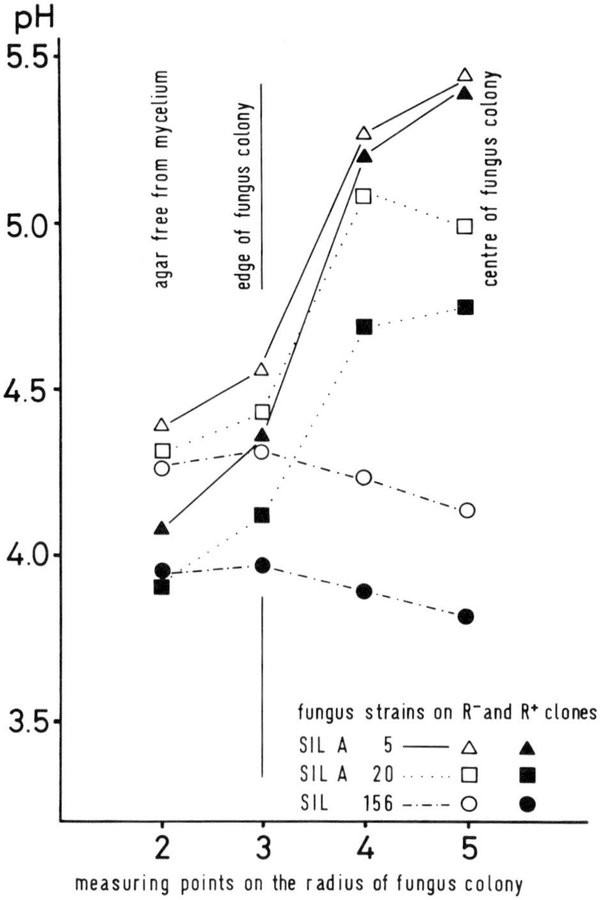

Figure 3. The pH-changing influence of 3 strains of *Lophodermium pinastri* on needle decoction agar. pH was measured along the radius of cultures growing in Petri dishes. Curves are means of 4 resistant (R^+) and 4 susceptible (R^-) non-infected *Pinus sylvestris* clones, respectively (from Scholz & Stephan, 1981).

In this connection Schütt (1964a) found that the colony diameter of *L. pinastri* decreased with increasing concentrations in the culture medium of needle extracts of resistant Scots pine clones. The diameter increased, however, when increasing concentrations of needle extracts of susceptible clones were added. Schütt (1964a) supposed biochemical components as possible influencing factors. Higher buffering capacity in resistant clones also could have contributed to this effect.

Corresponding effects should be expected in the living host plant cell. The host plant should maintain pH at its own intracellular lower optimum by its inherent buffering capacity or by additionally synthesized buffering substances.

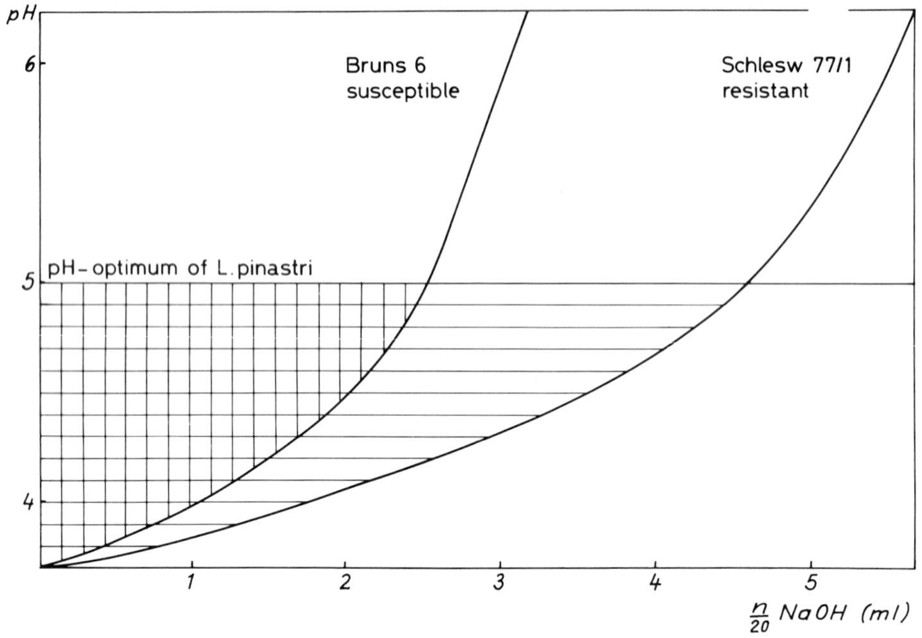

Figure 4. Buffering capacity in needle extracts from non-infected *Pinus sylvestris* clones with different degrees of resistance to *Lophodermium pinastri* (after Scholz & Stephan, 1974).

For procedural reasons we can test the inherent buffering capacity only in vitro, that is, in aqueous extracts of the particular plant organ that the fungus infects, in the present case the needles. Since in the course of pathogenesis nuclei of infected plant cells degenerate and compartmentation within the single plant cell is disintegrated, this state is presumably similar to that found by the fungus in the in vitro extracts.

In aqueous needle extracts from 8 clones of *Pinus sylvestris* of different degrees of resistance the pH was artificially influenced by adding OH^- This titration resulted in curves (Fig. 4) which showed that a susceptible clone had lower buffering capacity than a resistant one. The buffering capacity of resistant clones was significantly higher than that of susceptible clones. An analysis of variance was based on clones (Table 1). Variance between ramets in clones was small. Buffering capacity and resistance were highly correlated (r = 0.84) (Scholz & Stephan, 1974).

These summarized investigations support the hypothesis given above.

EXAMPLES OF RELATION OF BUFFERING CAPACITY AND RESISTANCE

The relation between buffering capacity and resistance was tested repeatedly for the *Pinus sylvestris/Lophodermium pinastri*-system with different pine populations, and investigated also for other host-parasite

Table 1. Analysis of variance of buffering capacity of different resistant clones of *Pinus sylvestris*.

Source of variation	df	F-value	Variance component $\hat{\sigma}^2$	Relative variance component (%)
between clones	7	22.01***	18953.7	83.4
betweem ramets	16	4.84***	2146.8	9.4
error	48	-	1675.3	7.4

*** significant at $p \leqq 0.001$.

systems. The results of the various authors are summarized in Table 2. Most of the investigations confirm the previous results. But there are also deviating results. They can be caused by various reasons several of which will be discussed in more detail, because they lead to a more thorough understanding of host-parasite interaction:

1. Procedural reasons related to statistical aspects of sampling, handling of plant organ samples before titration, and titration methods all can influence the results. Differences caused by these reasons can be excluded by standardization of methods and will not be discussed further.
2. Physiological processes are influenced by diurnal or seasonal variation, or both, and may change with the age of the host plant. Thus, results of investigations on the relation between resistance and buffering capacity in the respective plant organ also can be influenced.
3. The resistance in the investigated host populations may be based on different or interacting resistance mechanisms.
4. Finally, the well-known great variability of parasite populations may cause differences in virulence in different biotypes or races. Hence resistance mechanisms which are effective against a certain fungus population may be less effective against another fungus population. Such effects are known to occur in some host-parasite systems, but have not been investigated extensively in forest trees.

EFFECTS ON THE RELATION BETWEEN BUFFERING CAPACITY AND RESISTANCE

Influence of age of the plant organ

Buffering capacity and pH in aqueous extracts of Scots pine needles vary during the ontogeny of the needles as shown by Dupate (1977), Martinsson (1979), Wind (1979) and Scholz & Stephan (1981). Buffering capacity was highest in young needles after needle elongation, decreased during summer, autumn and winter of the first vegetation period, and stayed at low values

Table 2. Investigations of various authors on buffering capacity in relation to resistance of several pine and spruce species to fungi.

Host	Pathogen	Age of host (years)	Type of material	Organ of host	Time of sampling	Correlation	Reference
Pinus sylvestris	*Lophodermium pinastri*	15-20	clones	needle	October	yes ($r^* = 0.84$)	Scholz, 1973 Scholz & Stephan, 1974
Pinus sylvestris	*L. pinastri*	25	clones	needle	several dates	yes and no (depending on date)	Scholz & Stephan, 1981
Pinus sylvestris *Pinus nigra* *Pinus rigida* *Pinus strobus*	*L. pinastri*	20	clones	needle	August	yes	Scholz & *Stephan, 1975*
Pinus sylvestris	*L. pinastri*	6	families	needle	January	no	Stephan & Scholz, 1981
Pinus sylvestris	*L. pinastri*	4	families	needle	June/Oct.	yes	Dupate, 1977
Pinus sylvestris	*L. pinastri*	5	families	needle	July/Oct.	no	Martinsson, 1979
Pinus sylvestris	*L. pinastri*	.	provenances	needle	.	yes	Matschke & E. Scholz, 1980
Pinus radiata	*Dothistroma pini*	5 and 40	single trees	needle	1 year old needles	no	Franich & Wells, 1977
Pinus nigra	*Scleroderris lagerbergii*	22	clones	bark	October	yes ($r_s = 0.52$)	Stephan & Scholz, 1979
Picea spp.	*Rhizosphaera kalkhoffii*	.	single trees	needle	.	yes	Kumi & Lang, 1979

in the beginning of the second year. During the same period pH increased.

Parallel to those changes the life cycle of the needle cast fungus *Lophodermium pinastri* takes place. The highest levels of ascospore production and release in late summer and autumn coincide, at least under central European conditions, with the low-value period of buffering capacity in current year pine needles.

During this infection period and the following latent phase of the pathogen resistance mechanisms of the host plant should be most effective. Results of several investigations showed clearly that only in late summer, autumn and winter did the resistant Scots pine clone group have significantly higher buffering capacity than the susceptible group (Scholz & Stephan, 1974, 1981). It is important, therefore, to time investigations of traits like buffering capacity to the period when the decisive physiological interaction between host and parasite takes place.

Resistance mechanisms vary with further ontogeny of a plant organ. For example, Scots pine needles older than 1 year are resistant to the parasite *Lophodermium pinastri*.

Also, differences in age of the host plant itself could influence the relative importance of different mechanisms in host-parasite systems. Dupate (1977) found a relation between buffering capacity and resistance in young Scots pine plants, but Martinsson (1979), who investigated other young Scots pine families, did not. Studies described above were conducted on 25-year old Scots pine clones (Scholz & Stephan, 1974, 1981). Furthermore, Franich & Wells (1977) found no relation between buffering capacity of *Pinus radiata* D. Don needles and resistance against *Dothistroma pini* Hulb. in 5-year old trees, but in 40-year old trees high buffering capacity and resistance to the parasite generally exist.

Influence of host population

For investigations on the relation between buffering capacity and resistance, host-plant materials of quite different character and origin have been used by different authors. In various regions of the natural distribution area of a forest tree species different coevolution with the respective parasite may have resulted in different resistance mechanisms. Thus relations between traits at one location can be different from those at another one.

A few examples may support this concept for the *Pinus sylvestris/Lophodermium pinastri*-system. Relative resistance of provenances may change when they are cultivated in a distant location in comparison to that of provenances from other regions or that of local provenances. When Scots pine provenances are grown under central European conditions, those from Scandinavian countries often show higher resistance to the needle cast fungus than those from southern France (Schütt, 1964b). In a Dutch trial German Scots pine provenances were killed by *Lophodermium pinastri* whereas some local provenances remained largely unattacked (Kriek & Bikker, 1973). In a south-west German trial of the Forstliche Versuchs- und Forschungsanstalt Baden-Württemberg the progenies of non-local Scots pine plus trees selected for needle cast resistance were heavily attacked by *L. pinastri* but the local provenance remained unattacked (data unpublished). These reactions were the result of interactions with environmental factors, but also differences in resistance mechanisms might have been involved. Perhaps such differences could explain the results of Martinsson (1979) who did not find a relation between buffering capacity of Swedish Scots pine families and resistance. In our own investigations on Scots pine full-sib families from crosses in and between Alpine and Mazurian provenances the Alpine crosses showed high buffering capacity and susceptibility to *L. pinastri*, whereas Mazurian crosses showed low values for both traits (Stephan & Scholz, 1981). Within the provenances, however, this relation was not general.

Resistance of *P. sylvestris* to *L. pinastri*, which is assumed to be polygenic, is certainly governed not by only one mechanism. This might explain differences between pine populations investigated for one resistance trait only.

Influence by parasite population

Finally the interaction between host and different populations from different locations should be considered. *Lophodermium pinastri*, for instance, shows great variability with respect to morphological, physiological and biochemical traits (Millar & Watson, 1971; Stephan, 1973a, b), which recently lead to the separation of different species (Minter et al., 1978) with apparently different life cycles and biology (Minter & Millar, 1980). Obviously variation in the virulence of the pathogen is to be expected. Therefore in various host-parasite populations different interactions and balances between resistance mechanisms and virulence mechanisms are probable.

In considering pH regulation as a possible mechanism of host-parasite interaction Stephan (1975) reported variation between *L. pinastri* strains isolated from different pine species. The ability of the fungal strains to change pH as well as the reaction of the host species showed remarkable variation in vitro. Such variation also has been found between fungal strains isolated only from Scots pine (Scholz & Stephan, 1981). These strains differ in the amount of their pH-changing influence on needle agar and even in the direction of this influence (Fig. 3).

The results of *L. pinastri* strains growing on needle-decoction agar were largely confirmed with cultures in liquid needle extracts of resistant and susceptible Scots pine clones (data unpublished). The pH normally changed either to higher or lower values, depending on the fungal strain used. There were negative correlations between the degree of resistance of respective Scots pine clones and the extent of pH change in vitro after

14 days of culture. The importance of the variability of the parasite in the relation between resistance and buffering capacity is underlined by the fact that 1 strain showed a different behaviour in culture. In this case the amount of pH change was positively correlated with degree of resistance of the pine clones. The behaviour of the pathogen population interacting with the respective host population will depend on its inherent requirement for an optimum pH and its ability to regulate pH.

INHERITANCE OF BUFFERING CAPACITY

Inheritance of buffering capacity is not yet well known. Preliminary results of investigations of crosses between and within 2 Scots pine provenances showed significant differences between progenies (Stephan & Scholz, 1981). Alpine crosses showed the highest, and Mazurian crosses the lowest means. Interprovenance crosses were intermediate for this trait.

Estimations of genetic parameters in families of Norway spruce (*Picea abies* (L.) Karst.) showed that the phenotypic variance of buffering capacity is governed mainly by genetic factors (Scholz & Reck, 1977). Heritability was estimated roughly at $h^2_{n.s.} = 0.78$.

CONSEQUENCES AND CONCLUSIONS

1. For suitable comparability of results sampling criteria and buffering capacity evaluation methods should be standardized. With the necessary phytosanitary restriction, the exchange of host plant material and cultures of parasites for such host-parasite interaction studies is also suggested.
2. A better understanding of virulence and resistance mechanisms is necessary including more detailed knowledge of pH regulation in host-parasite interaction and its relation to other resistance mechanisms. This should be investigated in various host-parasite systems.
3. Up to now buffering capacity cannot be recommended generally as a trait for use in indirect selection. Before generalizing about the importance of a resistance mechanism, one should test it in different host populations.
4. Investigations on pH regulation can help in understanding the interaction between host and parasite at the level of biochemistry as well as at the level of the populations involved.

REFERENCES

Dupate, A.I., 1977. Investigations on resistance of Pinus sylvestris L. to Lophodermium pinastri (Schrad.) Chev. In: Ökolog: Besonderheiten der niederen Pflanzen des sowjetischen Ostseeraumes. Material zum 8. Symposium der balt. und weissruss. Mykologen und Lichenologen. Wilna 1977. p. 74-75.

Franich, R.A. & L.G. Wells, 1977. Infection of Pinus radiata by Dothistroma pini: effect of buffer capacity of needle homogenates. New Zealand Journal of Forest Science 7: 35-39.

Kriek, W. & G. Bikker, 1973. German and Dutch provenances of Scots pine in the Netherlands. Nederlands Bosbouw Tijdschrift 45: 154-161.
Kumi, J. & K.J. Lang, 1979. The susceptibility of various spruce species to Rhizosphaera kalkhoffii and some cultural characteristics of the fungus in vitro. European Journal of Forest Pathology 9: 35-46.
Martinsson, O., 1979. Testing Scots pine for resistance to Lophodermium needle cast. Studia Forestalia Suecica 150: 1-63.
Matschke, J. & E. Scholz, 1980. Biologische Grundlagen der Forstpflanzenzüchtung. Beiträge für die Forstwirtschaft 14: 116-121.
Millar, C.S. & A.R. Watson, 1971. Two biotypes of Lophodermium pinastri in Scotland. European Journal of Forest Pathology 1: 87-93.
Minter, D.W. & C.S. Millar, 1980. Ecology and biology of three Lophodermium species on secondary needles of Pinus sylvestris. European Journal of Forest Pathology 10: 169-181.
Minter, D.W., J.M. Staley & C.S. Millar, 1978. Four species of Lophodermium on Pinus sylvestris. Transactions of the British Mycological Society 71: 295-301.
Raven, J.A. & F.A. Smith, 1973. The regulation of intracellular pH as a fundamental biological process. In: W.P. Anderson (Ed.): Ion transport in plants. Academic Press, London/New York, p. 271-278.
Scholz, F., 1973. Biochemische Methoden in der Forstgenetik. Forstarchiv 44: 171-174.
Scholz, F. & W. Knabe, 1976. Investigations on buffering capacity in spruce clones of different resistance to air pollution. XVI. IUFRO World Congress, Oslo June 1976. Working Party S 2.09.04 - Physiological effects. Mimeograph. 10 p.
Scholz, F. & S. Reck, 1977. Effects of acids on forest trees as measured by titration in vitro, inheritance of buffering capacity in Picea abies. Water, Air and Soil Pollution 8: 41-45.
Scholz, F. & B.R. Stephan, 1974. Physiologische Untersuchungen über die unterschiedliche Resistenz von Pinus sylvestris gegen Lophodermium pinastri. I. Die Pufferkapazität in Nadeln. European Journal of Forest Pathology 4: 118-126.
Scholz, F. & B.R. Stephan, 1975. Zur pH-Regulation als Faktor der Resistenz von Pinus spp. gegen Lophodermium pinastri. Angewandte Botanik 49: 55-63.
Scholz, F. & B.R. Stephan, 1981. Physiological studies on variation in resistance of Pinus sylvestris to Lophodermium pinastri. II. Variation of buffering capacity in needles of different age with relation to resistance. European Journal of Forest Pathology 11 (in press).
Schütt, P., 1964a. Eine mykologische Methode der Resistenzprüfung, entwickelt für den Lophodermium-Befall der Kiefer. Phytopathologische Zeitschrift 51: 62-72.
Schütt, P., 1964b. Der Schüttebefall der Kiefer in Abhängigkeit von Herkunft und Anbauort. Forstwissenschaftliches Centralblatt 83: 129-192.
Smith, F.A. & J.A. Raven, 1976. H^+ transport and regulation of cell pH. In: U. Lüttge & M.G. Pitman (Eds): Encyclopaedia of plant physiology, New Series, 2. Transport in plants II, Part A, Cells. Springer, Berlin/Heidelberg/New York. p. 317-346.
Stephan, B.R., 1973a. Untersuchungen zur Variabilität von Lophodermium pinastri. I. Kulturvarianten. European Journal of Forest Pathology 3: 6-12.
Stephan, B.R., 1973b. Untersuchungen zur Variabilität von Lophodermium pinastri. II. Unterschiede im physiologischen Verhalten. European Journal of Forest Pathology 3: 112-120.
Stephan, B.R., 1975. Variability of Lophodermium pinastri in culture. Mitteilungen der Bundesforschungsanstalt für Forst- und Holzwirtschaft Hamburg-Reinbek 108: 69-78.
Stephan, B.R. & F. Scholz, 1979. Weitere Untersuchungen zur unterschiedlichen Anfälligkeit von Pinus nigra-Klonen gegenüber Scleroderris lagerbergii. European Journal of Forest Pathology 9: 46-51.
Stephan, B.R. & F. Scholz, 1981. Preliminary results on Lophodermium needle cast resistance and buffering of pH in Scots pine crosses. Proceeding of the meeting of the IUFRO Working Party S 2.06.04 - Needle diseases. Sarajevo, Yugoslavia, Sept. 1980. 6 p. (in press).
Wind, E., 1979. Pufferkapazität in Koniferennadeln. Phyton 19: 197-215.

Histoenzymological approach for detecting host-parasite interactions in forest trees caused by heart-rot diseases

Zuei-ching Chen & Paul D. Manion

Department of Botany, National Taiwan University, Taipei, Taiwan, Republic of China, and Department of Forest Botany and Pathology, State University of New York, College of Environmental Science and Forestry, Syracuse, New York, USA 13210

ABSTRACT

A progressive change of cell-wall degrading enzyme activities in living aspen *(Populus tremuloides)* sapwood inoculated with basidiospores of *Phellinus tremulae* were investigated by a routine enzymological method and a modified histoenzymological method. Peroxidase activity was inhibited moderately in the wounded tissue and severely in the inoculated tissue during the early period of discoloration. The activity of laccase in the inoculated tissue was stimulated immediately after treatment and increased 4 to 6 times over that in the sound tissue and about twice over that in the wounded tissue. More cellulase activity was noted in the wounded than in the inoculated tissue. A clearer picture of host-parasite interactions was obtained by the histoenzymological method. The sensitivity of the histoenzymological method was about 12-1018 times higher than of the routine extraction method.

INTRODUCTION

Basidiospores of *Phellinus tremulae* (Bond.) Bond. & Borisov germinate only on fresh aspen (*Populus tremuloides* Michx.) sapwood wounds during the summer (Manion & French, 1968; Chen, 1971). Recently Shigo's succession theory (Shigo, 1963, 1967) cast doubt on the pathogenicity of principal decay Hymenomycetes in the initial stage of decay. *Phellinus tremulae*, the white-rotter of aspen, possesses an enzyme system which can attack all components of the wood cell wall including lignins, resulting in a general thinning of the cell wall, even in the early stage of decay (Basham, 1958; Cowling, 1961; Wilcox, 1968). Cellulases of the white-rot fungi attack the crystalline and amorphous cellulose of the host cell wall at a similar rate, and act from the surface of the wall. Lignin-depolymerizing enzymes of the white-rot fungi are able to penetrate rapidly the secondary wall of the wood cells and act on the compound middle lamella. This results in the rapid utilization of lignin by the fungus in the earliest stage of decay. Enzymatic degradation of the cell wall by white-rot fungi, therefore, involves both a lignin- and a cellulose-degradation system (Cowling, 1961). Lignin-degradation enzymes involve polyphenol oxidases of the laccase type (Aufsess et al., 1968; Fåhraeus, 1952; Kirk et al., 1968a, 1968b) or the peroxidase type (Ishikawa & Oki, 1964; Lyr & Ziegler, 1959). The present investigation attempts to determine the decay capability of *P. tremulae* in fresh sapwood wounds of living aspen trees after germination of basidiospores. Histoenzymology is introduced into this study in a comparison with the routine enzyme extract method for the evaluation of the magnitude of host-parasite interactions in this heart-rot system.

MATERIALS AND METHODS

Basidiospore inoculum was collected on a clean glass slide placed in a plastic box attached to a bigtooth aspen (*Populus grandidentata* Michx.) bearing sporulating fruiting bodies. Spores were collected during a 16- to 18-hour-period starting at 3-5 p.m. Spores from a single fruiting body were washed from the slides with sterilized distilled water prior to inoculation. The concentration of spores was approximately 2-6 million per ml of solution. On June 1, 1969, wounds 2 cm deep and 0.5 cm in diameter were made at a height of 1.4 m on opposite sides of trembling aspen trees 10-15 cm in diameter located in the Heiberg Memorial Forest, Tully, New York, USA. After the wounds were made, one bore hole was inoculated with 1 ml of the basidiospore suspension of *P. tremulae* and the other hole in the opposite side was injected with 1 ml of sterilized distilled water. Collections from treated aspen trees were made monthly after inoculation until October. Each month 15-20 trees were cut and 1-meter sections were removed and stored at -20 °C. One half of each collection was used for investigating symptom development, fungal flora, cytological changes, and enzyme assays. The rest was used for investigating pH change, ash and mineral content. The stem sections were cut vertically through the bore-hole into halves to observe the development of discoloration and the change in morphology. Pieces of wood about 1 × 1 × 2-3 mm in size were removed aseptically from different parts of the discolored column and cultured on 2 % malt agar medium at 25 °C in order to recover *P. tremulae*.

For enzyme preparation extracted from wood meal, frozen sapwood blocks 2.5 cm × 1.0 cm × 0.5 cm were removed near the wound and ground in a Wiley Mill over a 100 mesh screen. One gram of wood meal was then suspended in 5 ml of 0.05 M phosphate buffer containing 0.25 M sucrose and 0.03 M ascorbic acid at pH 7.0. The mixture was ground in a mortar for 2 minutes at 5 °C. The mixture was then centrifuged at 4 °C for 30 minutes at 15 000 × *g*. To the supernatant extract ammonium sulfate was added to 0.8 saturation and the resultant enzyme precipitate was collected by centrifugation at 4 °C for 10 minutes at 2000 × *g*. The precipitated enzyme was dissolved in 5 ml of distilled water and dialysed in cold water for 12 h. The dialysed enzyme solution was used for measuring peroxidase, laccase, catalase, and cellulases.

For enzyme preparation from thin sections of sapwood tissue, frozen wood blocks about 0.5 cm × 0.5 cm × 0.5 cm in size were sectioned radially to a thickness of 20 μm with a cryostat. About 20 sections were added to each enzyme reaction solution. After the reaction was measured, the sections were collected on filter papers and dried at 105 °C overnight before their dry weight was obtained.

Peroxidase activity was determined by the method described by Jennings et al.(1969). The enzyme solution was diluted 1:5 with water and 0.5 ml was used to determine peroxidase activity. The reaction mixture consisted of 1.5 ml of 0.025 M citrate-phosphate buffer at pH 5.0, 0.5 ml of 0.02 M guiacol, and 0.5 ml of 0.06 M hydrogen peroxide. Activity was recorded as the change in absorbance per minute at 483 nm. The reaction was initiated by the addition of hydrogen peroxide. For the quantitative histochemical method, the tissue sections were added directly to the reaction solution. The mixture was incubated for 30 minutes at 30 °C. A preliminary test with various incubation times from 1 to 240 minutes revealed that 30 minutes was the best choice for obtaining the linear relation between enzyme activity and incubation time. The reaction was terminated by swift transfer of the reaction solution into ice. The reaction solution was filtered through Whatman No. 1 filter paper and filtrates transfered to cuvettes. For the control, tissue sections killed by autoclaving at 120 °C for 20 minutes, or living tissue sections incubated in the reaction solution without hydrogen peroxide, were used.

Laccase activity was determined by the method described by Jennings et al.(1969). The reaction solution contained 1.5 ml of 0.02 M citrate-phosphate buffer at pH 6.0, 0.5 ml proline (5 mg/ml), and 0.5 ml catechol (2 mg/ml). The mixture was aerated through a glass capillary for 2 minutes before addition of catechol, which initiated the reaction. For each assay 0.5 ml of undiluted enzyme solution was used.

Catalase activity was assayed by the method of Beer & Sizer (1952) as modified by the Worthington Biochemical Corp.(1963). The decomposition of hydrogen peroxide was determined in 3 ml of reaction mixture containing 0.1 ml of enzyme preparation, 1 ml of 0.059 M hydrogen peroxide in 0.05 M phosphate buffer at pH 7.0, and distilled water. The change in absorbance at 240 nm was recorded every 10 seconds for 70 seconds. A unit of catalase activity was defined as that amount which catalyzes the decomposition of 1 μmole of hydrogen peroxide per minute under the reaction condition described above.

Cellulases activity was determined by the method of Mandels & Reese (1964) as modified by the Worthington Biochemical Corp.(1963). This activity was demonstrated by a decrease in viscosity or by the formation of glucose in cellulose powder and carboxymethyl cellulose (CMC).

RESULTS

The amount of enzyme activity detected by the extracted enzyme preparation method was so low that only the activity at the third month of treatment is presented (Table 1). The inoculated tissue showed the highest catalase activity and the wounded the lowest. In the contrary, the wounded tissue had the highest peroxidase activity. Its activity was 2 and 7.5 times higher than that of the inoculated and sound tissue, respectively. Laccase activity in the inoculated tissue was the highest among the 3 treatments. Cellulase activity was negligible in the sound tissue but higher activities were found in both the inoculated and the wounded tissues. In general, all enzyme activities measured by this method tended to indicate low activity in aspen sapwood, even after the wound inoculation. This method also was unable to show the low level of enzyme activity taking place at the site of cell-wall decay in the host plants.

The histochemical method showed progressive changes of enzyme activities in the host tissues. The peroxidase activity in both the inoculated and the wounded tissues was inhibited during the first 2 months (Table 1). In the third month, the activity increased sharply and exceeded the activity in the sound tissue. Although the activity in the inoculated tissue was lowest in the first month, it became the highest in the third month. Inoculation with *P. tremulae* basidiospores caused a strong inhibition of peroxidase activity in the host tissue in the early stage of infection. Activity of laccase in the sound tissue was in the range of $\Delta A=0.15$ to 0.2. Contrary to the influence of host tissue on peroxidase activity, laccase activity in both the inoculated and the wounded tissues was higher than in the sound tissue. In comparisons between the inoculated and the wounded tissues, activity in the inoculated was slightly higher ($\Delta A=0.80$) than in the wounded tissue ($\Delta A=0.62$) one month after inoculation and reached $\Delta A=1.14$, which was twice as much as that in the wounded tissue ($\Delta A=0.56$) in the third month. A slight cellulase activity was detected in the sound tissue. The activity in normal aspen sapwood ranged from the lowest in September (0.08) to the highest in mid-summer (0.52). The wounded tissue showed the highest activity of the 3 types. Almost a linear increase of 'C_1' enzyme activity in the wounded tissue was noticed through the second month. Activity in the inoculated tissue increased in the second month but was only 2/7 of the activity in the wounded. A steady activity was observed in both the inoculated and the wounded tissues during the second and the third months while there was a sharp drop of cellulase activity in the normal aspen sapwood.

Table 1. Wood-degrading enzyme activities of living aspen sapwood 0.5 cm from the inoculation point[1].

Source of enzyme preparation	Enzymes	Months[6] after inoculation	Inoculated	Wounded	Sound
Extracted	catalase[3]	3	1.56	0.27 a	0.48 a
	peroxidase[3]	3	1.87	3.62	0.48
	laccase[3]	3	0.112 a	0.071 a	0.064
	"C_1" cellulase[5]	3	0.65 a	0.49 a	0.07
	"C_x" cellulase[4]	3	0.038	0.076	0
Wood sections	peroxidase[2]	0	-	-	0.37 a[7]
		1	0.08	0.16 b	0.37 a
		2	0.18 b	0.16 b	0.31
		3	0.55	0.43	0.35 a
	laccase[2]	0	-	-	0.20 bc
		1	0.80	0.62	0.25 b
		2	0.65	0.42	0.16 c
		3	1.14	0.16 c	0.20 bc
	cellulases	0	-	-	0.32 c
	"C_1"[4]	1	0.40 c	1.60	0.52
		2	0.80 b	2.86 a	0.52
		3	0.76 b	2.82 a	0.08

ΔA: Change of absorbance.
DWT: Dry weight of aspen sapwood.
1. Activity is based on the average of 3 replicated experiments.
2. ΔA/min/10 mg DWT.
3. ΔA/min/g DWT.
4. µg/mg DWT.
5. Unit/min/g DWT.
6. Date of inoculation: 1 June 1969.
7. Values followed by a common letter are not significantly different in a 5 % Duncan's multiplerange test. Each enzyme was analysed separately.

DISCUSSION AND CONCLUSION

There are 2 theories on the enzyme system responsible for lignin degradation. One theory stresses the role of laccase (Aufsess et al., 1968; Fåhraeus, 1952; Kirk et al., 1968a, 1968b). The other stresses peroxidase as more suitable for the degradation of lignin in the semi-anaerobic environment of wood tissue (Ishikawa & Oki, 1964; Lyr & Ziegler, 1959). In vivo, peroxidase activity was inhibited moderately in the wounded tissue and severely in the inoculated tissue during the first 2 months. At the end of the third month, the activity in the inoculated tissue recovered and finally exceeded the levels in the wounded and the sound tissues. The mechanism(s) operating in the early inhibition and late reactivation of peroxidase in the inoculated tissue is not understood. An increase of laccase activity was noted from the very beginning of wounding. Laccase activity in the inoculated tissue was 5.7 times higher than in the sound tissue and 2 times higher than in the wounded tissue 3 months after the inoculation.

The impact of high activity of polyphenol oxidases could affect lignin catabolism and phenol metabolism in the host. Pigmentation of the host tissue after wounding is considered to be the result of oxidation of phenol compounds catalyzed by many oxidative enzymes (Kellin & Mann, 1939) including the peroxidase system of contaminating bacteria (Shortle et al., 1978). The ability to utilize these oxidative products could be a crucial advantage for those organisms that grow and survive in the wound tissue (Lyr, 1962). Organisms with high polyphenol oxidase activity would be better able to dominate and survive in the phenol-rich tissue. High phenol oxidase activity in the discolored tissue might contribute to the degradation of aromatic inhibitors of peroxidase in the tissue and permit reactivation in the late stage of discoloration. It seems that peroxidase is less important than laccase in the initial stage of discoloration and decay of the wounded aspen sapwood by *P. tremulae*, because of the evidence of changes that occurred in the inoculated tissue.

In field inoculations, cellulase activity in the inoculated tissue was significantly higher than in the sound tissue but lower than in the wounded tissue. In vivo, cellulase activity in the sound sapwood decreased drastically after the summer growth season. Inoculation with basidiospores of *P. tremulae* altered cellulase activity in the inoculated tissue and activity was lower than in the normal sound sapwood in the first month, then gradually increased from the second month. The cause of this decrease in the early stage of wound inoculation is not known. Shortle et al.(1978) reported that bacteria found in discolored wood might alter the cellulase activity of white-rot fungi to some degree. In the present investigation, bacteria were isolated in about the same frequency from the wounded and the inoculated tissues. Therefore, the cause of inhibition of cellulase activity in the inoculated tissue probably is related to *P. tremulae* itself.

Only slight enzyme activities in the host tissue 3 months after wound inoculation were detected by the enzyme extraction method. In contrast, the histoenzymological method was 12-1018 times as sensitive. The results indicated that the majority of enzymes exuded from living hyphae in the wood decay ecosystem were tightly bound with their substrates such as lignin and cellulose components of the wood cell wall. The method modified in this way resulted in a very sensitive enzyme reaction and the outcome of the experiment was very satisfactory. A clearer picture of host-parasite interaction in terms of the progressive change of enzyme activity was obtained by this method. Histoenzymology, therefore, proved to be a dependable research technique.

REFERENCES

Aufsess, H. von, H. von Pechmann & E. Graessle, 1968. Fluoreszenzmikroskopische Beobachtungen an pilzbefallenem Holz. Holz als Roh- und Werkstoff 26: 50-61.

Basham, J.T., 1958. Decay of trembling aspen. Canadian Journal of Botany 36: 491-505.

Beer Jr, R.F. & I.W. Sizer, 1952. A spectrophotometric method for measuring the breakdown of hydrogen peroxide by catalase. Journal of Biological Chemistry 195: 133.

Chen, Z.C., 1971. Host response of aspen after wound inoculation with Fomes igniarius basidiospores. Ph.D. Thesis. SUNY College of Forestry at Syracuse, New York. 100 p.

Cowling, E.B., 1961. Comparative biochemistry of the decay of sweetgum sapwood by white-rot and brown-rot fungi. U.S. Department of Agriculture, Technical Bulletin 1258.

Fåhraeus, G., 1952. Formation of laccase by Polyporus versicolor in different culture media. Physiologia Plantarum 5: 284-291.

Ishikawa, H. & T. Oki, 1964. The oxidative decomposition of lignin I. The enzymatic degradation of softwood lignin and related aromatic compounds by peroxidase. Journal of Wood Society of Japan 10: 207-213.

Jennings, P.H., B.L. Brannaman & F.P. Zscheile Jr., 1969. Peroxidase and polyphenol oxidase activity associated with Helminthosporium leaf spot of maize. Phytopathology 59: 963-967.

Kellin, D. & T. Mann, 1939. Laccase, a blue copper-protein oxidase from the latex of Rhus succedanea. Nature 143: 23-24.

Kirk, T.K., J.M. Harkin & E.B. Cowling, 1968a. Oxidation of guaiacyl-and veratryl-glycerol-B-guaiacyl ether by Polyporus versicolor and Stereum frustulatum. Biochemica et Biophysica Acta 165: 134-144.

Kirk, T.K., J.M. Harkin & E.B. Cowling, 1968b. Degradation of the lignin model compound syringyl-glycol-B-quaiacyl ether by Polyporus versicolor and Stereum frustulatum. Biochemica et Biophysica Acta 165: 145-163.

Lyr, H. & H. Ziegler, 1959. Enzymausscheidung und Holzabbau durch Phellinus igniarius Quel. und Collybia velutipes Curt. Phytopathologische Zeitschrift 35: 173-200.

Lyr, H., 1962. Detoxification of heartwood toxins and chlorophenols by higher fungi. Nature 195: 289-290.

Mandels, M. & E.T. Reese, 1964. Fungal cellulases and the microbial decomposition of cellulosic fabric. Developments in Industrial Microbiology 5: 5.

Manion, P.D. & D.W. French, 1968. Inoculation of living aspen trees with basidiospores of Fomes igniarius var. populinus. Phytopathology 58: 1302-1304.

Shigo, A.L., 1963. Fungi associated with the discolorations around rot columns caused by Fomes igniarius. Plant Disease Reporter 47: 820-823.

Shigo, A.L., 1967. Successions of organisms in discoloration and decay of wood. International Review of Forest Research 2: 237-299.

Shortle, W.C., J.A. Menge & E.B. Cowling, 1978. Interaction of bacteria, decay fungi, and live sapwood in discoloration and decay of trees. European Journal of Forest Pathology 8: 293-300.

Wilcox, W.W., 1968. Changes in wood microstructure through progressive stages of decay. U.S. Forest Service Research Paper, FPL 70. 46 p.

Worthington manual, 1963. Enzymes. Worthington Biochem. Corp., Freehold, New Jersey, USA.

Preliminary work on the relation between resistance to Fomes annosus and the monoterpene composition of Sitka spruce resin

G.I. Forrest

Forestry Commission, Northern Research Station, Roslin, Midlothian, Scotland, UK

ABSTRACT

A preliminary study was carried out in a Sitka spruce population heavily infected by *Fomes annosus*. Significant differences in monoterpene composition of the root cortical oleoresin were found between infected and uninfected trees, as judged by the presence or absence of *Fomes* in radial cores taken from the lower stem. In particular, *Fomes* was absent from nearly all those trees in which α-pinene formed at least 37 % of the root oleoresin monoterpene fraction.

INTRODUCTION

Several of the monoterpenes which commonly occur in the oleoresin of conifers have been shown to possess fungitoxic properties (e.g. Hintikka, 1970; Gibbs, 1972; de Groot, 1972), and the degree of resistance of certain species to attack by *Fomes annosus* (Fr.) Cooke has in some instances been connected with the monoterpene composition of their oleoresin (e.g. for *Pinus sylvestris* L. by Chudnyi et al., 1972; Fedorov & Manukov, 1972). Sitka spruce (*Picea sitchensis* (Bong.) Carr.) is the main forestry species planted commercially in Britain, and is frequently subjected to serious attacks from *Fomes annosus*. In a preliminary attempt to elucidate any connection between its degree of resistance to infection and the chemical composition of its oleoresin, a study was carried out in a heavily infected stand growing in central Scotland.

EXPERIMENTAL METHODS

The area sampled formed part of Compartment 86A of Drummond Hill forest, Tayside Region, planted in 1943. A 4 mm diameter wood core was taken from the lower trunk of each of 102 trees. A sample of bark and wood was taken from the main root system of each tree for resin analysis. This was usually taken from the upper surface of one of the main roots, about 30 cm from the trunk base. The sample was square, and measured about 5 cm square and 1 cm deep.

Each wood core was examined for the presence of *Fomes annosus*, both by direct microscopic examination of the core surface after suitable incubation, and by culturing thin slices from the core on agar medium at room temperature.

Cortical oleoresin was extracted from the root bark samples either by direct suction into glass capillary tubing from resin canals where possible, or alternatively by macerating small quantities of the root cortex in n-pentane and evaporating the solution down to a small volume. Oleoresin was analysed by gas chromatography. Monoterpenes were quantified as the percentage contribution of each peak to the total monoterpenes present.

RESULTS

The presence of *Fomes* was confirmed in 41 out of the 102 cores either by fruiting on the core surface or by growth in agar culture. The monoterpene compositions of the 40 % of the population in which infection had been confirmed and the 60 % showing no evidence of infection could then be compared. The mean monoterpene composition of each division, together with standard errors, is shown in Table 1.

Of the three monoterpenes showing highly significant differences between the two groups, the most outstanding was α-pinene; a scatter-diagram of the data showed that there was a quite clearly defined cut-off point at about 37 % α-pinene, above which very few infected samples occurred. The distributions were quite different in the two groups: of the 28 trees in

Table 1. Mean monoterpene percentage composition of root cortical oleoresin related to incidence of *Fomes* in wood cores from main trunk of Sitka spruce trees.

Monoterpene	*Fomes* present	*Fomes* absent	t value	Significance of difference
α-pinene	25.6 ± 1.6	33.4 ± 1.3	3.67	***
camphene	0.1 ± 0.0	0.2 ± 0.0	2.89	**
β-pinene	20.1 ± 0.7	18.7 ± 0.4	1.70	-
3-carene	2.1 ± 0.7	0.6 ± 0.2	2.28	*
limonene	4.3 ± 0.7	1.8 ± 0.4	3.34	**
β-phellandrene	46.9 ± 1.2	44.5 ± 1.0	1.50	-
terpinolene	1.0 ± 0.2	0.8 ± 0.1	0.81	-

Significance levels: *** = $P \leqq 0.001$; ** = $P \leqq 0.01$; * = $P \leqq 0.05$.

Table 2. Percentage representation of monoterpene pattern types in infected and in uninfected Sitka spruce trees. The predominance of α-pinene-dominated types in uninfected trees is well shown, while infected trees were characterised by a predominance of β-phellandrene.

Dominant monoterpenes	Sub-dominant monoterpenes	Percentage representation	
		Fomes present	*Fomes* absent
β-phellandrene	α-pinene	24	16
β-phellandrene	β-pinene	10	2
β-phellandrene	α-pinene, β-pinene	10	3
β-pinene	β-phellandrene, α-pinene	2	-
α-pinene	β-phellandrene	29	67
α-pinene	β-phellandrene, β-pinene	2	-
β-phellandrene, α-pinene	β-pinene	20	10
β-phellandrene, β-pinene	α-pinene	2	2

which α-pinene exceeded 37 % of the total monoterpene component, *Fomes* was detected in only three of them. The frequency distributions of the two groups were also different below the 40 % mark, with infected trees having generally lower levels of α-pinene.

Camphene, although forming only a very small percentage of the total monoterpenes, showed a relatively striking difference between infected and uninfected trees: its frequency distribution was bimodal, and the higher mode, at 0.4 %, was represented only in uninfected trees.

Each tree may be characterised genotypically by its gas chromatogram pattern-type, which takes into account the quantitive relationships between the different monoterpenes. A system for allocating each tree to a definite type by visual inspection of its chromatogram has been devised for Sitka spruce (Forrest, 1980), and the results of classifying the infected and uninfected groups in this way are shown in Table 2.

CONCLUSIONS

The main conclusion to be drawn from this preliminary study was that *Fomes* was absent from nearly all those trees in which α-pinene formed at least 37 % of the root cortical oleoresin monoterpene fraction. There are two possible reasons for this effect. First, a high α-pinene content may confer a genuine resistance to attack by *Fomes* by virtue of the known fungitoxic properties of this monoterpene. Secondly, the invasion of the root system by *Fomes* may induce a decrease in the percentage representation of α-pinene in the cortical oleoresin. In the latter case, the relative increases in the synthesis of 3-carene and limonene following infection may be of biological significance.

Further work is necessary to decide between these two possibilities. The main aim will be to measure the resin composition both before and at various periods after artificial inoculation of tree roots with *Fomes*. If the composition is found to be unaltered by infection, then the possibility exists that α-pinene is the main fungitoxic monoterpene in Sitka spruce root cortical oleoresin. Breeding programmes could then be modified to incorporate a screening procedure favouring the preferential selection of high-α-pinene trees, in order to build up a breeding population with high resistance to *Fomes* attack.

ACKNOWLEDGEMENTS

I am grateful to Dr Redfern and his staff for assistance with sampling and culturing.

REFERENCES

Chudnyi, A.V., R.A. Krangauz & E.I. Gundaeva, 1972. Effect of oleoresin composition on the resistance of Scots pine to Fomes annosus. Lesnoe Khozyaistovo 7: 60-61. (In: Forestry Abstracts 34, No. 4654.)

Fedorov, N.I. & E.N. Manukov, 1972. Composition of gum turpentine and wood turpentine of Scots pine trees infected by Fomes annosus. Lesovedenie i Les. Kh.-vo, Minsk, USSR no. 5: 128-133. (In: Forestry Abstracts 34, no. 690.)

Forrest, G.I., 1980. Geographic variation in the monoterpene composition of Sitka spruce cortical oleoresin. Canadian Journal of Forest Research 10: 458-463.

Gibbs, J.N., 1972. Tolerance of Fomes annosus isolates to pine oleoresins and pinosylvins. European Journal of Forest Pathology 2: 147-151.

de Groot, R.C., 1972. Growth of wood-inhabiting fungi in saturated atmospheres of monoterpenoids. Mycologia 64: 863-870.

Hintikka, V., 1970. Selective effects of terpenes on wood-decomposing hymenomycetes. Karstenia 11: 28-32.

Biochemical aspects of the resistance of Pinus sylvestris to Fomes annosus

E.I. Ladejtschikova & G.M. Pasternak

Ukrainian Scientific Research Institute of Forestry and Agroforestry Amelioration, Kharkow, USSR

ABSTRACT

Diagnostic traits were studied in order to select resistant trees of Scotch pine in disease centres. The difference between resistant trees and trees with low resistance (both infected and non-infected) was established for the following parameters: resistance of wood to decay through the pathogen in vitro; thickness of the walls of tracheids; content of hemicelluloses in wood in relation to one unit of cellulose; qualitative composition of extractives from phloem tissue and their inhibitory effect on growth of the pathogen in pure culture; qualitative composition of extractives from wood, being extracted by acetone, and their contribution to the resistance of wood to decay through the pathogen in vitro; content of trace elements (e.g., Mn, B, Cu) in the phloem tissue of the stem. No relation was found between the composition of monoterpenes and the resistance of Scotch pine to *Fomes annosus*.

INTRODUCTION

The range of *Fomes annosus* (Fr.) Cooke indicates that this problem has evolved primarily due to man-made disturbance of ecological conditions. There is much evidence of the relation between the resistance of coniferous stands and ecological conditions (Laatsch et al., 1969; Ladejtschikova et al., 1975).

Along with resistance of stands, individual resistance has been established, but its nature has not yet been determined. There is evidence of the formation of a reaction zone in Scotch pine (Shain, 1967) as a manifestation of active protective reactions.

Some individuals of Scotch pine maintain their viability in disease centres (= openings due to *F. annosus* - attack in stands) during 20-30 years or more. They may be completely non-infected (these cases are extremely rare) or they may have local, minor infection of roots. Research results bear witness to a stability of their physiological vigour, which is typical for Scotch pine in given conditions: provision with the inorganic nutrients, nuclein and carbohydrate metabolism, tissue moisture, growth and the value of electrophysiological parameters (Ladejtschikova et al., 1975; Pasternak, 1975).

MATERIALS AND METHODS

The sample trees were selected in 40-60 years old stands, in disease centres infected by *F. annosus* 20-30 years ago. They were selected on the basis of their location in an infested stand, their external features, and the value of an electrophysiological parameter (impedance) of the stem phloem tissue complex in contact with the cambium (measured with an instrument designed by A.G. Korobtschenko), when horizontal roots were carefully dug up. Upon completion of the biochemical tests, all the roots were completely dug up. This made it possible to compare the values of biochemical and biophysical traits with the actual vigour of the root system.

The experimental model consisted of trees with 3 different categories of vigour: category I-Ia, consisting of trees with high and medium resistance growing in the very disease centre; category II, consisting of healthy trees growing beyond the boundaries of a disease centre serving as an example of a stand with low resistance - sometimes these trees could have infection of the initial phase (category IIa); category III, consisting of diseased trees growing in the periphery of a disease centre. This category of trees has features of low resistance and, in addition, the effect of the pathological process is reflected in it.

Resistant trees were selected in the very disease centre among trees with well-developed, pointed, dense crowns, with dark-green needles of normal length, and with normal shoot growth. The value of the stem impedance of such trees should exceed the average value of the impedance of those trees with equal diameter, growing in a healthy part of the stand. For this purpose the value of the impedance of 15 trees having equal diameter with the resistant tree was measured. A healthy tree with impedance value equal to the average of all trees in the stand was selected for the experiment (category II). Diseased trees were selected along the periphery of the disease centre. They were characterized by reduced growth and thin crowns with light green needles, usually shortened. Their impedance value was 60 % of the value of the sound tree or less.

A characteristic feature of the applied technique is that only healthy roots are analysed irrespective of the category of vigour of the tree, and that not only phloem tissue and wood of the roots, but also those of the stem are used as diagnostic tissues. Phloem tissue and wood of the stem chosen for analysis were not directly affected by the mycelium of the pathogen.

Furthermore, it has to be taken into account that wood of the stem provides more comparable samples in respect to methology than does wood of the root. Samples of the latter are characterized by uneven structure of the wood and by eccentricity. For the tests with the decay of wood this is of methological importance. Related conclusions of the diagnostic value of the phloem tissue of the stem were made by other scientists, too (Alcubilla et al., 1971).

The content of structural substances in the wood of roots has been studied (Melikyan, 1959), as well as the resistance of wood to decay in vitro (Naumov, 1932), the content of extractives soluble in chloroform and acetone (Ivanov, 1946) and their importance for the resistance of wood to decay (Shain, 1967), the composition of extractives in the phloem tissue of the stem (Ivanov, 1946), and the composition of turpentine oils. Biotests with phloem tissue or extractives from phloem tissue and pure cultures of *F. annosus* were also conducted (Laatsch et al., 1969).

RESULTS AND DISCUSSION

Research has shown that the stem wood of resistant 50-year old trees is decayed less than the wood of healthy trees beyond the disease centre (by 14 % less) and also less than the wood of diseased trees (by 23 % less). Earlier there had already been stated an opinion that the resistance of a living tree to wood-attacking fungi covers the wood, too (Rypaček, 1957).

A question arises: what are the reasons for the greater resistance of the wood of resistant trees? In this respect information about the anatomic structure and composition of the structural substances of the cell membrane of wood is of interest, as well as information about the content of different groups of extractives. It turned out that wood of the roots and the stems of resistant trees has a great quantity of late tracheids and a thicker structure of cell walls (Ladejtschikova et al., 1973).

The following phase in research was to find out, what kind of structural substances determine the formation of thick cell walls in the wood of resistant trees.

When the investigations were outlined it was supposed that this phenomenon is connected with lignin. It turned out, however, that the content of lignin in the healthy roots of diseased trees may even be higher (Table 1). A correlation was found between the vigour of the tree and the content of hemicelluloses both in roots and in stem: trees with low resistance have a very low content. This kind of regularity is quite evident when the content is related to dry weight of wood and, especially, to a unit of cellulose. In the roots and stem of Scotch pine the content of the analysed substances in sapwood is not different from the content in the central part of the root.

It may be assumed that results of the chemical analysis of wood of the

Table 1. The content of structural substances in cells of the wood of roots and stems of trees in different vigour categories.

Category of tree vigour	Tree organ	Sapwood			Central part		
		cellulose	hemicellulose	lignin	cellulose	hemicellulose	lignin
In % of the dry weight of wood							
I	healthy roots	46.3	22.0	28.3	45.9	22.0	29.5
II	healthy roots	47.4	23.7	26.1	47.7	23.4	26.4
III	healthy roots	50.0	18.9	27.6	50.0	17.9	29.0
I	stem	47.6	23.9	26.0	44.7	23.8	28.3
II	stem	46.7	24.3	25.6	43.1	25.2	27.9
III	stem	47.6	21.3	27.1	44.7	21.7	29.0
Per unit of cellulose							
I	healthy roots	-	0.47	0.61	-	0.50	0.64
II	healthy roots	-	0.50	0.55	-	0.50	0.55
III	healthy roots	-	0.38	0.55	-	0.36	0.58
I	stem	-	0.50	0.55	-	0.53	0.63
II	stem	-	0.52	0.55	-	0.58	0.65
III	stem	-	0.45	0.57	-	0.49	0.65

central part of root and stem characterize the content of structural substances in the roots of 10-15 year old Scotch pine, and, furthermore, that wood of the root and stem was formed at this age before the infection of the tree by *F. annosus*. Apparently low content of hemicelluloses in some trees is one of the reasons for their low resistance to *F. annosus*.

Correlation between the resistance of wood to decay and the composition of extractives is confirmed by the results of experiments in vitro with wood, which was consistently treated with different solvents (Shain, 1967). As was already mentioned above the wood of resistant trees is decayed less than the wood of both healthy and diseased trees. When a group of extractives soluble in chloroform was separated from wood, the degree of decay of the wood in all categories of vigour was increased by 20-24 %, but the former regularity is still maintained: the wood of resistant trees was decayed less.

Further treatment of wood with acetone increased the degree of decay only of the resistant trees (by 6 %). Consequently, the great resistance of the wood of resistant trees is evidently related to a group of extractives soluble in acetone. Extractives soluble in chloroform produce a certain level of resistance in the wood of Scotch pine irrespective of the vigour of the tree. The greater degree of decay of the wood of diseased trees with low resistance is also related to a lower content of hemicellulose and lignin.

Composition of extractives in phloem tissue and their inhibitory effect may also be a criterion for resistance, as the results of a study on *F. annosus* showed (Alcubilla et al., 1971). However, the phloem tissue of Scotch pine, unlike that of spruce, did not have an inhibitory effect on the growth of *F. annosus* in pure culture (Table 2). This is due to the fact that the content of the main group of extractives having an inhibitory effect on the growth of *F. annosus* - phenol compounds - is approximately 5 times smaller in the phloem tissue of Scotch pine than in spruce, while the content of carbohydrates is practically equal. Now, the biological effect of the phloem tissue of the stems of resistant trees is the same as that of healthy trees, but the phloem tissue of diseased trees is a less favourable substrate for the growth of mycelium of the pathogen.

Table 2. The content of extractives (M ± m) in phloem tissue of the stems of Scotch pine and their effect on the growth of *F. annosus* depending on the vigour of trees (n=10). LSD: Least significant difference.

Index	In disease centre		In intermediate zone	LSD
	resistant	diseased	healthy	
The content of extractives soluble in:				
chloroform (%)	6.50±0.40	8.6±0.72	5.1±0.40	1.41
acetone (%)	3.0±0.58	4.3±0.42	2.8±0.33	1.24
The biomass of mycelium:				
a. in experiments with extracts:				
chloroform extracts (mg)	71.9±10.1	44.9±4.8	91.6±8.2	21.71
acetone extracts (mg)	63.0±8.6	67.5±7.7	87.8±8.5	20.09
b. in experiments with phloem tissue (mg)	162.0±10.0	128.0±9.7	151.0±9.4	27.33
Inhibitory effect of extractives from phloem tissue soluble in:				
chloroform (%)	51.6	65.0	39.4	
acetone (%)	61.1	47.3	41.9	

Chloroform and acetone extracts of the phloem tissue of Scotch pine contain substances which inhibit the growth of the pathogen. The inhibitory effect of chloroform extracts increases directly with the quantitative content; whereas the effect of acetone extracts depends on their qualitative composition. While the contents of extractives soluble in acetone are equal in the phloem tissue of resistant and healthy trees (3.0 % and 2.8 % respectively), the inhibitory effect on the growth of the mycelium of *F. annosus* was greater in the case of resistant trees (61.1 % versus 41.9 %). The pathological process leads to a remarkable increase in the content of the group of extractives soluble in acetone (4.3 %), but it is not followed by an increase of their inhibitory effect, which remained on the level of healthy trees and was much lower compared with resistant trees (47.3 %).

The chromotographic (solvents: H-butyl alcohol: acetic acid: water 4:1:5) division of acetone extracts from phloem tissue made it possible to detect the difference in the composition depending on the category of tree vigour: in resistant trees unidentified extractives were discovered with Rf values 0 - 0.40 and 0.55 which were lacking in healthy and diseased trees.

Experiments in our laboratory have established that analyses of phloem tissue give us information for drawing conclusions about the relation between provision with inorganic nutrients and vigour of trees in an infected stand (Pobegailo & Ladejtschikova, 1975). This regularity is also valid for the provision of Scotch pine with trace elements. Resistant trees are dis-

Table 3. The average content of basic monoterpenes and its variability in trees with different vigour.

Category of tree vigour	α-pinene	β-pinene	Δ-3-carene
In disease centre			
I high degree of resistance (n=1)	52.7	1.7	22.0
Ia resistant (n=15)	53.1	6.3	20.5
	36.2-77.6	1.0-26.2	2.6-38.6
III diseased (n=7)	50.3	4.7	28.5
	28.2-62.8	0.9-26.6	16.5-40.5
Beyond the disease centre			
II healthy (n=4)	45.4	6.2	31.8
	34.6-56.5	0.7-21.8	21.7-51.5
III initially diseased (n=5)	58.1	1.6	24.9
	33.4-76.7	0.1-2.7	5.0-55.9

tinguished by a markedly higher content of trace elements, such as Mn, B, and Cu (respectively 34 %, 32 %, and 33 %), whereas diseased trees have a lower content (respectively 52 %, 43 %, and 53 %) compared with healthy trees.

In genetic research with coniferous species - including that where resistance of Scotch pine to *F. annosus* is selected - great importance is attached to composition of turpentine oils (Table 3). Studies of the turpentine of 32 trees with different degrees of resistance and a complete digging up of roots made it possible to compare the composition of monoterpenes with the degree of disease (Ladejtschikova et al., 1975).

Among 15 resistant trees there was only 1 which was totally free of *F. annosus* (an example of a high level of resistance). The content of Δ-3-carene (a monoterpene to which resistance to the pathogen is related) in such a tree turned out to be lower (22 %) than what was observed in resistant trees having local signs of disease (38.6 %) or in diseased trees (40.5 %). Beyond the disease centre among a group of healthy trees, trees with signs of initial disease were discovered in which the content of Δ-3-carene reached its maximum value: 55.9 %.

CONCLUSIONS

Resistant trees, which are a product of natural selection in heavily infectious surroundings, are characterized by the following features distinguishing them from trees with low resistance (both diseased and not diseased): higher resistance of wood to decay in vitro; greater thickness of the walls of thracheids; higher content of hemicelluloses in wood per unit of cellulose; specific composition of extractives soluble in acetone, which are found in phloem tissue and wood, with a higher inhibitory effect on the growth of *F. annosus* in pure culture; and higher content of trace elements (e.g., Mn, B, Cu) in the phloem tissue of stems.

Obviously part of these traits are genetically predetermined. Investigations have been started in the field of genetic resistance.

REFERENCES

Alcubilla, M, M.P. Diaz-Palacio, K. Kreutzer, W. Laatsch, K.E. Rehfuess & G. Wenzel, 1971. Beziehungen zwischen dem Ernährungszustand der Fichte (Picea abies Karst.), ihrem Kernfäulebefall und der Pilzhemmwirkung des Basts. Eur. J. For. Path. 1: 100-114.

Ivanov, N.N., 1946. Methods of physiology and biochemistry of plants. State Publishing House of Agricultural Literature, Moscow-Leningrad. 494 p.*.

Laatsch, W., M. Alcubilla, G. Wenzel & H. von Aufsess, 1969. Beziehungen zwischen dem Standort und der Kernfäule-Disposition der Fichte (Picea abies Karst.). Forstwiss. Cbl. 87: 193-203.

* Titles translated to English by the authors.

Ladejtschikova, E.I., A.G. Tschernych & A.I. Pobegailo, 1973. Anatomical and chemical characteristics of pine roots wood conditioned by various degree of Fomes annosus injury. In: Silviculture and Afforestation Amelioration, 33, Urozhay, Kiev. p. 137-142*.
Ladejtschikova, E.I., A.I. Pobegailo, G.D. Byely, I.A. Alekseyev, A.G. Tschernych, G.M. Pasternak, A.G. Shekhovtsov & A.G. Korobchenko, 1975a. Reasons of predisposition of pineta on old arable lands to diseases. In: Prapor, Kharkov. p. 22-30*.
Ladejtschikova, E.I., A.I. Pobegailo, G.M. Pasternak, A.G. Korobchenko & L.F. Ladnych, 1975b. Estimation of Pinus silvestris morphological and physiological-biochemical indices for plant-breeding on resistance to Fomes annosus. In: Silviculture and Afforestation Amelioration, 40, Urozhay, Kiev. p. 3-15*.
Melikyan, N., 1959. Structural changes and accumulation of lignin depending on the conditions of the environment. Publishing House of Yerevan University, Yerevan. 200 p.*.
Naumov, N.A., 1936. Methods of microscopical researches in phytopathology. State Publishing House of Agricultural Literature, Moscow-Leningrad, 224 p.*.
Pasternak, G.M., 1975. Effect of Fomes annosus injury on nucleic acids content in pine phloem tissue. In: Silviculture and Afforestation Amelioration, 40, Urozhay, Kiev. p. 84-90*.
Pobegailo, A.I. & E.I. Ladejtschikova, 1975. A way of diagnostics of inorganic nutrients of stands, injured by Fomes annosus. Author's Certificate N 475546. USSR Council of Ministers State Committee Official Bulletin in Discoveries and Inventions, N 24. p. 90*.
Rypaček, V., 1957. Biologie dřeokazných hub. Nakladatelstvi Československe Akademie věd. Praha. 275 p.
Shain, L., 1967. Resistance of sapwood in stems of loblolly pine to infection by Fomes annosus. Phytopathology 57: 1034-1045.

Susceptibility of European black pine (Pinus nigra) to the European pine shoot moth (Rhyacionia buoliana): variations of susceptibility at the provenance and individual level of the pine and effect of terpene composition

P.J. Charles[1], A. Delplanque[1], Anne Marpeau[2], Colette Bernard-Dagan[2] & M. Arbez[3]

ABSTRACT

The number of attacks of shoot moth (*Rhyacionia buoliana*) on individual trees were noted during 5 consecutive years in a provenance trial of European black pine (*Pinus nigra*). Results show that great differences exist among provenances in the level of attacks. Undamaged branch samples were taken from 5 heavily-attacked and 5 non-attacked trees within each of 4 provenances. Two provenances of the subspecies *nigricans* were heavily attacked, and the other 2, from the Corsican group of the subspecies *laricio* and from the subspecies *pallasiana*, were slightly attacked. The cortical tissues of these branch samples were peeled, macerated in pentane, and the terpene concentrations measured by gas chromatography. The results show that it is difficult to distinguish between terpene profiles of heavily-attacked trees and non-attacked trees within very susceptible provenances, but that a clear difference seems to exist between heavily-attacked trees and non-attacked trees when the provenances are partially resistant. Judged only from a statistical relationship, it seems that phellandrene, 3-carene and camphene could play a role in the biochemical mechanism of resistance. If confirmed, such a result indicates the promising possibility of making indirect selections of black pines which would be less susceptible to the European shoot moth and consequently developing fewer forks.

1. INRA, Station de Recherches de Zoologie et de Biocoenotique forestière, Ardon, 45160 Olivet, France
2. Université de Bordeaux I, Laboratoire de Physiologie cellulaire végétale, 33405 Talence, France
3. INRA, Laboratoire d'Amélioration des Arbres forestiers, Pierroton 33610 Cestas, France

INTRODUCTION

In spite of large geographic variation, excellent straightness of the trunk is a typical characteristic of the European black pine. Nevertheless, and depending on the environmental conditions, this average growth habit can change and a certain amount of forking can appear. From results of some exploratory investigations, it seems that at least 2 main factors could explain the incidence of forked trees: the occasional occurrence of a short second terminal shoot when the location is well provided with water, and damage to the terminal bud caused by *Rhyacionia buoliana* Schiff.

Provisionally disregarding environmental effects, we shall consider the susceptibility of the black pine to the European shoot moth. In 1 plantation, the infestation rate by *Rhyacionia* varied among provenances (geographic origins) and among individuals within provenances (Charles, 1974; Mallet, 1975). Mallet has already noted significant differences among provenances of *Pinus nigra* Arn. from various geographical origins. These observations were made in a provenance trial located in the French southern Piedmont of the Alps.

For a given year of observations (1975) the total number of attacks per tree decreased from the subspecies *nigricans* (Laragne, a French introduced population, and Tzarvaritza, a natural Bulgarian one) to the Corsican branch of the subspecies *laricio* (Vizzavona, from the central part of Corsica) and the subspecies *pallasiana* (Mugla, a provenance from south-western Turkey). Such conclusions remained true when the incidence of damaged terminal shoots was considered in place of the total number of attacks per tree.

Besides this, we have also noted clear differences in *Rhyacionia*-susceptibility among individuals within provenances.

These preliminary observations were promising enough to encourage us to continue more intensive studies on the same provenance experiment during 5 consecutive years.

Previous research (Arbez et al., 1974) also has shown that large variation in the monoterpene composition of the cortical tissue exists among the different subspecies of *P. nigra*. Because of such conditions it seemed of interest to investigate whether a relationship between terpene composition and shoot moth susceptibility existed. Of course, such a relationship could be highly valuable in making indirect selections for shoot moth resistance.

EXPERIMENTAL SCHEME

Observations were carried on in the same provenance trial. Nine different provenances were involved in this test planted in the spring of 1971, 6 of them belonging to the subspecies *nigricans*, 2 belonging to the subspecies *laricio*, and 1 to the subspecies *pallasiana* (Table 1). The main design

Table 1. Description of the populations used in the provenance test.

Abreviations	Provenances	Country	Subspecies
VIZZ	Vizzavona	France (Corsica)	*laricio*
COSE	Cosenza	Italy (Calabria)	*laricio*
LEDE	Ledererkogel	Austria	*nigricans*
STUD II	Studenica II	Yugoslavia (Serbia)	*nigricans*
VGRA II	Visegrad II	Yugoslavia (Bosnia)	*nigricans*
TZAR	Tzarvaritza	Bulgaria	*nigricans*
POLV	Popova-Livada	Yugoslavia (Macedonia)	*nigricans*
LARA	Laragne	France (artificial stand)	*nigricans*
MUGL	Mugla	Turkey	*pallasiana*

was a balanced lattice with 4 replications and 56 trees per plot. The experiment is located in the Pelleng State Forest near Montmeyan (Var Prefecture); the elevation is 550 m, the aspect is flat, and the soil overlies a hard limestone. The planting is surrounded by a natural oak forest (*Quercus pubescens* Willd.). Trees had been attacked by the European shoot moth since 1972. Damage was important from 1973 to 1975 but decreased every year since 1973. The number of attacks per tree were counted as well as attack of the terminal shoot. Finally these observations were combined for the first 5 consecutive years.

With a view toward eventually establishing more precisely a relationship between terpene composition and shoot moth resistance, in 1978 we decided to choose a sample including 4 provenances: 2 heavily and regularly attacked (Tzarvaritza (TZAR) and Laragne (LARA)) and 2 with light damage (Vizzavona (VIZZ) and Mugla (MUGL)). In each provenance, we chose 10 individuals: 5 heavily and regularly attacked during the last 5 years and 5 not or only slightly attacked during the same period.

Unattacked 1-year-old branch samples were taken from each selected tree. After maceration and extraction in pentane of cortical tissues of these individual samples, the extracts were submitted to gas chromatography analysis.

The relative concentrations of 9 terpenes (expressed in percent of total volatile terpene hydrocarbons) were then compared by a hierarchical desing for analysis of variance (provenances and individuals within provenances) and a principal component analysis. The 9 terpenes involved in the gas chromatography study were: α- and β-pinene, camphene, myrcene, 3-carene, limonene, β-phellandrene, terpinolene and caryophyllene. The first 8 are monoterpenes, the last one is a sesquiterpene.

RESULTS

Provenance and individual variation in shoot moth susceptibility (5 years' performance)

Observations made during 5 consecutive years show that some trees sustained repeated attacks, whereas others were never attacked by the European shoot moth. The percentages of trees without attacks and those of trees heavily attacked are given in Table 2.

It appears that large variation in susceptibility occurred, depending on geographic origin of the pine. Differences among trees, as well as among provenances were large immediately after planting, when trees were still small. The 'silhouette effect', often used when moths are concerned to explain differences in female attraction during the egg laying process does not apply in our case. Such differences in susceptibility to *R. buoliana* have been observed previously in other pine species such as *P. contorta* (Esjberg & Feilberg, 1971).

Relationship between intensity of attack and terpene composition of the cortical tissues

In every provenance, non-attacked trees and heavily-attacked trees (4 to 5 times during 5 consecutive years) were analysed separately. Subpopulations with an odd number correspond to heavily attacked trees and those with an even number to non-attacked trees (Fig. 1).

Terpene compositions of cortical tissues (percent of total volatile terpenes) for the 4 provenances and the 8 subpopulations are presented in

Table 2. Percentages of trees never attacked or heavily attacked depending on provenance level of susceptibility.

Provenances	Trees never attacked during a 5-year period	Trees attacked 3 or 4 times during a 5-year period	Trees attacked 5 times during a 5-year period
MUGL	21	1	0
VIZZ	30	4	0.5
COSE	3	4	0.5
POLV	8	9	1
STUD II	15	9	2
LEDE	10	17	2
VGRA II	7	21	3
LARA	5	25	1.5
TZAR	4	25	2

Table 3. Terpene composition of cortical tissue of 4 provenances and 8 subpopulations. A = heavily attacked trees, NA = non-attacked trees.

	α-pin.	camph.	β-pin.	myrc.	3-car.	limon.	phell.	terpi.	cary.
provenances									
LARA (I)	72.59	1.17	10.47	2.17	0.20	13.05	0.24	0.20	3.77
TZAR (II)	63.81	1.19	5.82	2.98	0.43	25.50	0.15	0.23	3.86
MUGL (III)	72.15	1.35	15.10	2.95	0.36	7.05	0.63	0.27	7.90
VIZZ (IV)	37.86	0.84	3.16	5.88	0.26	42.58	6.53	0.89	1.23
subpopulations									
LARA 1 (A)	78.49	1.18	3.80	2.24	0.25	13.91	0.12	0.16	5.37
LARA 2 (NA)	66.68	1.16	17.13	2.09	0.15	12.19	0.36	0.25	2.17
TZAR 3 (A)	66.99	1.17	7.50	2.87	0.40	20.86	0.10	0.22	2.26
TZAR 4 (NA)	59.04	1.23	3.31	3.14	0.46	32.46	0.23	0.25	6.25
MUGL 5 (A)	66.47	1.05	19.64	2.85	0.62	8.59	0.56	0.25	9.17
MUGL 6 (NA)	77.83	1.64	10.57	3.06	0.11	5.50	0.70	0.28	6.62
VIZZ 7 (A)	26.45	0.59	3.19	5.69	0.34	58.30	3.40	0.80	0.97
VIZZ 8 (NA)	49.27	1.10	3.13	6.07	0.18	26.85	9.66	0.97	1.49

Table 3.

At a provenance level, the results are in good agreement with the preliminary conclusions of Arbez et al. (1974). Provenances belonging to the subspecies *nigricans* and *pallasiana* were similar, with a relatively high level of α- and β-pinene, a small amount of myrcene and a relatively scant amount of limonene. The Corsican pine belonging to the subspecies *laricio* was characterized by a low level of pinenes, a typically high amount of limonene and phellandrene and a moderately high level of myrcene.

In consideration of these results and the quite different terpene compositions which occurred for the 2 highly resistant provenances, Vizzavona and Mugla, it appears that the single point of similarity in terpene composition was the higher amount of phellandrene. That was also the case for terpinolene concentration but the differences between resistant and susceptible provenances were smaller. In any case, only the provenance Vizzavona

	PROV I		PROV II		PROV III		PROV IV	
	LARA		TZAR		MUGL		VIZZ	
Subpopulation:	1(A)	2(NA)	3(A)	4(NA)	5(A)	6(NA)	7(A)	8(NA)

Figure 1. Experimental design. A = heavily-attacked trees, NA = non-attacked trees.

was significantly different from the others.

At the subpopulation level, it appears that within the same provenance the non-attacked individuals had the greater amount of phellandrene; although the differences in concentrations between heavily-attacked individuals and non-attacked individuals were sometimes slight, this relationship was consistent.

Analysis of the correlation matrix with provenance means of terpene concentrations shows that myrcene, limonene, phellandrene and terpinolene concentrations were highly correlated.

A principal component analysis was performed with the means of the 8 subpopulations.

The first 2 principal components explain 78 % of the total variance.

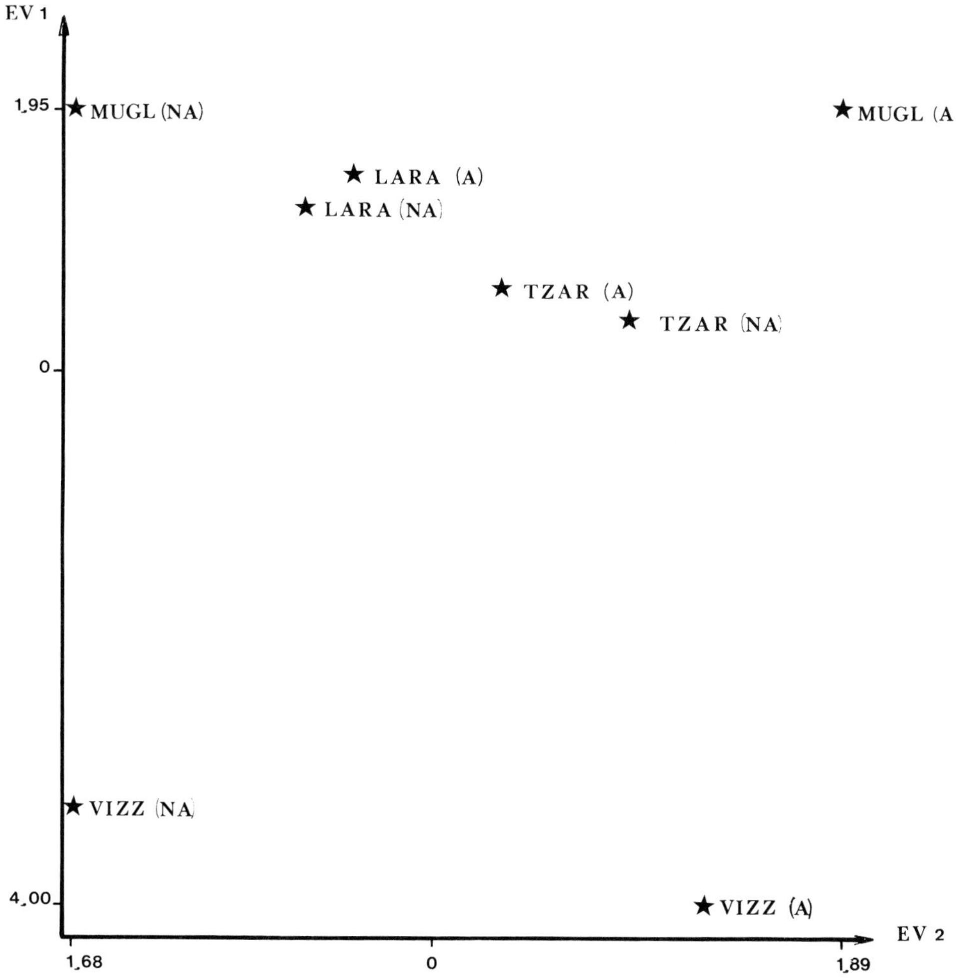

Figure 2. Projections of points of attacked (A) and non-attacked (NA) groups of individuals from the 4 provenances: eigen vectors EV_1 en EV_2.

Except for 3-carene, almost all the other terpenes involved in this study were correlated with the first principal component, but mainly α-pinene positively, and myrcene, limonene and terpinolene negatively.

The third principal component is mainly correlated with β-pinene and phellandrene concentrations. The graphical representation of the 8 subpopulations in the first 2 plans of the eigen vector EV_1, and EV_2 is shown in Fig. 2.

The most striking fact is that the differences between attacked and non-attacked individuals are clear mainly for the slightly-attacked provenances Vizzavona and Mugla. In Fig. 2, such differences can be seen only on the second eigen vector EV_2 (3-carene+, camphene-). This is also the case with the third eigen vector EV_3 (β-pinene+, phellandrene+).

If we disregard the details, it seems that attacked and non-attacked individuals have similar terpene profiles within highly-susceptible provenances, but within slightly-attacked provenances their terpene compositions are clearly different.

This could mean that, in explanation of differences between individuals, terpenes (directly or through some biochemical precursors) can be a sufficient barrier against *Rhyacionia* attacks only when the resistance level of the population is already high. This leads us to suppose that terpenes act probably in association with more important factors of resistance.

If confirmed with different provenances, in different environments, such results could make possible indirect selection for Corsican pine that would be less sensitive to the European shoot moth, with a subsequent decrease in the incidence of forking.

REFERENCES

Arbez, M., C. Bernard-Dagan & C. Fillon, 1974. Variabilité intraspécifique des monoterpènes de Pinus nigra Arn. Bilan des premiers résultats. Annales des Sciences forestières 31: 57-70.

Charles, P.J., 1974. Incidence des attaques de la tordeuse des pousses Rhyacionia (Evetria) buoliana Schiff. sur le choix des espèces de pins suceptibles d'être plantées en région méditerranéenne. Comte rendu IVèmes journées Phytiatrie-Phytopharmacie Circum-méditerranéennes, Montpellier. p. 468-471.

Esbjerg, P. & L. Feilberg, 1971. Infestation level of the European pine shoot moth (Rhyacionia buoliana Schiff.) on some provenances of lodgepole pine (Pinus contorta Loud.). Danish Forest Experiment Station Report no. 254, 32 (4): 345-358.

Mallet, B., 1975. Contribution à l'étude de la variabilité infraspécifique de la sensibilité à Rhyacionia buoliana Schiff. chez Pinus ponderosa et Pinus nigra. Rapport D.E.A. (limited distribution).

Genetic basis for variation in fungi

C.E. Caten

Department of Genetics, University of Birmingham, Birmingham B15 2TT, UK

ABSTRACT

Fungal species are highly variable in respect of both qualitative and quantitative characters. Most of this phenotypic variation is genetically determined and originates from nuclear-gene mutations and from the repeated shuffling of these into new combinations by heterokaryosis and sexual and parasexual hybridization. Changes in the mitochondrial genome and the effects of mycoviruses may also produce variation. All these processes occur in nature but the contribution of each to the observed variation is difficult to assess and probably variable. Among the factors involved, incompatibility systems acting at sexual reproduction or between vegetative cells are particularly important in controlling the flow of genetic determinants between strains. The variability of fungal species extends to their pathogenicity, which is a complex character with qualitative and quantitative components. The qualitative component (virulence) is primarily involved in host specificity and is controlled by major genes, while the quantitative component (aggressiveness) is polygenically determined.

INTRODUCTION

The genetic basis for variation in fungi encompasses much of fungal genetics and cannot be comprehensively treated in the space available. This paper concentrates on 3 topics: (i) the variability of fungal populations, (ii) the origin of variation and (iii) the genetic control of pathogenicity. For basic information on the genetics of fungi and for discussion of other aspects of fungal variation and pathogenicity the reader is referred to the texts by Day (1974), Burnett (1975) and Fincham et al. (1979).

VARIABILITY OF FUNGAL POPULATIONS

Variation may be conveniently divided into qualitative and quantitative. With the former, the differences are discrete and any individual can be unambiguously assigned to a particular class, while with the latter there is a continuous gradient of types and discrete classes cannot be recognised. Physiologic specialization provides the best documented example of qualitative variation in fungi. Isolates are classified into discrete races on the basis of their reaction with host cultivars carrying different sources of resistance and pathogen populations may be polymorphic with respect to their racial composition. Likewise, Spieth (1975) showed that populations of *Neurospora intermedia* Tai are frequently polymorphic for electrophoretically-distinct forms of individual proteins. The presence of quantitative variation in fungal populations is also widely appreciated. Brasier (1970) found that the growth rate of 77 dikaryons of *Schizophyllum commune* Fr. sampled from a single locality varied continuously and that the isolates could not be placed into discrete classes on this basis. Extensive quantitative variation occurs naturally for a range of other biologically important characters including fruiting (Brasier, 1970), fungicide resistance (Hollomon, 1978) and pathogenity (Clifford & Clothier, 1974). These comparative studies of isolates reveal only the phenotypic variation present in natural populations. However, many cases of qualitative and quantitative phenotypic variation have been shown to have a genetic basis (e.g., Flor, 1956; Croft & Jinks, 1977; Caten, 1979) and it is reasonable to conclude that this is generally the case.

ORIGIN OF VARIATION

Genetic variation is due to the presence of 2 or more different forms (alleles) of a gene in a population. Since even such simple organisms as fungi contain sufficient nuclear DNA for more than 10 000 genes it is clear that a large number of different genotypes are possible, even allowing for the fact that not all genes are polymorphic. Electrophoretic surveys of a range of organisms have shown that a high, although somewhat variable, proportion of genes is naturally polymorphic (Nevo, 1978). These different alleles arise by mutation which is therefore the ultimate basis for the genetic variation in fungi and all other organisms. However, mutation is a rare event and the contribution of new mutants to the variability of a population in any one generation is very small. Most of the observed variation results from prior mutations and from the repeated shuffling of these into new combinations. Laboratory studies have shown that in fungi reassociation and recombination of nuclear genes occurs through heterokaryosis, sexual hybridization and parasexual hybridization. The mechanisms underlying these processes are well understood in a few species (Burnett, 1975, 1976;

Fincham et al., 1979) but their contribution to the variability of fungal populations is far from clear. In addition to the nuclear system, fungi, in common with other eukaryotes, possess a second genetic system located in the cytoplasm. This cytoplasmic system does not conform to the laws of Mendelian genetics that govern the nuclear system. Nevertheless, cytoplasmic genes are subject to mutation and may be an additional source of variation.

I have identified mutation, heterokaryosis, sexual and parasexual hybridization and cytoplasmic inheritance as potential sources of genetic variation in fungi. To what extent do these processes occur in nature and what are their contributions to the observed variability of fungal populations? No definitive answers to these questions are possible but I aim to identify some of the major factors involved, to point out some of the gaps in our knowledge and to draw a few tentative conclusions.

Mutation

Several factors suggest that current mutation is of more significance for the variation and evolution of fungi than it is for higher organisms. Firstly, the dominant nuclear phase of most fungi is haploid and hence new mutations are expressed once they are segregated into a spore or hypha. Secondly, although spontaneous mutation rates are very low the large size of many fungal populations ensures that mutant alleles arise continuously. Slootmaker (cited by Day, 1974) calculated that a hectare of barley infected with powdery mildew may produce 10^{13} conidia per day. Even with a conservative estimate of the mutation rate to virulence to a particular resistance gene of 10^{-8}, 10^{5} virulent conidia may be produced per hectare per day and double mutants virulent to 2 resistance genes are a real possibility (Person et al., 1976). The frequency of these virulent alleles in the population may be very low (too low to be detected) but their practical significance is enormous because of the large selective advantage they have on resistant hosts. Thirdly, the high asexual multiplication rates of many fungi ensures that an initially rare allele can under selection increase in number and frequency very rapidly.

Heterokaryosis

The role of heterokaryosis in the variability of fungi has been the subject of considerable speculation (e.g., Snyder, 1961). However, knowledge of heterokaryosis derives largely from laboratory studies utilizing rather artificial techniques and confirmed examples of natural heterokaryons are rare (Caten & Jinks, 1966). (The dikaryons of heterothallic Basidiomycetes are invariably heterokaryotic but constitute a special case as part of the sexual cycle of these fungi.) The natural significance of heterokaryosis will depend not only upon its occurrence but also upon the diversity of the associated genomes. Attempts to form heterokaryons between independent isolates revealed the presence of heterokaryon (vegetative) incompatibility systems in many species (Caten & Jinks, 1966; Esser & Blaich, 1973). Strains carrying different alleles at one or more specific heterokaryon-incompatibility genes (*het*) are restricted or blocked in heterokaryon formation (Esser & Blaich, 1973; Croft & Jinks, 1977). Consequently heterokaryons form readily only between strains of similar genotype. Rayner & Todd (1979) have recently shown that nuclear exchange between dikaryons of wood-decaying Basidiomycetes is blocked by a similar incompatibility system which they term intraspecific antagonism. These considerations of the incidence of heterokaryosis and its genetical control suggest that its natural significance is less than has frequently been assumed in the past.

Sexual hybridization

Where it occurs sexual hybridization is undoubtedly the major source of variability in fungal populations. In rusts, many different races may be produced by crossing 2 contrasting parent races (Flor, 1955). With quantitative characters even hybridization of phenotypically similar strains may give a wide spectrum of variation (e.g. Fig. 1 in Caten & Jinks, 1976). Despite the obvious effect of sexual hybridization in these examples it should not be forgotten that many fungal species are asexual and furthermore that the sexual stage of certain perfect species occurs rarely or not at all in nature.

The amount of variation released by hybridization depends upon the diversity of the interacting genomes which is regulated by 2 independent and opposing sexual incompatibility systems, termed homogenic and heterogenic incompatibility (Esser, 1971). Homogenic incompatibility is the control exerted by the mating-type genes and encourages outbreeding by blocking matings between individuals of like genotype. The occurrence, genetic control and biological significance of homogenic incompatibility in fungi is widely appreciated (Raper, 1966; Burnett, 1976; Fincham et al., 1979) and will not be considered further here. Heterogenic incompatibility, in contrast, stops matings between strains of different compatibility genotype, thereby reducing outbreeding and the release of genetic variation. Knowledge of heterogenic incompatibility derives largely from investigations by Esser and associates into mating between races of *Podospora anserina* (Ces.) Niessl. where, in 1 case analysed in detail, compatibility/incompatibility was determined by complex interactions between 4 unlinked genes (Esser & Blaich, 1973). The homothallic Ascomycete, *Aspergillus amstelodami* (Mangin) Thom & Church, provides a second example (Caten, unpublished). In homothallic species the genetic significance of sex depends upon the frequency of hybridization as opposed to selfing. Genetic control of this character was sought by comparing the proportion of hybrid asci (outcrossing index) in mixed cultures of related und unrelated strains. Although all the strains hybridized, related strains did so more

Tabel 1. Heterokaryon formation and outcrossing between related and unrelated strains of *Aspergillus amstelodami*.
+ = heterokaryon compatible, - = heterokaryon incompatible. The figures indicate the proportion of hybrid asci among total asci. Strains with the same number are related, strain 13w is related to 9b.

		White-spored strains				
		9w	37w	115w	164w	13w
Brown-spored strains	9b	+ 0.24	- 0.03	- 0.01	- 0.01	+ 0.22
	37b	- 0.06	+ 0.24	- 0.01	- 0.05	- 0.07
	115b	- 0.01	- 0.01	+ 0.10	- 0.04	- 0.01
	164b	- 0.01	- 0.06	- 0.12	+ 0.33	- 0.01

readily than unrelated giving a higher outcrossing index (Table 1). The related strains were all heterokaryon-compatible while the unrelated strains were incompatible, implying that low hybridization is related to heterokaryon incompatibility. This was investigated by determining the outcrossing index for pairs of strains differing at known *het* genes. Of 5 *het* genes tested, 1 (*hetB*) gave a 5-fold reduction in outcrossing when heterozygous while the others had no effect. Clearly in *A. amstelodami* some, but not all, *het* genes also act as heterogenic sexual incompatibility factors.

Parasexual hybridization

The occurrence in fungi of recombination by processes other than sexual reproduction was first clearly demonstrated in *Aspergillus nidulans* (Eidam) Wint. by Pontecorvo and associates (Pontecorvo, 1956). The sequence of events was termed the parasexual cycle and is now extensively documented and widely appreciated (for review see Roper, 1966; Caten, 1981). More recently it has been shown that parasexual recombination in dikaryons of heterothallic Basidiomycetes can occur through processes other than the parasexual cycle (Leonard et al., 1978; Day, 1978; Frankel, 1979). At least one of these processes, meiotic-like recombination, produces both inter- and intra-chromosomal recombinants at a frequency much higher than expected from the parasexual cycle (Frankel, 1979).

Evidence that parasexuality occurs outside the laboratory has accumulated over recent years. Somatic diploid strains have been isolated from nature in several habitually-haploid species (Caten & Day, 1977; Caten, 1981) and new genotypes of rusts and smuts have been shown to arise in mixtures of dikaryons under near-natural conditions (Bartos et al., 1969; Megginson & Person, 1974). The origin of these variants is unclear, although diploid clones of *Puccinica graminis* Pers f. sp. *tritici* Eriks. & E. Henn arise in axenic cultures (Maclean et al., 1974) and produce variants (Green et al., 1978). Even though parasexual processes occur naturally their contribution to variation may be reduced by heterokaryon incompatibility since this restricts nuclear association to strains of similar genotype. It is significant in this context that many naturally-occurring somatic diploids appear homozygous (Caten, 1981). Further work is required to establish how common somatic diploids are in nature and the extent and nature of the genetic variation they carry.

Cytoplasmic inheritance

The cytoplasm is important for the variability of fungi in 2 respects: (i) the presence of a genetic system in the mitochondria and (ii) the occurrence of stable double-stranded RNA (ds RNA) molecules with genetic activity.

Eukaryotic mitochondria carry DNA genomes which can undergo mutation, recombination and segregation in manners analogous to their nuclear counterparts (Gilham, 1978). The number of genes in the mitochondrial genome is limited (approximately 50 in yeast) but they are essential for the functioning of the mitochondria. The mitochondrial genetic diversity present in natural populations remains to be established but analyses with restriction enzymes have revealed sequence heterogeneity within species (Sanders et al., 1977). It is particularly interesting that 2 long-recognised degenerative conditions in fungi, senescence and vegetative death, have recently been shown to result from gross changes in the mitochondrial genome (Cummings et al., 1979; Lazarus et al., 1980).

Double-stranded RNA molecules have been detected in many fungal species and they are believed to reflect mycovirus infections, although virus-like particles are not always visible (Ghabrial, 1980). Not all strains within a species carry any or the same ds RNA molecules and there may or may not be phenotypic differences associated with the various complements. Double-stranded RNA determines the production of, and sensitivity to, the toxins responsible for the 'killer' phenotype in *Ustilago maydis* (DC.) Cda. and is almost certainly responsible for the hypovirulent phenotype in *Endothia parasitica* (Murr.) P.J. & H.W. And. (Day & Dodds, 1979). These phenotypic effects raise the possibility of using mycoviruses to control fungal pathogens. Since cell-free mycovirus preparations are unable to infect intact fungal cells the normal route of transmission is via hyphal anastomosis and will be affected by those factors which control hyphal fusion and heterokaryon formation (Ghabrial, 1980). For example, transmission of hypovirulence in *Endothia* is restricted by heterokaryon incompatibility but sufficient infection occurs to give curing of cankers following inoculation with hypovirulent strains (Day & Dodds, 1979).

GENETIC CONTROL OF PATHOGENICITY

When a single host genotype is inoculated with independent isolates of a phytopathogenic fungus and the level of disease is assessed quantitatively it is typically found that the isolates differ in pathogenicity and that this variation is continuous ranging from isolates which give no obvious symptoms to others which produce severe disease (Schwarzbach & Wolfe, 1976; Fig. 1). On many host genotypes there is an excess of isolates which give only low levels of disease implying that the overall picture reflects 2 overlapping distributions, one qualitative and the other quantitative. In considering the genetic control of pathogenicity it is conventional to distinguish between these qualitative and quantitative components, although this distinction may not always be clear in practice. The low-disease producing isolates are generally classified as avirulent to that host genotype in contrast to the virulent isolates which produce significant levels of disease (Fig. 1). This classification provides the qualitative component of

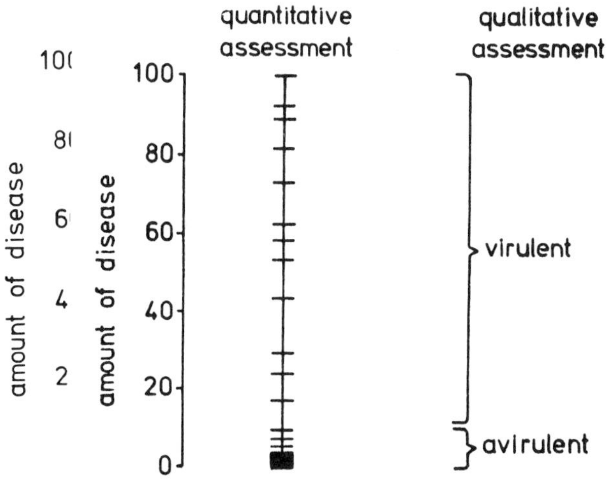

Figure 1. Assessment of the variation in pathogenicity shown by independent isolates inoculated onto a single host cultivar. Each horizontal bar represents the amount of disease produced by an individual isolate.

pathogenic variation and will be referred to as virulence. It is particularly important in denoting the host range of pathogen isolates. However, as Fig. 1 shows, not all virulent isolates produce the same level of disease and this represents the quantitative component of pathogenic variation which will be referred to as aggressiveness.

Virulence

The classical studies of Flor (1942, 1956) with flax rust established that avirulence/virulence to a particular resistance gene was controlled by alternate alleles at a corresponding major gene in the pathogen. This one-for-one relationship applied to several resistance genes, such that for every gene determining resistance in the host there was a specific complementary gene determining virulence in the pathogen. It has subsequently been demonstrated in several other host-parasite systems (Person & Sidhu, 1971; Day, 1974) and is formalized as the gene-for-gene hypothesis. However, Lawrence et al. (1981a) have recently found that pathogenicity of flax rust to each of the M^1, L^1, L^7, L^{10} and L^x resistance alleles is determined by 2 interacting pathogen genes, indicating that, although of wide validity, the gene-for-gene relationship does not fully account for all qualitative host-parasite interactions. The avirulent and resistant alleles in pathogen and host respectively are generally dominant over their alternate alleles (Person & Sidhu, 1971) and are thought to be the physiologically-active forms of these genes, i.e. host-pathogen incompatibility is the active response (Ellingboe, 1978). In a number of cases, resistance factors against the same pathogen are closely linked and may be allelic (Shepherd & Mayo, 1972). The little information available on the chromosomal location of the pathogen genes indicates that they are not clustered, although in flax rust the avirulence determinants to the P^1, P^2 and P^3 resistance alleles are tightly linked and may be allelic (Lawrence et al., 1981b).

Aggressiveness

The genetic control of aggressiveness has rarely been examined despite the fact that this component may determine the severity of an epidemic or whether a newly-arisen race spreads or not. As a quantitative character aggressiveness would be expected to be polygenically determined and the few analysed examples indicate that this is the case. The enhanced pathogenicity of the aggressive strain of *Ceratocystis ulmi* (Buism.) C. Moreau is determined by an unknown number of genes which interact in a complementary fashion (Brasier, 1977). Strains of *Ustilago hordei* (Pers.) Lagerh. differ in the % infection they produce on compatible barley cultivars. This variation in aggressiveness is determined by a polygenic system (Emara & Sidhu, 1974) and, at least in one strain, the genes involved show additive, dominance and epistatic effects (Caten, Groth, Person and Dhahi, unpublished).

CONCLUDING REMARKS

Fungi possess a variety of mechanisms which can generate, store and release genetic variation and it is not surprising that natural populations are highly variable. The relative importance in nature of these different mechanisms is not clear and is likely to vary between species, between localities and over time, precluding any general conclusions. The mechanisms are themselves subject to genetic control and may evolve to higher or lower efficiency in response to selection pressures and we might expect related species inhabiting contrasting environments to utilize different mechanisms and to exhibit different patterns of variation. Many agricultural chemicals have genetic effects on fungi (Kappas et al., 1974; Bignami, 1977) and they may enhance the activity of the genetic mechanisms and thereby increase the general variability of fungal populations. Just as important as the genetic mechanisms are the selection pressures operating in nature. These will markedly influence the fate of new mutations and the rate of change of gene and genotype frequencies. Clearly the host population is the major selective force affecting pathogen populations, but many detailed questions remain to be answered. For example, are virulent and avirulent alleles equally fit in the absence of the matching host resistance and what will be the response of pathogens to heterogeneous host populations? It is hoped that study of the population genetics of pathogens will lead both to a better understanding of host-parasite interactions and to the development of practical control measures (Person et al., 1976; Wolfe & Barrett, 1980).

REFERENCES

Bartos, P., G. Fleischmann, D.J. Samborski & W.A. Shipton, 1969. Studies on asexual variation in the virulence of oat rust, Puccinia coronata f. sp. avanae and wheat leaf rust, Puccinia recondita. Canadian Journal of Botany 47: 1383-1387.

Bignami, M., 1977. Mutagenic and recombinogenic action of pesticides in Aspergillus nidulans. Mutation Research 46: 395-402.

Brasier, C.M., 1970. Variation in a natural population of Schizophyllum commune. American Naturalist 104: 191-204.

Brasier, C.M., 1977. Inheritance of pathogenicity and cultural characters in Ceratocystis ulmi; hybridization of protoperithecial and non-aggressive strains. Transactions of the British Mycological Society 68: 45-52.

Burnett, J.H., 1975. Mycogenetics. Wiley, London. 375 p.

Burnett, J.H., 1976. Fundamentals of mycology. 2nd ed. Edward Arnold, London. 673 p.

Caten, C.E., 1979. Quantitative genetic variation in fungi. In: J.N. Thompson, Jr. & J.M. Thoday (Eds): Quantitative genetic variation. Academic Press, New York. p. 35-59.

Caten, C.E., 1981. Parasexual processes in fungi. In: K. Gull & S. Oliver (Eds): The fungal nucleus. Cambridge University Press, Cambridge (in press).

Caten, C.E. & A.W. Day, 1977. Diploidy in plant pathogenic fungi. Annual Review of Phytopathology 15: 295-318.

Caten, C.E. & J.L. Jinks, 1966. Heterokaryosis: Its significance in wild homothallic Ascomycetes and Fungi imperfecti. Transactions of the British Mycological Society 49: 81-93.

Caten, C.E. & J.L. Jinks, 1976. Quantitative genetics. In: K.D. Macdonald (Ed.): Second international symposium on the genetics of industrial micro-organisms. Academic Press, London. p. 93-111.

Clifford, B.C. & R.B. Clothier, 1974. Physiologic specialization of Puccinia hordei on barley hosts with non-hypersensitive resistance. Transactions of the British Mycological Society 63: 421-430.
Croft, J.H. & J.L. Jinks, 1977. Aspects of the population genetics of Aspergillus nidulans. In: J.E. Smith & J.A. Pateman (Eds): Genetics and physiology of Aspergillus. Academic Press, London. p. 339-360.
Cummings, D.J., L. Belcour & C. Grandchamp, 1979. Mitochondrial DNA from Podospora anserina II. Properties of mutant DNA and multimeric circular DNA from senescent cultures. Molecular and General Genetics 171: 239-250.
Day, A.W., 1978. Chromosome transfer in dikaryons of a smut fungus. Nature 273: 753-755.
Day, P.R., 1974. Genetics of host-parasite interaction. Freeman, San Francisco. 238 p.
Day, P.R. & J.A. Dodds, 1979. Viruses of plant pathogenic fungi. In: P.A. Lemke (Ed.): Viruses and plasmids in fungi. Marcel Dekker, New York. p. 201-238.
Ellingboe, A.H., 1978. A genetic analysis of host-parasite interactions. In: D.M. Spencer (Ed.): The powdery mildews. Academic Press, London. p. 159-181.
Emara, Y.A. & G. Sidhu, 1974. Polygenic inheritance of aggressiveness in Ustilago hordei. Heredity 32: 219-224.
Esser, K., 1971. Breeding systems in fungi and their significance for genetic recombination. Molecular and General Genetics 110: 86-100.
Esser, K. & R. Blaich, 1973. Heterogenic incompatibility in plants and animals. Advances in Genetics 17: 107-152.
Fincham, J.R.S., P.R. Day & A. Radford, 1979. Fungal genetics. 4th ed. Blackwells, Oxford. 636 p.
Flor, H.H., 1942. Inheritance of pathogenicity in Melampsora lini. Phytopathology 32: 653-669.
Flor, H.H., 1955. Host-parasite interaction in flax rust - its genetics and other implications. Phytopathology 45: 680-685.
Flor, H.H., 1956. The complementary genic systems in flax and flax rust. Advances in Genetics 8: 29-54.
Frankel, C., 1979. Meiotic-like recombination in vegetative dikaryons of Schizophyllum commune. Genetics 92: 1121-1126.
Ghabrial, S.A., 1980. Effects of fungal viruses on their hosts. Annual Review of Phytopathology 18: 441-461.
Gillham, N.W., 1978. Organelle heredity. Raven Press, New York. 602 p.
Green, G.J., P.G. Williams & D.J. Maclean, 1978. Variability of uredial cultures of Puccinia graminis tritici derived from vegetative axenic mycelium. Canadian Journal of Botany 56: 855-861.
Hollomon, D.W., 1978. Competitive ability and ethirimol sensitivity in strains of barley powdery mildew. Annals of Applied Biology 90: 195-204.
Kappas, A., S.G. Georgopoulos & A.C. Hastie, 1974. On the genetic activity of benzimi-dazole and thiophanate fungicides on diploid Aspergillus nidulans. Mutation Research 26: 17-27.
Lawrence, G.J., G.M.E. Mayo & K.W. Shepherd, 1981a. Interactions between genes controlling pathogenicity in the flax rust fungus. Phytopathology 71: 12-19.
Lawrence, G.J., K.W. Shepherd & G.M.E. Mayo, 1981b. Fine structure of genes controlling pathogenicity in flax rust, Melampsora lini. Heredity 46 (in press).
Lazarus, C.M., A.J. Earl, G. Turner & H. Küntzel, 1980. Amplification of a mitochondrial DNA sequence in the cytoplasmically inherited 'ragged' mutant of Aspergillus amstelodami. European Journal of Biochemistry 106: 633-641.
Leonard, T.J., R.F. Gaber & S. Dick, 1978. Internuclear genetic transfer in dikaryons of Schizophyllum commune. II. Direct recovery and analysis of recombinant nuclei. Genetics 89: 685-693.
Maclean, D.J., I.C. Tommerup & K.J. Scott, 1974. Genetic status of monokaryotic variants of the wheat stem rust fungus isolated from axenic culture. Journal of General Microbiology 84: 364-378.
Megginson, F.G.A. & C.O. Person, 1974. Somatic recombination in Ustilago hordei during the parasitic phase on barley. Canadian Journal of Genetics and Cytology 16: 851-855.
Nevo, E., 1978. Genetic variation in natural populations: patterns and theory. Theoretical Population Biology 13: 121-177.
Person, C. & G. Sidhu, 1971. Genetics of host-parasite interactions. In: Mutation breeding for disease resistance. International Atomic Energy Agency, Vienna. p. 31-38.
Person, C., J.V. Groth & O.M. Mylyk, 1976. Genetic change at the populational level in host-parasite systems. Annual Review of Phytopathology 14: 177-188.
Pontecorvo, G., 1956. The parasexual cycle. Annual Review of Microbiology 10: 393-400.

Raper, J.R., 1966. Life cycles, basic patterns of sexuality and sexual mechanisms. In: G.C. Ainsworth & A.S. Sussman (Eds): The fungi, volume II. Academic Press, New York. p. 473-511.

Rayner, A.D.M. & N.K. Todd, 1979. Population and community structure and dynamics of fungi in decaying wood. Advances in Botanical Research 7: 333-420.

Roper, J.A. 1966. The parasexual cycle. In: G.C. Ainsworth & A.S. Sussman (Eds): The fungi, volume II. Academic Press, New York. p. 589-617.

Sanders, J.P.M., C. Heyting, M.P. Verbeet, F.C.P.W. Meylink & P. Borst, 1977. The organisation of genes in yeast mitochondrial DNA. III. Comparison of the physical maps of mitochondrial DNAs from three wild-type Saccharomyces strains. Molecular and General Genetics 157: 239-261.

Schwarzbach, E. & M.S. Wolfe, 1976. Consideration of the relationship between Erysiphe graminis f. sp. hordei and its host, barley, in the exploitation of host resistance. In: H. Gaul (Ed.): Barley genetics III. Karl Thiemig, Munich. p. 426-432.

Shepherd, K.W. & G.M.E. Mayo, 1972. Genes conferring specific plant disease resistance. Science 175: 375-380.

Snyder, W.C., 1961. Heterokaryosis as a natural phenomenon. In: Recent advances in botany, vol. I. Proceedings of the 9th international botanical congress, Montreal, 1959. University of Toronto Press, Toronto. p. 371-374.

Spieth, P.T., 1975. Population genetics of allozyme variation in Neurospora intermedia. Genetics 80: 785-805.

Wolfe, M.S. & J.A. Barrett, 1980. Can we lead the pathogen astray? Plant Disease 64: 148-155.

Genetics of pathogenicity in Ceratocystis ulmi and its significance for elm breeding

C.M. Brasier

Forest Research Station. Alice Holt Lodge, Farnham, Surrey, UK

ABSTRACT

The relationship between research on variation in *Ceratocystis ulmi* and elm breeding is discussed. *C. ulmi* falls into 2 major sub-groups, a highly pathogenic 'aggressive strain' and a more weakly pathogenic 'non-aggressive strain'. The 2 strains also differ in most other major characteristics including colony morphology, growth rate, optimum temperature for growth and mating system. Each strain shows a wide range of variation. The 2 strains show genetic isolation and probably do not freely interbreed. Laboratory forced crosses between the aggressive and non-aggressive strains indicate that the genetic systems governing their pathogenic abilities are complex, polygenic and qualitatively different. The aggressive strain divides into 2 further sub-groups, the Eurasian and North American races, which differ in a number of characteristics including colony morphology and pathogenicity. They show a degree of reproductive isolation, and a fairly discrete geographical distribution. The terms 'strain' and 'race' as used in *C. ulmi* are discussed and defined.

INTRODUCTION

The elm is one of the most important shelter and amenity trees in the Northern Hemisphere, comprising some 30 taxa extending from North America through Europe and Western Asia eastwards via the Himalayas and the central Asian uplands to China and Japan. Broadly speaking, North American elms tend to include the most susceptible species to Dutch elm disease, the Asiatic elms the most resistant species, and many of the European elms lie somewhere in between.

Currently, 3 main centres at Wisconsin USA, Wageningen the Netherlands and Volgograd USSR are carrying out the breeding and selection of elms. Dutch elm disease is by far the biggest threat to the products of these breeding programmes, and all 3 centres are attempting to incorporate disease resistance from highly resistant Asiatic elms such as *Ulmus pumila* L. into their breeding material.

Tree breeding is obviously a long-term process and it is consequently important to have clear aims. In breeding for resistance to disease it is important to know what one is breeding for resistance to i.e. the extent of and the potential for variation in the pathogen. For a number of reasons, knowledge of pathogenic variation in forest pathogens has been lagging behind that in pathogens of arable crops. Indeed, until quite recently elm breeders were very much in ignorance of what they were breeding for resistance to. It was only in the early 1970s that the existence of a highly pathogenic 'aggressive' strain of the Dutch elm disease fungus, *Ceratocystis ulmi* (Buism.) C. Moreau, was recognised, and in consequence the elm breeding programmes in the USSR and in the Netherlands were severly shaken (Kryukova, 1972; Gibbs & Brasier, 1973; Heybroek, 1976). The breeding programme at Wisconsin survived the trauma largely through geographical accident, since they had, unknowingly, already been using the aggressive strain in their resistance testing procedures.

Clearly we as forest pathologists and fungal geneticists have a duty to the tree breeders to attempt to keep knowledge of variation in forest pathogens at least one, and hopefully several steps ahead of what is going on in their breeding programmes. Our failure to recognise the existence of several pathogenic forms within *C. ulmi* has had a number of repercussions.

Firstly, as just mentioned, it has undermined the basis upon which the selection of resistant elm material was carried out. Secondly, it has resulted in a lack of data on interactions between the products of the elm breeding programmes - the different elm species and their progenies - and the various strains and races of the fungus. This in turn must have resulted in the loss of valuable information on the genetic nature of host resistance and of fungal pathogenicity. Thirdly, it has allowed 2 new epidemics of Dutch elm disease to progress virtually uncharted and unrecognised across much of Europe and South West Asia during the past 10-20 years (Brasier, 1979).

Evidently continued negligence of variation in our forest pathogens is a short sighted policy. It should therefore be our constant aim to maximise information on the pathogenic potential of these organisms, and to attempt to understand the genetic systems controlling their pathogenicity by surveying and monitoring the extent of variation in the field and by experimenting on potential variation in the laboratory.

In the case of *C. ulmi*, an attempt to identify the geographical centre of origin of Dutch elm disease, still rather vaguely thought to be 'somewhere in Eastern Asia', is a third and vital requirement.

PRESENT KNOWLEDGE OF STRAINS AND RACES WITHIN CERATOCYSTIS ULMI

C. ulmi falls into 3 genetically distinct sub-groupings: a highly pathogenic aggressive strain which itself divides into 2 distinct races, and a more weakly pathogenic non-aggressive strain (Gibbs & Brasier, 1973; Brasier, 1979). The non-aggressive strain is found from North America through Europe to Iran. Of the 2 races of the aggressive strain, termed the North American or NAN and Eurasian or EAN races, the NAN is distributed throughout North America. In the late 1960s it was introduced into Britain and it has since spread to adjoining parts of north-west Europe as far east as Italy and northern Yugoslavia. The EAN race occurs mainly to the east of a line from Denmark to Italy, extending to Volgograd and the Caspian Sea area of Iran. The present situation in Europe is rapidly changing as the NAN race of the aggressive strain spreads eastwards and southwards and the EAN race spreads westwards.

The aggressive and non-aggressive strains have many differing properties (Table 1). In agar culture they differ in growth rate, optimum temperature for growth, upper temperature limit for growth and in colony morphology. They also differ in the frequency of their mating types in nature, and a barrier to hybridization exists between them. Moreover, as shown below, crosses provide evidence of considerable genetic isolation between them.

The 2 races of the aggressive strain also differ in cultural characteristics and some other features (Table 1). Barriers to their hybridization also exist but these are less marked than those occurring between the aggressive and non-aggressive strains.

DEFINITION OF 'STRAIN' AND 'RACE' IN C. ULMI

The terminology applied to sub-units below the species level in fungal populations is in a rather confused state, reflecting perhaps the youthful state of the art! The field of fungal pathogenic polymorphism is particularly difficult, and there certainly appears to be no established terminology in the literature appropriate to major sub-groups of the sort encountered in *C. ulmi*. In the event the somewhat neutral term 'strain' was chosen to apply to the aggressive and non-aggressive sub-groups of *C. ulmi* (Gibbs & Brasier, 1973) despite the fact that the same word is sometimes used rather loosely by mycologists when they may really mean 'isolate' (the difference between 'strain' and 'isolate' has been clearly defined by Ainsworth (1961)). Subsequently it has been necessary to find a suitable word to apply to sub-groups of lower rank than strain in *C. ulmi*, and in this case the word 'race' sensu Stebbins (1966) has been chosen (Brasier, 1979). Following Stebbins' hierarchy 'strain' as used in *C. ulmi* should probably assume a rank roughly equivalent to his 'sub-species', and indeed

Table 1. Differences between the aggressive and non-aggressive strains of Ceratocystis ulmi. Based on the data of Brasier (1977, 1978, 1979); Brasier & Afsharpour (1979); Brasier & Gibbs (1975a, 1975b, 1976); Brasier et al. (1981); Gibbs & Brasier (1973); Gibbs et al. (1979); Brasier (unpublished).

	Non-aggressive strain	Aggressive strain	
		NAN race	EAN race
Growth rate (common range in mm/day)			
20 °C	2.0-3.1	3.2-4.8	3.1-4.4
33 °C	1.1-2.8	0.1-0.5	0(-0.1)
Optimum temperature for growth (°C)	30	20-22	20-22
Upper limit for growth (°C)	35	33	32
Colony morphology	From a smooth waxy surface to a lawn of moderately dense relatively undifferentiated aerial mycelium. Weak diurnal zonation.	Commonly fibrous-striate petaloid. Strong diurnal zonation	Commonly less striate and petaloid, often lobed and uneven. Strong diurnal zonation.
Mating type frequency in nature	A- and B-types equal	B-type predominant	B-type predominant
Fertility reaction (when fertilised by other strains/races)	Universal acceptor - accepts non-aggressive and both races of aggressive strain.	Accepts both races of aggressive strain equally. Strongly rejects the non-aggressive strain	Partially rejects NAN race. Accepts EAN race. Strongly rejects the non-aggressive strain.
Pathogenicity (common range for % defoliation on clonal *Ulmus procera*)	10-35 typically recovers	80-100 no recovery	60-100 recovery occasional

this seems appropriate for the aggressive and non-aggressive strains.

The following definitions of 'strain' and 'race' as applied to *C. ulmi* are offered:

Strain. (Aggressive and non-aggressive strains). Major sub-groups within the *C. ulmi* population which do not freely interbreed and which differ in almost all important physiological and pathological characteristics (see Table 1). Each strain shows a wide range of variation comparable to that often found in a single species. Inter-strain hybrids have yet to be found in nature.

Race. (North American and Eurasian races). Sub-groups within the aggressive strain population which do not freely interbreed and which differ in certain important physiological and pathological characteristics, but in which their similarities broadly outweigh their differences (see Table 1). The population of each race shows a wide range of variation. No inter-race hybrids have yet been found in nature.

DIFFERENCES IN PATHOGENICITY

The strains and races of *C. ulmi* are best differentiated on elms of moderate disease resistance such as *Ulmus procera* Salisbury or certain clones of *U. japonica* (Rehd.) Sarg. Highly susceptible or highly resistant elms such as *U. americana* L. or *U. pumila* are, for obvious reasons, unsuitable.

Fig. 1 shows the differences in pathogenicity between population samples of the non-aggressive strain and of the EAN and NAN races of the aggressive strain when inoculated into clonal *U. procera*. Fig. 1a is the result of a pathogenicity test carried out in 1977 comparing 13 non-aggressive isolates originating from 7 countries from the USA to Iran. After 10 weeks, all the isolates had caused only relatively low levels of defoliation ranging from 14 to 39 % with a mean of 28.0 %. This is a typical result for the non-aggressive strain.

The existence of the North American and Eurasian races of the aggressive strain was only recognised as recently as 1978, when it was supposed that they both had comparably high levels of pathogenicity (Brasier, 1979). However, a pathogenicity test carried out in 1979 showed otherwise. The test (Fig. 1b, c) comprised a sample of 14 EAN race isolates originating from 7 countries for Ireland to Iran, and 14 NAN race isolates originating from 7 countries between the USA and Austria. Two non-aggressive control isolates (arrow) gave 16 % mean defoliation. As in previous tests the NAN race isolates caused very high levels of defoliation ranging from 81 to 100 %. With the EAN aggressive, however, defoliation ranged from 38 to 100 %, with a mean of 82 %, significantly below the NAN mean of 95 %.

Thus the EAN aggressive showed an unexpected range of variation in pathogenicity, with some isolates significantly less pathogenic than the

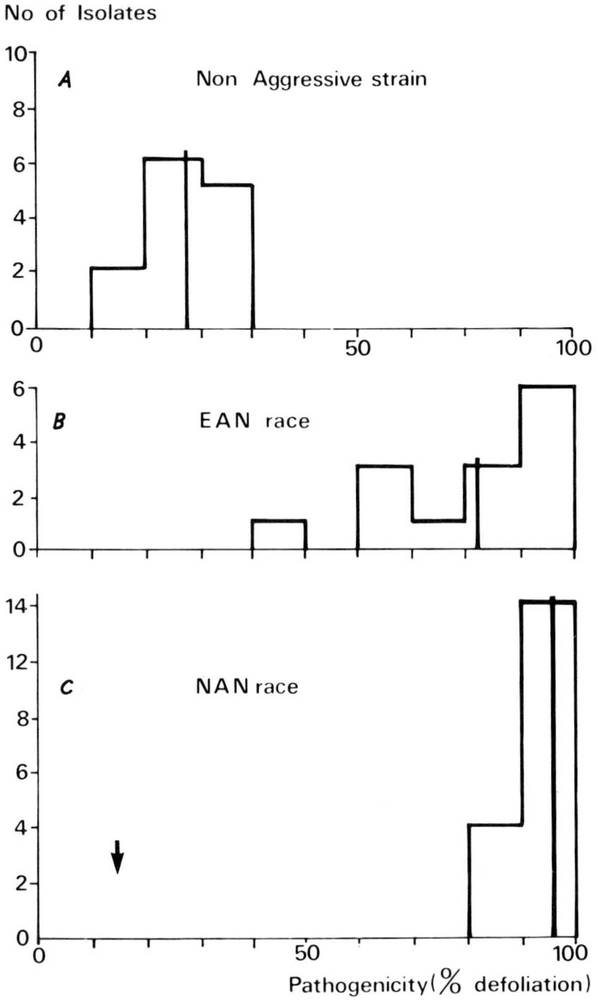

Figure 1. Pathogenicity of the non-aggressive and of the EAN and NAN races of the aggressive strain of *C. ulmi*. *A*, sample of 13 non-aggressive isolates inoculated into 3-year-old clonal *U. procera* in 1977, showing % defoliation after 10 weeks. *B, C*, samples of 14 EAN and 14 NAN aggressive isolates inoculated into 5-year-old clonal *U. procera* in 1979, showing % defoliation after 13 weeks. The arrow indicates the mean of 2 non-aggressive control isolates. Heavy bars represent sample means. There were 3 replicate trees for each isolate.

NAN race, though still more pathogenic than the non-aggressive strain.

Further important differences between the strains are revealed in disease progress curves. Fig. 2 shows the disease progress curves for the above pathogenicity test. Typically, with the non-aggressive strain on *U. procera* a burst of disease occurs in the first few weeks after inoculation,

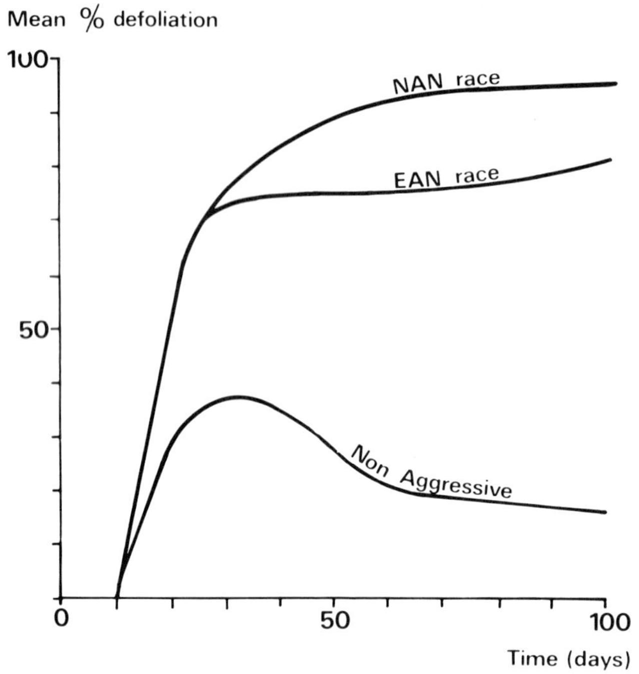

Figure 2. Disease progress curves (mean % defoliation against time from inoculation) for the non-aggressive strain and the EAN and NAN races of the aggressive strain of *C. ulmi* on *U. procera*. Note final differences in mean pathogenicity levels of the EAN and NAN races, and the recovery of trees inoculated with the non-aggressive strain. The tests were carried out on 5-year-old clonal *U. procera* in 1979. The curves are the means of 14 EAN, 14 NAN and 2 control non-aggressive isolates. There were 3 replicate trees for each isolate.

but thereafter disease progress is halted and the trees go into a recovery phase with the diseased shoots budding-up and eventually flushing. There is normally no re-occurrence of the disease in the following season.

With both the EAN and NAN races of the aggressive strain there is a greater initial burst of disease activity. With the NAN race disease usually progresses until defoliation levels of 90-100 % have been reached and most trees die within the same season. If they do not, disease re-occurrence and death usually follow in the second season.

With the EAN race the most highly pathogenic isolates behave similarly to the NAN isolates, but with the less pathogenic isolates some recovery may take place. A full recovery may occur in the following season when disease re-occurrence appears to be a relatively uncommon phenomenon.

The NAN race is the form of the aggressive strain first described in 1973 by Gibbs & Brasier. From its pathogenic behaviour the EAN race is clearly a more distinct genetic entity than was originally supposed. Its present status as a race within the aggressive strain (alongside the NAN race) rather than that of a third strain in its own right may yet be called into question. Its biological characteristics, distribution and history of spread are, together with those of the NAN race and the non-aggressive strain, still undergoing detailed survey and investigation.

GENETIC CONTROL OF PATHOGENICITY: CROSSES BETWEEN THE AGGRESSIVE AND NON-AGGRESSIVE STRAINS

Although the aggressive and non-aggressive strains may not hybridise freely in nature, crosses can be forced between them in the laboratory. Numerous crosses have now been undertaken using aggressive and non-aggressive isolates from 3 continents, and the results have been broadly the same.

In terms of the inheritance of growth rate, the progeny of aggressive × non-aggressive crosses show considerable variation, many growing significantly faster than the aggressive parent or slower than the non-aggressive parent. Fig. 3 shows the result of one such cross between non-aggressive

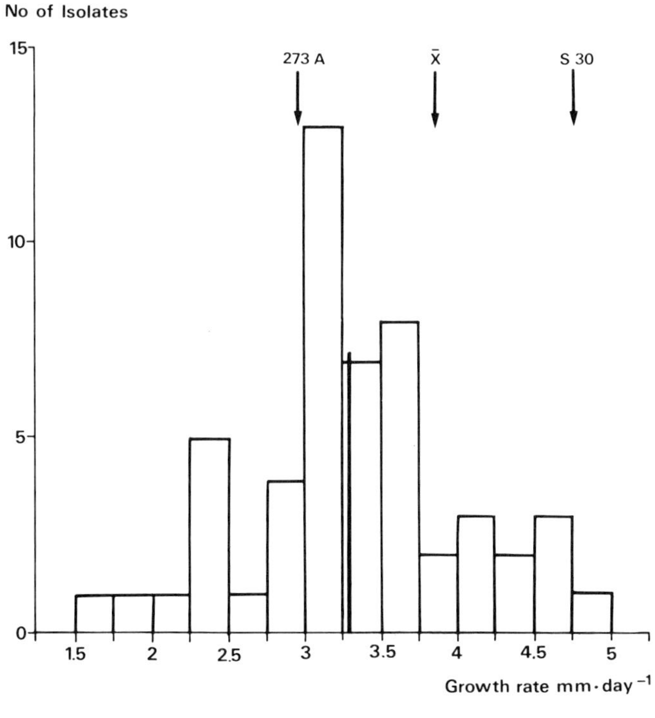

Figure 3. Growth rate distribution of F_1 progeny from a cross between a NAN aggressive (S30) and a non-aggressive (273A) isolate of *C. ulmi*. The heavy bar shows the progeny mean, and the arrows the parental growth rates and the parent mid-point value. The test was carried out at 18 °C and there were 2 replicates per isolate.

parent 273A and aggressive parent S30. The progeny distribution indicates that a large component of the variation is simply additive, but a degree of negative interaction between the parents can often be detected, as here, with the mean of the progeny (heavy bar) lying significantly below the mid-point of the 2 parents (middle arrow) (see Brasier, 1977).

In colony morphology, progeny of aggressive × non-aggressive crosses are always extraordinary variable (see Brasier & Gibbs, 1976, Plate 1). Usually very few of the progeny can be characterised in terms of the striate-petaloid colony morphology of the aggressive parent, or the flat relatively undifferentiated morphology of the non-aggressive parent, giving a further indication of segregation of extensive genetic differences between the 2 parents.

In their pathogenic behaviour the progeny of aggressive × non-aggressive crosses are particularly interesting. Fig. 4 shows the results of 2 such crosses, one between non-aggressive isolate M35 and aggressive isolate O27 (left) the other between non-aggressive isolate W10 and aggressive isolate G36 (right). Above (Fig. 4a, c) is the result when a sample of their progenies were inoculated into the moderately resistant *U. procera*. Points to note are as follows:

- The aggressive parents gave characteristically high and the non-aggressive parents characteristically low levels of defoliation.
- The majority of the progeny are weak pathogens. With the exception of 1 isolate (in the cross W10 × G36, Fig. 4c) none approached the aggressive strain in pathogenicity. Statistically, only a few progeny were significantly more pathogenic than their non-aggressive parent.
- There is a strong negative interaction: in both crosses the progeny mean (heavy bar) fell significantly below the mid-point of the parents (middle arrow).

Inoculation of the progeny of W10 × G36 into the more susceptible *U. glabra* Hudson (Fig. 4d) produced a greater spread in their pathogenic distribution thus providing a clearer picture of their pathogenic behaviour. Again, none of the progeny approached the aggressive parent in pathogenicity, and the progeny mean (heavy bar) lay significantly below the parent mid-point value (middle arrow).

The behaviour of the progeny M35 × O27 on the even more susceptible *U. laevis* Pallas (Fig. 4b) amplifies the picture even further. On this species even the non-aggressive parent (left arrow) caused 57 % defoliation, with the aggressive parent (right arrow) causing 97 % defoliation. The progeny showed a broad distribution. Not only was the progeny mean (heavy bar) significantly below the parent mid-point (middle arrow) but statistically many progeny were significantly less pathogenic than the non-aggressive parent, 1 progeny isolate causing only 0.5 % defoliation. A few progeny approached the pathogenicity level of the aggressive parent, the highest giving 90 % mean defoliation (compared with the aggressive parent value of 97 %).

The picture given by the progeny is therefore one of a large negative interaction between the genetic systems of the aggressive and non-aggressive strains (i.e. a drop in pathogenicity of the progeny relative to that of the parents), together (as indicated by the *U. laevis* data) with some additive component. The interaction cannot be accounted for by cytoplasmic

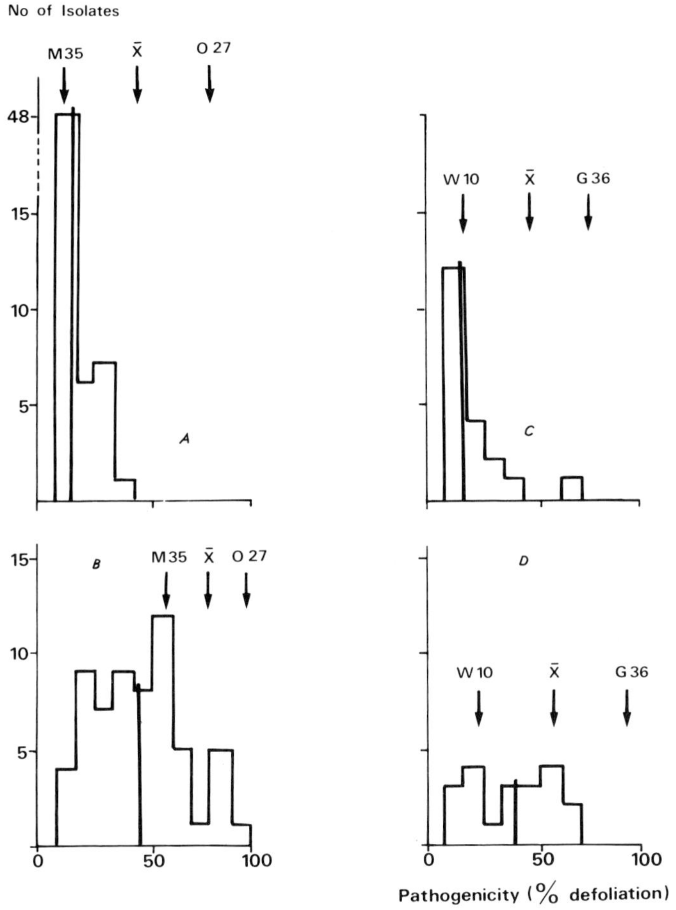

Figure 4. Pathogenicity of F_1 progeny from crosses between the NAN aggressive and non-aggressive strains of *C. ulmi*. *A*, *B*, progeny sample of non-aggressive × aggressive cross M35 × O27 on 5-year-old clonal *U. procera* and 3-year-old seedling *U. laevis* respectively. The inoculations were carried out in 1973 and data are for 10 weeks after inoculation. *C*, *D*, progeny sample of non-aggressive × aggressive cross W10 × G36 on 3-year-old clonal *U. procera* and 3-year-old seedling *U. glabra* respectively. The inoculations were carried out in 1974, and data are for 12 weeks after inoculation. There were 3 replicate trees per isolate. The heavy bars represent the progeny means and the arrows the pathogenicity levels of the parents and the parent mid-point values.

effects: reciprocal aggressive × non-aggressive crosses give the same results (see Brasier & Gibbs, 1976; Brasier, 1977).

The following explanation is proposed for the negative interaction:
- The genetic systems governing pathogenicity in the aggressive and non-aggressive strains are complex, polygenic, and qualitatively different.
- The high level of pathogenicity in the aggressive strain is conferred by special gene combinations which on hybridization with the non-aggressive strain are dispersed, resulting in a large number of weakly pathogenic genotypes (Fig. 4).
- It is suggested that the special gene combinations may involve gene sequences acting as functional units, such as a series of structural genes governed by an operator gene. Thus the physiology of the host-parasite relationship in this vascular wilt disease is highly complex, and several highly specialised enzyme systems are likely to be involved. Their control by operon-type genetic systems is a distinct possibility. Substitution of an operator allele in the aggressive strain with the equivalent locus in the non-aggressive strain might lead to half of the progeny being non-functional at the operon concerned. Substitution of alleles at the structural loci might also impair the efficiency of the metabolic pathway involved (see Brasier, 1977).

CONCLUSIONS

To summarize, it is clear that there are extensive genetic differences between the aggressive and non-aggressive strains of *C. ulmi*. These are exemplified firstly by their differing properties of growth-rate, their differing colony morphologies, and in their remarkably different temperature-growth responses. Secondly, they are shown in their different mating systems, and in the fertility barrier that operates between them. Thirdly, they are highlighted by the complex patterns of inheritance that occur among their progeny, and by the fact that although thousands of wild isolates of *C. ulmi* have been examined, no indication of hybridization between the 2 strains in nature has been found.

Many of these factors point to a considerable degree of genetic isolation or evolutionary divergence. Indeed, we appear to have here 2 discrete virtually non-hybridizing populations which approximate to the level of sub-species. Whether the 2 strains have evolved as a consequence of sympatric or allopatric evolution has been investigated. Experimental evidence points to a sympatric divergence, with the aggressive strain having evolved from the non-aggressive strain (Brasier, in preparation).

Turning to the 2 races of the aggressive strain, they have a number of features in common indicating common ancestry, but it is also true that ever greater genetic differences are being found between them as shown in the pathogenicity data presented here. The present status of the EAN aggressive as a race rather than a separate strain could readily be called into question as mentioned above. Further light may be thrown on this problem when crosses between the EAN and NAN races, now being carried out, have been analysed in detail.

Thus it is clear that there are 3 distinct genetic groupings within the wild population of *C. ulmi*. With the geographical centre of origin of Dutch elm disease still to be determined, the possible existence of further discrete pathogenic forms of the fungus cannot be ruled out. This possibility represents a continuing threat to the products of our expensive elm breeding programmes. Such a threat can only be alleviated by further detailed mycological surveys, in particular in the several remaining uncharted areas of elm diversity such as the Himalayas and parts of China.

REFERENCES

Ainsworth, G.C., 1961. Dictionary of the fungi. Commonwealth Mycological Institute, London. 547 p.

Brasier, C.M., 1977. Inheritance of pathogenicity and cultural characters in Ceratocystis ulmi: hybridization of protoperithecial and non-aggressive strains. Transactions of the British Mycological Society 68: 45-52.

Brasier, C.M., 1978. Mites and reproduction in Ceratocystis ulmi and other fungi. Transactions of the British Mycological Society 70: 81-89.

Brasier, C.M., 1979. Dual origin of recent Dutch elm disease outbreaks in Europe. Nature 281: 78-80.

Brasier, C.M. & F. Afsharpour, 1979. The aggressive and non-aggressive strains of Ceratocystis ulmi in Iran. European Journal of Forest Pathology 9: 113-122.

Brasier, C.M. & J.N. Gibbs, 1975a. A highly fertile form of the aggressive strain of Ceratocystis ulmi. Nature 257: 128-131.

Brasier, C.M. & J.N. Gibbs, 1975b. Variation in Ceratocystis ulmi. Significance of the aggressive and non-aggressive strains. In: D.A. Burdekin & H.M. Heybroek (Eds): Dutch elm disease. Proceedings of IUFRO Conference, Minneapolis - St. Paul, 1973. USDA Northeastern Forest Experiment Station, Upper Darby, Pennsylvania. p. 53-66.

Brasier, C.M. & J.N. Gibbs, 1976. Hybridization of the aggressive and non-aggressive strains of Ceratocystis ulmi. Annals of Applied Biology 83: 31-37.

Brasier, C.M., J. Lea & M.K. Rawlings, 1981. Aggressive and non-aggressive strains of Ceratocystis ulmi have different temperature optima for growth. Transactions of the British Mycological Society 76: 213-218.

Gibbs, J.N. & C.M. Brasier, 1973. Correlation between cultural characters and pathogenicity in Ceratocystis ulmi isolates from Britain, Europe and North America. Nature 241: 381-383.

Gibbs, J.N., D.R. Houston & E.B. Smalley, 1979. Aggressive and non-aggressive strains of Ceratocystis ulmi in North America. Phytopathology 69: 1215-1219.

Heybroek, H.M., 1976. Chapters on the genetic improvement of elms. In: Better trees for metropolitan landscapes. USDA Forest Service General Technical Report NE 22. p. 203-214.

Kryukova, E.A., 1972. The Dutch elm disease pathogen of elms in the droughty zone of South East European Russia. In: Protection of Forests against Pests and Diseases. Kolos, Moscow. p. 153-161.

Stebbins, G.L., 1966. Processes of organic evolution. Prentice Hall, Englewood Cliffs, New Jersey. 179 p.

Pathogenic variability within the genus Cronartium

H.R. Powers, Jr.

Chief Plant Pathologist, Southeastern Forest Experiment Station, USDA, Forestry Sciences Laboratory, Athens, Georgia 30602, USA

ABSTRACT

In the genus *Cronartium*, each pathogen that has been investigated has shown a high degree of pathogenic variability on both pine hosts and alternate host. Thus, constant monitoring of the pathogen will be required to detect shifts in pathogenicity, and to determine the most effective placement of resistant materials. In addition, an extremely broad genetic base of resistance will be required in each of the host species if tree rusts are to be controlled by the development of resistant strains of pines.

INTRODUCTION

The rust fungi have earned a reputation for variability. This characteristic was first demonstrated in rusts of cereals, particularly the black stem rust organism (*Puccinia graminis* Pers.) which has several hundred physiologic races. Research on pathogenic variability of rust fungi on forest trees began much later. One reason is that the uredial, or repeating stage, of cereal rusts occurs on the economically important host. With the tree rusts, this stage is usually on the less important hosts.

WHITE PINE BLISTER RUST

White pine blister rust, caused by *Cronartium ribicola* J.C. Fisch. ex Rabenh., was the first tree rust to be closely studied. Widespread in Asia and Europe, it was introduced into the United States early in this century. The first 50 years of research on white pine blister rust yielded little evidence of pathogenic strains for this fungus. The fungus attacks many soft pines, ranging from the very susceptible sugar (*Pinus lambertiana* Dougl.) and western white (*P. monticola* Dougl.) pines to the highly resistant Swiss stone pine (*P. cembra* L.) of Europe. All species of *Ribes* can serve as alternate hosts, but they also vary widely in susceptibility to infection. Mielke et al. (1937) showed variation both between and within 4 species of *Ribes*. Some highly resistant individuals were found in the generally susceptible *R. petiolare* Dougl. Anderson & French (1955) reported variation in response to infection on *R. hirtellum* Michx. when collections of *C. ribicola* from 3 different pine species were used as inocula.

Yokota et al. (1975) reported a further step in variability--, a strain of the organism from Japan that was able to utilize both *Pedicularis* and *Ribes* as alternate hosts. Subsequently, Yokota & Uozumi (1976) made a tentative revision of *Cronartium* rusts on white pine (*P. strobus* L.) in Japan: *C. ribicola* f. sp. *ribicola* on *P. strobus* and *Ribes* sp.; *C. ribicola* f. sp. *pedicularis* on *P. strobus* and *P. pumila*, with both *Pedicularis* and *Ribes* as alternate hosts. In Canada, Hiratsuka & Maruyama (1976) found a strain of *C. ribicola* that was able to infect both *Ribes* and *Castilleja miniata* Douglas. This was the first report of *C. ribicola* infecting a scrophulariaceous host in North America.

In addition, Saho & Takahashi (1976) reported a *Peridermium* species on *P. pumila* Regel in Japan which they consider to be an autoecious form of *C. ribicola*. That is, a pine-to-pine rust with an intermediate host not being required to complete the life cycle. This would be pathogenic variability of the first magnitude.

Some of the most extensive work indicating variation within *C. ribicola* has been done on western white pine by McDonald & Hoff (1975). They have shown extensive variation in pigmentation of spots developing on inoculated pine seedlings. Their results demonstrate the existence of 4 seedling types: those with only red spots, those with only yellow spots, those with both red and yellow spots, and those with no spots. They hypothesize that these reactions indicate races of *C. ribicola*, and that the pine hosts are differentially resistant. Further, they feel resistance to these races is simply inherited. Seedlings with red needle spots are resistant to the rust shown to produce yellow spots. Conversely, the seedlings with yellow spots are resistant to the strain producing red spots. Seedlings with spots of both colors are, therefore, susceptible to both strains, and seedlings with no spots are resistant to both strains.

Recently, McDonald & Andrews (1980) found highly significant variation in development of the pathogen on *Ribes*. They found differences in development of uredial infections after single aeciospore inoculations on leaf disks, and in the percent of these infections that produced teliospores within 35 days. These variations could influence pathogenic capabilities of the fungus.

Two recent investigations provide some of the strongest evidence found to date on pathogenic variability of *C. ribicola* on pine. Kinloch & Comstock (1981) found that sugar pines with single major gene resistance to blister rust suddenly became susceptible to infection. These trees had been free of disease under extremely hazardous conditions for up to 14 years in field progeny tests. Their sudden susceptibility demonstrates a new race of *C. ribicola* that overcomes this gene for resistance in sugar pine. These authors indicate that necrotic flecks on inoculated seedlings indicate resistance while yellow or occasional red lesions indicate susceptibility. G.I. McDonald (personal communication) also has found a new, more virulent strain of *C. ribicola* on *P. monticola*. Compared to standard isolates of *C. ribicola*, the new strain (Champion Mine) more readily infected previously tested seedlings from one specific area of the Pacific Northwest.

HARD PINE RUSTS

Red and yellow needle spots also occur in Italy on seedlings of a hard pine (*P. pinaster* Ait.) inoculated with *C. flaccidum* (Alb. et Schw.) Wint. These spots indicate 2 races of the pathogen and differential resistance of maritime pine seedlings to these 2 races (Raddi, 1976). The host-parasite interaction, in this case on a hard pine, closely resembles that of soft pines with *C. ribicola*.

Studies on pathogenic variability of hard pine rusts in the United States have primarily dealt with the *C. quercuum* complex. Kais & Snow (1972) and Powers (1972) indicated sharp differential responses between jack (*P. banksiana* Lamb.) and Virginia (*P. virginiana* Mill.) pines when inoculated with cultures of *C. quercuum* derived from each species. Each pine species was susceptible to its isolate of the organism, but resistant to the isolate from the other species. This work demonstrates a relationship for this host-parasite complex comparable to that with the wheat rust. That is, the isolates of *C. quercuum* from a given pine species were similar to the *formae speciales* on the various cereals. These relationships within the *C. quercuum* complex were clarified subsequently by Burdsall & Snow (1977). Four *formae speciales* (*banksianae*, *virginianae*, *echinatae*, and *fusiforme*) were established to accommodate the rust on jack, Virginia, shortleaf (*P. echinata* Mill.), and the combination of loblolly (*P. taeda* L.) and slash (*P. elliottii* var. *elliottii* Engelm.) pines.

Snow & Kais (1970) provided evidence of pathogenic variability within the fusiform rust organism (*C. q.* f. sp. *fusiforme*). They found differential reactions between resistant slash pine progenies and rust isolates from 5 states. Later, Snow et al. (1975) found a significant inoculum source × family interaction among rust isolates collected on a east-west transect (Florida-Texas) and 2 north-south transects (Mississippi and Florida-Georgia). Among 8 individual galls and 3 resistant slash pine families a significant inoculum × family interaction was also found, as was substantial pathogenic variability among collections from individual galls. An absence of active galls on 1 resistant family when inoculated with 2 of the rust isolates was the first observation of an immune reaction between this pathogen and slash pine. In addition, 2 of the 10 families were resistant to all of the inocula from 5 states. The authors hypothesized that this phenomenon might be comparable to horizontal resistance in various agronomic crops.

Pathogenic variability of *C. quercuum* f. sp. *fusiforme* on loblolly pine was studied by extensively sampling the pathogen across the Southern United States (Powers et al., 1978). Isolates from 56 individual rust galls from 7 states from Louisiana through North Carolina were used to inoculate 3 loblolly pine families rated as resistant, intermediate, and susceptible to fusiform rust. The results of these inoculations showed highly significant variability at every level examined. The proportion of seedlings infected was influenced by the host family, the state in which the rust was collected, interactions between host family and state of rust collection, the individual rust collection within states, and interactions between host families and individual collections within states. Several interactions between specific rust collections within states and families demonstrate the need to label highly virulent strains and make tree breeders aware of the fact that these isolates can cause heavy damage on some of the most resistant clones or selections available. Extensive planting based on only 1 source of resistance could lead to a rapid build-up of such virulent strains of rust. For example, 1 of the isolates from South Carolina was highly virulent on resistant family 11-20, which was selected in South Carolina. In fact, it produced a higher percentage of infection on the resistant family than on the susceptible family. This study also produced evidence of the classic reversal of infection types between families and rust isolates, similar to differential reactions on wheat host lines that result in the classification of physiological races of the wheat stem rust fungus. In both slash and loblolly pine, differences among infection levels caused by individual rust collections from within a state were sometimes greater than the differences among collections from different states. Thus, a broad genetic base of resistance is needed for planting, and tests for resistance must include range of rust isolates from different geographic areas.

A recent study included 7 loblolly pine families rated as resistant to intermediate in resistance, 2 susceptible families, and 5 rust isolates (Powers & Matthews, 1979). Again, there were highly significant differences in resistance among host families and in infection levels produced by the various rust isolates. All resistant families had relatively low levels of infection, but only 1 was resistant to all rust isolates. Highly virulent rust isolates caused particularly heavy infection on pine families originating in the same geographic area. The South Carolina rust collection again caused heavy infection on family 11-20. Families from Texas and Louisiana were most susceptible to rust isolates from their own states.

Snow et al. (1976) demonstrated increasing virulence of rust isolates developing on resistant pine selections. They found a striking increase in virulence produced by collections of *C. quercuum* f. sp. *fusiforme* gathered from one specific resistant slash pine family. Because such increases challenge programs of selecting and breeding for resistance to the rust pathogen, additional studies with loblolly pine were carried out (Powers et al., 1977). They used 3 families: a resistant family from which the rust collections were obtained, an additional, non-related resistant family, and a susceptible check. Collections of individual rust galls from field plantings of the resistant family were compared with galls representing the general rust populations from nearby trees. The rust isolates from the resistant family produced 9 % more infection on seedlings of that family than did isolates from the general population. This increase may be biologically significant, even though it was not statistically significant. There was no increase in virulence on the unrelated rust resistant family. The relatively modest increase in virulence of these family-specific rust isolates and maintenance of resistance to all isolates by the unrelated resistant loblolly family were encouraging.

To determine if a trend for increasing virulence has been developing in recent years, isolates of the fusiform rust organism originating in approximately 1945 were compared with those originating around 1970 (Powers & Dwinell, 1978). No differences in virulence were found between the 2 groups, but the more recent isolates were much more variable.

Pathogenic variability has also been found among 10 single-spore isolates originating from 1 isolated rust gall (Powers, 1980). Three half-sib families of loblolly pine previously determined to be resistant, intermediate, and susceptible to fusiform rust were inoculated with each of the isolates. Isolates differed significantly in percentage of seedlings infected; the range, 42 % to 56 %, was similar to that produced by unrelated galls within an individual county or plantation.

In populations of slash and loblolly pines, resistance to *C. quercuum* f. sp. *fusiforme* has been found in single, random trees. In loblolly pine there are also specific geographic zones, primarily around the periphery of its natural range, where bulk collections of seed are moderately resistant (Wells & Wakeley, 1966). To compare seeds from 6 of these geographic areas and rust collections from each area, a study was designed to cross-inoculate seedlings from each seed source with each rust collection (Powers & Matthews, 1980). Seed and rust spores were collected from Arkansas, east Texas; Livingston parish, Louisiana; the eastern shore of Maryland; and Marion county, Florida. Seeds from central Georgia were included as a susceptible control. The Maryland seed source was significantly more resistant than any other source. Virulence of rust isolates varied by state of collection, but the range in percentage of seedlings infected (< 10 %) was too small to be biologically meaningful. Interactions between seed sources and rust collections from different geographic areas were significant. The range of responses was from 48 % infection on the Arkansas seed

Table 1. Incidence of infection on loblolly pine seedlings from 6 geographic sources after inoculation with rust collections from each geographic area.

Seed sources	Seedlings (%) with galls 9 mo. after inoculation with spores from:						Host mean[1]
	Ga.	Md.	Ark.	Fla.	Tex.	La.	
Maryland	53	*69*	49	52	56	55	56a
Arkansas	54	48	*72*	61	68	61	61b
Florida	69	74	49	*76*	61	73	67c
Texas	58	53	80	68	*84*	83	71c
Louisiana	78	73	81	81	81	*82*	79d
Georgia control	*88*	89	81	82	80	87	84e
Mean	66a	68ab	69abc	70abc	72bc	73c	

1. Infection percentages followed by the same letter do not differ signifi cantly ($P \leqq 0.01$ for host means and $P \leqq 0.05$ for spore means) as determine by Duncan's multiple range test.

sources produced by the Maryland rust, to 89 % infection on the Georgia check lot by the Maryland rust. In all cases except the control, the resulting infection level on a specific seed source was highest when inoculated with the rust isolates from the same geographic area (Table 1). This and earlier results indicate that genes for virulence tend to develop where the complementary genes for resistance are found, and that while this virulence is relatively common in these areas, such genes are not common throughout the ranges of the host and its parasite.

REFERENCES

Anderson, R.L. & D.W. French, 1955. Evidence of races of Cronartium ribicola on Ribes. Forest Science 1: 38-39.

Burdsall, H.H., Jr. & G.A. Snow, 1977. Taxonomy of Cronartium quercuum and C. fusiforme. Mycologia 69: 503-508.

Hiratsuka, Y. & P.J. Maruyama, 1976. Castilleja miniata, a new alternate host of Cronartium ribicola. Plant Disease Reporter 60: 241.

Kais, A.G. & G.A. Snow, 1972. Host response of pines to various isolates of Cronartium quercuum and Cronartium fusiforme. In: R.T. Bingham, R.J. Hoff & G.I. McDonald (Eds): Biology of Rust Resistance in Forest Trees, USDA Miscellaneous Publication No. 1221. p. 495-503.

Kinloch, B.B. & M. Comstock, 1981. Virulent race of Cronartium ribicola to major gene resistance in sugar pine. Plant Disease 65: 604-605.

McDonald, G.I. & D.S. Andrews, 1980. Influence of temperature and spore stage on production of teliospores by single aeciospore lines of Cronartium ribicola. USDA

Forest Service Research Paper 256. 9 p.
McDonald, G.I. & R.J. Hoff, 1975. Resistance to Cronartium ribicola in Pinus monticola: an analysis of needle-spot types and frequencies. Canadian Journal of Botany 53: 2497-2505.
Mielke, J.L., T.W. Childs & H.G. Lachmund, 1937. Susceptibility to Cronartium ribicola of the four principal Ribes species found within the commercial range of Pinus monticola. Journal of Agriculture Research 55: 317-346.
Powers, H.R., Jr., 1972. Testing for pathogenic variability within Cronartium fusiforme and C. quercuum. p. 505-509. In: R.T. Bingham, R.J. Hoff & G.I. McDonald (Eds): Biology of rust resistance in forest trees. USDA Forest Service Miscellaneous Publication No. 1221.
Powers, H.R., Jr., 1980. Pathogenic variation among single-aeciospore isolates of Cronartium quercuum f. sp. fusiforme. Forest Science 26: 280-282.
Powers, H.R., Jr. & L.D. Dwinell, 1978. Virulence of Cronartium fusiforme stable after 25 years. Plant Disease Reporter 62: 877-879.
Powers, H.R., Jr. & F.R. Matthews, 1979. Interactions between virulent isolates of Cronartium quercuum f. sp. fusiforme and loblolly pine families of varying resistance. Phytopathology 69: 720-722.
Powers, H.R., Jr. & F.R. Matthews, 1980. Comparison of six geographic sources of loblolly pine for fusiform rust resistance. Phytopathology 70: 1141-1143.
Powers, H.R., Jr., F.R. Matthews & L.D. Dwinell, 1977. Evaluation of pathogenic variability of Cronartium fusiforme on loblolly pine in the Southern USA. Phytopathology 67: 1403-1407.
Powers, H.R., Jr., F.R. Matthews & L.D. Dwinell, 1978. The potential for increased virulence of Cronartium fusiforme on resistant loblolly pine. Phytopathology 68: 808-810.
Raddi, P., 1976. Indagine sui tipi di macchie fogliari e sul loro significato in discendenze di Pinus pinaster infette da Cronartium flaccidum. Rivista di Patologia Vegetale 12: 91-98.
Saho, H. & I. Takahashi, 1976. A preliminary report on a Peridermium species found on Pinus pumila Regel in Japan. European Journal of Forest Pathology 6: 187-191.
Snow, G.A., R.J. Dinus & A.G. Kais, 1975. Variation in pathogenicity of diverse sources of Cronartium fusiforme on selected slash pine families. Phytopathology 65: 170-175.
Snow, G.A., R.J. Dinus & C.H. Walkinshaw, 1976. Increase in virulence of Cronartium fusiforme on resistant slash pine. Phytopathology 66: 511-513.
Snow, G.A. & A.G. Kais, 1970. Pathogenic variability in isolates of Cronartium fusiforme from five southern states. Phytopathology 60: 1730-1731.
Wells, O.O. & P.C. Wakeley, 1966. Geographic variation in survival, growth, and fusiform-rust infection of planted loblolly pine. Forest Science Monographs II. 40 p.
Yakota, S. & T. Uozumi, 1976. New developments in white pine blister rusts in Japan. XVI IUFRO World Congress. Proceedings Division II. Oslo, Norway. p. 330-337.
Yokota, S., T. Uozumi, K. Endo & S. Matsuzaki, 1975. A Cronartium rust of strobe pine in eastern Hokkaido, Japan. Plant Disease Reporter 59: 419-422.

Relative virulence of Cronartium quercuum f. sp. fusiforme on loblolly pine from Livingston parish

G.A. Snow, W.L. Nance & E.B. Snyder

USDA Forest Service, Forestry Sciences Laboratory, Gulfport, Mississippi, USA

ABSTRACT

Seedlings of 43 controlled-pollinated pine families were inoculated with 5 sources of *Cronartium quercuum* f. sp. *fusiforme*. The seedlings were grown from seed produced in a diallel crossing experiment with 10 loblolly pines in Livingston Parish. Data were assessed as to percentage of trees with galls and the form of galls on infected trees. Differences in pine families, inocula, and a pine family × inocula interaction were identified with both variables. The most precise separation of inocula was achieved with gall-form. The variable response of full-sib pine families to diverse inocula emphasizes the need to evaluate potential breeding stock with several different rust inocula. Progeny from some parent trees were relatively stable to the varying sources of inocula. Such trees may be valuable as testers in future breeding programs.

INTRODUCTION

Resistance to fusiform rust (*Cronartium quercuum* (Berk.) Miyabe ex Shirai f. sp. *fusiforme*) has been found in certain geographic seed sources of loblolly (*Pinus taeda* L.) pine (Wells & Wakeley, 1966; Wells & Switzer, 1971). Since large quantities of this seed is readily available, it has been used extensively. Livingston Parish, La., is an important source for loblolly seed. Trees grown from this seed perform well in growth and resistance to fusiform rust. Moreover, the seed has performed well throughout much of the southern pine region. As a result, the Livingston Parish area will likely remain important as a source of resistant trees for seed orchards.

Powers et al. (1977) recently reported that Livingston Parish (LP) seedlings grown from bulk seed collections varied in susceptibility to different sources of rust inocula. At least 1 of 10 rust collections from each of 7 states was relatively virulent on bulk stock. Inocula from

Mississippi were generally more virulent than inocula from other states. In subsequent tests, differences in virulence of the pathogen were demonstrated on a half-sib LP family (Powers & Matthews, 1979). These findings indicate a need to evaluate potential breeding stock from the Livingston Parish area with an array of different inocula to identify and maintain adequate genetic resistance.

This study seeks to evaluate the interaction among LP pine families with varying levels of resistance and rust inocula with varying levels of virulence. Understanding this interaction is necessary to maximize resistant traits that are stable to the diverse rust population.

METHODS

Host material: Forty-three control-pollinated pine families from 10 randomly selected loblolly pines in Livingston Parish, La., were tested. The trees had been intercrossed in all combinations without regard to which was used as the male or female parent of a given cross. The result was a half diallel with 45 possible crosses. Trees from this experiment had been planted in replicated plots on the Harrison Experimental Forest in Mississippi in 1968. Extra seed was available from all but 2 of the crosses. In this experiment, pine seedlings from each cross were grown in 2.5 cm × 12.7 cm plastic tubes containing a 1:1 mixture of vermiculite and peat moss. They were transplanted to nursery beds 4-6 weeks after inoculation.

Inocula: Five different inocula of *C.q.* f. sp. *fusiforme* were used. These were chosen to be diverse in virulence on the stock. Each inoculum was derived from 1 loblolly pine gall, the collection points for all inocula were widely separated, and 2 inocula were from LP trees. The inocula from LP trees were expected to be more virulent than the other inocula (Snow et al., 1975; Powers et al., 1978). Inoculum A was from a tree growing in Livingston Parish; inocula B, C, and D were collected near Jasper, Texas; Kiln, Mississippi; and Brewton, Alabama, respectively, and were taken from trees thought to be native to the collection points. Inoculum E was collected from a tree in the diallel planting on the Harrison Experimental Forest.

Artificial inoculations: Ten 5-week-old seedlings from each of the 43 pine families were inoculated with each of the inocula in 1978. The entire procedure was repeated in 1979 to obtain a second replication. The order of inoculation, with respect to both family and inoculum, was determined randomly to provide a complete factorial arrangement. A forced air inoculation system (Snow & Kais, 1972) was used throughout the experiment with basidiospore density maintained at 15-25/mm^2.

Six months after inoculation, the seedlings were scored for presence of galls, gall length and diameter. Several derived variables were studied graphically to identify those most repeatable from replication to replication and that best separated families and inocula. The variables selected were percentage of plants with galls and gall-form (gall length/gall diameter). Gall-form is a quantitative measure of shape; round galls produce gall-form values near 1; elongated, or spindle-shaped galls yield larger values. To interpret the results, it was assumed that resistance of an individual pine family could be expressed as a low percentage of plants with galls or by low gall-form values. Low gall-form values would indicate a reduction in longitudinal growth of the fungus in the host tissues. Analysis of variance was used to test for significance of differences among families and inocula and for the family × inocula interaction. Significance was appraised at the 0.01 level.

RESULTS

Analyses of percentage of plants galled and gall-form showed significant differences among both inocula and pine families, and significant pine family × inocula interactions:

	Df	F values	
		Percent galled	Gall form
Replication	1	11.24	12.31
Inocula	4	3.75	21.69
Pine family	42	8.12	7.75
Inocula × family	168	1.70	1.54
Error	214		
Total	429		

The F values for pine family and the inocula by pine family interaction were slightly higher for percent galled than gall-form. However, the F value for inocula was much larger for gall-form, indicating that more precision in separation of inocula was achieved with the gall-form variable. This resulted from (1) less replication-to-replication variation in the gall-form data than occurred in the percentage data and (2) the uniformly high percentage levels obtained with most of the family × inocula combinations.

More than 70 percent of the trees developed galls in 45 of the 50 half-sib family × inoculum combinations (Fig. 1). Families 3, 5, and 7 had the lowest percentages and families 3 and 7 showed the most variation with inocula source. Families 4, 5, 8, and 9 showed much less variation with inocula than the other families. Inoculum A ranked first or second highest on 5 of 6 more variable families (nos. 1, 3, 6, 7, and 10). The rankings

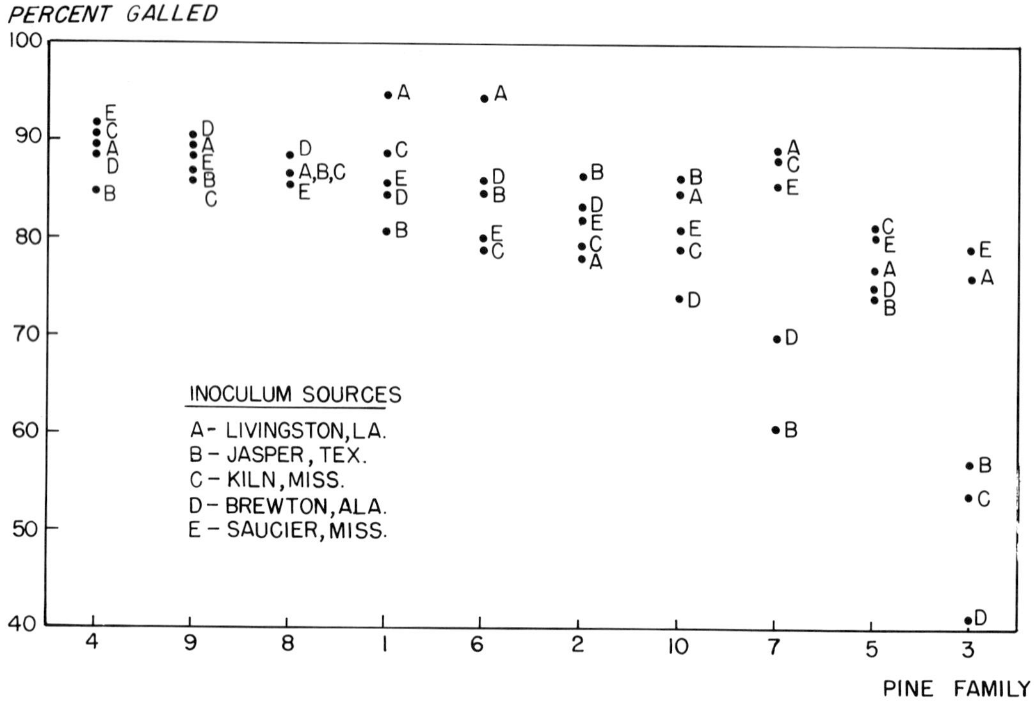

Figure 1. Percentage of trees with galls for 10 half-sib pine families 6 months after exposure to 5 different rust inocula. The half-sib family values are an average of 8 or 9 full-sib families with a common parent; family numbers correspond with the common parent numbers. Families are ranked in decreasing order of susceptibility. Each entry represents the mean percentage infection for all trees in a particular family × inocula combination.

of the other inocula changed on these pine families and none ranked consistently high or low.

Rankings of the 5 inocula were much more consistent with regard to gall-form values (Fig. 2). Inoculum A ranked first or second highest on all half-sib families. Inoculum E also ranked high with the exception of family 10. Inocula B and C ranked lowest on most families, indicating that galls caused by these inocula were generally rounder than galls caused by inocula A, E, and D.

Gall-form was generally related to percent galled in that families 3, 5, and 7 with the lowest percentages (Fig. 1) also had the lowest gall-form values (Fig. 2). Conversely, families 4 and 8, which ranked first and third for highest percentages, had the largest gall-form values. A noticeable exception was family 9 which was second highest in percent galled and fourth lowest for gall-form.

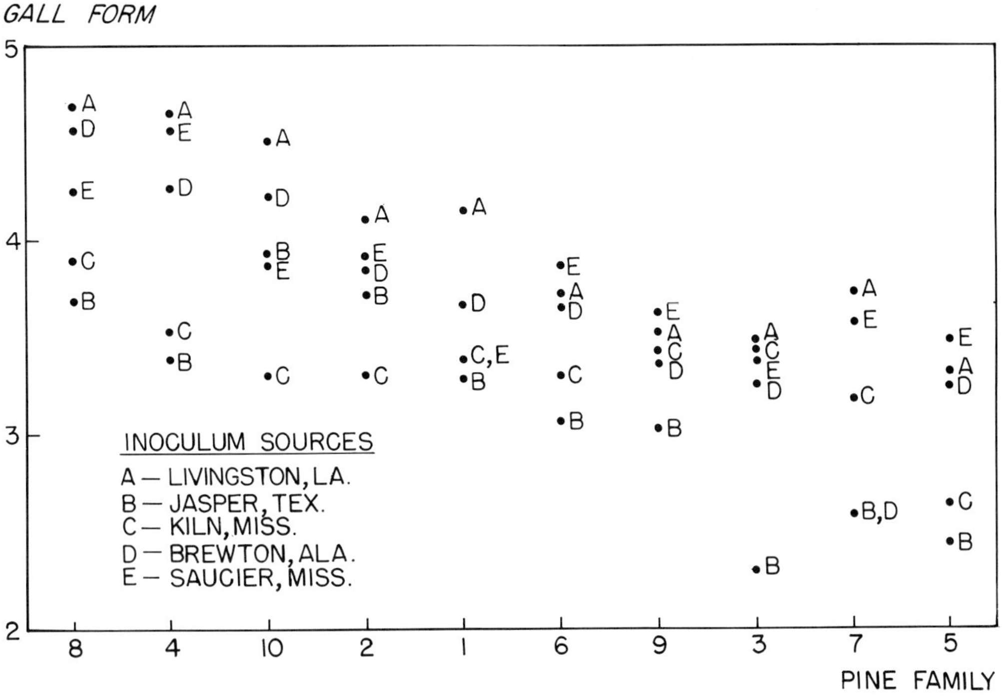

Figure 2. Form (gall length/gall diameter) of galls on 10 different rust inocula. The half-sib family values are an average of 8 or 9 full-sib families with a common parent; family numbers correspond with the common parent numbers. Families are ranked in decreasing order for average gall-form values. Each entry represents the mean gall-form for all galled trees in a particular family × inocula combination.

The interaction between full-sib families and inocula was strong. Graphic analysis of this interaction revealed that crosses from parents 5 and 8 showed much less tendency to interact with inoculum source than the crosses from other parents. Gall-form values for the full-sib crosses of parent 5 are shown in Fig. 3. Cross 7 × 5 ranked lowest with inocula B, D, and E, and intermediate with inocula A and C; families 8 × 5, and 4 × 5 were intermediate to high on all inocula. In contrast, the ranking of family 3 × 5 changed from low to intermediate to high, depending on the inocula employed. Similar changes in rank with inocula occurred for the crosses of parent 8 (Fig. 4). Family 9 × 8 was one of the most variable while 5 × 8 ranked uniformly low and 4 × 8 was uniformly high with most inocula. The range between families with the lowest and highest gall-form values on each inocula was greater for the families of parent 8 than the families of parent 5.

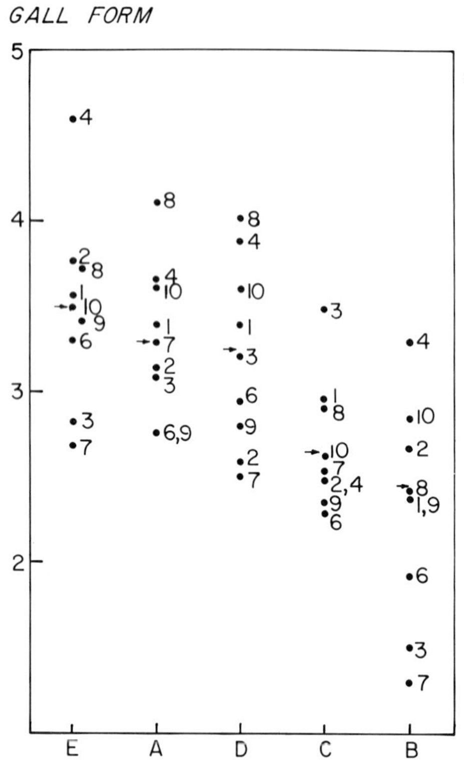

Figure 3. Form of galls on 9 full-sib pine families with common parent No. 5 six months after inoculation with 5 sources of *C.q.* f. sp. *fusiforme.*

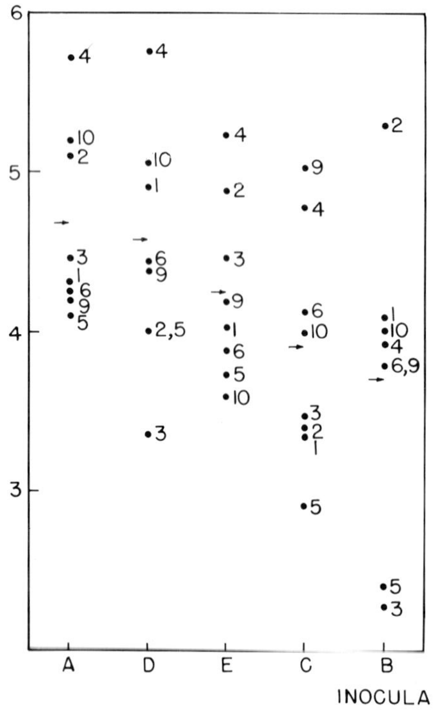

Figure 4. Form of galls on 8 full-sib pine families with common parent No. 8 six months after inoculation with 5 sources of *C.q.* f. sp. *fusiforme.*

DISCUSSION

Infection levels of inoculated seedlings were much higher than expected. In the field planting at the Harrison Experimental Forest only 5.2 % of all the trees had branch or stem galls at age 10, while another seed source of loblolly pine included in the planting had 57 % infection. It appears that all components of the diallel have a high level of resistance. There is some reservation therefore about using the percentage data from the artificial inoculations to make interferences about the field performance of these trees. However, the results with the gall-form data are encouraging and it is hoped that meaningful correlations can be made between field and greenhouse results using form data. Infection levels were too low in the field planting to make these correlations. The diallel was re-established in a high rust hazard area in January 1980, and we hope sufficient infection will occur for more valid comparisons in 2 or 3 years.

It is not known whether the mechanism that limits longitudinal growth of the rust fungus in the host is different from that which limits its establishment. Until the biological relationship between the host responses is better understood, variables that quantify both events seem necessary to assess resistance. The gall-form variable appears to be an adequate expression of longitudinal growth and has at least 3 advantages over percentage data: (1) Data are from infected plants only and experimental error associated with inoculation procedures is reduced; (2) fewer plants are required to obtain repeatable results, and (3) there is no upper limit to gall-form as in percentage data. The disadvantage of gall-form is that infected plants that do not develop galls are ignored.

It was not surprising to find that inocula A and E, which were collected from LP trees, were usually more virulent than the other inocula. Although a larger sample of inocula would be needed to establish that LP stock select virulent forms of the rust population, the results are in line with those of Powers & Matthews (1980) who recently reported inocula from Livingston Parish tend to be more virulent on LP stock than inocula from other areas.

The differences in virulence of the 5 cultures of *C.q.* f. sp. *fusiforme* observed in this experiment are similar to those that have been observed by Powers et al. (1977), and Powers & Matthews (1979) in other experiments with loblolly pine. Use of full-sib pine families and the gall-form variable have provided a different and perhaps closer look at the effects of diverse inocula on pines than previous research has indicated.

With respect to selection and breeding rust resistant trees, the results thus far are in general agreement with other experiments of this nature (Powers et al., 1977; Snow et al., 1975). That is, resistance in LP stock is relative to the virulence of the pathogen and efforts should be made toward finding resistance that is stable to as many forms of the pathogen as possible. The stability of progeny from 2 of the pine parents (8 and 5) is encouraging. These trees seem to impart stable response to varying sources of inocula when used in combinations with other parents in this study. Such parents could prove valuable as top-cross testers in breeding programs.

REFERENCES

Powers, H.R., Jr. & F.R. Matthews, 1979. Interactions between virulent isolates of Cronartium quercuum f. sp. fusiforme and loblolly pine families of varying resistance. Phytopathology 69: 720-722.

Powers, H.R., Jr. & F.R. Matthews, 1980. Comparison of six geographic sources of loblolly pine for fusiform rust resistance. Phytopathology 70: 1141-1143.

Powers, H.R., Jr., F.R. Matthews & L.D. Dwinell, 1977. Evaluation of pathogenic variability of Cronartium fusiforme on loblolly pine in the southern USA. Phytopathology 67: 1403-1407.

Powers, H.R., Jr., F.R. Matthews & L.D. Dwinell, 1978. The potential for increased virulence of Cronartium fusiforme on resistant loblolly pine. Phytopathology 68: 808-810.

Snow, G.A. & A.G. Kais, 1972. Technique for inoculating pine seedlings with Cronartium fusiforme. In: R.T. Bingham, R.J. Hoff & G.I. McDonald (Eds): Biology of rust resistance in forest trees. USDA Miscellaneous Publication 1221. p. 325-326.

Snow, G.A., R.J. Dinus & A.G. Kais, 1975. Variation in pathogenicity of diverse sources of Cronartium fusiforme on selected slash pine families. Phytopathology 65: 170-175.

Wells, O.O. & P.C. Wakeley, 1966. Geographic variation in survival, growth, and fusiform-rust infection of planted loblolly pine. Forest Science Monograph II. 40 p.

Wells, O.O. & G.L. Switzer, 1971. Variation in rust resistance in Mississippi loblolly pine. Eleventh Southern Forest Tree Improvement Conference Proceedings, 1971. p. 25-30.

Variation in pathogenicity and virulence in Fomes annosus and Armillaria mellea

Robert D. Raabe

Department of Plant Pathology, University of California, Berkeley, California 94720, USA

ABSTRACT

There are varying reports regarding the variability in pathogenicity and virulence of *Fomes annosus*. The fact that there are reports of the fungus differing in these characteristics suggests that the fungus is somewhat variable. It also suggests that this is an area where more research needs to be done. Reports that crosses have been made in culture and that fruiting bodies have been produced in culture should lead to studies on the inheritance of many characteristics of the fungus, including its pathogenicity and virulence. In contrast, *Armillaria mellea* is an extremely variable fungus in many characteristics including its growth in culture media, pathogenicity and virulence. It also is variable in mating types and recent research has shown that based on this phenomenon, these are 3 distinct 'biologic species' in Finland and 10 in the United States. Matings between 5 from Europe and the 10 from the United States have shown there to be some interfertility between isolates from the 2 continents. The complications resulting from such 'biologic species' suggest additional research needs to be done concerning various aspects of this fungus, and particularly in regard to pathogenicity and virulence. Although fruiting bodies have been produced in culture, a consistent method for this is not available. When developed, it will be of great value in the study of many inherited characteristics including those regarding pathogenicity and virulence.

INTRODUCTION

A review of the literature regarding *Fomes annosus* (Fr.) Cooke and *Armillaria mellea* (Vahl ex Fr.) Kummer reveals an extensive amount of research concerning these fungi and the diseases resulting from infection by them. Very little, however, has been published regarding the genetic variability of these fungi, particularly in regard to pathogenicity and virulence. One could speculate as to the reasons for this. Inoculations of seedlings are easy but inoculations of larger trees are more difficult. This might be complicated by the fact that trees of given species may vary in susceptibility depending upon the age when exposed to the pathogen. For example, peach (*Prunus persica* Batsch) is susceptible at any age when exposed to *A. mellea*. In contrast, *Pinus radiata* D. Don is very susceptible as a young plant and can be killed by the fungus in a little over a month in inoculations in a glasshouse (Raabe, 1967). In its native area of coastal California, this plant is rarely infected and more rarely killed by the fungus. When tested for susceptibility in an open plot, it was found to be very resistant (Raabe, 1979). Another reason might be that knowledge of the biology and genetical constitution of pathogens frequently is correlated with programs for breeding or selecting for resistance. Such research programs involve long term research projects with few rewards regarding research publications and therefore researchers are reluctant to be involved. One other reason is that the difficulty of producing basidiocarps of these fungi under controlled conditions is a deterrent to the study of any genetic variability of the organisms.

There may be other reasons for the lack of research in this area but the purpose of this paper is to bring together the information available regarding variability in these fungi. To do this most efficiently, each fungus will be discussed separately unless information may apply to both fungi. Information concerning general variability will be included though the main emphasis will be variability in pathogenicity and virulence.

Pathogenicity here, refers to the ability of an organism to produce a disease. Virulence, as used here, refers to the disease-producing ability of an organism as measured by the severity of the disease.

FOMES ANNOSUS

Although found mostly in temperate zones throughout the world, *F. annosus* generally is not considered to be an exceedingly variable fungus. Roll-Hansen (1940) noted the absence of distinct strains in eastern and western Norway. Dimitri (1974) reported minor differences in the virulences of different races of the fungus. Kuhlman (1969c) used 8 isolates of the fungus from 2 hosts from southern United States to inoculate seedlings of *Pinus taeda* L. in a glasshouse and found no variations in virulence. Lane & Witcher (1974) used 19 isolates to inoculate pine seedlings and they varied only slightly in virulence.

Possibly related to pathogenicity and virulence was the report of Bassett et al. (1967), who found a toxin produced by *F. annosus*. Of the 9 isolates used, 7 produced the toxin. The authors concluded that toxin production was not related to pathogenicity or virulence. Gibbs (1972) subjected 10 isolates from pine and 10 from non-pine hosts to volatile components of pine oleoresin and to various concentrations of pinosylvin and pinosylvin monomethyl esther. No detectable differences were found between the 2 groups in regard to their reactions to the substances tested.

In contrast to reports of little variation in pathogenicity and/or virulence are some reports of much variation in these characteristics. Kuhlman (1970), in further experiments, used 23 isolates of the fungus collected from various parts of the world and from various hosts. He found the isolates varied considerably in their virulence on seedlings of *P. taeda* and in addition, found they varied in virulence on different hosts. James (1977) used 5 isolates to inoculate large roots of *P. ponderosa* Laws. and found considerable differences in the percentages of infection and the amounts of colonization. He also used 10 isolates to inoculate seedlings of pine and found considerable variations in percentages of infection, numbers of trees killed and colonization rates. In another experiment, using mass conidial inoculum from 2 isolates, he found that the conidia from the more virulent isolate proved to be more virulent that those from the less virulent isolate. Negrutsky (1978) also reported differences in aggressiveness to affected plants by different forms of the fungus.

Platt et al. (1965) studied the wood-rotting ability of different isolates on excised branches and roots, and found considerable differences existed between the isolates used. They also found that single basidiospore isolates varied in their ability to rot wood and that they were generally less vigorous than tissue isolates. Similar results were reported by Kaufmann & Wellendorf (1978), who reported a procedure to by-pass the years involved in testing trees for resistance. They found that the fungus, when grown on heartwood sawdust in 1 % agar, varied in growth rates dependent not only on the source of the wood but also on the strain of the fungus used. If this could be used to determine accurately the resistance of trees, great strides could be made in solving the *F. annosus* problem.

Of interest are the researches of Korhonen (1978a) and Worrall (1979). The former, in Finland, divided the fungus into 2 groups. One, called S, from spruce was found mostly on *Picea abies* (L.) Karst. This type also was found to attack *Pinus sylvestris* L. saplings near infected spruce stumps. It was not isolated from mature pines and there were few records of the S group on plants other than spruce or pine in Finland. Another group of isolates, from pine, was called the P group. These were common on pine of all ages and some also were found on *Juniperus communis* L., *Betula* sp., *Alnus* sp., *Calluna* sp., and other plants. In further studies, the S group was found to be present in Norway, Denmark, Germany, Italy, India, and Japan. The P group was found to occur in Norway, Denmark, Scotland, France, Italy, Madeira, Canada, and the United States. In addition, cultures were received from Australia, New Zealand, the Fiji Islands and one from Japan that did not belong in either group. Worrall (1979) found in California that isolates from pine were more virulent on pine than on fir and that though most isolates from fir were more virulent on fir than on pine, some isolates from fir were more virulent on pine. He suggested that the pine isolates might be excluded from fir whereas the fir isolates could infect both pine and fir.

An exciting contribution by Korhonen (1978a) was the finding that mating in *F. annosus* can take place in culture. The fungus showed a bipolar mating system and heterocaryosis was indicated by the presence of clamp connections. This information with that of Sychev & Negrutsky (1978), in which they reported the production of fruiting bodies in pure culture, should lead to many studies regarding the genetics of *F. annosus* and its variability in many characteristics including pathogenicity and virulence.

The fact that *F. annosus* produces a conidial stage must also be considered. Little is know of the importance of the conidia in the disease cycle. It has been reported they are not found frequently in nature (Risbeth, 1951) though Risbeth (1957) and Kallio (1971) reported natural production of conidia on stumps covered by branches. Morris & Knox (1962), Kallio (1967) and Nuorteva & Laine (1968) also have reported the production of conidia on stumps. If they are important, this might account, in part, for the lack of variability. The fact that the numbers of nuclei in the conidia may vary from 1 to 4 or more (Griffin & Wilson, 1967; Ahrberg, 1975) also might have an effect upon the variability of the fungus.

It is known that under laboratory conditions, the conidia can infect seedlings of pine and spruce (Hüppel, 1970). Kuhlman (1969b) also used conidia for infecting roots of pines under natural conditions. It is also known that conidia can be used to infect stumps (Kuhlman & Hendrix, 1964; Ross, 1968; Hunt et al., 1974, and others). In their experiments, Kuhlman and Hendrix found basidiospores were more than 600 times as effective as conidia in colonizing stumps.

Of interest is the fact that the conidia are more resistant to heat than basidiospores (Ross, 1969) and are able to retain their viability up to 10 months in soil under field conditions (Kuhlman, 1969a), suggesting the possibility that as more resistant structures, they might have an important part in the disease cycle.

Although not connected with virulence or pathogenicity, the variability of cultural characteristics when grown on media may give an indication of the general variability of a fungus. In regard to *F. annosus*, there are varying reports regarding the variability of the fungus in culture. Bega & Hendrix (1962), Bega (1963), de Azevedo et al. (1974), Fedorov & Staichenko (1974), and Negrutsky (1978) reported considerable variations in cultural characteristics. Based on cultural characteristics, Kaufamnn & Wellendorf (1978) separated isolates of the fungus from Denmark and Germany into 2 - 5 strains. In contrast, Etheridge (1955) reported that cultures originating in Europe could not be distinguished from those originating in North America by colony appearance, growth rate, pH optimum or cellulitic activity. He divided them into 3 growth rate groups and noted that though some variability occurred, all could be included in these groups. Courtois (1974), based upon cultural characteristics, reported 2 ecotypes of the fungus. Roll-Hansen (1940) and Cowling & Kelman (1964) reported little effect of temperature on the growth of different isolates in culture. The variability of this fungus in culture thus is somewhat in question.

Of interest are the reports of Dimitri (page 260 of this book) and von Weissenberg (in prep.). Dimitri reported that the fungus is known to vary considerably in in-vitro characteristics. However, using a total of 10 isolates of the fungus in inoculation studies, he found that with his method of using only 3 isolates to inoculate a single tree, there were no differences in the growth of the different isolates in any given tree. Von Weissenberg used single basidiospore isolates, crosses of single basidiospore isolates and hyphal tip isolates from Finland. He also found they varied considerably in in-vitro characteristics but did not vary much in their growth rates when used to inoculate trees. Delatour (page 268 of this book) showed various isolates of the fungus grew differently in culture and he found that there was a negative correlation between the growth of the fungus on malt agar and the growth of the isolate in trees from where they had been isolated.

ARMILLARIA MELLEA

Although there is a present tendency to use the binomial *Armillariella mellea* (Vahl) Karst., in this paper the suggestion of Thiers & Sundberg (1976) will be followed, thus retaining the name *Armillaria mellea* (Vahl ex Fr.) Kummer.

In contrast to *F. annosus*, *A. mellea* is an exceptionally variable fungus. Isolates from infected plants throughout California, single-spore isolates from a single sporophore (Raabe, 1966a), and single-spore isolates from various California crops (Maclean, 1950) showed considerable variations in many characteristics when grown on potato dextrose agar. These variations were much more pronounced than those reported by Hintikka (1973) where the colonies produced mainly an aerial mycelium with usually no brown crustose formation.

The first report of variation in pathogenicity in *A. mellea* was that of Childs & Zeller (1929) who reported an oak strain which was pathogenic and a fir strain which was non-pathogenic. Van Vloten (1936) found variations in the ability of different isolates to attack potato tubers. The fact that the isolate most virulent on potato also infected *Ligustrum* sp., *Pinus* sp., and *Rosa* sp., led him to state that specialization by the fungus was not apparent. Bliss (1946) also noted differences in the pathogenicity of isolates of *A. mellea*. Raabe (1955, 1967) reported distinct differences in the pathogenicities and virulences of field isolates of *A. mellea* when 3 different host species were inoculated under glasshouse conditions. These differences were also shown to occur when single-spore isolates from a single sporophore were used to inoculate the same 3 host species (Raabe, 1972). In that experiment, the single-spore isolates were compared with the parent isolate and none were as virulent as the parent and only 1 was as pathogenic as the parent. The results of the inoculations using field isolates and those using single-spore isolates were tabulated (Raabe, 1969) and in general, the single-spore isolates were less pathogenic and much less virulent than isolates from infected plants. This apparent decrease in vigor is also found when the isolates are grown in culture. The same has been shown for *F. annosus* (Kuhlman & Hendrix, 1964; Platt et al., 1965). This may be due to genetic variability alone, the small sample of isolates selected or may be due to the fact that hyphae in a haploid mycelium are less vigorous than when in the diploid condition. Also, it is important to consider that in comparing single-spore isolates with those from infected plants, the latter were selected because of their pathogenicity, and possibly because of their virulence.

In 1973, Hintikka showed that mating of single-spore isolates of *A. mellea* could take place when pairing compatible isolates in culture. With a compatible reaction, the resultant new diploid produced a distinctive different growth pattern in culture and therefore was detected easily. This has led to the determination that in Finland, there are 3 distinct intersterile 'biologic species' in *A. mellea* (Korhonen, 1978b) and that in North America, there are at least 10 'biologic species' (Anderson & Ullrich, 1979). The relationships between the 10 North American 'biologic species' and 5 'biologic species' from Europe also have been studied (Anderson et al., 1980).

The discovery of different 'biologic species' of *A. mellea* in rather limited areas raises questions as to how many other 'biologic species' might be found throughout the world. It also raises questions as to the pathogenicities and virulences of these many groups and the individuals in these groups. The discovery of an easy laboratory method for determining the degrees of virulences and perhaps the degrees of pathogenicities would be most helpful. An attempt to do this has not been successful (Raabe, 1969) but if found, such methods would be most helpful in a testing program for selecting resistant plants. Presently, in California, a program is in progress to find resistant species of ornamental plants and fruit trees (Raabe, 1966b, 1979). In such experiments, plants are grown in areas known to be infested with the fungus. In addition, inoculum of 2 pathogenic and virulent isolates of the fungus are put near the test plants when they are planted. Plants are rated as to the length of time before they are killed. Those still alive after 10 years are dug and the roots examined and rated as to the amounts of infection.

A most important aspect in the research regarding this fungus would be the discovery of a reliable method of producing basidiocarps easily. The closely related *A. tabescens* (Scop. ex Fr.) Emel. produces basidiocarps readily in culture and successful matings in culture have produced basidiocarps (Tang & Raabe, 1973) but *A. mellea* fruits only occasionally and reluctantly, though there are reports of it fruiting in culture (Reitsma, 1932; Raabe, 1972). If a consistent and reliable method for this could be found, many studies on this important pathogen could be made including those regarding pathogenicity and virulence.

REFERENCES

Ahrberg, H.E., 1975. Untersuchungen zur Cytologie und zum Wachstum verschiedener Stämme von Fomes annosus (Fr.) Cooke. European Journal of Forest Pathology 5: 287-303.

Anderson, J.B. & R. Ullrich, 1979. Biological species of Armillaria mellea in North America. Mycologia 71: 402-414.

Anderson, J.B., K. Korhonen & R. Ullrich, 1980. Relationships between European and North American biological species of Armillaria mellea. Experimental Mycology 4: 87-95.

Azevedo, N.F. dos S. de & M. de F.M.A. Moniz, 1974. Influence of temperature, pH and nutrients on the growth of Fomes annosus isolates. Proceedings 4th International Conference on Fomes annosus: 163-168.

Basset, C., R.T. Sherwood, A. Kepler & P.B. Hamilton, 1967. Production and biological activity of fomannosin a toxic sequeterpene metabolite of Fomes annosus. Phytopathology 57: 1046-1052.

Bega, R.V., 1963. Symposium on root diseases of forest trees: Fomes annosus. Phytopathology 53: 1120-1123.

Bega, R.V. & F.F. Hendrix, 1962. Variation of monobasidiospore isolates of Fomes annosus. Phytopathology 52: 3 (abstract).

Bliss, D.E., 1946. The relation of soil temperature to the development of Armillaria root rot. Phytopathology 36: 302-318.

Childs, L. & S.M. Zeller, 1929. Observations on Armillaria root rot of orchard trees. Phytopathology 19: 869-873.

Courtois, H., 1974. About the occurrence of ecotypes of the species Fomes annosus (Fr.) Cooke. Proceedings 4th International Conference on Fomes annosus: 125-132.

Cowling, E.B. & A. Kelman, 1964. Influence of temperature on growth of Fomes annosus isolates. Phytopathology 54: 373-378.

Dimitri, L., 1974. Resistenzforschung bei der Fichte gegenüber dem Fomes annosus. Proceedings of the 4th International Conference on Fomes annosus. p. 76-80.

Etheridge, D.E., 1955. Comparative studies of North American and European cultures of the root rot fungus, Fomes annosus (Fr.) Cooke. Canadian Journal of Botany 33: 416-428.

Fedorov, W.I. & N.I. Staichenko, 1974. [Rate of growth and activity of oxidizing enzymes of different Fomitopsis annosa strains]. Proceedings 4th International Conference on Fomes annosus. p. 106-113.

Gibbs, J.N., 1972. Tolerance of Fomes annosus to pine oleoresins and pinosylvins. European Journal of Forest Pathology 2: 147-151.

Griffin, B.R. & C.L. Wilson, 1967. Asexual nuclear behavior and formation of conidia in Fomes annosus. Phytopathology 57: 1176-1177.

Hintikka, V., 1973. A note on the polarity of Armillaria mellea. Karstenia 13: 32-39.

Hunt, R.S., W.W. Wilcox & F.W. Cobb, 1974. Resistance of stump tops to colonization by Fomes annosus. Canadian Journal of Forest Research 4: 140-142.

Hüppel, A., 1970. Inoculation of pine and spruce seedlings with conidia of Fomes annosus. Proceedings 3rd International Conference on Fomes annosus. p. 54-56.

James, R.L., 1977. The effects of photochemical air pollution on the epidemiology of Fomes annosus. Ph.D. Thesis, University of California, Berkeley.

Kallio, T., 1967. Havaintoja maannousemasientesta kevaalla 1967. Metsätal. aikakauslehti 7-8: 228-230, 234.

Kallio, T., 1971. Incidence of the conidiophores of Fomes annosus (Fr.) Cooke on the logging waste of spruce (Picea abies (L.) Karst.). Acta Forestalia Fennica 124. 9 p.

Kaufmann, U. & H. Wellendorf, 1978. Genetic variation in Picea abies (L.) Karst and Fomes annosus (Fr.) Cooke shown by growth of the fungus on wood dust. European Journal of Forest Pathology 8: 226-239.
Korhonen, K., 1978a. Intersterility groups of Heterobasidion annosum. Seloste: Juurikäävän risteytymissuhteet. Communicationes Instituti Forestalis Fenniae 94(6): 1-25.
Korhonen, K., 1978b. Interfertility and clonal size in the Armillaria mellea complex. Karstenia 18: 31-42.
Kuhlman, E.G., 1969a. Survival of Fomes annosus spores in soil. Phytopathology 59: 198-201.
Kuhlman, E.G., 1969b. Number of conidia necessary for stump root infection by Fomes annosus. Phytopathology 59: 1168-1169.
Kuhlman, E.G., 1969c. Inoculation of loblolly pine seedlings with Fomes annosus in the greenhouse. Canadian Journal of Botany 47: 2079-2082.
Kuhlman, E.G., 1970. Seedling inoculations with Fomes annosus show variation in virulence and in host susceptibility. Phytopathology 60: 1743-1746.
Kuhlman, E.G. & F.F. Hendrix Jr., 1964. Infection, growth rate and competitive ability of Fomes annosus in inoculated Pinus echinata stumps. Phytopathology 54: 556-561.
Lane, C. & W. Witcher, 1974. Studies on Fomes annosus and the infection of pine seedlings. Proceedings 4th International Conference on Fomes annosus: 227-230.
Maclean, N.A., 1950. Variation in monospore cultures of Armillaria mellea. Phytopathology 40: 968 (abstract).
Morris, C.L. & K.A. Knox, 1962. Fomes annosus: A report on the production of conidia in nature and other studies in Virginia. Plant Disease Reporter 46: 340-341.
Negrutsky, S.F., 1978. [Position and perspectives on investigations about Fomitopsis annosa (Fr.) Karst in USSR and in countries of Eastern Europe.] Proceedings 5th International Conference on Problems of Root and Butt Rot of Conifers: 19-26.
Nuorteva, M. & L. Laine, 1968. Über die Möglichkeiten der Insekten als Überträger des Wurzelschwamms (Fomes annosus (Fr.) Cooke). Annales Entomologici Fennici 34: 113-135.
Platt, W.D., E.B. Cowling & C.S. Hodges, 1965. Comparative resistance of coniferous root wood and stem wood to decay isolates of Fomes annosus. Phytopathology 55: 1347-1353.
Raabe, R.D., 1955. Variation in pathogenicity of isolates of Armillaria mellea. Phytopathology 45: 695 (abstract).
Raabe, R.D., 1966a. Variation of Armillaria mellea in culture. Phytopathology 56: 1241-1244.
Raabe, R.D., 1966b. Testing plants for resistance to oak root fungus. California Agriculture 20: 12.
Raabe, R.D., 1967. Variation in pathogenicity and virulence in Armillaria mellea. Phytopathology 57: 73-75.
Raabe, R.D., 1969. Cultural variations of Armillaria mellea not related to pathogenicity and virulence. Proceedings 1st International Citrus Symposium 3: 1263-1272.
Raabe, R.D., 1972. Variation in pathogenicity and virulence in single-spore isolates of Armillaria mellea. Mycologia 64: 1154-1159.
Raabe, R.D., 1979. Resistance or susceptibility of certain plants to Armillaria root rot. University of California Division of Agricultural Sciences Leaflet 2591. 11 p. (review).
Reitsma, J., 1932. Studien über Armillaria mellea (Vahl) Quel. Phytopathologische Zeitschrift 4: 461-522.
Risbeth, J., 1951. Observations on the biology of Fomes annosus with particular reference to East Anglian pine plantations. II. Spore reduction, stump infection and saprophytic activity in stumps. Annals of Botany, New Series 15: 1-22.
Risbeth, J., 1957. Some further observations on Fomes annosus Fr. Forestry 30: 69-89.
Roll-Hansen, F., 1940. Underskelseer over Polyporus annosus Fr. saerlig med henblikk pa dens forekomst i sonnafjelske Norge. Meddelelser fra Det Norske Skogforsøksvesen 7: 1-100.
Ross, E.W., 1968. Duration of stump susceptibility of loblolly pine to infection by Fomes annosus. Forest Science 14: 206-211.
Ross, E.W., 1969. Thermal inactivation of conidia and basidiospores of Fomes annosus. Phytopathology 59: 1798-1801.
Sychev, P.A. & S.F. Negrutsky, 1978. [The formation of fruit bodies by different strains of Fomitopsis annosa (Fr.) Karst and their interactions in culture.] Mikologiya i Fitopatologiya 12: 148-151. (Review of Plant Pathology 58:547.)
Tang, H. & R.D. Raabe, 1973. Sporophore production and heterothallism in Clitocybe tabescens. Phytopathology 63: 1218 (abstract).
Thiers, H.D. & W.J. Sundberg, 1976. Armillaria (Tricholomataceae, Agaricales) in the

western United States including a new species from California. Madrano 23: 448-453.
Vloten, H. van, 1936. Onderzoekingen over Armillaria mellea (Vahl) Quel. Fungus, Wageningen 8: 20-23.
Weissenberg, K. von. Between-isolate variation in pathogenicity of Heterobasidion annosum to Picea abies (in preparation).
Worrall, J., 1979. Ecology of Fomes annosus (Fr.) Karst on Abies concolor (Gond.) Lindl. ex Hildebn. I. Development of host specialization to pine and fir in Fomes annosus isolates from California. Report for Masters Degree, University of California, Berkeley.

Some host/parasite relationships between Norway spruce (Picea abies) and Fomes annosus

L. Dimitri

Hessische Forstliche Versuchsanstalt, 3510 Hann. Münden, BRD

ABSTRACT

Research on the root rot caused by *Fomes annosus* has been underway for more than 100 years. However, experiments on host-parasite interactions and the norm of reaction (i.e., the resistance) against *Fomes* have been initiated only quite recently. The paper presents some results of extensive work (Dimitri, 1980) on the interrelationships between host/parasite as well as host/environment/parasite, in the case of Norway spruce inoculated with *F. annosus*. In spite of considerable differences in morphological and physiological features of special strains of the parasite *F. annosus* in vitro, we have so far been unable to find any significant difference in the spread of these strains in the host plants in vivo. We should, therefore, not imply that strains show the same parasitic activity in vivo as in vitro. On the basis of present results we can assume a horizontal resistance of Norway spruce to *F. annosus*. Environment seems to influence both height-growth of the host and spread of the mycelium in the trees. Most provenances show the same trend: the better the height-growth of the host, the higher the mycelium-height in the trees. But in each provenance a certain portion of individuals is found, which combine outstanding height-growth with high phenotypic resistance against the spread of the mycelium. Thus, in view of resistance to *F. annosus*, the selection of Norway spruce individuals will be more efficient than that of provenances.

INTRODUCTION

Stem rot is the most important and most serious fungal disease which afflicts Norway spruce (*Picea abies* L. Karst.), the silviculturally and economically most significant tree species in the Federal Republic of Germany. The causes of this damage and the organisms which induce it can be very different. When the lower part of the stem is wounded, the ensuing wound-rot is caused mainly by 2 *Stereum* species. Heart-rot, on the other hand, which spreads up from the root zone, is usually caused by the root and butt rot fungus, *Fomes annosus* (Fr.) Cooke. A substantial reduction of wound-rot definitely appears to be possible if surface wounds receive an early treatment with suitable preservatives (Dimitri & Schumann, 1975; Schönhar, 1979; Bonneman, 1979). Several preventive measures against heart-rot are also possible (Greig, 1980).

One possibility for reducing damage, which has so far received comparatively little attention and study, is the utilization of this tree species' natural resistance to *F. annosus*. Since this resistance is dependent upon several factors (virulence of the causative organisms, environmental conditions, etc.), it was investigated within the scope of an extensive project (Dimitri, 1980). Only a few results of the experiments, which have been described in detail in the above-mentioned paper, are specified here.

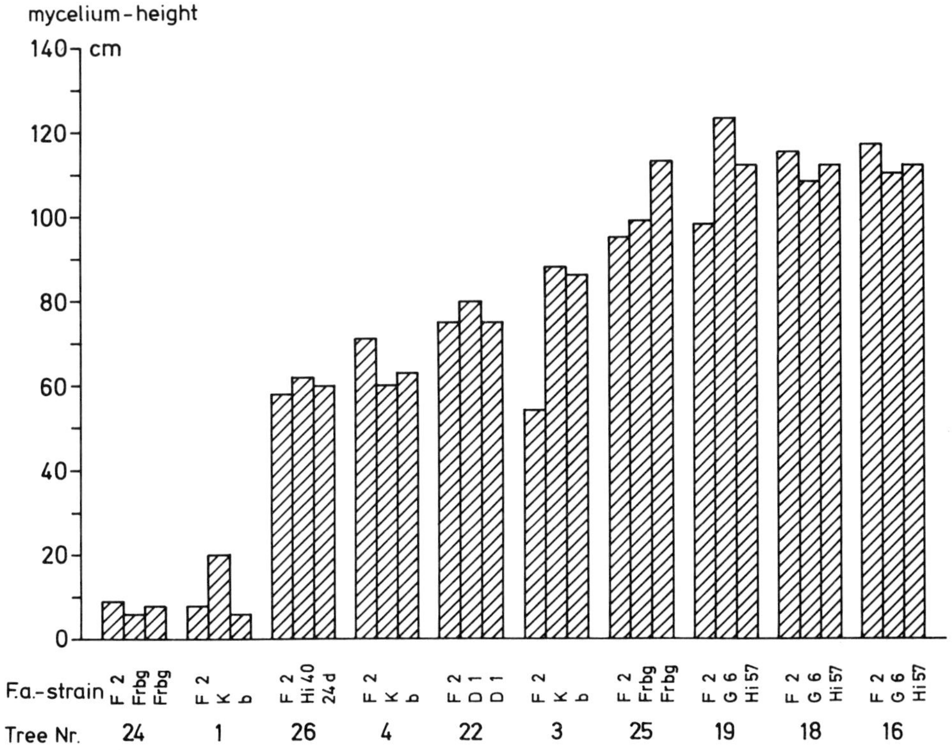

Figure 1. Spread of the mycelium of different *F. annosus* strains, 3 of which had been inoculated simultaneously into each of ten 60-year-old Norway spruce trees 12 months earlier.

RELATIONSHIP BETWEEN THE HOST PLANT AND THE PARASITE UNDER IDENTICAL SITE CONDITIONS

It is known that in laboratory tests different *F. annosus* strains vary distinctly in their morphological and physiological properties (Courtois, 1980; Dimitri, 1980). More important than the behaviour in vitro is the parasitic productive capacity of the parasite strains in vivo. Neither in simultaneously inoculated 8-year-old spruce clones nor in older spruce trees, which had been inoculated simultaneously with 2 *F. annosus* strains (obtained from different parts of West Germany) as well as a third check strain (F2-culture, obtained from a basidiospore) could any significant differences in the mycelium height be determined. As examples of the differing virulence of individual *F. annosus* strains in vivo, Figures 1 and 2 illustrate the results obtained in 60-year-old and 67-year-old spruce stands.

In both figures the dominating effect of the host plant upon the vertical spread of the mycelium can clearly be seen: if a tree has a high phe-

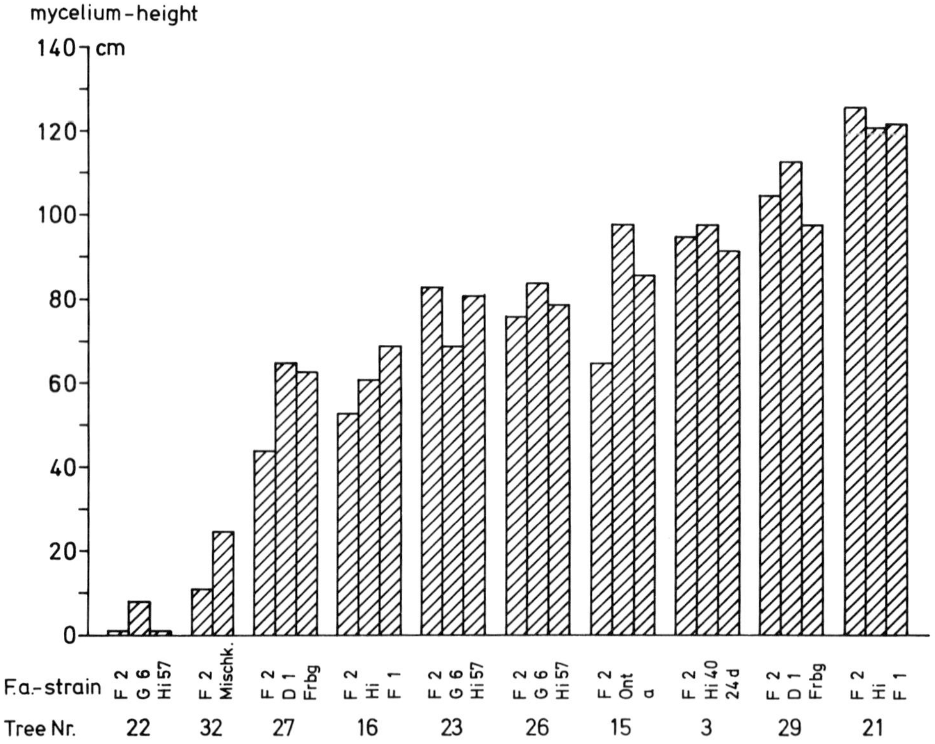

Figure 2. Spread of the mycelium of different *F. annosus* strains, 3 of which had been inoculated simultaneously into each of ten 67-year-old Norway spruce trees 12 months earlier.

notypic resistance, the spread of all *F. annosus* strains is checked to almost the same extent (e.g., trees no. 1 and 24 in Fig.1, as well as trees no. 22 and 32 in Fig. 2). Vice versa, all parasite strains can develop very well if the host tree has a low resistance (e.g., trees no. 16 and 18 in Fig. 1, as well as trees no. 21 and 29 in Fig. 2). If one compares the individuals in these 2 populations of spruce, each of which had been inoculated with the same *F. annosus* strains, the picture of Table 1 develops.

On the basis of these results, which are still to be substantiated by means of further experiments, it can be assumed that one cannot readily draw conclusions from morphological characteristics or from physiological behaviour in test cultures about the parasitic activity of different *F. annosus* strains in nature.

From the fact that older spruce trees react to several *F. annosus* strains in completely the same way (that is, either vulnerable or resistant), a horizontal (i.e. generalized) resistance can be deduced.

Table 1. Resistance and reaction of trees to *Fomes annosus* strains.

Tree No.	Strain	Resistance	Reaction
In Fig. 1			
1	F 2	high	
	K		different
4	b	medium	
24	F 2	high	
	Frbg		different
25	Frbg	low	
16	F 2	low	
18	G 6	low	identical
19	Hi 57	low	
In Fig. 2			
16	F 2	medium	
	Hi		different
21	F 1	low	
22	F 2	high	different
23	G 6	medium	identical
26	Hi 57	medium	

For the investigation of these interrelations, 3086 trees from 10 different Norway spruce provenances were inoculated with the F_2 *F. annosus* strain, half of these within, the other half outside of the vegetative period. These 10 provenances grew with approximately the same number of stems on 6 different sites. A description of the provenances, the sites, the methods of inoculation and investigation, as well as the individual results can be found in Dimitri (1980). Again, 2 diagrams will serve to illustrate these interrelations.

Fig. 3 shows the mean values for the height of the Norway spruce (HWL) and the height of the mycelium (MH) which were attained in trees of individual provenances on respectively the same sites after inoculation outside the vegetative period. As expected, the mean HWL of all provenances ($\bar{x}$ h) varied greatly from site to site, which is an indication of the differing site quality of the experimental sites for the spruce plantations. Among

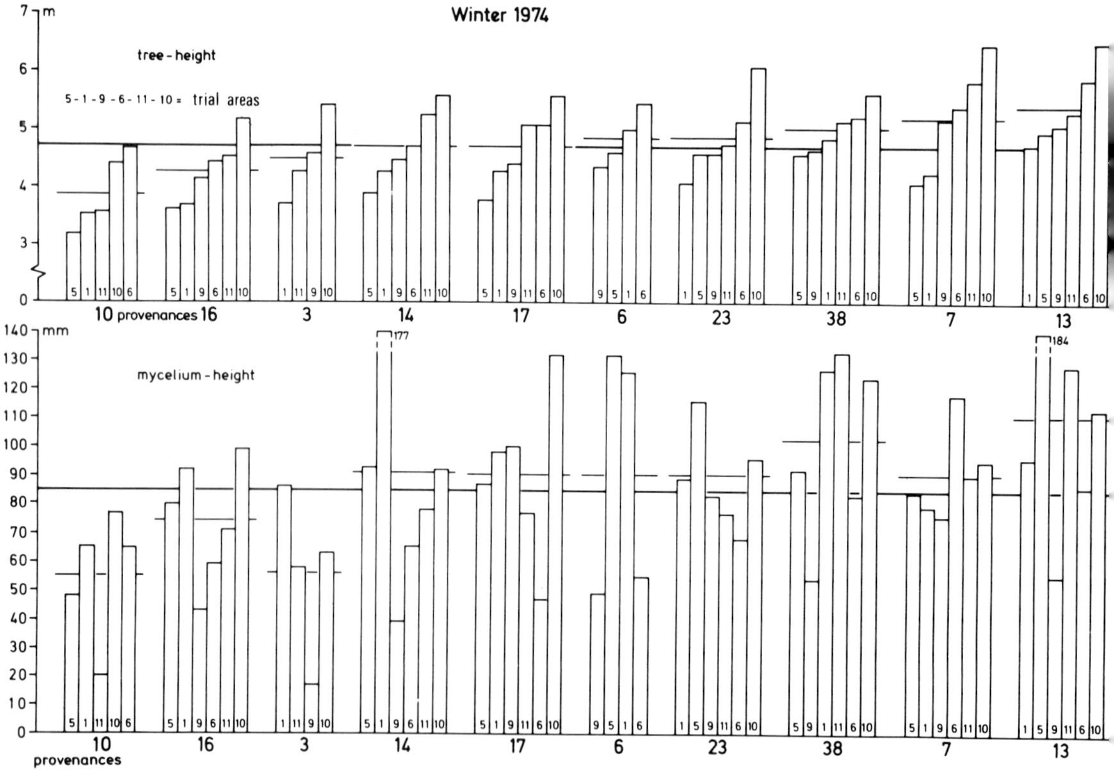

Figure 3. Average height of different Norway spruce provenances ($\bar{x}$ h) and average mycelium height ($\bar{x}$ m) in trees of these provenances in provenance-order inoculated outside the vegetative period 12 months earlier.

the provenances on the same site of each test plot it was possibly to ascertain substantial and statistically significant differences in the mean HWL, which clearly points out the differences in the hereditary production potential of the provenances.

Both the individual values and the respective means of the mycelium height (MH) have a much greater deviation than those of the HWL. Among different sites there can be considerable variation in the mean mycelium height ($\bar{x}$ m) of all provenances. Statistically significant differences could be determined for this property ($\bar{x}$ m) as well.

The very substantial differences among the mean values for the mycelium height of individual provenances on the same site (e.g., between provenances 10 and 38 in Nentershausen, site no. 11) could not be proved statistically significant by means of analysis of variance, due to the extremely large deviation of the single values.

Also in an analysis of covariance (with height as covariant) there were no significant differences between the tested variables; among the sites, however, significant differences in the mycelium height were found.

Sites on which almost all tested Norway spruce provenances have attained an above-average HWL but a mycelium height far below average (e.g., Bieber, site no. 6) are especially suited for the cultivation of spruce.

If one compiles the data for the characteristics 'height' and 'mycelium height', which were attained on different sites after inoculation of the respective provenances outside of the vegetative period, the following observations can be made (Fig. 4).

- There seems to be a positive correlation between height and mycelium height: all provenances with a mean HWL ($\bar{x}$ h) which reaches or surpasses the overall average ($\bar{x}$ H) also consistently have a mycelium height ($\bar{x}$ m) surpassing the overall average ($\bar{x}$ M). In a calculation of correlation according to Pearson, the interrelations of HWL and MH proved to be statistically significant.
- The best-growing provenance (No. 13) had the lowest resistance to the spread of *F. annosus* mycelium.
- The mean HWL of provenance No. 7 ranks close behind that of provenance No. 13, however, the average mycelium spread was distinctly lower (by 20 %).
- Individual provenances appear to be suited for different sites: thus, for example, provenance 23 had a below-average HWL in Königstein (5) but an above-average MH, whereas this relationship was exactly the opposite in the forest district Bieber (6). (Compare also provenance 3 in Reinhardshagen (1) versus in Ober-Ramstadt (10).).

Even if the mean values of the HWL and the ML for each provenance and repetition are taken as a ratio of the respective sample plot mean (= 100 %) (relative tree and mycelium height), it can be seen that all factors with a positive influence on the HWL (nutrient supply, moisture

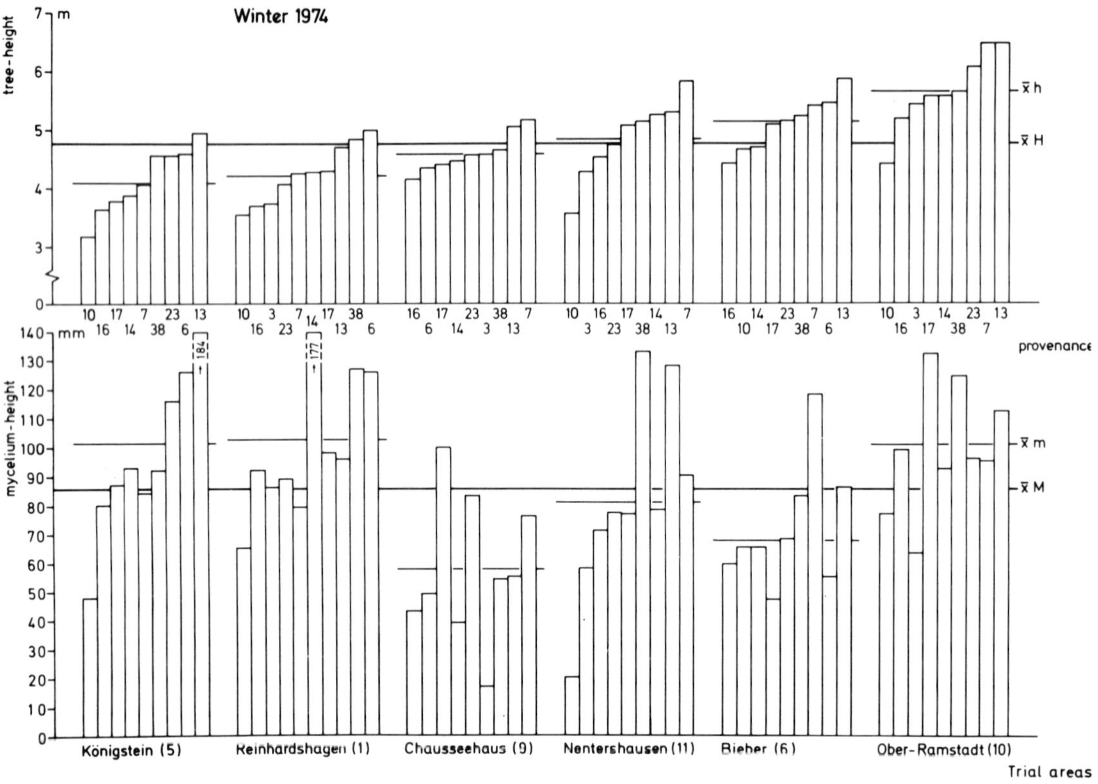

Figure 4. Average height of several Norway spruce provenances ($\bar{x}$ h) and average mycelium height ($\bar{x}$ m) in the trees of these provenances in site-order inoculated outside the vegetative period 12 months earlier.

content, temperature, etc.) also appear to influence the development of decay in the same direction. Analyses of regression between tree height and mycelium height for each site and provenance could establish only a low correlation coefficient (r) for this tendency, in other words, no close relationship was found.

It has been made evident that a selection of individuals appears to promise more advantages, also for *F. annosus* resistance, than selection of provenances. The proportion of individuals with above-average yield in growth (HWL 50 % above the mean of the repetition) and above-average phenotypic resistance to the spread of mycelium (MH 50 % below the mean of the repetition) is much higher (6.4 to 10.3 %) in the best-growing provenances (Cottbus, Chausseehaus, Rothenkirchen, Westerhof) than in the others (0.4 to 4.0 %). The influence of site in this respect is less clearly recognizable; the proportion of positive individuals on the whole was nearly the same on all tested sites (approx. 4.5 %).

REFERENCES

Bonnemann, I., 1979. Untersuchungen über die Entstehung und Verhütung von "Wundfäulen" bei der Fichte. Dissertation and der Forstl. Fakultät der Univ. Göttingen. 173 p.

Courtois, H., 1980. Pathogen biology (Fomes annosus). In: L. Dimitri: Proceedings of the Fifth International Conference of Root and Butt Rot in Conifers. p. 43-53.

Dimitri, L., 1980. Untersuchungen über die Widerstandsfähigkeit der Fichte (Picea abies Karst) gegenüber dem Wurzelschwamm (Fomes annosus (Fr.) Cooke). Schriftenreihe Forstl. Fak. Univ. Göttingen Nr. 66. 126 p.

Dimitri, L. & G. Schumann, 1975. Die Wundfäule - ein aktuelles Problem der Forstproduktion. Holz-Zentralblatt 101: 803-806.

Greig, B.J.W., 1980. Chemical, biological and silvicultural control of Fomes annosus. In: L. Dimitri: Proceedings of the Fifth International Conference on Problems of Root and Butt Rot in Conifers. p. 75-84.

Schönhar, S., 1979. Erprobung von Wundschutzmitteln auf der Fichte. Forst- u. Holzwirt 34: 12-14.

Behaviour of Fomes annosus in the stem of Norway spruce and in the laboratory

C. Delatour

CNRF, Laboratoire de Pathologie forestière, Champenoux, 54280 Seichamps, France

ABSTRACT

The growth rate of *F. annosus* in the stem of Norway spruce is connected with in vitro characters such as mycelial growth rate on malt-agar (negative correlation). This demonstrates that fungus variations are at least partially responsible for variations of the disease in nature. Such variations have to be taken into account for breeding experiments. Observations indicate that *F. annosus* may encounter limitations in height extension in stems, thus leading to the Potentially Colonizable Column concept. The significance of data achieved by the stem inoculation method are discussed and experiments comparing natural and artificial infections are advocated.

INTRODUCTION

Very little is known about the behaviour of *F. annosus* (Fr.) Cooke inside trees, e.g. in Norway spruce (*Picea abies* (L.) Karst.). It is known that Norway spruce sapwood can exhibit strong reaction against diametrical growth of *F. annosus* in the stem (Shain, 1971), but does *F. annosus* encounter limitations in its height development in trees? This is probably true because some workers (Weissenberg, 1975; Dimitri & Kliefoth, 1978) have found that *F. annosus* does not grow equally in all the inoculated trees, allowing the hypothesis of an internal resistance. But is this resistance the same in all parts of the stem wood (excluding resistant sapwood) and during the whole life of the tree?

On the other hand the parasite itself exhibits variations studied by some authors (e.g. Courtois, 1972; Kaufmann & Wellendorf, 1978) in the laboratory, but we do not know exactly their significance in nature. I performed in Nancy some observations and experiments in this field that are the subject of this paper.

GROWTH RATE OF F. ANNOSUS IN THE STEM OF NORWAY SPRUCE

Growth in natural conditions

My observations have been performed in a second generation stand of Norway spruce, 60 years old, growing on calcareous soil 60 km S.W.of Nancy. In 1973 and again in late 1977 I determined the height of *F. annosus* in the stems of 46 trees by the increment borer method, and thus measured the growth rate of the fungus in every individual tree.

Results: Variations are very large in height and also in growth rate (Table 1). These variations are not clearly explained by tree size (height and diameter). The height reached by the fungus is probably related, at least partially, to the infection age which is not known. The growth rate might be connected with the height of *F. annosus* in the stem as we can expect slower growth at great heights. This assumption was not exactly confirmed within our 46 trees: indeed the correlation between these 2 characters is not significant ($r = -0.11$).

Growth following inoculation at different heights in the stem

To achieve a clearer view of this height effect on the growth rate of the fungus I performed trial inoculations of some trees. The inoculation method was similar to Dimitri's, i.e. an inoculated dart of wood introduced in a bore hole. Four trees, 25 years old, were inoculated in April 1979 from the soil level up to the living crown. One strain was used, collected from an infected *Abies grandis* Lindl. in Dec. 1976 at the Arboretum des Barres (Center of France), then not stored on artificial medium but in living spruce. This strain belongs to the S intersterility group. The trees were felled 6 months later, discs were cut into them and observed for *F. annosus*.

In spite of the very small number of trees studied some interesting phenomena were observed (Fig. 1). An unquestionable result is the identical growth rate of *F. annosus* in every tree at 6 m high, which is approximately the living crown level. Below, the growth rate of the fungus is very erratic, although we generally observe greater growth rates at lower levels. Incidentally, we can emphasize the annual growth of *F. annosus* inside the inoculated stems is frequently greater than the 'natural growth', which is a very common feature for inoculations ranging over a period of 12 months or less.

BEHAVIOUR OF F. ANNOSUS IN THE LABORATORY

From every tree in which I measured the growth of *F. annosus*, was collected in pure culture the corresponding strain of the fungus (at the top of the decay column). So it was possible to undertake some studies on variations of the parasite in connection with its natural performance, including (1) the mycelial growth on malt-agar, (2) the decay capacity, (3) the cellulolytic capacity, and (4) the intersterility group of each strain. Mycelial growth took place in petri dishes (diam. 8.5 cm) on Difco malt-agar (3 %) at 23 °C in darkness during 8 days. Decay capacity was

Table 1. Height growth (cm) of *F. annosus* in the stems of 46 Norway spruces.

Tree No.	Height of *F. annosus*		Height increase between	Annual growth rate	Tree No.	Height of *F. annosus*		Height increase between	Annual growth rate
	1973	1977	1973-1977			1973	1977	1973-1977	
1	380	605	225	56	24	260	460	200	50
2	270	407	137	34	25	285	447	162	41
3	312	437	125	31	26	335	410	75	19
4	85	272	187	47	27	210	335	125	31
5	210	260	50	13	28	110	272	162	41
6	582	644	62	16	29	160	335	175	44
7	480	617	137	34	30	110	272	162	41
8	382	482	100	25	31	110	222	112	28
9	425	475	50	13	32	260	435	175	44
10	582	694	112	28	33	235	285	50	13
11	182	269	87	22	34	310	460	150	38
12	282	319	37	9	35	410	522	112	28
13	25	137	112	28	36	185	447	262	66
14	290	407	117	29	37	160	372	212	53
15	330	592	262	66	38	260	385	125	31
16	230	380	150	38	39	185	347	162	41
17	210	210	0	0	40	210	247	37	9
18	360	522	162	41	41	185	247	62	16
19	235	445	210	53	42	185	222	37	9
20	460	572	112	28	43	135	297	162	41
21	235	347	112	28	44	315	427	112	28
22	260	360	100	25	45	295	470	175	44
23	360	522	162	41	46	265	352	87	22

General values:	Min.	Mean	Max.
height in 1973	25	268	582
height in 1977	137	396	694
height increase	0	128	262
annual growth rate	0	32	66

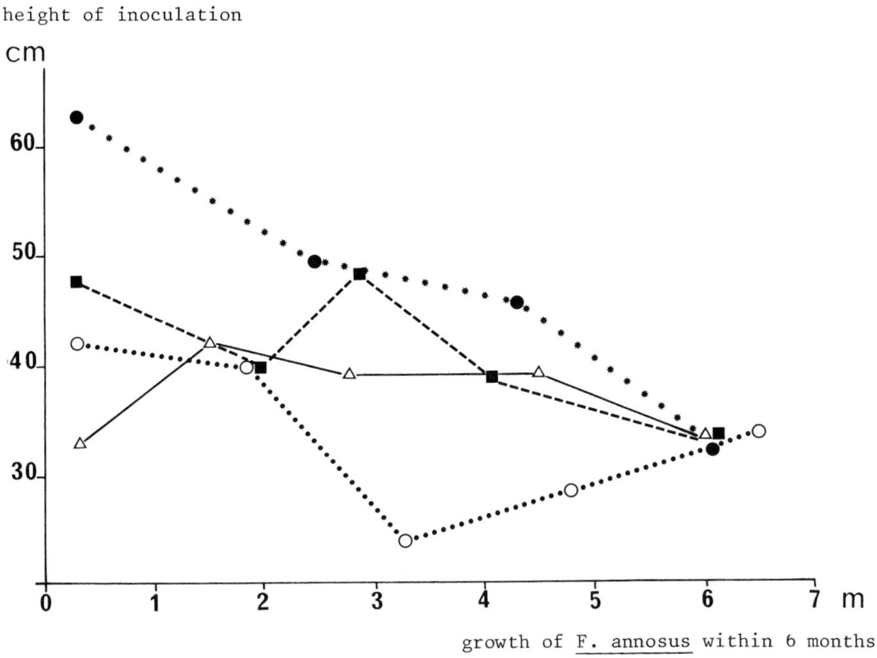

Figure 1. Growth of *F. annosus* in 4 Norway spruces inoculated at different heights.

studied on wood blocks in test tubes (Delatour & Sylvestre-Guinot, 1978). Cellulolytic capacity was studied following a classical method used by Eghbaltalab et al. (1976). Intersterility groups were defined following the Korhonen method (1978).

Results: Nearly all the strains belonged to the P intersterility group (No. 22 and 29 belonged to the S group), so that no comparison between these 2 different groups is possible. The correlation between the growth rate of the fungus in the trees and the 3 in-vitro characters was calculated. No significant correlation was found either with decay capacity ($r = -0.19$) or with cellulolytic capacity ($r = 0.01$). In contrast, a significant negative correlation exists with mycelial growth ($r = -0.49$), which means the strongest strains in the trees generally grow more slowly on malt-agar (Fig. 2).

DISCUSSION

The most conspicuous result is the connection between growth rate of the fungus in the trees and on malt-agar.

Nevertheless some criticism exists. A very important question is: what really was measured in the stems between 1973 and 1977? In fact it is the growth of the decay columns after holes had been drilled in the stems in

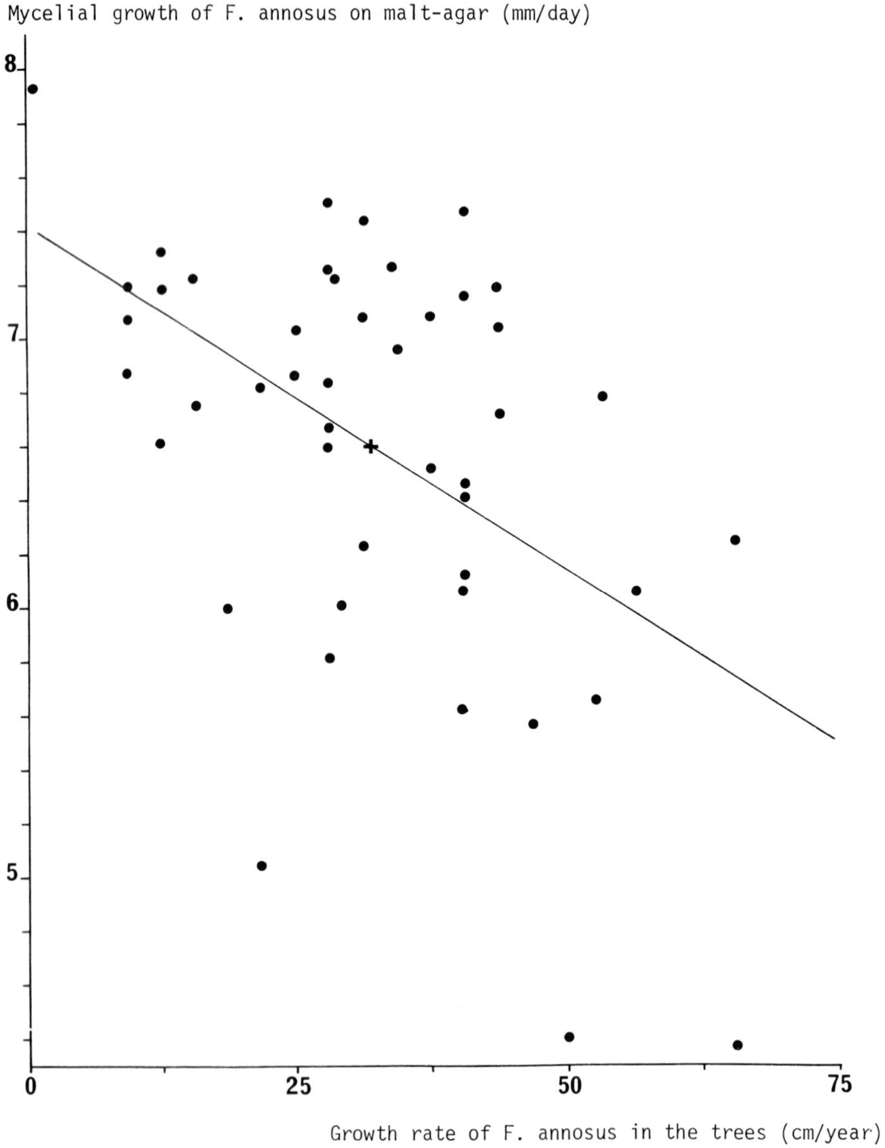

Figure 2. Relationship between growth rate of *F. annosus* in the stem of Norway spruce and mycelial growth on malt-agar. The line is the regression. Regression equation: $y = -0.025x + 7.42$; $r = -0.49$; level of significance < 0.01.

1973. In this way there are certainly introduced some disorders in the internal wood and then in the fungus growth, but one doesn't know at all the extent of them. So it is possible that the measured growth rate of the fungus is not exactly the 'natural growth' in the stems; but one can expect any error that exists is the same from one tree to another. Furthermore one may expect that over a 4 year period the trees overcame the disorders. In spite of these restrictive remarks it may be emphasized that variations of the fungus itself are, at least to some extent, responsible for the variations of disease in nature. Unfortunately in vitro variation of the fungus in this study was only based on mycelial growth on malt-agar, which is quite a rough character. It certainly would be more interesting to observe correlation with more precise physiological characters. In a breeding project it appears necessary not only to compare the trees vs. one individual strain of *F. annosus*, as is currently done, but also to take into account the performance of the fungus itself.

As concerns liability of the trees to the spread of *F. annosus*, some tentative hypotheses may be proposed. My observations show that *F. annosus* probably does not grow equally fast in all parts of the heartwood in Norway spruce. Generally speaking, the slowest growth occurs at the top of the stem; that is not a surprising fact and one may suppose theoretically there exists a height limit to *F. annosus* extension in the stem at every moment of its lifetime. So one may conceive of a Potentially Colonizable Column in the stem of Norway spruce. This potential column is likely related to the existence of 'heartwood', and its evolution depends on many factors such as ecological, silvicultural, and genetic conditions. Moreover the potential column is not necessarily homogeneous as regards fungal growth: some pre-existing factors in it may adversely affect the fungus such as wood structure, composition of the ligno-cellulosic compound, or presence and chemical evolution of phenolics. The nearby living tissues (i.e.: sapwood, cambium) may also play a role that is in connection with the current physiological status of the tree.

I believe that this Potentially Colonizable Column concept may be useful to consider, especially when inoculating young trees in which the potential column is probably not yet largely developed. For instance, inoculations inside the potential column or above it may not have the same significance, as the origin of the resistance is probably not the same. It may be postulated that inoculations for breeding purposes have to be performed inside the Potentially Colonizable Column because it is the natural area of development of the fungus. Another consequence of this concept is the following question: is it utopic to imagine that one can reduce the height of the potential column by genetic means or silvicultural practices?

Finally, another field of discussion is the significance of data collected after artificial inoculations. What is measured in this way? When inoculating trees by drilling holes in the stem it is well known that deep perturbations occur within it. I believe that the growth of *F. annosus* measured by this method is really the growth in a traumatic area where resistance factors of the stem seem to have been weakened. Indeed several unpublished observations show that the mycelial extent of *F. annosus* is at first large after inoculation and then decreases more or less constantly; for example, I observed at 6 months the extent is greater than at 12 months; a large extension 3 years after inoculation then decreased till 5 years (Weissenberg, pers. commun. 1978). So the tree builds up its resistance again. Possibly the phenomenon is different, depending on whether inoculation occurred in the Potential Column or outside it.

Unfortunately the connection, if any exists, between resistance measured by means of stem inoculations and that occurring in nature is not known at all. It would certainly be very important to study this connection to determine the effectiveness and limits of breeding based on artificial inoculations.

This knowledge will be achieved in a long term experiment only, by promoting natural infection of clones planted in an inoculum-managed area. Results may be compared then to those achieved in a stem inoculation experiment.

REFERENCES

Courtois, H., 1972. Das Keimverhalten der Konidiosporen von Fomes annosus bei Einwirkung verschiedener Standortfaktoren. European Journal of Forest Pathology 2: 152-171.

Delatour, C. & G. Sylvestre-Guinot, 1978. Microflore interne des tissus ligneux de l'épicéa commun sur pied. II - Etude in vitro de la microflore naturelle. Annales des Sciences Forestières 35: 285-298.

Dimitri, L. & R. Kliefoth, 1978. Untersuchungen über die Widerstandsfähigkeit verschiedener Fichtenklone gegenüber dem F. annosus. Proceedings of the 5th International Conference on Problems of Root and Butt Rot in Conifers, Kassel, 1978. p. 85-93.

Eghbaltalab, M., J.C. Debaud & G. Bruchet, 1976. Etude in vitro des activités pectinolytique et cellulolytique de Fomes annosus et de Rosellinia quercina. Bulletin de la Société Linéenne de Lyon 2: 85-96.

Kaufmann, U. & H. Wellendorf, 1978. Genetic variation in Picea abies and Fomes annosus shown by growth of the fungus on wood dust. European Journal of Forest Pathology 8: 226-239.

Korhonen, K., 1978. Intersterility groups of Heterobasidion annosum. Communicationes Instituti Forestalis Fenniae 94(5): 1-25.

Shain, L., 1971. The response of sapwood of Norway spruce to infection by F. annosus. Phytopathology 61: 301-307.

Weissenberg, K.v., 1975. Variation in relative resistance to spread of F. annosus in four clones of Picea abies. European Journal of Forest Pathology 5: 112-117.

Detection of disease resistance in Douglas-fir seedlings and variation in pathogenicity in Phellinus weirii by monitoring water stress after inoculation

Chas. H. Driver, D.R. Morse & R.L. Edmonds

College of Forest Resources, University of Washington, Seattle, Washington, USA

ABSTRACT

Natural resistance to effects of disease in Douglas-fir seedlings inoculated with *Phellinus weirii* was detected by water stress reaction monitoring. A series of four patterns were discussed relative to moisture stress reactions exhibited by inoculated seedlings. In addition, it was demonstrated that a synthesized karyotype of the pathogen exhibited virulence equal to that expressed by naturally occurring wood decay tissue isolate dikaryons. Homokaryons were found to be of lower virulence than dikaryons. These results indicate that it should be possible to develop a program for the detection and breeding of Douglas-fir for improved natural resistance to the serious root rot pathogen *P. weirii*.

PART I

Introduction

Poria root rot of Douglas-fir (*Pseudotsuga menziezii* (Mirb.) Franco) induced by *Phellinus weirii* (Murr.) Gilbertson is a major problem in the intensive management of this tree species. It is estimated that the disease results in an annual loss of 1 000 000 m^3 of Douglas-fir wood volume in the states of Washington and Oregon in the USA (Childs & Shea, 1967). Such impact is thought to increase as the young growth forest stands become cultured more intensively.

Present methods of limiting the impact of this disease may not be economically feasible under more limited management priorities. The use of Douglas-fir seedlings, with improved natural disease resistance to the pathogen, in artificially regenerating *P. weirii* infected sites would be a sound silvicultural practice.

Since Douglas-fir is an indigenous tree of the Pacific Northwest forest of the United States and the pathogen *P. weirii* is also known to be native to the area, sources of natural disease resistance should have been developed through natural selection. To date, to our knowledge, no method has been reported for the detection in Douglas-fir seedlings of natural disease resistance to infection and disease effects induced by *P. weirii*.

Therefore, a program of research was initiated to determine if seedlings of Douglas-fir exhibited natural resistance to infection by *P. weirii*. A second part of this effort was to investigate variation in pathogenicity of homokaryotic, dikaryotic and tissue isolates of this heterothallic pathogen. An initial step in the program was to develop a method of detection of natural resistance in seedlings of Douglas-fir upon being infected by *P. weirii*. The following details will describe the results of such efforts.

Methods

Root vascular pathogen activities have been demonstrated to increase the resistance to flow of water in the transpirational stream in many plants (Dimond, 1970; Helms et al., 1971). Waring & Clary (1967) demonstrated that the Scholander pressure chamber could be used to measure plant water stress in Douglas-fir infected by *P. weirii*. In addition Ritchie & Hinckley (1971) found that needles instead of twigs could be used with this system for monitoring seedling water stress. Such techniques were used to monitor seedling water stress changes in our research.

The second step in the program was to accomplish a standard pathogen exposure system to allow the expression of disease resistance. A standard infection technique was followed that incorporated the methods of Wallis & Reynolds (1962), Kuhlman (1969) and Nelson (1972). It consisted of introducing a 20 to 25 g infected red alder block segment into seedling tissue wounded by removing the bark at the root collar area. Such a procedure, as reported by Nelson (1972), resulted in a method always yielding better than 60 % initial infection.

Utilizing these methods of seedling inoculation and monitoring of water stress change a system was developed for following the results of infection by the pathogen and host resistance reactions.

Results and discussion

The following are the results when some 200 randomly selected nursery-run 2-year-old Douglas-fir seedlings were tested for the expression of natural disease resistance.

Seedlings were inoculated with known karyotype isolates of *P. weirii*. Using the Scholander pressure chamber, plant water stress was monitored on second year needles randomly selected from the leader of each test seedling. Pressure chamber evaluations were conducted during predawn morning hours beginning 1 week after inoculation.

Four general patterns of moisture stress were detected.

- No abnormal water stress pattern was exhibited different from the control treatment seedlings. This was interpreted to indicate no infection had resulted from exposure to the pathogen.

- Within 16 days after exposure to the pathogen a significant increase in water stress was exhibited by many seedlings (Fig. 1). By 22 days many seedlings exhibited over -30 bars (1 bar = 0.985 atm., Slatyer, 1967) of water stress and never recovered. Seedlings, which exhibited -30 bars reactions for a period of 10 days, were considered dead. Tissue from many of these seedlings yielded the pathogen upon isolation from sites distant from the place of inoculation.
- Some seedlings continued to exhibit abnormally high water stress indications for a period of 100 to 200 days (Fig. 2) before exceeding the -30 bars point of no recovery. Such seedlings also were found to yield the pathogen on isolation. This type of reaction could be interpreted as some indication of relative disease resistance. Further work, however, is required in this respect.
- A small number of seedlings (7 %) exhibited a most interesting water stress reaction (Fig. 3).

A constantly higher-than-normal water stress was detected for a period of 120 to 240 days before the seedlings return to what appeared to be

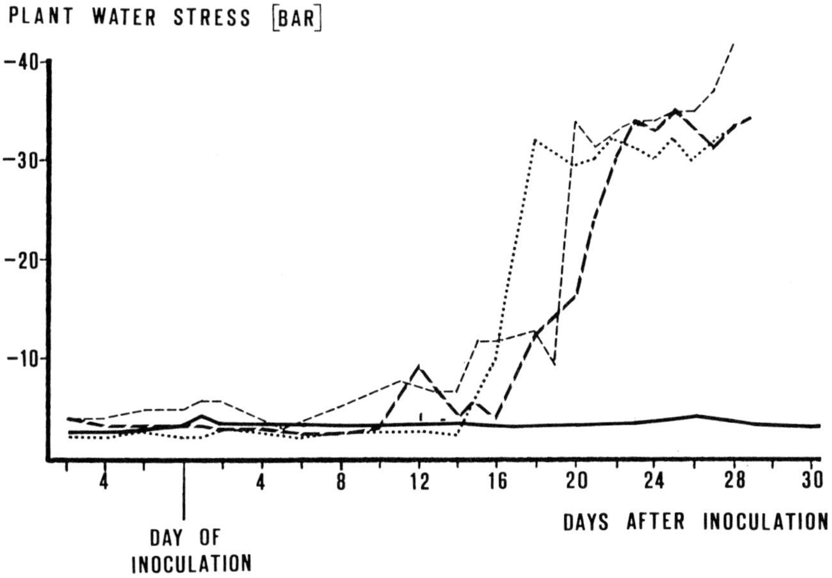

Figure 1. Lethal plant water stress of representative 2-year-old seedlings of Douglas-fir exhibiting typical disease reactions to infection by *Phellinus weirii*. The seedlings were inoculated with either synthesized dikaryon isolates D-1 or D-2, or a Douglas-fir decayed wood tissue isolate (DF). (······) D_1-seedling no. 4; (— —) D_2-seedling no. 3; (– – –) DF-seedling no. 17; (——) average of 20 control seedlings.

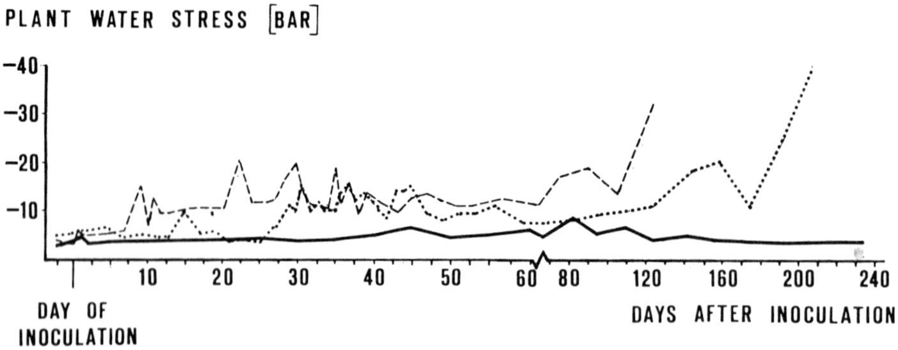

Figure 2. Latent, lethal plant water stress disease symptoms exhibited by 2-year-old Douglas-fir seedlings infected with *Phellinus weirii*. The seedlings were inoculated with either a synthesized dikaryon (D-1) isolate or a Douglas-fir decay tissue isolate (DF). (– – –) D_1-seedling no. 13; (......) DF-seedling no. 9; (——) average of 20 control seedlings.

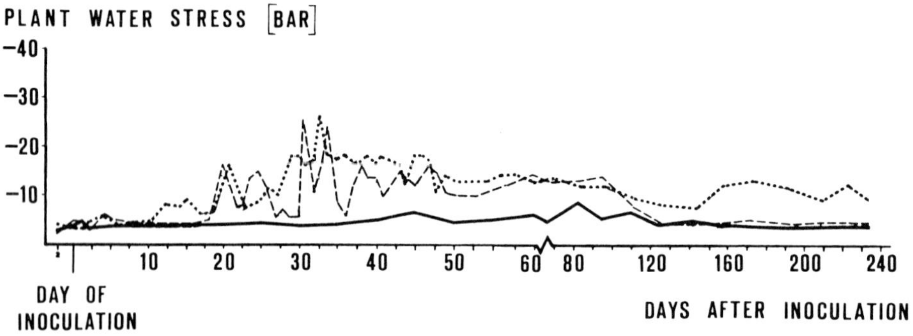

Figure 3. Water stress patterns typical of 2-year-old seedlings of Douglas-fir exhibiting natural resistance to *Phellinus weirii*. The seedlings were inoculated with either a synthesized dikaryon isolate (D-2) or a homokaryon isolate (H-3). (......) D_2-seedling no. 11; (– – – –) H_3-seedling no. 4; (——) average of 20 control seedlings.

comparable to the control seedlings water potential. These seedlings were exposed to a second complete series of inoculations and water stress evaluations that yielded similar results. Most of these seedlings remained alive after 1 year and exhibited vigorous growth. Such seedlings were thought to exhibit some form of relative disease resistance after exposure to *P. weirii*.

In conclusion it was determined that:

- needle water stress monitoring evaluations can be used to detect infection of Douglas-fir seedlings infected by *P. weirii*;
- Douglas-fir seedlings after inoculation with *P. weirii* that exhibit -30

bars never recover from the effects of the disease;
- and finally, Douglas-fir seedlings after being exposed to the pathogen and exhibiting the water stress patterns of Fig. 3 are thought to demonstrate some form of natural disease resistance to *P. weirii*.

PART II

Introduction

A second phase of our program of investigation was to search for the expression of variation in pathogenicity by *P. weirii*. Utilizing the above described methods for detecting disease reactions by the host a series of investigations were conducted using known karyotype cultures of the pathogen.

The majority of the infection centers in present day young growth forest are thought to be attributable to the vegetative persistence of the pathogen from the previous stand's infected roots. To date no asexual spore stage is known and sexual spores have not been demonstrated to be effective in the spread of the pathogen (Childs, 1970).

A wide range of morphological variability has been recognized by various workers (Buckland et al., 1954; Clark, 1958; Childs, 1963) studying vegetative cultures under laboratory conditions. Genetic characteristics of this prominent forest pathogen, however, were not described until recently (Gillette & Driver, 1974; Gillette, 1975; Driver et al., 1979). These workers and Hansen (1979) have demonstrated *P. weirii* to be heterothallic. In addition, dikaryotic cultures have been demonstrated to induce higher rates of white rot type of wood decay than homokaryotic cultures of this pathogen (Charlermpongse, 1976).

In respect to further cultural variation, Childs (1963) indicated that isolates from decayed wood from trees from a given infection center within a forest stand yielded a single vegetative type, which was designated as a clone. In addition, it was found that 2 different clones could be isolated, each from separate infection centers, within the same stand. These were demonstrated to be different clones by the development of a distinct zone reaction line between the different clones at the contact area between the mycelial mats when cultured together on a plate. Since Childs' (1963) cultures were initially derived from infected tree tissue, it could probably be assumed that most of his cultures were dikaryotic, therefore, making genetic interpretation difficult.

This limited review of the literature establishes the existence of cultural strains in *P. weirii* that logically lead to pose the question of the existence of variability in pathogenicity.

Raabe (1967, and this book, page 251) showed that strains of the root rot inducing fungus *Armillaria mellea* (Vahl ex Fr.) Kummer exhibited significant variation in pathogenicity. In addition, Day (1960) points out that heterokaryosis often leads to variation in ability to attack host species. These indications for strain variability in pathogenicity and the in-depth work on breeding southern pine species for resistance to strains of *Cronartium fusiforme* Hedgc. & Hunt (Snow et al., 1975; Powers et al., 1977; and this book, page 236 and 427) indicate the need to consider the potential for strain differences in *P. weirii* for pathogenicity. In this case the heterothallic sexuality of *P. weirii* offered an initial opportunity to conduct such studies.

Methods

Heterokaryotic strains were synthesized in the laboratory and were demonstrated to be compatible dikaryons (D-1 and D-2) by the ability to produce fertile fruiting bodies in culture. Homokaryotic strains (H-3 and H-5) were derived from monosporic cultures, and were used to produce the dikaryotic strain, D-1, used in these studies.

In addition to these known karyotype strains, tissue isolates were obtained from decayed wood from infected Douglas-fir (DF) and from western hemlock (WH) (*Tsuga heterophylla* (R.) Sarg.) trees. The tissue isolates were used to function as the wild-types occurring in nature. Such isolates were usually found to give rise to fertile fruiting bodies under laboratory cultural conditions and were therefore determined to be compatible dikaryotic heterokaryons.

The thus obtained cultures were used to prepare inoculum employed in the seedling infection and detection of natural resistance techniques described in the previous part of this paper.

To accomplish a comparative pathogenicity evaluation induced by the described strains of the pathogen a disease rating system was employed. The disease ratings were based on the following seedling reactions after inoculation with the known karyotype cultures of *P. weirii*.

Disease rating values were designated as follows:

Pathogenicity rates	Symptom characteristics
0	absence of abnormal growth conditions,
2	latent mortality (after 200 days),
3	average mortality (approximately 120 days),
4	or rapid mortality (before 20 days).

Mortality was indicated when seedlings exhibited above -30 bars of water stress for at least 10 days.

Results and discussion

These results (Table 1) indicates no statistically significant differ-

Table 1. Karyotype of cultures of *Phellinus weirii* as related to abilities to induce disease reactions in seedlings of Douglas-fir. 0.0 = No disease reactions, 3.6 = rapid mortality.

Isolate	Disease reaction ratings
Synthesized dikaryon D-1	3.6
Synthesized dikaryon D-2	3.2
Douglas-fir tissue isolate (DF)	3.2
Western hemlock tissue isolate (WH)	3.1
Homokaryon H-3	2.1
Homokaryon H-5	0.0

ences in pathogenicity between the synthesized dikaryons (D-1 and D-2) and tissue isolates (DF and WH). Each of the dikaryotic isolates and the tissue isolates yielded significantly higher disease ratings than the homokaryotic isolates (H-3 and H-5). In addition, it is of interest that the homokaryotic isolate H-3 was significantly more pathogenic than the H-5 isolate. The H-5 isolate failed to induce abnormal water stress patterns in inoculated seedlings. It is of special interest to note that the synthesized dikaryon isolate D-1 produced the highest disease rating of all isolates tested, even including the naturally occurring tissue isolates. In conclusion, it can be stated that true differences in pathogenicity can be demonstrated between karyotypes of *P. weirii*.

In addition, the results of these initial trials indicate that it should be possible to maintain a standard inoculum culture equal in pathogenicity to naturally occurring strains; or, at least possible to synthesize a karyotype to represent those combinations found in nature.

We feel that this latter attribute offers a rare opportunity for the plant breeder of improved natural disease resistance in forest trees to maintain a standard inoculum of known pathogenicity. In addition, it appears that since *P. weirii* has been reported to be spread mainly by vegetative systems, little unexpected changes in virulence should occur; thereby offering assurance of maintaining established lines of dependable disease resistance in artificially produced forest tree progeny.

REFERENCES

Buckland, D.C., A.C. Molnar & G.W. Wallis, 1954. Yellow laminated root rot of Douglas-fir. Can. J. Botany 32: 69-81.

Childs, T.W., 1963. Poria weirii root rot. Phytopathology 53: 1123-1127.

Childs, T.W., 1970. Laminated root rot of Douglasfir in western Oregon and Washington. PNW Forest Range Exper. Sta. USDA For. Serv. Res. Paper PNW-102.

Childs, T.W. & K.R. Shea, 1967. Annual losses from disease in Pacific Northwest forest. USDA For. Serv. Resources Bull. PNW-20.
Clark, J.W., 1958. The occurrence and investigation of physiological strains of Poria weirii Murr. Ph.D. Thesis, Yale University.
Charlermpongse, A., 1976. A comparative study of the decomposition capabilities of homokaryotic and dikaryotic cultures of Phellinus weirii on Douglasfir sapwood. M.S. Thesis. Univ. Wash. Seattle, Wash.
Day, P.R., 1960. Variation of phytopathogenic fungi. Ann. Rev. Microbiology 14: 1-16.
Dimond, A.E., 1970. Biophysics and biochemistry of the vascular wilt syndrome. Ann. Rev. Phytopath. 8: 301-322.
Driver, C.H., W.D. Gillette & Yi-Mei Kao, 1979. Heterothallism of Phellinus weirii influenced by genes at one or more loci. Abst. Proceedings 52nd Ann. Meeting. Northwest Sci. Assoc.
Gillette, W.D., 1975. Biology of Poria weirii: sexuality and effect on height growth of Douglas-fir M.S. Thesis. Univ. of Wash. Seattle, Wash.
Gillette, W.D. & C.H. Driver, 1974. Sporophore formation by Poria weirii in culture. Abst. Proceedings 66th Annual Meeting of the Amer. Phytopath. Soc.
Hansen, E.M., 1979. Sexual and vegetative incompatibility reactions in Phellinus weirii. Can. J. Botany 57: 1573-1578.
Helms, J.A., F.W. Cobb Jr. & H.S. Whitney, 1971. Effect of infection by Verticicladiella wagnerii on the physiology of Pinus ponderosa. Phytopathology 61: 920-925.
Kuhlman, E.G., 1969. Inoculation of loblolly pine seedlings with Fomes annosus in the greenhouse. Can. J. Botany 47: 2079-2082.
Nelson, B.D., 1972. Effects of fertilization with urea on root rot diseases of Douglas-fir and western hemlock caused by Poria weirii and Fomes annosus. M.S. Thesis Univ. of Wash.
Powers, H., F.R. Matthews & L.D. Dwinell, 1977. Evaluation of pathogenic viability of Cronartium fusiforme on loblolly pine in southern USA. Phytopathology 67: 1403-1407.
Raabe, R.D., 1967. Variation in pathogenicity and virulence in Armillaria mellea. Phytopathology 57: 73-75.
Ritchie, G.A. & T.M. Hinckley, 1971. Evidence for error in pressure-bomb estimates of stem xylem potentials. Ecology 52: 534-536.
Slatyer, R.O., 1967. Plant-water relationships. Academic Press, New York.
Snow, G.A., A.J. Dinus & A.G. Kais, 1975. Variation in pathogenicity of diverse sources of Cronartium fusiforme on selected slash pine families. Phytopathology 65: 170-175.
Wallis, G.W. & G. Reynolds, 1962. Inoculation of Douglasfir roots with Poria weirii. Can. J. Botany 40: 637-645.
Waring, R.H. & B.D. Cleary, 1967. Plant moisture stress: evaluation by pressure bomb. Science 155: 1248-1254.

Variability of host and pathogen in the pitch canker complex

L.D. Dwinell & J.B. Barrows-Broaddus

Principal Plant Pathologist and Plant Pathologist, U.S. Department of Agriculture, Forest Service, Southeastern Forest Experiment Station, Carlton Street, Athens, Georgia 30602, USA

ABSTRACT

Since 1974, there has been considerable research on the variability of host and pathogen in the pitch canker complex. There is tremendous variation in the susceptibility of the hosts and pathogenicity of the causal fungus, *F. moniliforme* var. *subglutinans*. However, researchers have only begun to assess the impact that environment has on this host-parasite system.

INTRODUCTION

Pitch canker is an important disease of planted pines in the Southern United States. The disease was originally attributed to *Fusarium lateritium* Nees emend. Snyd. & Hans. f. sp. *pini* Hepting (Snyder et al., 1949), but recent research has demonstrated that the causal agent is *Fusarium moniliforme* Sheld. var. *subglutinans* Wr. & Reink. (Kuhlman et al., 1978). This species of *Fusarium* occurs on many hosts and in many areas of the world (Booth, 1971).

In the pitch canker complex, the plasticity of the pathogen and the variability of the hosts make the problem of evaluating the host/parasite interaction extremely difficult. In a study of *Fusarium moniliforme* from corn, Leonian (1932) noted that even the most vigorous strains exhibited their pathogenicity in cycles; at one time they were able to infect the host, at another time and under identical conditions they failed to do so. The pitch canker fungus frequently demonstrates this same type of variability on pine.

VARIATION IN THE HOST

Several species of pines, including slash (*Pinus elliottii* Engelm. var. *elliottii*), south Florida slash (*P. elliottii* var. *densa* Little and Dorman), loblolly (*P. taeda* L.), Virginia (*P. virginiana* Mill.), shortleaf (*P.*

echinata Mill.), longleaf (*P. palustris* Mill.), pitch (*P. rigida* Mill.), Table Mountain (*P. pungens* Lamb.), Monterey (*P. radiata* D.Don), scotch (*P. sylvestris* L.), eastern white (*P. strobus* L.), pond (*P. serotina* Michx.), and sand (*P. clausa* (Chapm.) Vasey), have been reported as hosts of *F. moniliforme* var. *subglutinans* in the United States (Hepting, 1971; Hepting & Roth, 1953). Pitch canker has also been reported on Cuban pine (*P. occidentalis* Sw.) in Haiti (Hepting & Roth, 1953). Of these pine species, slash and loblolly pines are the most important economically in the south-eastern United States.

The incidence and symptoms of pitch canker are highly variable and depend on the inherent susceptibility of the pine species and the environmental conditions in which the trees are growing. In Florida, for example, shoot dieback is the principal symptom on planted slash pines (Dwinell & Phelps, 1977; Phelps & Chellman, 1976). However, elsewhere in the South, bole cankers are the predominant symptom of the disease on slash pines. In slash pine seed orchards, cankers are frequently associated with the use of tree shakers for cone removal. In loblolly pine, pitch canker is rarely observed in plantations and natural stands, but can cause serious dieback in seed orchards, markedly reducing seed crops (Dwinell et al., 1977). Bole and branch cankers are the common symptoms on Virginia pine, both in seed orchards and plantations. On eastern white pine, bole cankers are normally found on trees planted in urban environments outside the species' natural range.

To delineate the relative susceptibility of southern pines, we recently inoculated 7 species with 4 isolates of *F. moniliforme* var. *subglutinans*. Based on shoot mortality from girdling cankers, Virginia pine was ranked as highly susceptible; slash, loblolly, shortleaf and pitch pines were rated as moderately susceptible; and pond and eastern white pines as highly resistant. The isolate × pine species interaction was not statistically significant. In a previous study, utilizing less quantitative procedures, shortleaf pine was rated as highly susceptible (Dwinell, 1978). It is apparent from greenhouse studies and field observations that pine species vary markedly in their susceptibility to pitch canker.

Within pine species, the incidence of pitch is frequently related to the geographic source of provenance of the host. In 3 loblolly pine seed orchards on the Coastal Plain, for example, incidence of pitch canker on loblolly pines was higher on Piedmont than on Coastal Plain seed sources (Dwinell et al., 1977). In Florida, Blakeslee & Rockwood (1978) found that pine clones from central Florida were more resistant than clones from other geographic areas.

Individual clones also vary greatly in their susceptibility to infection by *F. moniliforme* var. *subglutinans*. This phenomenon has been noted for loblolly, slash, longleaf, shortleaf and Virginia pines grown in southern seed orchards (Dwinell et al., 1977; Phelps & Chellman, 1976). In a Virginia pine progeny test, Barnett & Thor (1978) also found clonal variation in the incidence of pitch canker.

There may be sufficient genetic variation in resistance to select and breed pines for control of the disease. Open-pollinated families of slash and loblolly pine varied considerably in their susceptibility to *F. moniliforme* var. *subglutinans* (Dwinell & Barrows-Broaddus, 1979). Shoot mortality for the 7 slash pine families ranged from 46-96 %, with a mean of 80 %. Recently, this study was expanded to include 43 open-pollinated families of slash pine collected from trees in natural stands in Florida. Of the 43 families, 3 were classed as highly susceptible, 33 as moderately susceptible, and 7 as highly resistant (unpublished data). These slash pine families are currently being field tested in Florida. For the 12 loblolly pine families, shoot mortality ranged from 26-78 % with a mean of 52 %. However, there was no correlation between the response of the families and the shoot dieback of the parent clones (unpublished data). In progeny tests, stress, insects, host nutrition, and physiology will probably have to be considered, along with environmental factors which influence the host: parasite interaction.

VARIATION IN THE PATHOGEN

Although *F. moniliforme* var. *subglutinans* occurs on a wide range of hosts and is found in many areas of the world (Booth, 1971), not all strains of the fungus are pathogenic to pines. In a study on the pathogenicity of *F. moniliforme* var. *subglutinans* from nonpine hosts, only isolates from *Gladiolus* corms grown in Florida were capable of infecting slash and loblolly pines. The other 17 isolates from corn, *Dracaena*, sycamore, pecan, lily, *Araucaria* and *Amaryllis* were avirulent (Dwinell & Nelson, 1978).

In studies on sources of inoculum, we have isolated the virulent variety of *Fusarium* from forest soil (Dwinell & Barrows, 1978), surfaces of pine needles, branches and boles, and from tree-shaker pads. Results of pathogenicity tests on slash and loblolly pines indicate that the fungus occurs naturally in mixed populations of saprophytic and pathogenic strains. For example, 55 % of the isolates of *F. moniliforme* var. *subglutinans* from seed orchard and forest soils were pathogenic on slash and loblolly pine seedlings. Similar results have been obtained with isolates from bark and needle washes and from tree-shaker pads (unpublished data). Isolates of *F. moniliforme* var. *subglutinans* are identified as the pitch canker strain if their pathogenicity is demonstrated.

Within pathogenic populations of the pine pitch canker fungus, there is considerable variation in virulence. However, the variation generally appears to be among isolates rather than among geographic sources or pine host sources of isolates (Barrows-Broaddus & Dwinell, 1979).

PATHOGENIC RACES

Most studies to date indicate that the relationship between the susceptibility of the host and the pathogenicity of the causal agent is additive (Barrows-Broaddus & Dwinell, 1979; Dwinell, 1978; Dwinell & Nelson, 1978; Dwinell & Barrows-Broaddus, 1979; Kuhlman et al., 1978). However, some exceptions have been noted. In a comparison of the pathogenicity of *F. moniliforme* var. *subglutinans* on slash and loblolly pine, one isolate from loblolly pine in South Carolina was highly virulent on loblolly pine, but avirulent on slash pine (unpublished data). This condition may be caused by the racial difference that Hepting noted in 1971. Edwards (1935) reported pathogenic specialization in *F. moniliforme* var. *subglutinans* on corn in the New South Wales. There may also be pathogenic races of the pine pitch canker fungus.

REFERENCES

Barnett, P.E. & E. Thor, 1978. Effects of site and inheritance on Fusarium incidence in Virginia pine. Proceedings Symposium for management of pine of the interior South. (Knoxville, Tenn., Nov. 1978). USDA Forest Service Southeast Area State and Private Forestry Technical Publication SA-TP2. p. 159-161.

Barrows-Broaddus, J.B. & L.D. Dwinell, 1979. Variation in virulence of diverse sources of Fusarium moniliforme var. subglutinans on Virginia and loblolly pine. Phytopathology 69: 525 (abstract).

Blakeslee, G.M. & D.L. Rockwood, 1978. Variation in resistance of slash pine to pitch canker caused by Fusarium moniliforme var. subglutinans. Phytopath. News 12: 207. (abstr.).

Booth, C., 1971. The genus Fusarium. Commonwealth Mycological Institute, Kew, Surrey, England. 237 p.

Dwinell, L.D., 1978. Susceptibility of southern pines to infection by Fusarium moniliforme var. subglutinans. Plant Disease Reporter 62: 108-111.

Dwinell, L.D. & J.B. Barrows, 1978. Recovery of the pine pitch canker fungus from pine plantation and seed orchard soil. Phytopathology News 12: 207. (abstract).

Dwinell, L.D. & J.B. Barrows-Broaddus, 1979. Susceptibility of half-sib families of slash and loblolly pine to the pitch canker fungus, Fusarium moniliforme var. subglutinans. Phytopathology 69: 527. (abstract).

Dwinell, L.D. & P.E. Nelson, 1978. Susceptibility of slash and loblolly pines to strains of Fusarium moniliforme and its variety subglutinans. Phytopath. News 12: 207. (abstr.).

Dwinell, L.D. & W.R. Phelps, 1977. Pitch canker of slash pine in Florida. Journal of Forestry 75: 488-489.

Dwinell, L.D., P.L. Ryan & E.G. Kuhlman, 1977. Pitch canker of loblolly pine in seed orchards. Proceedings 14th Southern Forest Tree Improvement Conference (Gainesville, Florida, June 1977). p. 130-137.

Edwards, E.T., 1935. Studies on Gibberella fujikuroi var. subglutinans the hitherto undescribed ascigerous state of Fusarium moniliforme var. subglutinans and on its pathogenicity on maize in New South Wales. Science Bulletin. N.S.W. 49. 68 p.

Hepting, G.H., 1971. Disease of forest and shade trees of the United States. USDA Agricultural Handbook 386. 658 p.

Hepting, G.H. &. E.R. Roth, 1953. Host relations and spread of the pine pitch canker disease. Phytopathology 43: 475. (abstract).

Kuhlman, E.G., L.D. Dwinell, P.E. Nelson & C. Booth, 1978. Characterization of the Fusarium causing pitch canker of southern pines. Mycologia 70: 1131-1143.

Leonian, L.H., 1932. The pathogenicity and the variability of Fusarium moniliforme from corn. West Virginia Agricultural Experiment Station Bulletin 248: 1-16.

Phelps, W.R. & C.W. Chellman, 1976. Evaluation of "pitch canker" in Florida slash pine plantations and seed orchards, 1976. USDA Forest Service, State and Private Forestry, Southeastern Area. 22 p.

Snyder, W.C., E.R. Toole & G.H. Hepting, 1949. Fusaria associated with mimosa wilt, sumac wilt, and pine pitch canker. Journal of Agricultural Research 78: 365-382.

Pest outbreaks as a function of variability in pests and plants

Nancy Lorimer

USDA-Forest Service, North Central Forest Experiment Station, 1992 Folwell Avenue, St. Paul, Minnesota 55108, USA

ABSTRACT

A model is proposed that describes pest outbreaks as a function of variability in both plant and pest populations. Plant variability selects for pest variability and pest numbers remain low. Environmental factors may reduce genetic and phenotypic variability in plant populations so that the plant population quickly selects a homogeneous pest population. Pest numbers then increase rapidly. Crashes occur because lack of variability in the pest population increases its vulnerability to elements of its environment, including its own predators and parasites. The model incorporates principles of genetics, ecology, and evolutionary biology in the attempt to understand the economically and biologically important phenomenon of pest outbreaks.

INTRODUCTION

The dynamical behavior of populations of some forest insect pests (*Malacosoma disstria* Hübner, *Choristoneura fumiferana* (Clem.), *Orygia pseudotsugata*, *Zeiraphera diniana* Guén. and others) can be characterized by quick changes in population size over time. Insect numbers, very low for some time, suddenly and rapidly increase, or outbreak. The outbreak stage continues for several years until the population rapidly decreases, or crashes. The outbreak stage is shorter than the period of time from crash to new outbreak.

Forest insects that outbreak and crash also show properties that facilitate rapid changes in variability (Lorimer, 1979d). Therefore, I proposed in an earlier paper that changes in pest variability may be related to changes in pest population size, with natural selection playing an important role (Lorimer, 1979d). In this paper I wish to extend the model of pest outbreaks by adding a third parameter, variability of the plant host populations. Some considerations for forest tree breeding and management emerge.

PEST AND HOST VARIABILITY

Population variability is defined as the following: the space and time differences in properties or traits within and among individuals in the population. Space refers to differences within parts of an individual, and differences from individual to individual over the physical area of the population habitation. Time refers to differences within an individual's life cycle, and differences from individual to progeny. Factors that determine differences between individual x and individual y in population P are genotype, age, and environment during development.

The genetics of forest insects is beginning to be studied (e.g. Stehr, 1955, 1959; Harvey, 1957; Campbell, 1966; Morris & Fulton, 1970; Morris, 1971; May et al., 1977; Lynch & Hoy, 1978; Stock et al., 1979; Stock & Guenther, 1979; Lorimer, 1979a, b, c), and the possibility of genetic changes in forest insect populations during rise and decline is being recognized (Franz, 1949; Baltensweiler, 1971; Lorimer, 1979d; Haukioja, 1980).

When individuals from 2 populations interact, like herbivores and plants, or pests and trees, variability is defined in terms of individual differences in properties or traits that affect the interaction. The important properties of plants in their interaction with herbivores in general and with insect pests in particular can be divided into 2 broad classes: 1. properties that satisfy pest needs, and 2. defense properties. Properties in the first class include nutritional content (i.e. protein, amino acids, vitamins, minerals), water content, shelter, and a large group of other properties that enable insects to survive and reproduce. Defensive properties are also many and varied and would include secondary compound production and physical features like thorniness and leaf toughness.

Plant populations are also variable among individuals. Individual trees x and y in population P will not have identical amounts of nutrients (Shaw & Little, 1977) or secondary compounds (Chew & Rodman, 1979) in their foliage, for example. Nor will they have the same amounts of these compounds at different times (Seigler & Price, 1976).

VARIABILITY AND NATURAL SELECTION WITHIN THE INTERACTION

Consider a heterogeneous plant population in opposition to a homogeneous herbivore population (Fig. 1A). Suppose that each herbivore can survive and reproduce successfully only on plants that 'match' it, meaning plants that meet the herbivore's needs and whose defenses the herbivore can overcome (Mattson et al., page 295 in this book). Thus y herbivores can survive and reproduce only on v plants and not on u and w plants. The offspring will be y.

Now consider the situation where the herbivore population has increased in variability (Fig. 1B). Herbivore x survives and reproduces on plant u,

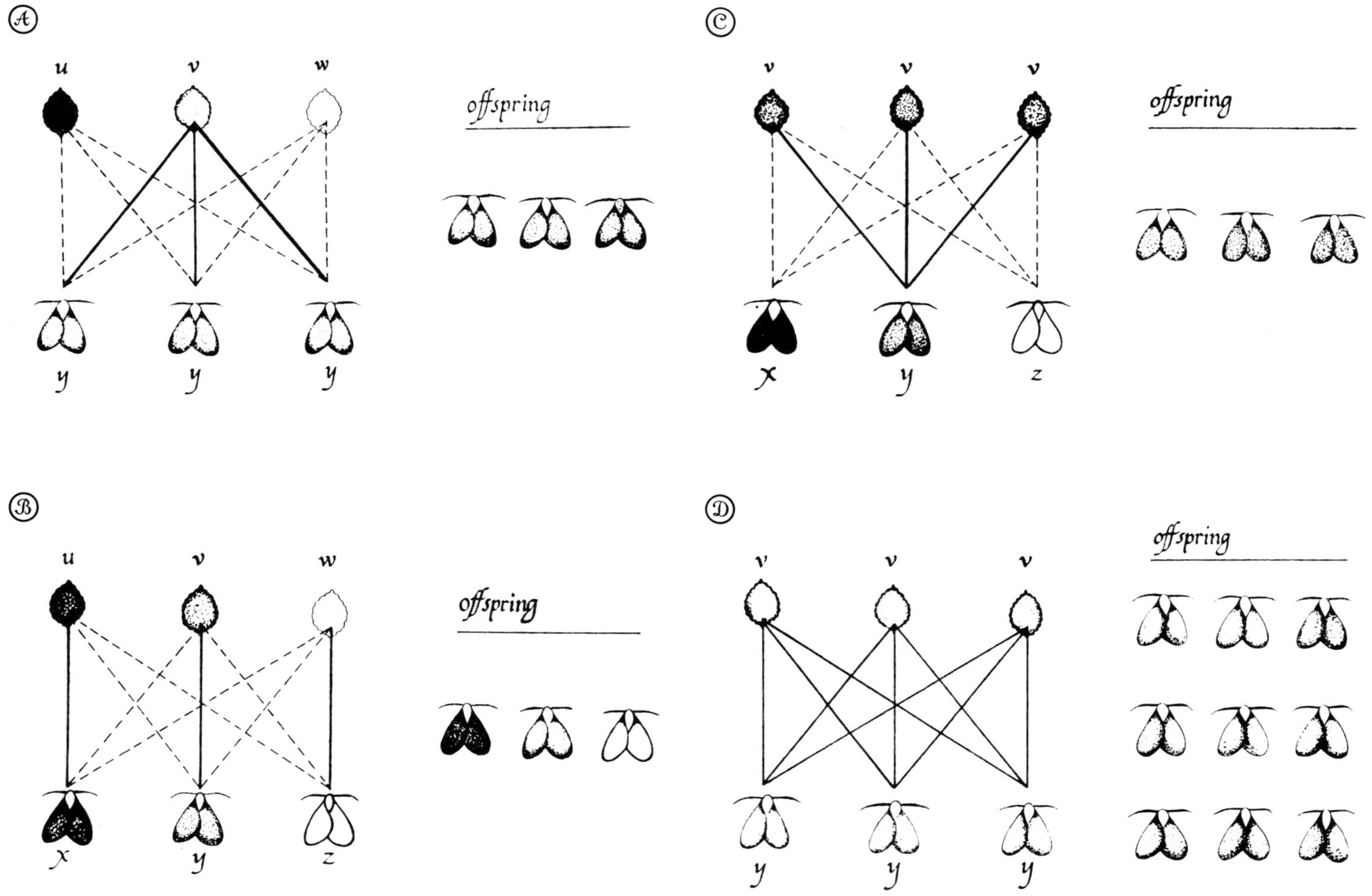

Figure 1. Variability and frequency dependent selection between populations of plants and pests, explained in the text.

but not on plants v and w. Similarly, y does well on v, but not on u or w, and so on. The progeny reflect the successful reproductions. Hence, heterogeneity in the plant population has maintained heterogeneity in the herbivore population by a process of frequency dependent selection (Clarke, 1976).

When the plant population is homogeneous (Fig. 1C), there is rapid selection for the herbivore genotypes that grow and reproduce successfully on these plants: in our example, herbivore y. The whole of the next generation will be y. Their offspring will show a marked increase in numbers (Fig. 1D). The increase in pest numbers has come about because a homogeneous plant population selected an adapted, homogeneous herbivore population. The herbivore/plant interaction will be governed by a high degree of uncertainty if the plant population maintains space and time variability. Variability keeps the number of matches low.

DECLINING VARIABILITY IN THE PLANT POPULATION

The degree of variability in a population of trees is first a function of the original genetic base, the variability of the founders of the stand. A population established from a homogeneous seed source is likely to be less variable than a population from a heterogeneous seed source. As populations age, variability may be reduced if natural selection or thinning eliminates some members from the population. Short-term genetic recomposition is impossible for populations of long-lived individuals.

Phenotypic variability in plant populations is more changeable than genetic variability and may be a function of stress. Stresses may reduce the variety of chemical defenses produced in plants (Rhoades, personal communication). Environmental stresses like temperature changes and drought have been documented as preceding insect outbreaks. But rather than affecting the insects directly, these occurrences may be affecting the insects indirectly by reducing variability in the tree population. Stresses may trigger simultaneous metabolic and physiological responses among individual trees, thereby reducing variability.

For example, drought stress decreased protein levels and increased levels of free amino acids in *Eucalyptus* prior to psyllid outbreaks (White, 1969). Although the data are not available to ascertain biochemical variability among the trees before and after the drought stress stimulus, similarity of response is indicated.

As a second example, occurrence of flowering in balsam fir increases with the age of the stand, and is promoted by certain weather conditions, so that nearly all of the trees flower at the same time (Morris, 1971). Spruce budworm outbreaks have been known to occur following heavy flowering (Greenbank, 1963; Kimmons, 1971). Although flower feeding appeared to enhance larval development, nutritional value of flowers was not a likely factor in the outbreaks (Greenbank, 1963). In the light of the hypothesis proposed here, massive flowering represents synchrony of response to stress. Decrease in the plant population's variability results in uniform selection of the pest population. The pest population then quickly increases in number (Fig. 1D).

Haukioja (1980) apparently observed a different pattern in outbreak areas of the autumnal moth, *Oporinia autumnata* Bkh. on birch, *Betula pubescens* Ehrh., in northern Finland, where poor climatic and edaphic conditions prevail. While these populations might be subject to more stresses than populations in moderate climates, interindividual variance in defense was noted as being 'high'. However, the defensive characters were not specified, and no direct comparison of defense variance with other populations was supplied.

The evidence for changes in variability in insect and plant populations before outbreaks is sparse for several reasons. (1) The role of the plant in the population dynamics of the herbivore has only recently been recognized and documented. With the realization that the plant is not merely a passive partner in the interaction but an active defender of its own tissue has come the necessity to re-evaluate former assumptions about the factors controlling population numbers. (2) Entomologists have tended to regard variability as noise. Life table studies have concentrated on means, and individuals in populations were assumed to be nearly identical. (3) The process of natural selection has been investigated mostly on an evolutionary, rather than an ecological, time scale. The process of natural selection as a force in the relationship between trophic levels has not been sufficiently recognized.

A CATASTROPHE APPROACH TO MODELLING PEST OUTBREAKS

Systems with variables that cannot be averaged over time and space can be visualized with catastrophe models (Thom, 1970). Thus catastrophe models may be useful in solving problems that deal with non-linear systems, phenomena involving sudden change, the kind of problems that abound at the interface of genetics and ecology. Population outbreaks exhibit the 5 qualities of the cusp catastrophe (Zeeman, 1976), so the model is applicable to this aspect of population behavior (Jones, 1975). The 5 features of cusp catastrophes (Zeeman, 1976) shared by outbreak phenomena (Lorimer, 1979d) are (1) bimodality of behavior: populations at very low or very high levels, (2) sudden jumps between modes: quick increases in numbers, or outbreaks, and quick decreases in numbers, or crashes, (3) hysteresis: population composition before the outbreak dissimilar to composition at the end of the crash, (4) inaccessibility: the population seldom at a moderate level, and (5) divergence: a gradually changing force causing an abrupt change in behavior.

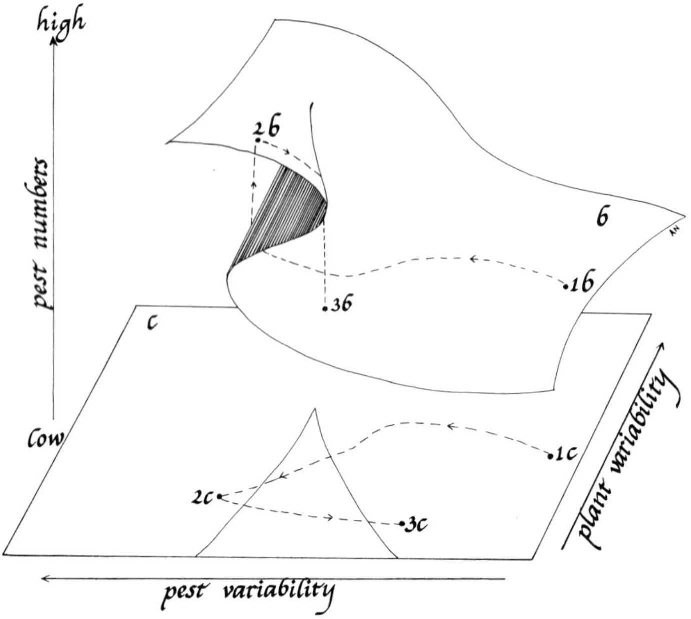

Figure 2. Control surface (C) and behavior surface (B) of a cusp catastrophe of pest outbreaks and crashes. The 2 control variables are pest variability and plant variability. The behavior variable is pest numbers. The model is described in the text.

The cusp is described by the function $x^4/4 + ax^2/2 + bx$, where a and b are control parameters, and x is the behavior variable. If plant and pest variability are made the 2 control dimensions in a cusp model, and pest numbers is the behavior dimension, an outbreak and crash cycle can be traced on both the behavior surface (Fig. 2B) and the control surface (Fig. 2C). At (1), high variability in the plant population and low variability in the pest population keep the pest population at a low level. As variability in pest population increases, the system passes into and out of the bifurcation set, jumping up to (2). Here variability is low in the plant population, but high in the insect population, and pest numbers are high. From (2) to (3) insect variability decreases, plant variability remains low, and the pest population, passing through the other side of the bifurcation set, crashes to a very low level.

Why does the population crash? The homogeneous tree population has selected for homogeneity in the pest population, which, in turn, makes the pests vulnerable to their enemies or the weather or other factors in their environment. The population is no longer genetically buffered against environmental influences. Secondly, in large numbers the pests are forced by density considerations to feed on other species of plant, which would have the same effect as increasing the variability in the original plant population (Fig. 1A).

MANAGEMENT PRACTICES AND THE PEST OUTBREAK MODEL

Management practices will clearly affect a system represented by the model proposed here. In natural systems, diversity maintains dynamical equilibria (Browning, 1974), but agricultural pests encounter built-in homogeneity in food crops and the response is outbreak (Marshall, 1977). The same danger exists in forestry. Besides reducing genetic variability in stands, human-applied stresses could lead to reduction in phenotypic variability and then to pest outbreak. Presenting a homogeneous phenotype to pests promotes rapid selection and ultimate destruction of the tree crop.

ACKNOWLEDGEMENTS

I thank Drs. W.J. Mattson and R.A. Leary for helpful discussions, Drs. H.O. Batzer, F. Chew, E. Haukioja, M.F. Stock, J.C. Schultz, and conference participants for critical comments.

REFERENCES

Baltensweiler, W., 1971. The relevance of changes in the composition of larch bud moth populations for the dynamics of its numbers. In: P.J. den Boer & G.R. Gradwell (Eds): Dynamics of populations. Proceedings of the Advanced Study Institute on the Dynamics of Numbers in Populations. Oosterbeek, 1970. Pudoc, Wageningen. p. 208-219.

Browning, J.A., 1974. Relevance of knowledge about natural ecosystems to development of pest management programs for agro-ecosystems. Proceedings of the American Phytopathological Society 1: 191-199.

Campbell, I.M., 1966. Genetic variation related to survival in Lepidopteran species. In: H.D. Gerhold, E.J. Schreiner, R.E. McDermott & J.A Winieski (Eds): Breeding pest-resistant trees. Pergamon Press. Oxford. p. 129-135.

Chew, F.S. & J.E. Rodman, 1979. Plant resources for chemical defense. In: G.A. Rosenthal & D.H. Janzen (Eds): Herbivores: Their interaction with secondary plant metabolites. Academic Press. N.Y. p. 271-307.

Clark, B., 1976. The ecological genetics of host-parasite relationships. In: A.E.R. Taylor & R. Muller (Eds): Genetic aspects of host-parasite relationships. Symposia of the British Society of Parasitology 14: 87-103.

Franz, J., 1949. Über die genetischen Grundlagen des Zusammenbruchs einer Massenvermehrung aus inneren Ursachen. Zeitschrift für Angewandte Entomologie 31: 228-260.

Greenbank, D.O., 1963. Staminate flowers and the spruce budworm. In: R.F. Morris (Ed.): The dynamics of epidemic spruce budworm populations. Memoirs Entomological Society of Canada 31: 202-218.

Harvey, G.T., 1957. The occurrence and nature of diapause-free development in the spruce budworm, Choristoneura fumiferana. Canadian Journal of Zoology 35: 549-572.

Haukioja, E., 1980. On the role of plant defenses in the fluctuation of herbivore populations. Oikos 35: 202-213.

Jones, D.D., 1975. The application of catastrophe theory to ecological systems. In: G.S. Innis (Ed.): New directions in the analysis of ecological systems. Simulation Councils Proceedings Series 5: 133-148.

Kimmons, J.P., 1971. Variations in the foliar amino acid composition of flowering and non-flowering balsam fir and white spruce in relation to outbreaks of the spruce budworm. Canadian Journal of Zoology 49: 1005-1011.

Lorimer, N., 1979a. Patterns of variation in some quantitative characters of Malacosoma disstria (Lepidoptera: Lasiocampidae). Annals of the Entomological Society of America 72: 275-280.

Lorimer, N., 1979b. Genetics of superoxide disumtase in the forest tent caterpillar and other organisms. Journal of Heredity 70: 199-204.

Lorimer, N., 1979c. The genetics of melanism in Malacosoma disstria Hübner (Lepidoptera: Lasiocampidae). Genetics 92: 555-561.

Lorimer, N., 1979d. Genetic causes of pest population outbreaks and crashes. In: M.A. Hoy & J.J. McKelvey, Jr. (Eds): Genetics in relation to insect management. The Rockefeller Foundation. p. 50-54.

Lynch, C.B. & M.A. Hoy, 1978. Diapause in the gypsy moth: environment specific mode of inheritance. Genetical Research 32: 129-133.

Marshall, D.R., 1977. The advantages and hazards of genetic homogeneity. In: P.R. Day (Ed.): The genetic basis of epidemics in agriculture. New York Academy of Sciences. p. 1-20.

May, B., D.E. Leonard & R.L. Vadas, 1977. Electrophoretic variation and sex-linkage in spruce budworm. Journal of Heredity 68: 355-359.

Morris, R.F., 1971. Observed and simulated changes in genetic quality in natural populations of Hyphantria cunea. Canadian Entomologist 103: 893-906.

Morris, R.F. & W.C. Fulton, 1970. Heritability of diapause intensity in Hyphantria cunea and correlated fitness responses. Canadian Entomologist 102: 927-938.

Seigler, D. & P.W. Price, 1976. Secondary compounds in plants: primary functions. American Naturalist 110: 101-104.

Shaw, G.G. & C.H.A. Little, 1977. Natural variation in balsam fir foliar components of dietary importance to spruce budworm. Canadian Journal of Forest Research 7: 47-53.

Stehr, G., 1955. Brown female, a sex-linked and sex-limited character. Journal of Heredity 46: 263-266.

Stehr, G., 1959. Hemolymph polymorphism in a moth and the nature of sex-controlled inheritance. Evolution 13: 537-560.

Stock, M.W. & J.D. Guenther, 1979. Isozyme variation among mountain pine beetle (Dendroctonus ponderosae) populations in the Pacific Northwest. Environmental Entomology 8: 889-893.

Stock, M.W., G.B. Pitman & J.D. Guenther, 1979. Genetic differences between Douglasfir beetles (Dendroctonus pseudotsugae) from Idaho and coastal Oregon. Annals of the Entomological Society of America 72: 394-397.

Thom, R., 1970. Topological models in biology. In: C.H. Waddington (Ed.): Towards a theoretical biology, Vol. 3. Aldine, Chicago. p. 89-116.

White, T.C.R., 1969. An index to measure weather-induced stress of trees associated with outbreaks of psyllids in Australia. Ecology 50: 905-909.

Zeeman, E.C., 1976. Catastrophe theory. Scientific American 234: 65-83.

Role of plant variability (trait vector dynamics and diversity) in plant/herbivore interactions[1]

William J. Mattson, Nancy Lorimer & Rolfe A. Leary

North Central Forest Experiment Station, 1992 Folwell Avenue, St. Paul, Minnesota 55108, USA

ABSTRACT

Plants and insects interact in a manner prescribed by their respective individual, population, and ecosystem level properties. At the individual level, an insect's properties must match those of its host plant because insect performance depends on the goodness of match. However, the plant's properties are not constant but ever changing, thereby influencing the quality of match and thus insect performance. Individual plants or their modular compartments each present a highly variable set of properties owing to their (a) rapid phenological and ontogenetic change, and (b) induced change in response to herbivory and variations in the environment. We propose that all plants, but especially long-lived ones (relative to their consumers) depend upon rapid changes or variability in their properties to thwart matching with phytophagous organisms.

INTRODUCTION

Each living organism may be characterized by the collection of properties or special traits it possesses. However, the list of traits assigned to an organism will vary depending on the organism which perceives it because each observer's perception is a function of its own set of traits and no two organisms have identical sets of traits. Ultimately then, the relationship between organisms or the manner in which they interact is prescribed by the set of traits that each possesses, modified, of course, by environment.

All plant/herbivore coevolutionary arguments begin with a large, important assumption; namely that the herbivore population is somewhat preadapted to the set of traits possessed by the plant population. In other

1. This research was supported in part by the Canada-United States (CANUSA) Spruce Budworms Program, USDA Forest Service, 870 Reed Road, Broomall, PA 19008.

words, the herbivore population must consist of at least a few individuals which each possess a set of traits that allow it to interact with a corresponding set of traits possessed by some members of the plant population. If this is not the case, the plant population is immune or incompatible with the herbivore population. Hence, no coevolutionary scenario is possible. We emphasize that this is not a trivial assumption because the matching of corresponding plant and herbivore traits is at the heart of understanding the nature of plant/herbivore interactions. We propose that the variability in the 'goodness of fit' or matching between the trait sets of plant and animal populations contributes significantly to their long-term coexistence, their population dynamics (Haukioja, 1980), and their evolution.

Traits are those individual properties of each individual organism that are important in its interactions with members of its own and the other trophic level. Examples of some plant traits are growth rate, phenology of development, compositions of amino-acids, carbohydrates, lipids, moisture, and allelochemics, physical characteristics (toughness, hairiness, etc.), and associations with members of other trophic levels such as predators and parasites of herbivores. Examples of some herbivore traits are growth rates, timing of emergence for feeding and mating, sensory apparata for detecting chemical and/or electromagnetic cues from the host plant, systems for sequestering or detoxifying harmful substances, and associations with members of other trophic levels such as gut micro-organisms.

TRAIT VECTORS

Each trait (T_i) can be measured at some time t and assigned a particular numerical value, ($T_{i(t)}$). The reference to time is exceedingly important because trait values are not invariant. Change occurs in response to internal (i.e. spontaneous) and external (i.e. inducing) factors. Particular traits may not only change in numerical value, they may even submerge or disappear; just as others may suddenly emerge or appear in an individual's trait collection. Such changes are deep or qualitative changes in contrast to ordinary or quantitative change (Bunge, 1977). Furthermore, each trait may have its own characteristic rate of change, and sometimes the rates of change of several traits may be correlated because traits are not entirely separate and independent (Gould & Lewontin, 1979). Traits are not necessarily independent of one another both in their origin and their effects on other organisms. For example, several traits may be linked to the same gene(s), and other traits may be influenced by more than one gene. Moreover, we want to emphasize that the interactions between organisms depend on the whole collection of traits, i.e. the set of traits possessed by each organism, not any individual traits of and by themselves.

Thus, the properties of each individual plant and herbivore can be described by a set of traits having specific trait values; in short, a trait vector $[T_{1(t)}, T_{2(t)}, T_{3(t)}, \ldots T_{n(t)}]$. Any 2 individuals within a population can differ with respect to the number of traits they posses as well as their specific trait values. As we have defined traits, some are clearly pervasive, that is they will occur in the trait vectors of all members of the population (growth rates, flushing phenologies, etc.). Whereas, others will not necessarily occur in all individual trait vectors.

FITNESS IN RELATION TO TRAIT MATCHING

We believe that the outcome of a particular plant/herbivore interaction is a function of the goodness of matching between their respective traits. A non-match occurs either when one or more of the instantaneous trait values (Act_i) in the plant's trait vector fall outside of the tolerance limits (Min_i and Max_i) for the corresponding traits of the herbivore, or when a plant has one or more traits for which the herbivore has no corresponding traits. On the other hand, a match occurs when the instantaneous values for all the individual traits of a plant's vector fall between the tolerance limit values for the corresponding individual traits of the herbivore. A match can vary from poor to good depending on how near the individual plant traits come to exceeding the tolerance limits for the corresponding herbivore traits. For example, as plant trait values (Act_i) approach the 'optimum' value (Opt_i) of the herbivore's corresponding trait tolerance range, the match becomes increasingly better. However, as the instantaneous values for the plant traits (Act_i) deviate from the 'optimum' to the limits (Min_i and Max_i) of the herbivore's trait tolerance range, the match becomes progressively worse. We suggest the following matching function:

$$M_{j,m,t} = \frac{a}{n} \sum_{i=1}^{n} 1-k_i \frac{Opt_i - Act_i}{Opt_i}^{b_i} \qquad \text{If and only if } Min_i \leqq Act_i \leqq Max_i, \text{ otherwise } M_{j,m,t} = 0.$$

where $M_{j,m,t}$ is the match between herbivore$_j$ and plant$_m$ at some time t; n is the total number of traits of concern, k_i and b_i are trait specific numerical constants and a is a numerical constant specific to the particular species of herbivore and plant.

We contend that the better the match ($M_{j,m,t}$) between the herbivore (H_j) and the plant (P_m), the higher the likelihood that the herbivore will be able to express its full reproductive potential (rpH_j) and therefore realized reproduction ($rrH_{j,m}$) will be maximal. On the other hand, a poor match brings about the opposite effect. The outcome of the interaction for the plant is usually the opposite of that for the herbivore; namely plant realized reproduction ($rrP_{m,j}$) approaches its reproductive potential (rpP_m)

when herbivore matching is very poor. The likely relation between these variables is formalized in the following equations:

$$rrH_{j,m} = rpH_j \left(1-e^{-b \int_{t_0 = \text{birth } H_j}^{t_n = \text{death } H_j} (M_{j,m,t})(E_{j,m,t})dt}\right)$$

$$rrP_{m,j} = rpP_m \left(1-e^{-c/ \int_{t_0 = \text{birth } P_m}^{t_n = \text{death } P_m} (M_{j,m,t})(E_{j,m,t})dt}\right)$$

where $E_{j,m,t}$ is an encounter function which is a unit step function that has value one when the encounter occurs and zero where there is no encounter, and b and c are numerical constants.

The equations explicitly consider the impact of matching "quality" over the entire life span (birth to death) of each organism in the determination of realized reproduction. If, however, reproduction normally ceases well before death occurs, one could integrate the matching function $(M_{j,m,t})dt$ just up to that point, unless the post-reproductive adult still contributes in some way to the survival of its progeny - hence its realized reproduction. These equations are most easily interpreted when both members of the interaction pair have similar life spans or generation times. However, if their life spans are different, then the evaluation and impact of matching become more complicated because the longer-lived member outlives the influence of the shorter-lived member. To adjust for this relational asymmetry, one can allow the shorter-lived member to reproduce (if it can after the initial encounter) so as to carry the interaction through time to the end of the life span of the longer-lived member.

Fitness (f) for both organisms is merely the realized reproduction (rr) of each, ranked relative to that of all other members of its own population. In other words, it is simply relative realized reproduction (rrr). The notion of fitness is important because evolutionary theory purports that those organisms with higher rrr will eventually supplant those with lower rrr. Given enough time, a population should finally evolve to consist of individuals which have nearly equal rrr. In the case of a plant/herbivore coevolutionary scenario, the ultimate survivors would be those individuals whose trait vector matching qualities and reproductive potentials resulted in nearly equal rrr. For example, such plants might on the one hand be individuals which have highly matching trait vectors and high rp, or on the other hand individuals which have poorly matching trait vectors and low rp. Thus, an organism's capacity to 'stay in the game', to be an effective competitor with its conspecifics, depends on the relationship which exists between the matching quality of its trait vector and its reproductive potential. Next we examine changes in matching quality of plant trait vectors.

TRAIT VECTOR DYNAMICS

Both the instantaneous values of traits and the magnitude and periodicity of their variation are important because the goodness of matching between herbivores and plants has to be evaluated not just at an instant, but over the entire life span of each organism. The variability of matching is really a special case of a more general issue of the variability or stability of favorableness of a habitat for particular occupants of that habitat (Southwood, 1977). In this special case, we propose that change in the quality of the herbivore habitat (i.e. plant trait vectors) is selected for as a consequence of herbivory.

Seasonal and ontogenetic change

Perhaps the most obvious (to humans anyway) changes in plant trait vectors are those that we casually observe and recognize as seasonal change: changes in leaf coloration, and leaf morphology, changes in woodiness of plant parts, fruit development, and the like. Although each plant has a trait vector containing n individual traits, we will focus on just 2 of them for a more detailed examination of seasonal change. Two which are extremely important to herbivores are foliar organic nitrogen content (T_1) and moisture content (T_2). The trajectory for these traits (Fig. 1) shows that plants begin with both high nitrogen and high moisture and then steadily decrease over the duration of the growing season. The pattern is the same for many kinds of plants (Scriber & Slansky, 1981). The importance of such change for herbivores becomes obvious when one plots the herbivore's corresponding trait ranges, that is the minimum (Min_i) and maximum (Max_i) tolerable limits for each particular plant trait (Fig. 1). For example, we know that certain kinds of herbivores must have diets containing at least 2.0 % nitrogen or survival is zero. On the other hand, as plant nitrogen levels approach the upper bounds, around 7.0 %, the survival of herbivores likewise approaches zero, but for different reasons (Mattson, 1980). One can make the same determinations for moisture content. The result when plotted is the domain of plant nitrogen and moisture values for which herbivory by a particular herbivore is possible. It appears as a rectangle in our two-dimensional plant state space. If the plant trait space were instead n-dimensional, then the herbivore's domain would be an n-dimensional volume. We call this rectangle or volume the herbivore's 'window' in the plant's state space. Different kinds of herbivores have different windows and hence, as plant trait vectors change through seasonal time, plants can pass in and out of the windows of different kinds of herbivores

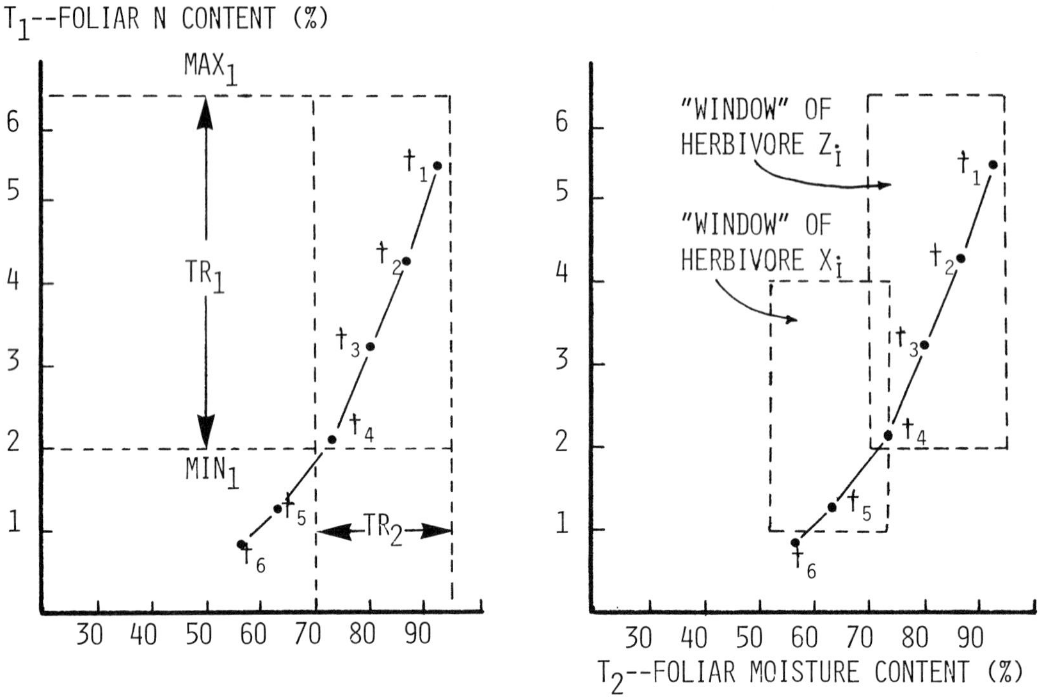

Figure 1. Seasonal change in foliar nitrogen (Trait 1) and foliar moisture content (Trait 2) in relation to the minimum and maximum tolerable limits or trait ranges (TR_i) for these traits by different herbivores. The region delimited by the herbivore's trait ranges for all plant traits is the herbivore's window on the plant's trait state space. Time (t_i) is represented parametrically.

(Fig. 1).

Just as there is trait vector change over seasonal time, there is similar change over the life span of the organisms, i.e. over ontogenetic time. Particularly important are the qualitative changes; i.e. the emergence of new traits and the submergence of old ones. Those which are most obvious (to human beings) are the emergence of reproductive traits during the course of an organism's ontogeny such as those associated with sex identification, and reproductive capacity. Other less obvious changes are those in an organism's immunity systems, growth capacities, and so on.

Induced change

The fact that plants are highly responsive to their abiotic and biotic environment has long been well known (Bradshaw, 1965). For example, it is now common knowledge that plants react in many ways to wounding by biotic agents. Typical changes are increased levels and kinds of various phenolic compounds, protease inhibitors, altered levels of photosynthesis and phytosynthate allocation, as well as altered phenology, morphology, electromagnetic spectra, growth rates, and structural form or complexity (Bradshaw, 1965; Knutson, 1979).

Just as significant as the biotic inducing agents are the abiotic ones, e.g. moisture, heat, nutrient, and light regimes. Changes in the abiotic flux can elicit large, significant changes in a plant's trait vector. Increasing levels of nitrogen, for example, often increase growth rates at the expense of other processes in the plant that compete for carbon and/or amino acid precursors, such as the production of polyphenols (Mattson, 1980). In short, each plant has a unique potential to react to various kinds of inducing agents with the result that it has a unique capacity to change its state or its trait vector (Bradshaw, 1965). Consequently, a plant's trait vector may change in its matching relationship with an herbivore. If matching becomes poorer, this would be advantageous because it would increase the plant's opportunity to fulfill its reproductive potential. On the other hand, some inducing factors could cause a plant's trait vector to change in the opposite direction, thereby making it better matching. Moreover, some inducing factors may restrict the plant's potential to change, thereby 'locking' it onto a relatively stable state which may be more susceptible or highly matching for a particular population of herbivores. Similarly, various spontaneous ontogenetic changes in the plant may also induce plants to move into more susceptible states or cause them to lose flexibility and thereby remain in susceptible states longer than usual.

If one expresses the duration of plant suitability (i.e. good matching) by D and the duration of the herbivore life cycle as d, then over the course of evolution between plants and herbivores, one might expect D/d to evolve toward values of 1. In other words, one would expect the plants to evolve labile trait vectors which would permit them to escape from their 'adapted' consumers through time by quickly passing through the herbivore's window, thereby minimizing herbivore impact on plant fitness attainment. On the other hand, because herbivores themselves have a capacity for change (cc), one might expect that plant/herbivore coevolution would eventually result in both organisms evolving nearly equal capacities for change: ccP/ccH = 1. They need not, however, employ the same basic mechanisms for generating change (e.g., mutations, recombination, trait vector dynamics, etc.).

TRAIT VECTOR DIVERSITY

Up to this point we have emphasized that the susceptibility of an individual plant to an herbivore population depends on the properties or trait vector of that individual with respect to the trait vectors of the individual herbivores, all with respect to some time, t. Thus, at a given time t, a particular plant may be a poor match for a very large fraction of the herbivore population, and therefore is defined to be minimally susceptible. As time passes, however, not only the plant's trait vector, but also the herbivore population's set of trait vectors can change. The 'direction' that the herbivore population changes is particularly important to the long term 'resistance' of the particular plant. This direction is, of course, due in part to both spontaneous factors as well as inducing factors in the herbivore's environment. Especially important is the selection pressure exerted on the herbivores by the rest of the plant population, i.e. the contemporaries of the particular plant in question. In the short term an individual plant's susceptibility or resistance depends primarily on its own particular properties, but over the longer term (i.e. over successive generations of the herbivore populations) its susceptibility depends also upon the properties of other plants in the same population because they have a marked influence on the direction of evolution or the trait vector change in the herbivore population.

The effect of a plant population on an individual herbivore can be viewed in terms of the number of different plant trait vectors that an individual herbivore encounters through attacks in its given life span. When an herbivore attacks a plant, each organism is changed as a result of the interaction. The change is more or less unique to the specific trait vectors of each interacting pair. Moreover, the order in which different herbivores and different plants interact is also important because each different sequence of plants that a particular individual herbivore attacks has unique cumulative effects on the survival and reproduction of that herbivore. The herbivore's attack sequence, then, really represents the cumulative selection pressure exerted by the plant population on an herbivore.

We can quantify the total possible number of different attack sequences or attack pathways that herbivores could trace or follow in a plant population given (1) the total number of different plants, that is those having different trait vectors, and (2) the usual number of plants attacked per herbivore. For example, if the plant population consists of m different individuals, all of which have different trait vectors, and the herbivores normally attack n plants each, then the total possible number of attack sequences and hence possible different selection pressures facing herbivores in the particular population of plants is simply m^n. If the entire plant population consists of only 3 different trait vectors such as (A, B, C), and the individual herbivore normally attacks only 2 individual plants during its life span, then there are only 3^2 or 9 potential attack sequences or selection pathways facing the herbivore populations: [AA, AB, AC, BA, BB, BC, CA, CB, CC]. On the other hand, as m increases the number of possible attack sequences increases exponentially. From the other direction, as the numbers of trait vectors in the plant population decrease to approach one, that of the hypothetical resistant plant, the selection pressures facing herbivores become increasingly homogeneous and therefore convergent instead of divergent. In a nutshell, then, as the rest of the plant population becomes more and more like the 'minimally susceptible' plant, then more and more selection pressure will direct the herbivore population's evolution toward better and better matching with the here-to-fore minimally susceptible individual and hence with the plant population as a whole.

This argument holds not only for within herbivore generations but also between generations. For example, if each individual herbivore attacks but one plant during its life span, it has the possibility of selecting any one of m different types in the plant population. Its progeny (let's say it produces a minimum of p of them) each also have the possibility of selecting any one of the m different types of plants in the plant population. Therefore, over the course of n different generations there are m^{pn} different attack sequences or pathways for a given herbivore's germ cell line.

In conclusion, the type and intensity of interaction between plants and herbivores depends on the collective goodness of matching between the trait vectors of individuals comprising the plant and herbivore populations. Variations in the goodness of match depend on changes which can occur in the trait vectors of both plants and herbivores. Minimal goodness of match will occur when individual plants show high rates of trait vector change with respect to capacity for change in the herbivores, and high trait vector diversity among individuals. On the other hand, maximal goodness of match will occur when individual plants show low rates of trait vector change with respect to capacity for change in herbivores, and low trait vector diversity among individual plants.

REFERENCES

Bradshaw, A.D., 1965. Evolutionary significance of phenotypic plasticity in plants. Adv. in Genetics 13: 115-155.

Bunge, M., 1977. Ontology I: The furniture of the world. Vol. 3 of Treatise on basic philosophy. D. Reidel Publishing Co., Boston. 352 p.

Gould, S.J. & R.C. Lewontin, 1979. The spandrels of San Marco and the panglossian paradigm: a critique of the adaptionist programme. Proc. R. Soc. Lond. B. 205: 147-164.

Haukioja, E., 1980. On the role of plant defences in the fluctuation of herbivore populations. Oikos 35: 202-213.

Knutson, D.M., 1979. How parasitic seed plants induce disease in other plants. In: J.G. Horsfall & E.B. Cowling (Eds): Plant disease. Vol. 4. Academic Press, New York/ London. p. 293-312.

Mattson, W.J., 1980. Herbivory in relation to plant nitrogen content. Ann. Rev. Ecol. Syst. 11: 119-161.

Scriber, J.M. & F. Slansky, Jr., 1981. The nutritional ecology of immature insects. Ann. Rev. Entomol. 26 (in press).

Southwood, T.R.E., 1977. Habitat, the template for ecological strategies? J. Anim. Ecol. 46: 337-365.

Stability in a multi-component host-pathogen model

E.G. Brittain

Botany Department, School of General Studies, Australian National University, P.O. Box 4, Canberra 2600, Australia

ABSTRACT

A 4 component model of 2 competing organisms, each with a specific parasite, has been exercised so as to investigate its equilibria. In addition to the equilibrium to which the model was already known to return after moderate disturbance, severe perturbation led to a new steady state which had the appearance of an equilibrium but which was not maintained in the long term. Stability of equilibria and what constitutes an equilibrium are also discussed.

INTRODUCTION

In 1972 Graham Chilvers and I (Chilvers & Brittain, 1972) published a model for a system of 2 species in competition with each other and each subject to attack by 1 host-specific parasite. We presented this model in general terms but in fact in developing it we thought in terms of 2 competing and closely related trees, each subjected to a specific leaf spot fungus disease, so that the equations used in the model are intended to be appropriate for that particular system.

In the course of discussing this model, we asserted that it was stable in the sense that for a given set of parameters, it would return the amounts of the 4 components to the same level from almost any set of initial values.

Since 1972 there have been a number of papers which have been concerned with stability of multi-component systems. Most of these have been mathematical in presentation and have been concerned with seeking analytical methods with which one might predict whether a given model should be stable or not. Discussing the problem from rather a different standpoint, May (1977) has considered the possibility that natural multi-species assemblies of plants and animals and also models of them, are likely to possess several different equilibrium points. The possibility that an ecosystem may possess more than one stable equilibrium is of considerable interest because it raises the thought that the familiar stable state of the system may have some degree of instability, sufficient that it may, if perturbed sufficiently change to some alternative and perhaps very unfamiliar equilibrium.

For these reasons, I have recently made an effort to exercise our 1972 model more vigorously than before in order to test more stringently our contention that it had only one equilibrium point (Brittain, 1980).

The equations which constitute the 4 component model are set out below. I do not propose to discuss the details of these now. The assumptions made and the arguments used in deriving the model are set out in full in the original paper. For the purposes of the present work, the model has been implemented in ACSL on a Univac 1100/42 computer at the Australian National University (ACSL - Advanced Continuous Simulation Language. Mitchell and Gauthier Associates Inc. P.O. Box 685 Concord, Mass. 01742).

I should mention that in the model, the environment is conceived to be capable of supporting a limiting quantity of biota to which we ascribed the value of unity, so that the amounts of each of the components as they vary within the environment are correctly described as proportions. However, I propose loosely to call these quantities, since proportions of components in a multi-component system has another meaning.

MODEL EQUATIONS

Host X:

$$dX/dt = R_X X(1-(X + Y)) - (D_x x + D_n X - D_x x \cdot D_n X) \qquad (i)$$

Host Y:

$$dY/dt = R_Y Y(1-(X + Y)) - (D_y y + D_n Y - D_y y \cdot D_n Y) \qquad (ii)$$

Parasite x:

$$dx/dt = R_x x(X-x)\ (1-x(A_X x+A_y Y)) - (D_x x+D_n x-D_x D_n x^2) \qquad (iii)$$

Parasite y:

$$dy/dt = R_y y(Y-y)\ (1-y(A_Y y+A_x X)) - (D_y y+D_n y-D_y D_n y^2) \qquad (iv)$$

Dead tissue p:

$$dp/dt = R_x x(X-x)\ (1-x(A_X x+A_y Y)) - (D_x x+D_n x-D_x D_n x^2) \qquad (v)$$

Where X = amount of one host

Y = amount of second host

x = amount of X specific parasite

y = amount of Y specific parasite

R_X, R_Y, R_x, R_y = growth constants

D_x, D_y = death rate constants for parasites

D_n = death rate due to other causes
A_X, A_Y = absorption constants for collection of inoculum
p = amount of dead tissue.

EXERCISING THE MODEL

The model was first run using the parameters and initial amounts of the 4 components (Table 1) until the steady state was reached. I shall refer to this as the first steady state. The values of the state variables at this point are in column 2 of Table 2.

Perturbation of this first equilibrium by reduction of either host with its attendant parasite from the equilibrium amount to its initial amount resulted predictably in a temporary upsurge of the other host-parasite pair due to reduction of competition, followed by recovery of the perturbed pair and, after some strongly damped oscillations, return to the first steady state. However, when the perturbation was more extreme, as when I reduced one host to $1/10^8$ of its equilibrium value, something rather different occurred. Removal of 1 host and its parasite to that low level allowed the other pair to surge upwards as before. They reached a new steady level 120 time steps after the perturbation. This new level was maintained for a further 880 time steps. I shall refer to this as the second steady state. Only after 1000 time steps had elapsed since perturbation did the perturbed component reappear in an amount sufficient to be distinguished from zero. Then after a period of more or less violent oscillation, the first steady state was restored.

Perturbation of the first equilibrium by reduction of 1 parasite to much smaller proportions, in all cases resulted in a temporary upsurge in the corresponding host which was thus relieved of its parasite load. However the parasite always came back again before any equilibrium could be attained so that after a few oscillations, the first steady state was restored. Only by reducing a parasite exactly to zero could any other equilibrium be demonstrated.

Table 1. Initial state variable and parameter values.

State variables	Growth constants	Other rate factors
X 0.01	R_X 0.04	D_x 0.014
Y 0.01	R_Y 0.035	D_y 0.014
x 0.0005	R_x 0.18	D_n 0.009
y 0.0005	R_y 0.19	A_X 0.05
		A_Y 0.05

Table 2. Values of state variables in the first and second steady states.

State variable	Steady state
$Xi = 0.01$, $Yi = 0.01$	
X	0.3353
Y	0.2243
x	0.2072
y	0.1028
$Xi = 0.01$, $Yi = 0.4 \times 10^{-8}$	
X	0.5137
Y	0.25×10^{-6}
x	0.3853
y	0.46×10^{-14}
$Xi = 0.4\ 10^{-8}$, $Yi = 0.01$	
X	0.21×10^{-6}
Y	0.4516
x	0.62×10^{-14}
y	0.3301

MODIFICATION OF THE MODEL

May (1977) has presented curves for the rate of change of vegetation biomass plotted against the amount of that biomass. The curve appears to be parabolic, which is a natural consequence of plotting the first derivative of data for a symmetrical sigmoid growth curve in this way. He has superimposed on this curve, a set of other curves for rate of removal of vegetation biomass by various fixed levels of grazing. These intercept the parabolic curve at either 1 or 3 points. At each intersection, grazing removes vegetation biomass at the same rate as it is produced so that an equilibrium appears to be indicated. May uses this diagram as a means of emphasising his point that multi-component systems may possess multiple equilibria.

It occurred to me that it would be fairly easy to modify our model and arrange its output in such a way as to resemble May's diagrams and so to check whether we in fact have the possibility of multiple equilibria in our model.

In order to simulate this situation it was first necessary to reduce it to 2 components and that was done by setting the other host-parasite pair to zero. Then in order to obtain a fixed level of parasitism to compare with May's fixed level of herbivory, I broke the feed-back link between parasite and host. This was readily done by changing the left hand side of equation (iii) from dx/dt to dp/dt. The initial value of x, amount of X-specific parasite, which appears in equation (i) was then retained as a fixed value. We thus had a fixed amount of parasite reacting with a growing host. The product of this interaction, which I called p, may be thought of as dead tissue, neither living host nor infective parasite. The rate of production of p was then equivalent to the rate of removal of host.

A set of values for different fixed amounts of parasite was investigated. These, I have not superimposed on the 1 parabolic curve for the amount of host because the amount of parasite present itself has an effect on the host curve. Therefore I have presented them separately.

A pair of curves for each of 4 different levels of parasite is shown in Fig. 1. The model operates dynamically so that it actually begins at the left with a small amount of host X and then progresses to the right. Even-

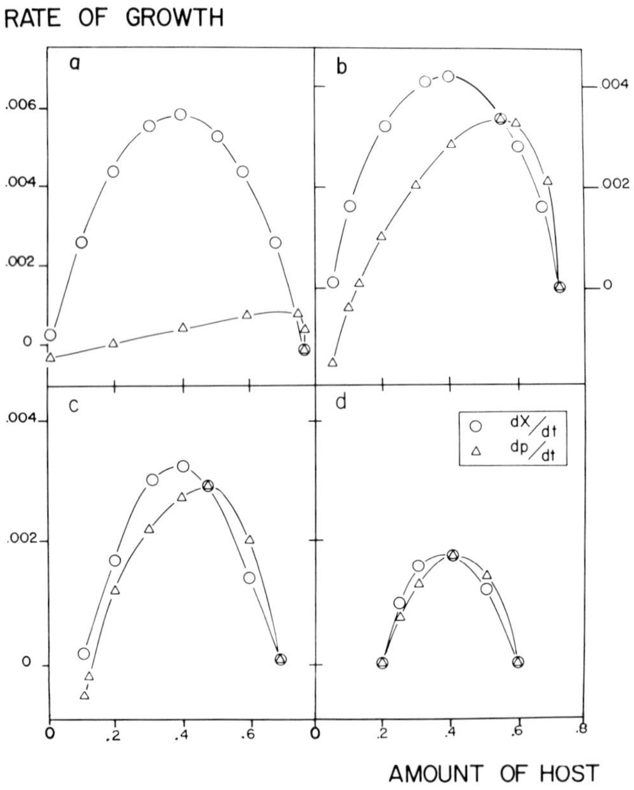

Figure 1. Rate of growth of host (O) and rate of production of dead tissue (Δ) plotted as a function of the amount of host present in a 2 component model. Four different fixed amounts of parasite were used: a 0.01, b 0.10, c 0.20, d 0.30.

tually the host curve passes through an intersection with the curve for the rate of formation of dead tissue, p, and then it continues to decrease along the descending limb of the approximately parabolic course. Removal of host tissue to dead tissue also continues past the intersection and the 2 curves cross but presently come together again at what I am calling a confluence. At this point the rates of increase of host and of dead tissue are virtually zero and we have what I believe to be the only equilibrium for the system. In no case was there ever more than 1 crossover point, contrary to the contention of May that there could be 3. Furthermore the crossover point was never an equilibrium. I have checked that deliberately by setting the quantities to those obtaining at the crossover but the model then moved away from that to the confluence.

At the crossover points the 2 first derivates are equal but nonzero. Put another way, the rate of removal of host equals the rate of growth of host but neither is zero. Yet these points are not equilibria. They are not, because although the rates are equal they are so only briefly. Both are changing more or less rapidly so that their courses intersect and continue beyond the intersection.

I use the term confluence to describe the circumstance where the 2 curves approach each other and come to an end while coinciding. They do this only when both the net rate of growth of host and the net rate of production of dead tissue become zero.

EQUILIBRIA AND STABILITY

I now want to discuss a simple physical model. Consider a cubic prism at rest on one of its faces on a stationary horizontal plane. We may certainly say that the cube is in equilibrium, is in a steady state and is stable. The cube can be perturbed by tilting it. Because it is highly stable it will return to its original state from quite large perturbations. However if the limit of its equilibrium is exceeded it will assume a second steady state, different from the first. We know that it can assume in all 6 such steady states, resting on each of its 6 faces. Thus the cube is a system which has multiple equilibria or steady states in each of which it is highly stable. The cube in fact possesses 20 other possible equilibrium positions which are unstable states, 12 on the edges and 8 on the apices; positions in which it might be balanced if one had sufficient patience and skill. Given a steady environment, a cube balanced in any of those 20 unstable positions might be said to be in a steady state or in equilibrium. But it would be highly unstable, the slightest perturbation causing it to revert to one of its more stable positions.

The point I want to make with this simple illustration is that stability is quantitative and that equilibria and steady states exist which have varying degrees of stability.

In our 1972 paper, Chilvers and I stated that 'the equilibria eventually arrived at are typical of the rate constants used in the model but are quite independent of the starting proportions ... of the components'. This statement is true only if the model has a unique equilibrium position.

In the present work some quite extreme perturbations were used and it was found that some of these produced a new state which appeared to be steady for a significant length of time. This second steady state was virtually indistinguishable from that which the system assumes when one of the hosts is reduced exactly to zero, from which of course neither it nor its parasite can ever recover. However slight changes were occurring, below the level of resolution of graphic output but clearly visible in the numerical results. These ultimately led to a return from the second state to the original equilibrium. It is interesting to consider whether that second state, which can be seen to have been transitory, should be called an equilibrium. Probably a decision on that would depend on one's own time scale. If this were a real system and the time steps were years, I believe we should all be happy to call it an equilibrium if we lived at any time between about year 200 and year 800. Yet in the model situation with an arbitrary time scale, we see that it is not a final equilibrium.

One might say that there is a risk that a steady state of such long duration might be mistaken for an equilibrium. Alternatively it might be suggested that a steady state deserves to be called an equilibrium if it endures for long enough. In either case the crucial point would involve a subjective decision about the length of the apparent steady state. The dictionary meaning of 'equilibrium' is 'state of balance' with no qualification as to the duration of this state. Yet there seems to be a tendency in the modelling literature to reserve the term for situations which have some element of permanency. Certainly if this were not done, if the term could be used for any fleeting condition, it would not be of much value.

CONCLUSIONS

I have argued that for the interacting pairs of rate curves of Fig. 1 it is not sufficient for the rates to be equal for an equilibrium to be defined. It is also necessary that the net rates of change be zero. Equilibrium occurs when the quantity of a substrate and the quantity of its product become stationary, which is to say that their rates of change become zero.

Whether or not a system or a model system possesses multiple equilibria depends at least to some extent on the definition of equilibrium. I have concluded that a necessary condition for equilibrium is that the first derivatives of the amounts of the components should be zero.

This must be regarded as an ideal state for in any real system or even in a computer model if it is operated within realistic constraints of time and cost, exactly zero rates of change seldom occur. Rather they may seem to approximate zero. Whether they will not continue to remain near zero or tend away from it again may only be resolved if a great deal of observation time is available. Meanwhile one may speak of the existence of an equilibrium although with the reservation that this may be transitory on some other time scale.

REFERENCES

Brittain, E.G., 1980. Equilibria in a model system of four species (in preparation).

Chilvers, G.A. & E.G. Brittain, 1972. Plant competition mediated by host-specific parasites - a simple model. Australian Journal of Biological Science 25: 749-756.

May, R.M., 1977. Thresholds and breakpoints in ecosystems with a multiplicity of stable states. Nature 269: 471-477.

Selection in host-parasite systems

D. Krusche

Bundesforschungsanstalt für Forst- und Holzwirtschaft Hamburg, Institut für Forstgenetik und Forstpflanzenzüchtung, Siekerlandstr. 2, D 2070 Grosshansdorf 2, BRD

ABSTRACT

A mathematical model for a natural host-parasite system with genetic coupling is developed. The genetic variances and the heritability needed for predicting genetic gain in the course of a selection programme are dependent on the actual genetic state of the parasite population. The results suggest that without detailed knowledge of the evolutionary state of the system, genetic improvement by artificial selection may not be successful.

INTRODUCTION

In forestry a widely accepted method for breeding for resistance to fungal diseases is started by selecting some parent trees with low infection within a heavily infected area. Controlled crosses with several tester trees are made and the progenies are grown in field tests and observed for symptoms after artificial infection (Becker & Marsden, 1972; Bingham et al., 1969). The measured variation is attributed to several sources, such as additive genetic variance, dominance variance, etc., and the heritabilities and genetic gains are calculated. An assumption underlying this breeding method is that the genetic background of the disease-producing parasite population does not change over time. But this assumption is unrealistic, especially in reference to forest trees whose parasite populations may reproduce up to several hundred times within one host generation. Thus there could be a considerable change in the genetic composition of the parasite population. Breeding for disease resistance differs therefore in a fundamental way from breeding for characters such as frost resistance or a high growth rate.

In this paper I present a model of a host-parasite system. The analysis shows how artificial mass selection acts on the dynamic behaviour of the system. The dependence of heritability and genetic gain of the character 'infection density' on the time and state of the system is demonstrated.

Even though this model may no correspond in detail to any real host-parasite system, it aims to clarify the inherent dynamics and may stimulate discussion in order to improve the theoretical basis of resistance breeding.

THE MODEL

The model is an improved version of a similar model from Krusche (1975), and is based on the work of Mode (1958), Person (1967), Groth & Person (1977) and the 'genetic feedback' concept of Pimentel (1961). Consider a closed host-parasite system with the following properties:

- The host and parasite are diploid and random mating with non-overlapping generations and alternate reproduction.
- The host population carries a locus with 2 alleles, R standing for resistance and r for susceptibility, R being dominant over r.
- The parasite population has 2 alleles on a corresponding locus, A standing for avirulence and a for virulence, A being dominant over a.
- The interaction of host and parasite is according to the gene-for-gene concept (Flor, 1956). Whenever a parasite with an A-gene comes into contact with a host possessing an R-gene, the parasite cannot reproduce. The remaining 3 possibilities result in successful infections of the host.
- The absolute fitness of the host genotypes is a decreasing function of the number of infections present on a host before reproduction.

In the following section R- denotes RR and Rr genotypes and A- denotes AA and Aa genotypes, and relative gene frequencies are in brackets.

The fitness of the parasite genotypes on different host genotypes is:

		Parasite	
		A-	aa
Host	R-	0	p_2
	rr	p_1	p_2

The total (net) fitness of the parasite population is

$$W_p = (A-)\,(rr)p_1 + (aa)p_2 \qquad (1)$$

The number of parasites after reproduction is $P_1 = W_p P_0$, with P_0 the number of parasites before reproduction. Let H_0 be the number of hosts in the present generation, then $\alpha = P_1/H_0$ is the number of parasites coming into contact with a single host, i.e. the parasite density.

The numbers of successful infections by different parasite genotypes on different host genotypes are:

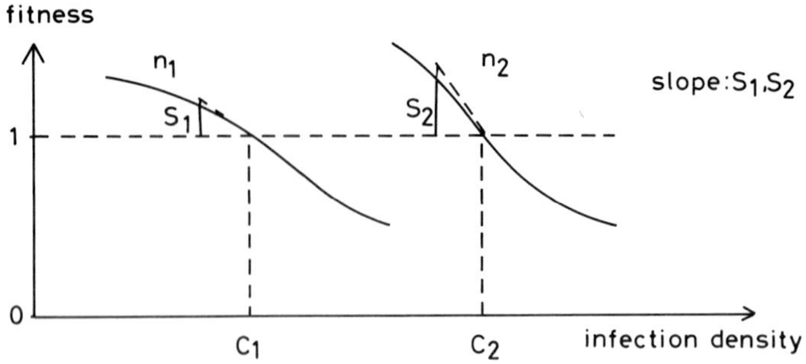

Figure 1. Fitness functions of the host genotypes.

		Parasite		
		A-	aa	total
Host	R-	0	α(aa)	α(aa)
	rr	α(A-)	α(aa)	α

We assume the fitness functions: n_1 for a R- host and n_2 for rr hosts as presented in Fig. 1.

The fitness of the host population is

$$W_H = (R-)n_1 + (rr)n_2 \quad (2)$$

THE EQUILIBRIUM OF THE SYSTEM

The equilibrium condition for the system is met when the fitness of all genotypes is 1. From (1) we have ($\bar{r}$, $\bar{a}$, etc., are the stationary values):

$$\bar{r}\bar{r}\, p_1 = 1,\ p_2 = 1,\ p_1 > 1$$

The stationary value of the susceptibility gene frequency is fully determined by the reproductive properties of the parasite. A requirement for the existence of an equilibrium is $p_1 > 1$, i.e. a fully differential fitness pattern of the parasite. The difference of the fitness of the A-parasites on R- and rr hosts must exceed the difference of the fitness of the aa parasite on R- and rr hosts (Person et al., 1976).

From (1) we have the change in gene frequency for the virulence gene:

$$\Delta a = a^2(1-a)(p_2-(rr)p_1)/W_p$$
$$= a^2(1-a)p_1((R-)-(\bar{R}\bar{-}))/W_p$$

Accordingly, the evolutionary speed of the virulence gene is proportional to p_1 and to the excess of the resistance gene frequency over the equilibrium value. The selective advantage of the virulence gene is positive when the related resistance gene exceeds its equilibrium, which has been termed the 'genetic feedback mechanism' (Pimentel, 1961).

The equilibrium condition for the R- gene is from (2): $n_1=n_2=1$. We linearize n_1 and n_2 in the neighbourhood of the equilibrium:

$$n_1 = (\alpha(aa) - c_1)s_1 + 1$$
$$n_2 = (\alpha - c_2)s_2 + 1$$

It follows that $\bar{\alpha} = c_2$ and $\overline{aa} = c_1/c_2$, $c_1 < c_2$.

The equilibrium frequency of the virulence gene is determined by the reproductive properties of the host population. The rr hosts have to support $\bar{\alpha}$ infections on the average. R- hosts suffer only a fraction $(aa)\bar{\alpha}$ infections, but the decrease of fitness with increasing α is lower than the decrease of fitness of the rr hosts.

A typical sample run is represented in Fig. 2. We notice that:
- The equilibrium gene frequencies for each population are determined by the reproductive properties of the other population.
- Polymorphisms of the R- and A- genes are maintained in the system.
- The gene-for-gene concept allows negative genetic feedback.
- The highest evolutionary change in the gene frequencies is for medium values, i.e. all genotypes are present in sufficient proportions.

HERITABILITY AND GENETIC GAIN

The selection of plants apparently low in infection as parents for the next generation in a mass selection programme will increase the R- gene frequency. As we do not know the exact state of the system, the consequences may be quite different.

In our model the character to be improved by selection in the negative direction is the number of infections per tree, i.e. the infection density. It is a direct measure of the relevant economic damage to the host individual, and easy to assess. The genotypic values of the infection density for the host genotypes are:

host genotype	RR	Rr	rr
infection density	$\alpha(aa)$	$\alpha(aa)$	α

Thus the R- genotypes are less infected than the rr genotypes, as we would expect. We calculate the additive genetic and dominance variance V_A and V_D and assume that there is environmental variance V_E for the infection density. h^2 denotes the heritability.

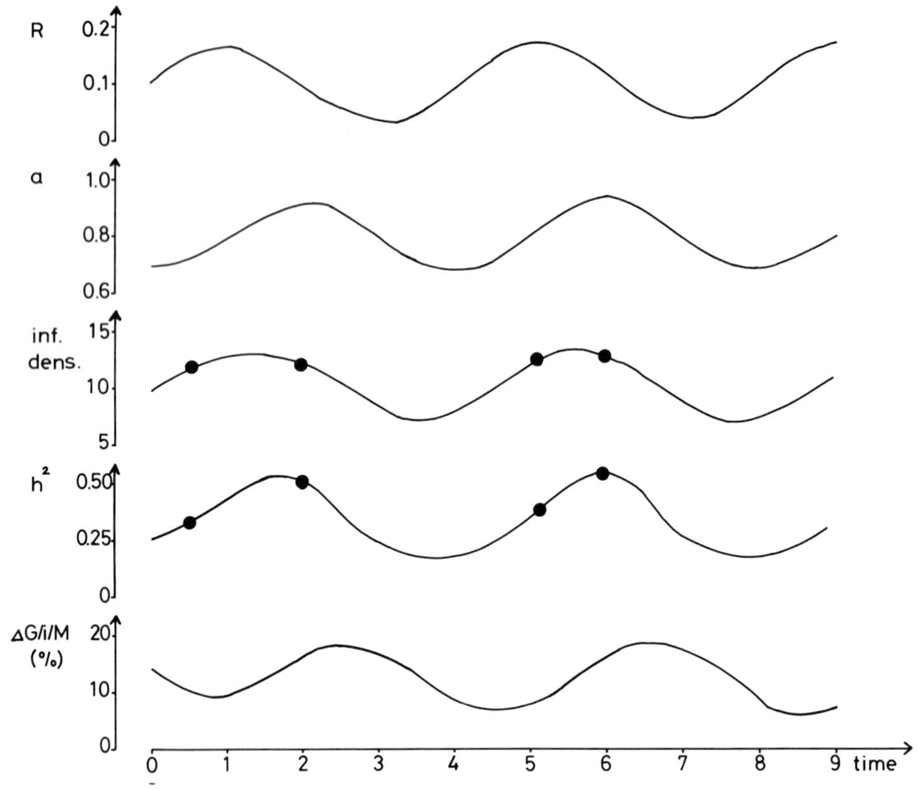

Figure 2. A typical sample run of the system (p_1=1.234, p_2=1, c_1=6.4, c_2=10). The time is measured in host generations.

$$V_A = 2r^3R\ \alpha^2(A-)^2$$

$$V_D = r^2R^2\ \alpha^2(A-)^2$$

$$h^2 = \frac{V_A}{V_A+V_D+V_E} = \frac{2r^3R\ \alpha^2(A-)^2}{r^2(R-)\ \alpha^2(A-)^2+V_E}$$

The phenotypic population mean of the infection density is

$$M = \alpha((R-)(aa) + (rr))$$

The genetic gain divided by the standardized selection differential i and the population mean M for mass selection is

$$\Delta G/i/M = h^2 V_p^{\frac{1}{2}} /M,$$

where V_p is the phenotypic variance.

The additive genetic variance and the heritability are large for intermediate gene frequencies of the R- gene, and tend toward 0 for values near 0 or 1. Very important is the dependence of V_A and h^2 upon the evolutionary state of the parasite population. For a very high virulence gene frequency, V_A and V_D are very low. The reason is that the difference between the high infection class (rr hosts) and the low infection class (R- hosts) decreases with increasing virulence gene frequency. Furthermore all genetic variances increase with the parasite density.

As a result there is a strong dependence of the genetic variances upon the actual state of the parasite population. In Fig. 2 it is shown that for the same phenotypic mean infection density M, different heritabilities and genetic gains can be realized. Without knowledge of the internal evolutionary state of the system, the normal heritability may be a poor predictor of the realized genetic gain from artificial selection.

CONCLUDING REMARKS

- The present model should be extended to several loci.
- A general quantitative genetic theory for the description of interacting host-parasite systems should be developed.
- The influence of the breeding systems of the interacting populations is very important for the dynamic behaviour of the system.

REFERENCES

Becker, W.A. & M.A. Marsden, 1972. Estimation of heritability and selection gain for blister rust resistance in western white pine. In: R.T. Bingham, R.J. Hoff & G.I. McDonald (Eds): Biology of rust resistance in forest trees. USDA Misc. Publ. No. 1221. p. 397-410.

Bingham, R.T., R.J. Olson, W.A. Becker & M.A. Marsden, 1969. Breeding blister rust resistant western white pine. V. Estimates of heritability, combining ability, and genetic advance based on tester matings. Silvae Genetica 18: 28-38.

Flor, H.H., 1956. The complementary genic systems in flax and flax rust. Advances in Genetics 8: 29-59.

Groth, J.V. & C.O. Person, 1977. Genetic interdependence of host and parasite in epidemics. Annals of the New York Academy of Sciences 287: 97-106.

Krusche, D., 1975. Some remarks about stability management in host-parasite systems. In: B.R. Stephan & C.S. Millar (Eds): Lophodermium in pines; Proceedings of the 5th European Colloquium for Forest Pathologists. Mitt. d. Bundesforschungsanstalt f. Forst- und Holzw. No. 108. p. 97-104.

Mode, C.J., 1958. A mathematical model for the co-evolution of obligate parasites and their hosts. Evolution 12: 158-165.

Person, C.O., 1967. Genetic aspects of parasitism. Canadian Journal of Botany 45: 1193-1204.

Person, C.O., J.V. Groth & O.M. Mylyk, 1976. Genetic change in host-parasite populations. Annual Review of Phytopathology 14: 177-188.

Pimentel, D., 1961. Animal population regulation by the genetic feedback mechanism. American Naturalist 95: 65-79.

Non-specific interaction based on polygenes

C. Person, R. Fleming[1] & L. Cargeeg

University of British Columbia, Vancouver, V6T 2B1, Canada

ABSTRACT

Polygenically-based systems of host-parasite interaction are expected by many researchers to exhibit: non-specificity; stability; and constant ranking of host and pathogen genotypes. This paper briefly describes a theoretical analysis that also leads to these expectations. It also describes experimental studies that have demonstrated the reality of polygenically-based pathogenicity for at least one parasitic system. For this particular system (*Ustilago hordei* and *Hordeum vulgare*) the experimental results suggest that polygenes are only part of a total genetic system that includes major genes with which the polygenes interact.

'Field resistance to disease is extremely common and appears to be the normal type of resistance to those diseases which normally infect our crops but do little damage. It appears to be evolutionarily stable, not likely to be upset by changes in the pathogen, and so should be the type of resistance used by plant breeders... Field resistance appears to be polygenically controlled, so that to overcome resistance, several gene changes may be necessary by the fungus.' These statements are taken from a paper written almost 20 years ago by C.M. Driver (Driver, 1962). The postulates they contain have since received a great deal of attention: they have been examined from the epidemiological (Vanderplank, 1963) and ecological (Robinson, 1976) points of view, and new terminologies have been introduced (loc. cit.). It is now widely accepted that field, or horizontal, resistance is stable, and that its stability arises from the fact that it is non-specific and polygenically controlled.

The evidence on which Driver's postulates were based was, and has

1. Present address: School of Civil and Environmental Engineering, Hollister Hall, Cornell University, Ithaca, N.Y., 14853, USA.

remained, largely empirical. Much of it originates from the history of the late-blight disease (*Phytophthora infestans* (Montagne) de Bary) of potato. To quote again from Driver's paper: 'A familiar example is the potato blight fungus on non-immune varieties. These varieties can be rated in the same order and with the same degree of infection as when first introduced even though some are over one hundred years old. This is true throughout the world, so that there is no evidence of potato blight varying the balance between it and the field resistance of commercial varieties over the past century'. This quotation also presents the concept of 'common ranking' which was subsequently amplified by Vanderplank and, later, by Robinson. Constant ranking is now regarded by many researchers as a distinguishing characteristic of polygenically-controlled resistance.

The mechanisms that make for stability and for constant ranking of polygenically controlled resistance are neither known nor understood. Fleming & Person (unpublished) have examined this problem in a theoretical way by assuming: (i) that resistance and pathogenicity were both non-specific and polygenically controlled; and (ii) that for each of the 2 interacting populations the variability was continuous and of the kind described by a normal curve. Since nothing is actually known of how these assumed polygenes would interact, 2 different 'models' of interaction were constructed (Fig. 1a, b). The normal curves in the upper part of each figure represent the polygenically-based potential for enabling the disease to develop. Each of the upper curves is divided (by 1, 2 and 3 standard deviations from the mean) into 6 classes to which are assigned arbitrary values (ranging from 0 to 5, inclusive) that numerically represent the potential for enabling the disease to develop. The numbers in the matrix represent the levels of disease that are assumed to develop when hosts and parasites interact. For the additive model (Fig. 1a) these 'realized' levels of disease are obtained by adding, and for the multiplicative model (Fig. 1b) by multiplying, the numbers that were arbitrarily assigned. The bottom curves describe the 'realized' levels of disease. For the additive model it was shown mathematically: (i) that the 'realized' levels of disease would also conform to a normal frequency distribution, and (ii) that a shift in the mean of the distribution curve for 'potential' disease development, occurring in either of the 2 interacting populations, would result in a relatively smaller shift in the mean of the curve that represents 'realized' disease development. This reduction in effect has been labelled 'phenotypic damping', it is thought to be a source for stability in polygenically-determined systems. The multiplicative model was less easy to analyze: the mean level of disease favoured host resistance and the effects brought on by assuming changes in either of the 2 interacting populations were more complicated. However, examination of the 2 contrasting models did support the following general conclusion: (i) polygenically-based systems should display constant ranking, and (ii) polygenically-based systems should be

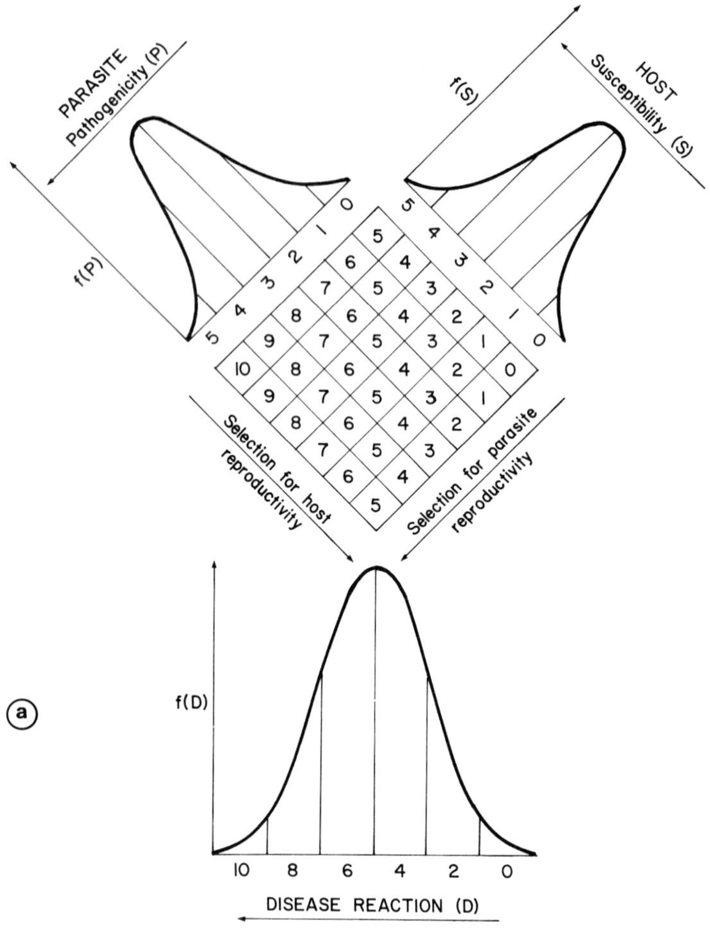

Figure 1. For these figures (1a and 1b), f(P) and f(S) represent the frequencies with which individuals with varying levels of pathogenicity (P) or susceptibility (S) occur in parasite and host populations. Because of their assumed polygenic determination these are represented as normal frequency distributions. For the additive model (Fig. 1a), where it is assumed that the amount of disease is D = A + S, the matrix records the severity of disease for the various indicated classes of host-parasite interaction. The resulting frequency distribution for D, f(D), is normally distributed. For the multiplicative model (Fig. 1b), where it is assumed that D = A × S, the shape of the resulting frequency distribution for disease reaction, f(D), depends on the parameters of the distributions for pathogenicity and susceptibility, and the distribution is one that would favor host resistance. (Fig. 1a was used, with our permission, by Robinson, 1976.)

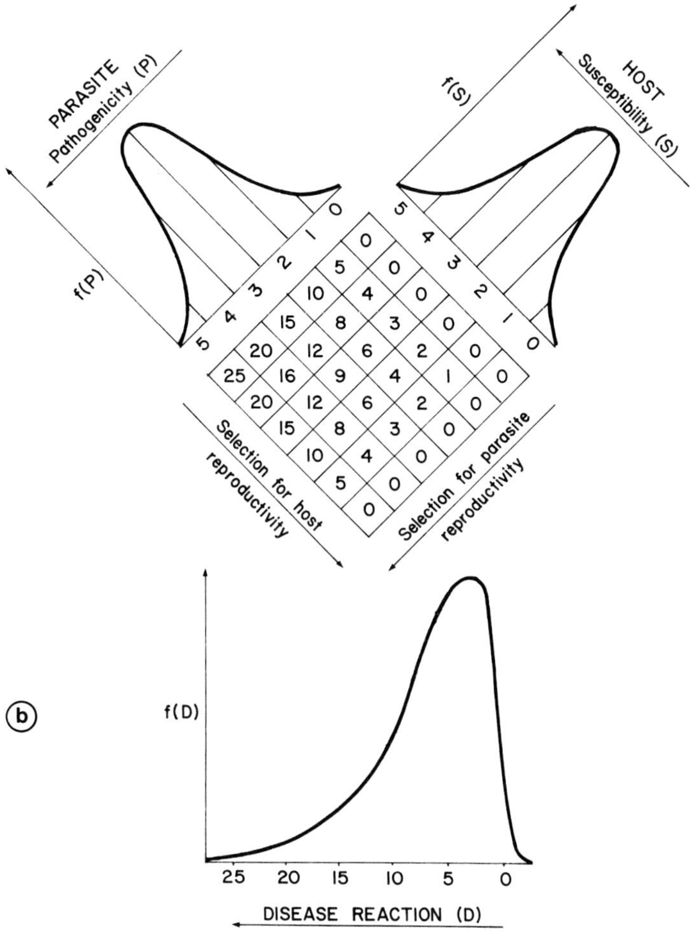

relatively stable. This second conclusion depends on the assumption that pathogenicity and resistance both correlate positively with reproductivity. Thus the directions taken by natural selection in the 2 interacting populations will be diametrically opposed. (This conclusion is in disagreement with Vanderplank's (1975) suggestion that the expected stability derives from stabilizing selection operating independently in each of the 2 interacting populations).

Although there have been numerous studies of polygenically-determined host resistance, relatively little is known of polygenically-determined pathogenicity. At the University of British Columbia, Emara (1972) and Emara & Sidhu (1974) have demonstrated that strains of *Ustilago hordei* (Pers.) Kell. & Sw. differ in their pathogenicity toward susceptible barley cultivars and that this variability is polygenically determined. These experiments have been continued in cooperation with Dr. Caten at Birmingham and, at least in 1 strain, the genes involved show additive, dominance and epistatic effects (unpublished data).

For the remainder of my talk I wish to describe another set of experiments that illustrate the *combined* action of major genes and polygenes in determining levels of disease development. These experiments also involve *U. hordei* and take advantage of the fact that *U. hordei* forms an ordered tetrad from which 4 'selfed' (and obligately parasitic) dikaryons can be established. A 'selfing' that results in a 3:1 ratio identifies the 2:2 ratio of '*V*' and '*v*' gametes in the tetrad (see Table 1), and the identified 2:2 gametic ratio can be readily verified in a test cross. A large number of gametes can thus be extracted from a single dikaryon, and for each of these the genetic content (*V* and *v*) in relation to the major gene can be determined. Where pathogenicity is also determined in part by polygenes it is also to be expected that the gametes formed by a single dikaryon will vary in their content of polygenes. This expectation leads to the prediction that the *V*- and *v*-containing gametes will vary in their pathogenic behaviour, and that this variability will be polygenically determined. To test this hypothesis the prediction was made that the *V*- and *v*-containing gametes would show constant ranking (Table 2).

Table 1. The 3 kinds of tetrad, and the progeny obtained via selfing from single teliospores, of a *Vv* dikaryon for which the original mating was $V+ \times v-$. Parental ditype (PD) and non-parental ditype (NPD) tetrads do not segregate when 'selfed'. Tetratype (*T*) tetrads yield 3:1 ($V-:vv$) segregations that lead to identification of *V* and *v* sporidia.

		Type of tetrad		
		PD	NPD	T
Sporidia	1	V^+	v^+	V^+
	2	V^+	v^+	v^+
	3	v^-	V^-	V^-
	4	v^-	V^-	v^-
Matings and 'selfed' progeny	1 × 3	*Vv*	*Vv*	*VV*
	1 × 4	*Vv*	*Vv*	*Vv*
	2 × 3	*Vv*	*Vv*	*Vv*
	2 × 4	*Vv*	*Vv*	*vv*[a]
Segregation	($V-:vv$)	4:0	4:0	3:1

a. This culture identifies the two *v*-containing and, by extension, the two *V*-containing sporidia of the tetrad.

Table 2. Expectation of 'constant ranking' among dikaryons: when plus and minus sporidia with varying numbers of polygenes (range 0 to 5) are arranged in ascending series the resulting dikaryons (having 0 to 10 polygenes) are expected to show constant ranking in respect to their pathogenicity.

Number of polygenes in '-' sporidium	Number of polygenes in '+' sporidium: 0	1	2	3	4	5
0	0	1	2	3	4	5
1	1	2	3	4	5	6
2	2	3	4	5	6	7
3	3	4	5	6	7	8
4	4	5	6	7	8	9
5	5	6	7	8	9	10

The results of a large experiment are shown in Table 3. Mathematical analysis of these data confirm the expectation of constant ranking. Within each of the 3 major genotypes (*VV*, *Vv* and *vv*) there is a considerable range of pathogenicity attributable to polygenes. The analysis of variance shows significant variation between the gametes in their contribution to the pathogenicity of dikaryons. However the analysis of variance also showed significant interactions, detected in the *VV* group and in one of the *Vv* groups, which indicate that the polygenes governing pathogenicity do not always act in an additive manner as assumed by the general theory of polygenes. These can also be seen by direct visual inspection of Table 3. With use of sporidium 21B3+ the performance of *VV* dikaryons was enhanced while that of *Vv* dikaryons was reduced, whereas with use of sporidium 22G3+ the opposite result was obtained. The genetic basis for these particular interaction effects is now being studied further.

So far as we are aware, the expectation of constant ranking among gametes represents an approach to the analysis of polygenes that is entirely new. The method is, of course, restricted to organisms such as *U. hordei* for which it is possible to establish gametic cultures. The method has yielded results which add support to the conclusions that were reached earlier by using the traditional approach.

To summarize: polygenically-based systems of host : parasite interaction are expected by many researchers to display 2 important characteristics, viz. constant ranking of host and pathogen genotypes, and stability. Although a good deal is known of polygenically-based host resistance, relatively little is known of polygenically-based pathogenicity, and almost

Table 3. Disease reactions (recorded as percent of infected plants) caused by *Ustilago hordei* on the barley cultivar Trebi. Data are for F_2 dikaryons all of which derive from a single parental F_1 dikaryon that was heterozygous for a major gene.

		Sporidial lines	Sporidia, mating type *a*(-)												Means			
			genotype *V*						genotype *v*						*VV*	*Vv*	*vv*	Over-all
			17 A 4⁻	21 C 4⁻	19 B 4⁻	20 A 4⁻	21 E 4⁻	20 C 1⁻	21 C 2⁻	21 B 4⁻	24 C 2⁻	24 A 4⁻	20 C 4⁻	20 A 3⁻				
Sporidia, mating type, *A*(+)	genotype *V*	18 D 1⁺	60	62	48	35	53	42	42	48	43	28	32	35	50	38		44
		21 B 1⁺	42	42	50	41	34	27	35	41	34	26	28	26	39	32		36
		21 B 3⁺	50	44	61	50	54	49	33	25	13	8	10	5	51	16		34
		20 A 2⁺	53	34	34	47	31	25	34	43	29	24	14	20	37	27		32
		23 C 3⁺	43	30	18	24	29	17	31	31	21	17	16	23	27	23		25
	genotype *v*	21 C 1⁺	50	48	50	56	45	38	16	16	11	4	10	3		48	10	29
		19 A 3⁺	55	47	55	60	48	32	14	0	4	6	5	9		50	6	28
		22 G 3⁺	47	35	27	32	27	23	12	17	18	14	8	8		32	13	22
		20 A 1⁺	48	46	36	36	1	7	13	2	0	1	4	0		29	3	16
		24 D 2⁺	45	31	33	25	13	19	5	2	3	1	2	2		28	3	15
		24 A 3⁺	38	38	21	16	8	15	5	2	0	2	1	1		23	2	12
		22 F 2⁺	24	16	24	8	10	2	8	4	4	1	3	1		14	4	9
Means		*VV*	50	42	42	39	40	32							41			
		Vv	44	37	35	33	22	19	35	38	28	21	20	22		30		
		vv							10	6	6	4	5	3			6	
		Overall	46	39	38	36	29	25	21	19	15	11	11	11				25

nothing is known of how polygenically-based systems of host and parasite would actually interact. Constant ranking has received very little study, and the mechanisms that would lead to stability have yet to be discovered.

Our experimental studies have demonstrated the reality of polygenically-based pathogenicity for at least one parasitic system. These studies suggest that polygenes are only part of a total genetic system that also includes major genes with which the polygenes interact.

Our theoretical analysis leads to the expectation that polygenically-based systems will show constant ranking, and that such systems will be relatively more stable than those that are based exclusively on major genes and specific gene-for-gene interaction. However, there is both practical and theoretical investigation to support the view that major genes could be used more efficiently than they have been in the past.

Our overall conlusion is that the question of major *vs* polygenes is still an open one that requires extensive study. At this particular point in time it seems that both kinds of resistance (if indeed they exist separately) are valuable, and that for both kinds of resistance the objective should be to breed for genetic heterogeneity rather than for a monoculture.

ACKNOWLEDGEMENT

The work described in this paper was financed by a grant from the Natural Sciences and Engineering Council of Canada.

REFERENCES

Driver, C.M., 1962. Breeding for disease resistance. Scottish Plant Breeding Station, Research Report. 1-11.

Emara, Y.A., 1972. Genetic control of aggressiveness in Ustilago hordei. I. Natural variability among physiological races. Canadian Journal of Genetics and Cytology 14: 919-924.

Emara, Y.A. & G. Sidhu, 1974. Polygenic inheritance of aggressiveness in Ustilago hordei. Heredity 32: 219-224.

Robinson, R.A., 1976. Plant pathosystems. Springer-Verlag, Berlin. 184 p.

Vanderplank, J.E., 1963. Plant disease: Epidemics and control. Academic Press, New York. 349 p.

Vanderplank, J.E., 1975. Principles of plant infection. Academic Press, New York. 216 p.

Monoculture versus mixture: interactions between susceptible and resistant trees in a mixed stand

Hans M. Heybroek

"Dorschkamp" Research Institute for Forestry and Landscape Planning, Wageningen, the Netherlands

ABSTRACT

The use of diverse genotypes is an effective way of spreading risks. For this purpose, a mosaic of pure stands, each consisting of a single genotype, may be adequate under certain conditions. Mixed stands may offer a bonus if and in so far as an interaction between resistant and susceptible trees in a stand will reduce damage by pest, disease, wind, etc. A survey of such interactions and the mechanisms involved is given with regard to diseases, including *Fomes annosus*, *Nectria cinnabarina* and *Lophodermium pinastri*. Beneficial effects for the susceptible partners in the mixture as well as harmful effects for the partially resistant partners are noted. The net effect for the stand is dependent on many factors, including the damage threshold. Consequences for testing designs are mentioned.

INTRODUCTION

In the decisions to either plant or avoid stands of mixed genotypes, many considerations must play a role. Pure stands of a single genotype often may cost less to establish, tend and harvest, they may promise to produce more timber of a higher value, they seem to require less silvicultural expertise. There may be no alternative to a pure stand in cases where only one resistant or locally adapted genotype is available.

Monocultures are, however, widely believed to attract diseases and pests and to be more vulnerable than mixed stands, especially in the long term. Multiclonal varieties have been regarded as a means of deploying not-too-resistant clones (Schreiner, 1965). Further, mixed stands may have the advantage of offering a more varied scenery and, under certain conditions, of giving a higher production (Heybroek, 1978).

Monocultures have health hazards that come in 3 forms: first they may constitute a large, undivided risk, and it may be better to spread risks; second, the concentration of susceptible plants per se might increase the disease rates of the individuals, while the interaction between neighbours of different susceptibility might reduce those rates; third, monocultures might stimulate the evolution of new, more virulent or aggressive forms of the parasite. This paper deals briefly with the first aspect and concentrates on the second.

Mixing of different species that serve as alternate hosts to the same rust (or aphids, etc.) has profound but evident effects and will not be discussed here.

The severity of some diseases is considerably affected by the microclimate in the stand. Mixing of species may influence disease severity in either direction by influencing the microclimate. This effect will not be discussed either.

SPREADING THE RISKS

The notion is widely accepted that the exclusive use of a single genotype (e.g. clone) over a large area entails an enormous risk: if the clone fails for any reason, the failure could be total over the entire area, thus causing almost insurmountable problems for the management of the forest and for the industries depending on it. It seems more acceptable to have 10 clones, each on one tenth of the area: even if the chance of failure of any of these is 10 times as great (supposing that each has the same chance of failure as the earlier monopolist clone) the prospective damage is only one tenth and can be much better absorbed. It is the wisdom of not putting all eggs in one basket. It is the philosophy of insurance: many small risks are less serious than a single big risk that could radically destroy continuity. Spreading the risk is a good common sense precaution.

From this limited point of view, there is little advantage in individual mixtures: planting 10 blocks of 1000 ha with 1 clone each would spread the risk just as effectively as planting 10 000 ha with the mixture of the 10 clones.

The individual mixture may have an advantage, however, if compensation occurs: compensation being the process in which the neighbours fill the gap caused by a failing tree, thus increasing their own production and more or less compensating for the loss. This mechanism may be particularly effective if the failure occurs early in the stands' development and if initial spacing was narrow. On the other hand, if 2-5 clones fail, the mixed forest would become defective over the entire area. That loss is less easy to handle than if the failing clones had been planted in pure stands, which could be salvaged and replanted.

These considerations might apply particularly to poplar planted at final wide spacing, which reduces the effect of compensation, and grown in short rotations. The latter silvicultural trait is among those mentioned by Kleinschmit (1979) as reducing the need for mixing.

A relatively safe situation may exist even when monoclonal stands are used, provided the number of clones used per region is not small, provided the clones have varying backgrounds, and provided they are backed up by a wide array of experimental clones which are being kept under test for different sites and which form a reserve from which old clones can be replaced when failing for some reason (Heybroek, 1981).

THE EFFECT OF INTERACTIONS

This section deals with the question of how far the damage to a susceptible plant is decreased if it is surrounded by resistant instead of susceptible plants. It seems important to study the mechanisms involved.

Trees with different levels of resistance are found mixed in:
- stands consisting of different tree species, differing in resistance;
- stands consisting of a seedling population of 1 species with variation in resistance;
- stands consisting of clonal mixtures, varying in resistance;
- stands in which trees of 1 species of different ages are mixed, while susceptibility to the disease in question is limited to a narrow age-class.

This means that experience gained with traditional 'mixed stands' can be used to understand and perhaps to predict what will happen in a clonal mixture. However, in spite of the customary interest by foresters in the advantages and disadvantages of mixed stands, literature does not abound with well-documented examples of diseases that are much more serious in pure than in mixed stands. A case often cited, having the charm of the exotic, concerns *Hevea brasiliensis* (Willd.) Muell.: this species is said to be severely damaged by leaf disease if grown in pure stands in South America, but to be mainly healthy in the neighbouring virgin forest where it occurs singly between other species (Boyce, 1954). The search for examples is complicated by the fact that in the older literature, examples were collected and used mainly to prove or disprove the idea that diseases etc. are bound to be more serious in 'unnatural' stands than in 'natural' ones (Boyce, 1954; Peace, 1962, p. 18).

The disease process can be divided in 2 phases, and in both neighbours can play a role. In the first phase, the individual host-tree must be reached by the parasite, and in the second, the parasite must multiply or spread on or in the host until the damage-treshold is reached.

Reaching the host

In the first phase, in which the host is reached by the parasite, 3 groups of cases can be distinguished.

1. In one group of cases, the host needs to be reached only once: once the

parasite has reached the tree, it can survive and spread in it or on it. This applies for example to perennial cankers, wood inhabiting fungi, etc.

The presence of resistant neighbour trees can have a delaying effect, which may or may not be of practical use, as illustrated in the following examples:

- The chestnut blight spread quickly through most of the area of *Castanea dentata* (Marsh.) Borkh., scarcely differentiating between pure and mixed stands. Some outlying stands or individuals escaped at first, but even isolated trees, planted in the Midwest of the United States, far outside the natural area of the species, were finally reached and succumbed.
- The perennial canker of beech, caused by *Nectria ditissima* Tul., was a rarity in the original mixed coppice-with-standards in north-eastern France. It increased to epidemic levels after these forests were converted to pure even-aged beech woods (Perrin & Vernier, 1979). Apparently, under the old regime, the presence of many non-hosts allowed even susceptible beeches to escape infection.
- Several tree diseases spread through live root grafts; for some, even mechanical root contact is sufficient (Epstein, 1978). Among them are oak wilt, Dutch elm disease and *Fomes* root rot. It seems plausible that planting tree species susceptible to these diseases in mixtures with resistant individuals, of the same or of other species, will greatly reduce this type of spread of the disease. Indeed it has been shown that both oak wilt (Epstein, 1978) and *Fomes* root rot (Rennerfelt, 1947) are much less severe in mixed than in pure stands. From the models in Fig. 1a-c it appears, that the use of 50 % resistant trees is enough to virtually stop the spread from one focus, provided transmission occurs between neighbours only. Roots may, however, extend much further than the crown of the tree, making root contacts possible among trees that are not direct neighbours (Stiell, 1970; Eis, 1978). If root contacts occur between diagonal neighbours in 50 % of the cases, reduction in infection is still considerable (Fig. 1): in the same time required to infect 20 trees in the pure stand, only 5 trees would be infected in the 50 % resistant stand. In the models of Fig. 1, systematic mixing was more effective than random mixing. The protective effect will clearly diminish as the root systems range wider, and tends to disappear when the initial infection is higher, that is, when the number of foci increases. There is also an age-effect as the relative range of the root system changes with age (Eis, 1978) and as mixtures of different species tend to be more difficult to maintain with increasing age. Perhaps the possibility of reducing the incidence of these diseases through mixed plantings should be explored further. It may be difficult to find non-hosts that can be mixed in without silvicultural problems, but even the admixture of partially resistant hosts, that is trees through which spread is slower, might give a worthwhile reduction of disease in the stand.

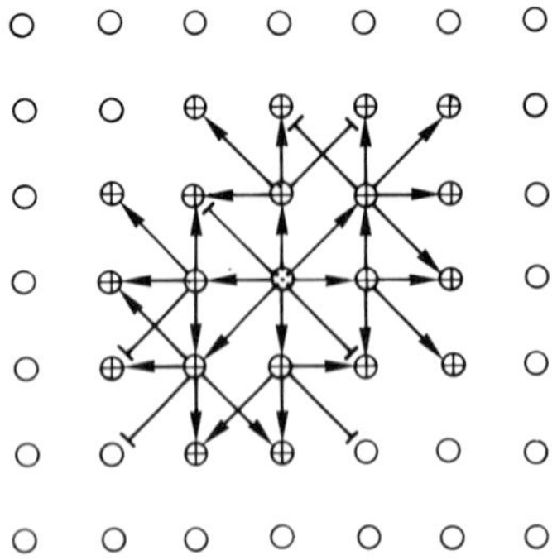

a. Spread in a homogeneously susceptible stand. 20 transmissions.

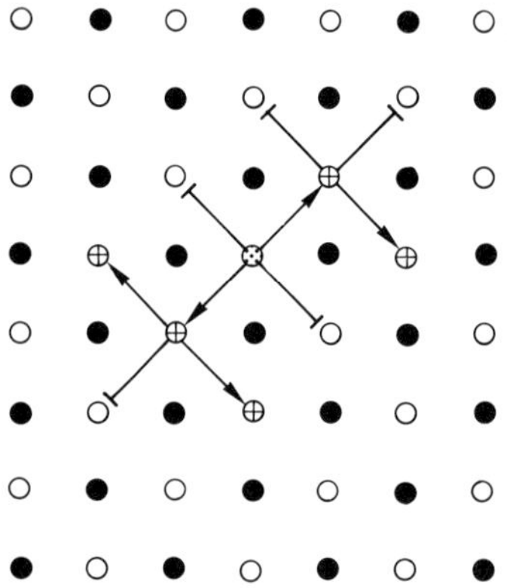

b. Spread in a stand with 50%, evenly distributed, resistant plants. 5 transmissions.

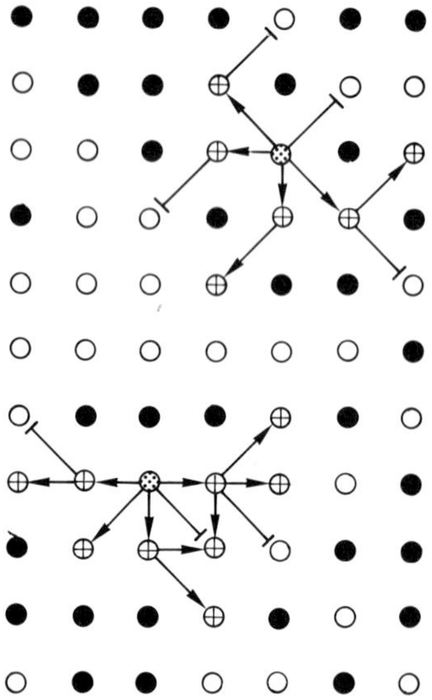

c. Spread in a stand with 50%, randomly distributed resistant plants. In two examples 6 and 9 transmissions are realized.

Legend:

- ● resistant plant
- ○ susceptible plant
- ——► spread of disease
- ——┤ unsuccessful attempt at spreading
- ⊕ infected plant
- ⊗ focus

Figure 1. Spread of disease through root contacts; assuming that all direct neighbourships lead to transmission, but only 50% of the diagonal neighbourships; over a time period sufficient for two diagonal transmissions.

2. In a second group of cases, diseases that require an annual reinfection from outside, the poplar rusts *Melampsora medusae* Thuem. and *M. larici-populina* Kleb. can be taken as examples. In these, the fungus often has to come from far. A mixture of susceptible poplars with some non-hosts might at best cause a minor delay in the first infection and the onset of the build-up of the epidemic. This effect may be even non-existent if the whole stand is infected at the same time by a cloud of spores. (The effect of mixing on the build-up of the epidemic is discussed in the next section.)

3. Between these 2 groups, *Nectria cinnabarina* (Tode) Fries provides an intermediate case. Pathogenic strains of this fungus form annual cankers that continue to sporulate for 1 or 2 years. For a successful infection, wounds must be present at the right time of the year and under the right weather conditions. Genetic resistance occurs. If genotype, condition and environment of the host are favourable for disease development, a single tree can reinfect itself year after year; if not, the tree will lose the parasite for few or many years until a new colonization occurs. Then, a larger closed population of host trees is needed to maintain a local population of the pathogenic strain of the fungus. A single such tree, surrounded by non-hosts, might remain free of infection for many years as the fungus did not happen to reach it at the right moment.

For this intermediate category, in which a tree may shed or lose the parasite for some time until the next recolonization, parallels can be found in the island-theory of MacArthur & Wilson (1976). These authors, comparing the numbers of animal species present of islands of different size and at different distances from the mainland, conclude that the chance for extinction of a species on an island increases as the island is smaller and as the possibility for recolonization decreases where other islands or the mainland are more remote. Chances for extinction also depend on the size of the population: all populations fluctuate in size over the years, a small population can easily disappear completely in such a fluctuation. Now, the groups of host-trees may be regarded as ecological islands in a 'sea' of non-hosts; local extinction of the parasite on the group of host trees might in critical cases depend on the size of the group, and on the distance from the 'mainland', the source of reinfection. It further depends on the intensity of attack on the host or the amount of parasite present on the single tree: the 'population size'.

Returning to *Nectria cinnabarina*, it can be expected that at low levels of disease, a mixed stand of host and non-host trees may lead to better health of the hosts. Matters are complicated, however, by the fact that pathogenic strains can survive for some time as saprophytes. Local extinction of *N. cinnabarina* is pursued artificially by the nurseryman who practices sanitation rigorously. Once the nursery is 'clean', many susceptible moments will not lead to infection because of the absence of the parasite.

In forestry practice, this mechanism of local extinction of and delayed recolonization by the parasite seems of limited value, as it may require the reduction of the proportion or number of host trees to a level that is not interesting economically.

The build-up of a parasite population

Once the parasite has found the host tree, generally a build-up of its population is needed before the disease can reach the damage treshold. Here again, resistant neighbours may have an effect on the outcome of the process.

The classical and successful example is the multiline in oats, as described by Browning & Frey (1969). It consists of a mixture of several separate host-parasite combinations, in which host A is compatible with parasite race 1, and B with 2, etc., but not A with 2, nor B with 1, etc. Thus the spores of 1 generated on A and landing on neighbour B, will be ineffective and lost for the epidemic on A, and vice versa. Thus the build-up of the population of parasite 1 on the component A will be based on a severely diluted spore cloud, so that the build-up is delayed considerably.

The concept cannot be simply copied for all host-parasite combinations. Success depends on whether a reduction in the density of the spore cloud will be sufficient to delay the build-up of the population, and whether this delay will be sufficient to reduce the damage to the host. Then, size is important: the individual cereal plant is small, long and narrow, so that spores produced on it have a fair chance of landing on neighbouring plants. In comparison, the crown of a single tree provides a large volume of leaves with the same genotype. In the crown, a small epidemic could develop independently, unaffected by the presence or absence of a resistant neighbour tree.

The multiline is developed to employ vertical resistance. In forest trees, however, within 1 species, mainly some degree of partial, horizontal resistance can be found. A priori, mixing such trees does not look promising. More effect might be expected from mixing highly divergent genotypes: hosts and non-hosts, different species.

Certainly, not any mixing of trees with different levels of resistance will be beneficial. Although it might be hoped that the presence of the more resistant would somehow protect the less resistant plants, the reverse may happen as well: the less resistant plants may act as disease spreaders, overcoming the resistance of the more resistant partners. An example of the latter possibility is given by Maslow (1970, page 62):

"White elm (*U. laevis*) suffered in the prefecture Rostov, Ukraine, much less from Dutch elm disease than field elm (*U. carpinifolia*), at least in the earlier years of the epidemic. In plantings of the first and second growth class, where up to 100 % of the trees [of field elm, see page 61] got infected, no more than 7 % of the white elms died.

The condition of white elm was dependent on the abundance of spore infection and of stem damaging insects, the latter being dependent on the presence of field elm in the immediate neighbourhood of white elm: *in all stands where white elm occurred in mixture with the severely diseased field elm, dying of white elm was of much higher significance, in cases even nearly total.*" (Italics by H.M.H.)

LOPHODERMIUM PINASTRI

Some idea of what can happen when hosts with varying degrees of partial resistance are mixed can be gained from data on needle cast caused by *Lophodermium pinastri* (Schrad. ex Hook.) Chev., collected in the Dutch programme on genetic improvement of Scots pine (Squillace et al., 1972, 1975; Kriek & Bikker, 1973).

In one experiment, 14 Dutch provenances were compared with 4 German provenances under severe disease-conditions. The experiment contained 3 replicate blocks: each single plot consisted of 5 rows, each of 60 plants; spacing 1.50 m between rows and 1.20 m in rows. In each block, provenances were randomized. Four and eight years after planting, mortality and growth were assessed. The provenances clearly separated into 2 groups: after 8 years, the German provenances suffered an average mortality of 45.5 %; the Dutch provenances varied slightly around an average of 20.1 %. Mortality could be primarily attributed to needle cast, though *Armillaria mellea* (Vahl.) Quél., present throughout the area, accelerated the dying of weakened trees.

Plots of the 'resistant' provenances,when situated next to a plot of the 'susceptible' provenances (Fig. 2) showed a slightly significant higher mortality. The influence of the susceptible plots was not evident beyond the width of one neighbour plot, that is 10.5 m. Unfortunately, the limited number of susceptible provenances did not allow assessment of the reverse effect, that is, a possible lower disease rating in susceptible plots bordered by resistant plots.

Fig. 2 is based on assessments by Kriek & Bikker (1973), following the method that Squillace et al. (1972) had used with earlier data. For each plot of a resistant Dutch provenance, the deviation of its disease rating from the average provenance rating was plotted against its distance from a susceptible German plot. This showed that on average, a plot of a resistant provenance had a 6.5 % higher disease rating than the provenance average if it was adjacent to a susceptible plot. The more remote plots, necessarily, had a slightly lower disease rating than the provenance average.

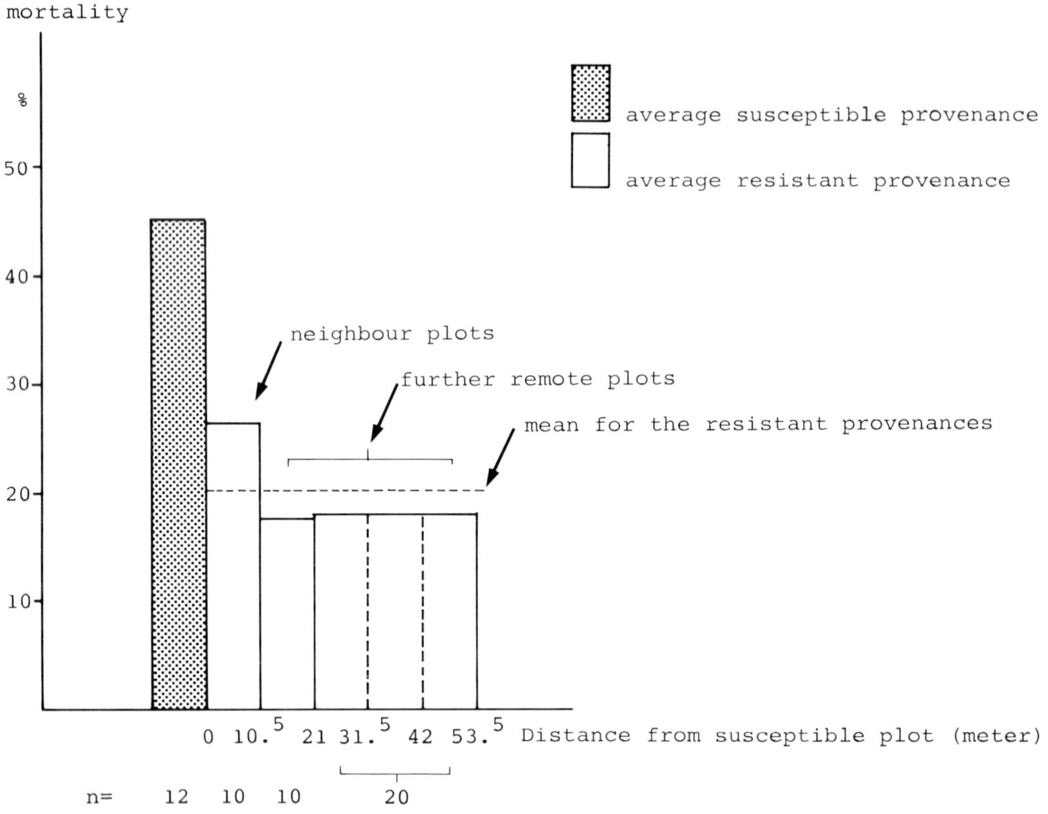

Figure 2. Disease ratings after attack by *Lophodermium pinastri* in plots of relatively resistant provenances of Scots pine, in dependence on their distance from the nearest plot of a susceptible provenance (based on data in Kriek & Bikker, 1973).

It was possible to study both effects, however, in a large scale half-sib test of 294 Dutch plus trees of Scots pine. Needle cast was assessed 7 years after establishment of the test, using a scale of 1 through 7, based on the presence of needle spots and on loss of foliage. The highest rating represented dead trees. The test consisted of six 7 × 7 lattice squares, each containing 4 replicates. One of the lattice squares, containing 49 families, was analyzed in detail for needle cast occurrence. Interaction between resistant and susceptible families was evident when the ratings of the 4 corner trees of the 4 × 4 tree plots were compared with those of the 4 inner trees. The inner trees were always surrounded by their own kind; the corner trees, however, by trees of other families. It was found that corner trees of resistant plots had higher average disease ratings than the inner trees, as they were mostly surrounded by trees of more susceptible other families; the reverse was true for the susceptible plots, while there

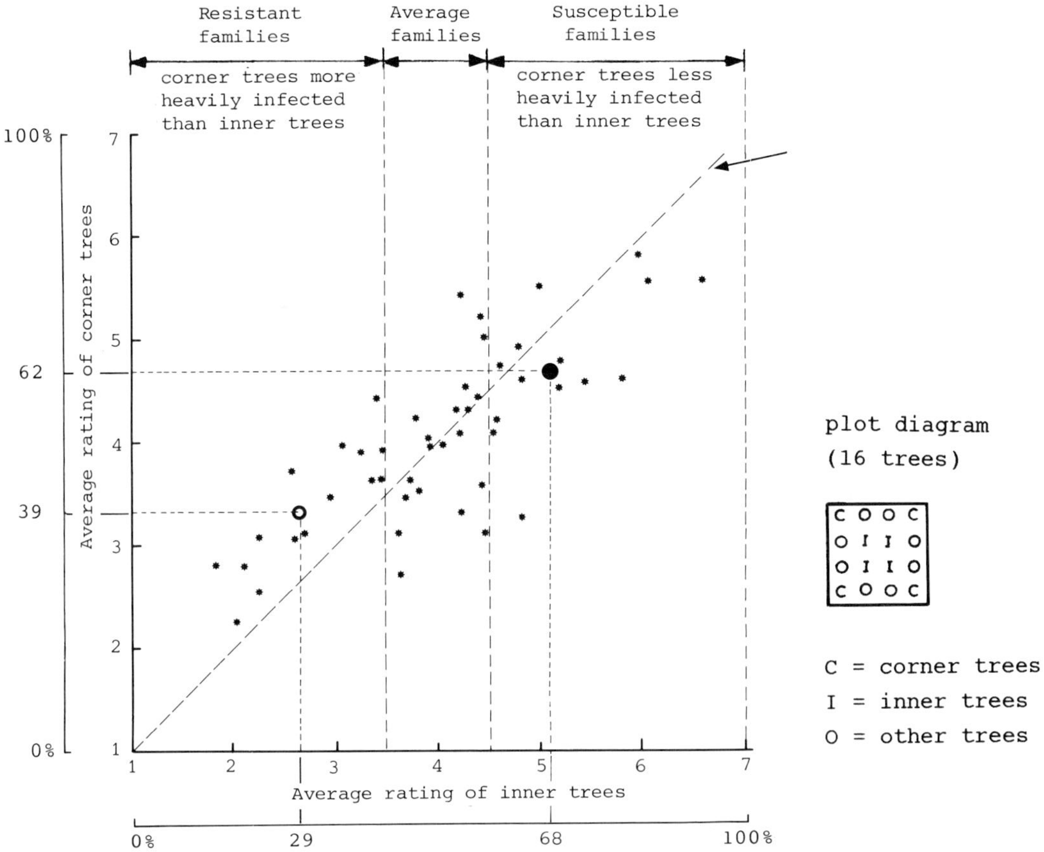

Figure 3. Average disease ratings (*Lophodermium pinastri* on *Pinus silvestris*) of 'corner trees' over 'inner trees' (see plot diagram) of the 4 plots of each of the 49 half-sib families in Experiment 2 of the progeny trial. The field ratings: 1 = healthy, 7 = dead, are converted to 0-100%. RL: hypothetical slope (= 1) if the differences between inner and corner trees were independent of resistance of families. o = median score for the (relatively) 'resistant families', • = median score for the 'susceptible families'. In the former group, the inner trees are less diseased, in the latter group, they are more diseased than the corner trees. After Squillace et al., 1972.

was no difference between corner- and inner trees in the average families (Fig. 3).

If the resistant families of Fig. 3 as a group are contrasted with the susceptible families as a group, the data of Fig. 3 can conveniently be represented in a diagram like Fig. 4. In this 'mixogram', disease ratings (ordinate) can be given for a resistant host in pure stand (left-hand

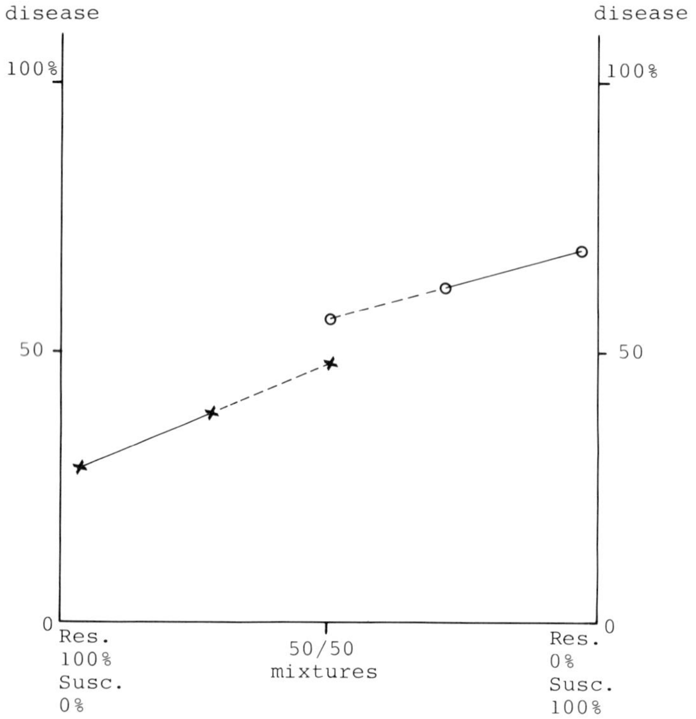

Figure 4. Disease rating (*Lophodermium pinastri*) for a group of resistant families (×) and susceptible families (o) of Scots pine grown in mixtures of different proportions (--- extrapolations).

side), for a susceptible host in pure stand (right-hand side) and for each of them grown in mixture in any proportion. The line for the resistant family group shows that the inner trees, which represent a more pure stand situation, are less diseased than the corner trees which represent a more mixed stand situation. The actual position of the line in the graph, and thus its slope, is dependent on some assumptions about the reach of this neighbour effect: even the inner trees may be affected to a certain extent by plots of different susceptibility in the periphery.

Fig. 4 is derived from Fig. 3 as follows. In Fig. 3 in both the relatively resistant and susceptible group, the median score was drawn. Then the median disease rating of the corner tree and of the inner tree of either group could be read on the ordinate and abscissa respectively. These were converted into percentage (1 = 0 %, 7 = 100 %), giving the values on the ordinate in Fig. 4.

Some assumptions had to be made in order to assign to the corner trees and inner trees certain values on the abscissa. It was assumed that the influence of a neighbour tree is proportional to the square root of its distance from the tree in question. This means that, if the surrounding trees are regarded as standing in concentric rings around the tree in question, the influence of rings farther out decreases quickly in spite of the fact that they contain more trees. The ratio of the influence of the first 3 rings (of 4, 4 and 12 trees) to that of the following 3 rings (of 16, 20 and 24 trees) is 9 to 4. This proportion is even increased if not only distance, but also interception or a filtering effect is considered to reduce the influence. It could be assumed, rather arbitrarily, that interception is 0 % for the first ring (direct neighbour), 20 % for the second ring, 40 % for the third, 60 % for the fourth, 80 % for the fifth, 90 % for the sixth. By doing so, the ratio of the influence of the first 3 rings to that of the following 3 rings is 7.4 to 1.1. Rings further out play a minute role. Under these assumptions, a tree surrounded by no more than 3 rings of its own kind, would still score $7.4/(7.4 + 1.1) \times 100\ \% = 87\ \%$ on the abscissa.

The half-sib test contained families of all levels of susceptibility, not just two. A corner tree, e.g. in a resistant plot, has in its first ring of 4 neighbours 2 of its own kind, and 2 'others' of 2 different families. These are samples of the entire population: 'resistant', 'average' and 'susceptible' families. In a simplifying approximation the 'others' are regarded as consisting of 50 % 'resistants' and 50 % 'susceptibles'. The situation is then reduced to a mixture between 'resistants' and 'susceptibles', or 'own kind' and 'the different kind'. The composition of the various rings of neighbours can thus be identified as to 'own' and 'foreign'; this is multiplied by the 'weight' or the relative influence of the ring, and values of the different rings added. Using higher interception-factors than in the above example (0 %, 50 %, 80 %, 90 %), the result is that corner trees would be influenced 72 % by their own kind and 28 % by the opposite kind, while for the inner trees the relation would be 97 % versus 3 %.

A size-factor is also apparent here: when the stand is fully grown, each plot of 16 trees will be replaced by 1 large tree, which will thus be situated at the 50-50 point on the abscissa of the mixogram. Although this tree could still be influenced by its neighbours, the entire situation of disease, susceptibility, spore flight and stand climate would be different, which would probably reduce the effect.

It is thus clear that a 'neighbour effect' does exist. From these data, it can even be predicted that in the more resistant families a further decrease in disease rating can be obtained by planting the tree in an individual mixture with a non-host, as was reported by Fischer (1957).

CONSEQUENCES FOR TESTING DESIGNS

A closer inspection of Fig. 4, and a tentative extrapolation of the lines shows several points of interest. Apparently, when grown in an individual 50:50 mixture, the differences in disease ratings between the resistant and susceptible families would even be smaller than now. On the other hand, when grown in pure stands, the resistant families would be less diseased, and the susceptible families more diseased than the average obtained in this test field. In a test field such as the one under consideration, the resistance of the relatively resistant families tends to be underestimated, and this effect will be greater when the plots are smaller. This bias is well-known in agricultural plant breeding (Parlevliet, 1975, 1979).

Squillace et al. (1972) discuss the merits of different test designs. As they found the disease may have a partially unexplained 'patchy' pattern, they would like randomized individual tree plots with many replications and wide spacing. Alternatively, to reduce interactions between families, they suggest rather large square plots (4 × 4 trees and more), in which only the interior trees are measured. Intermediate blocks would be undesirable. Isolation rows of a relatively resistant family between plots could also reduce interactions between families.

A similar but reverse effect is recognized in the testing for growth rate in forest tree breeding, as differences in growth rate between clones or families tend to be overestimated when plots are small.

The use of mixograms

In Fig. 4 it can be seen that the negative influence of the mixture on the resistant partner is similar to the beneficial influence on the susceptible partner. How would this change if a more resistant partner would be involved? Such plants would be less affected by the spores produced by the susceptible plants, their line in the graph would drop and become more horizontal. At the same time, as the resistant trees would contribute less to the spore cloud, the line of the susceptible group would tend to descend more steeply. The parallellism in Fig. 4 would be lost, the new graph would approach that of a mixture between a host and a non-host. On the other hand, if the susceptibility of the susceptible partner was increased, its line in the graph would rise and tend to become more horizontal as mixing with resistant material would tend to have less effect.

Thus the 'mixogram' may be a useful tool for visualizing the effects of mixing genotypes. Mixograms were tentatively sketched for some of the examples of host-disease combinations given above (Fig. 5). Note that in the case of the multiline, the pathogens for the 2 hosts are different.

A mixogram by itself, giving only biological information, cannot tell whether a certain mixture is advantageous or not. This will depend not only on the economic or other value of the partners in the mixture, but also on the level of the damage treshold (Fig. 6). If the damage threshold is high, at A, the 50:50 mixture is very advantageous as the disease rating of the susceptible host is reduced to below that level. If the damage threshold is

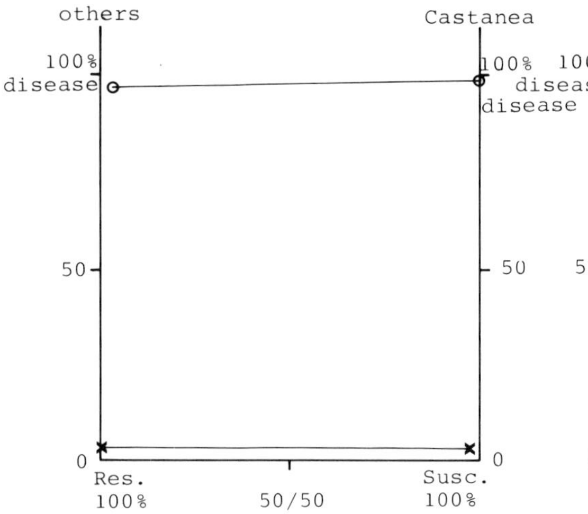

Chestnut blight on *Castanea dentata* (o) and other tree species (×) in the eastern American forest. Lines nearly horizontal: hardly any interaction.

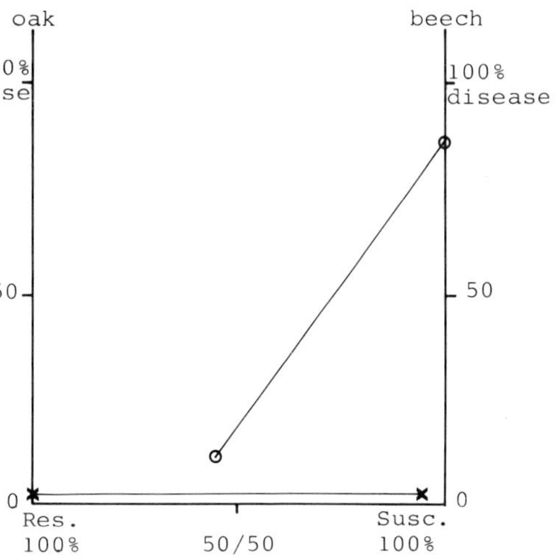

Nectria ditissima in NE France (after Perrin). Susceptible beech (o) is heavily affected in pure stands, little if mixed with resistant oak (×).

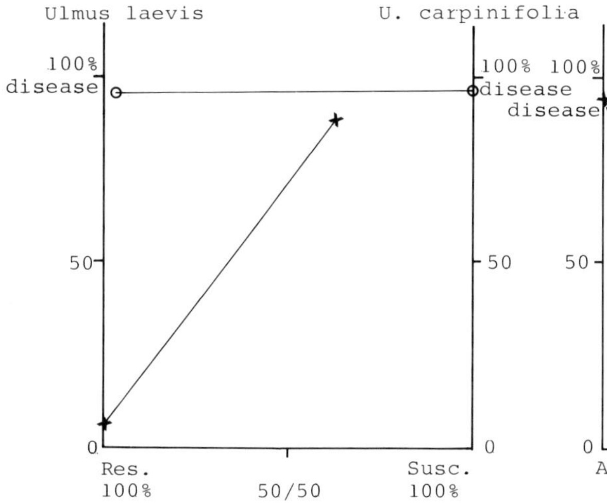

Dutch elm disease in Ukraine (after Maslov). If susceptible *U. carpinifolia* (o) is admixed in stands of 'resistant' *U. laevis* (×), the latter is badly affected.

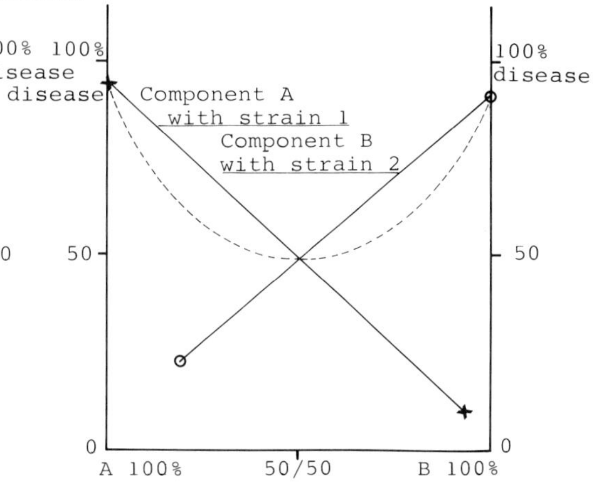

Simplified multiline with two 'lines' (with their pathogens). The dotted line would give the average disease rating of the mixed stand at different compositions.

Figure 5. Sketches of 'mixograms' of four different host-host-parasite combinations.

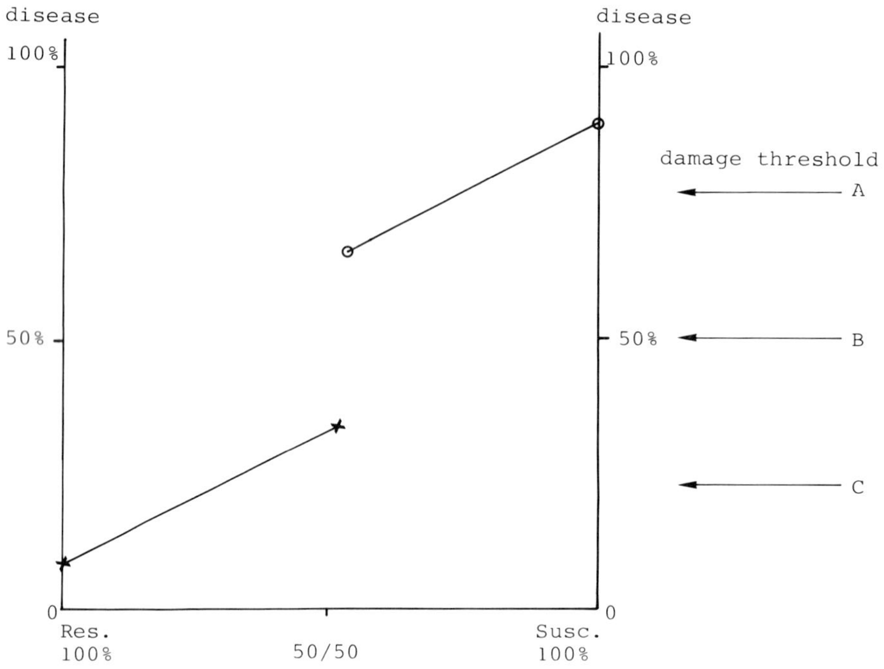

Figure 6. The effect of different levels of the damage threshold on the economic acceptability of mixtures of hosts with partial resistances. See text.

intermediate at B, mixing is of little use as the resistant partner remains good, the susceptible poor. If the damage threshold is low at C, the mixture is unadvisable as damage can only be avoided by growing the resistant partner in a pure stand.

CONCLUSIONS

1. In discussing the health hazards of pure stands, arguments stressing the danger of the undivided risk should be kept separate from those stressing that disease incidence must be higher in pure stands.
2. Broad generalizations on the effect of mixing of genotypes on the health of a stand or its components are dangerous. The effect may be different for each disease, site, host or case. Mixing may even be detrimental.
3. The mixing of a host with a non-host, which often means mixing different species, may have more effect than mixing hosts with partial or horizontal resistance, but it may entail greater silvicultural problems.
4. Mixing should be done consciously, after a study of disease and host has allowed the prediction that mixing is advantageous. Blind mixing will only in rare cases lead to useful reduction of disease incidence.

5. Mixing of genotypes is a poor substitute for breeding for resistance in forest trees.

REFERENCES

Boyce, J.S., 1954. Forest plantation protection against diseases and insect pests. FAO Forest Development Paper 3. 41 p.

Browning, J.A. & K.J. Frey, 1969. Multiline cultivars as a means of disease control. Annual Review of Phytopathology 7: 355-382.

Eis, S., 1978. Natural root forms of western conifers. Symposium on Root Form of Planted Trees, 1978. p. 23-27.

Epstein, A.H., 1978. Root graft transmission of tree pathogens. Annual Review of Phytopathology 16: 181-192.

Fischer, W., 1957. Zur Föhrenschütte Lophodermium pinastri. Schweizerische Zeitschrift für Forstwesen 108: 260-270.

Heybroek, H.M., 1978. Primary considerations: multiplication and genetic diversity. Unasylva 30 (119/120): 27-33 and 49-50 (a corrected reprint is available from the author).

Heybroek, H.M., 1981. The right tree in the right place. Unasylva (in press).

Kleinschmit, J., 1979. Limitations for restriction of the genetic variation. Silvae Genetica 28: 61-67.

Kriek, W. & G. Bikker, 1973. German and Dutch provenances of Scots pine in the Netherlands. Ned. Bosb. Tijds. 45: 154-161. Also: Dorschkamp Mededeling No. 131.

MacArthur, R.H. & E.O. Wilson, 1967. The theory of island biogeography. Princeton.

Maslov, A.D., 1970. [Insects noxious for elm species and their control]. Moscow. 76 p.

Parlevliet, J., 1979. Wederzijdse beïnvloeding van proefveldjes en de gevolgen hiervan voor de evaluatie van partiële resistentie. Zaadbelangen 33: 329-332.

Parlevliet, J.E. & A. van Ommeren, 1975. Partial resistance of barley to leaf rust, Puccinia hordei. II. Relationship between field trials, microplot tests and latent period. Euphytica 24: 293-303.

Peace, T.R., 1962. Pathology of trees and shrubs, with special reference to Britain. Oxford. 723 p.

Perrin, R. & F. Vernier, 1979. La chancre du hêtre. Influence des conditions stationelles sur la gravité de la maladie. Revue Forestiere Française 31: 286-297.

Rennerfelt, E., 1947. Om rotrötan (Polyporus annosus Fr.) i Sverige. Dess utbretning och sätt at upträda. Meddel. från Statens Skogsforskningsinstitut. Band 35 för 1946 (8) 1-88.

Schreiner, E.J., 1965. Maximum genetic improvement of forest trees through synthetic multiclonal hybrid varieties. NE Forest Tree Improvement Conference Proceedings 13 p. 7-13.

Stiell, W.M., 1970. Some competitive relations in a red pine plantation. Canadian Forest Service. Publ. 1275, 10 p.

Squillace, A.E., J.G.A. la Bastide & C.L.H. van Vredenburch, 1972. Breeding of superior Scots pine in the Netherlands. Dorschkamp, Intern Rapport nr. 10.

Squillace, A.E., J.G.A. la Bastide & C.L.H. van Vredenburch, 1975. Genetic variation and breeding of Scots pine in the Netherlands. Forest Science 41: 341-352. Also: Dorschkamp, Mededeling No. 155.

What is a safe number of clones per plantation?

W.J. Libby

Departments of Genetics, and of Forestry & Resource Management, University of California, Berkeley, California, 94720, USA

ABSTRACT

This paper discusses the following concepts: that a forest plantation is at risk from a great variety of possible physical and biotic events, many of which cannot be anticipated, and each has its own different probability distribution of damaging a particular population of trees; that management, through its decisions on planting density, thinning schedule, and salvage operations, can influence the amount of damage or mortality that a plantation can sustain and still be managed economically; and, that cross-adaptation of a short-generation pest may be more likely among many clones than among few unrelated clones. Some of the (perhaps unexpected) conclusions from analyses employing these concepts are: that monoclonal plantations are frequently the best strategy; that mixtures of 2 or 3 clones are frequently the worst strategy, and are rarely or never the best; that mixtures of a large number of clones are as safe as seedling plantations; and, if planting density and subsequent silviculture allow damage or mortality at levels comfortably above reasonable expected risk, then mixtures of modest numbers (7-25) of clones appear to provide a robust and perhaps optimum strategy.

INTRODUCTION

The question is simple and clear. The answer, it seems, is not. This paper will address some aspects of the problem of assessing risk when clones are used in production plantations.

Many thoughtful people have considered that a forest plantation of a single clone is less 'safe' than a plantation of seedlings (Goodman, 1975; Marshall, 1977; Singer, 1979). If the clone is susceptible to some physical event, such as drought, cold or wind, then a single such event may destroy or seriously damage the entire plantation. Worse, fastbreeding diseases or insects, over several of their generations, may adapt more closely to a single widely-planted clone than to the greater diversity of phenotypes in a seedling plantation (Day, 1974). Janzen (1970) used part of this argument to explain species-level diversity in tropical forests.

Planting a single clone over large areas has been done, and with some success. It provides operational simplicity, and may also maximize genetic gain (Appendix Comment 1). Perhaps the most notable forestry example of monoclonal plantings is provided by poplar clone I-214. But even this unusually successful clone has encountered serious difficulties, not all of which were perceived at the time of planting (Heybroek, 1978; Herpka & Guzina, 1979).

There has been, particularly in poplar culture, some resistance to planting mixtures of clones (Kolster, 1978; but see Schreiner, 1971 and earlier). However, seedling plantations of forest trees are usually mixtures of many different genotypes. Thus, the idea of intimately mixed genotypes in a plantation is hardly new or untried in forest practice.

THE PROBLEM

The relationship between risk and genetic variation in a plantation or forest is frequently presented as in Fig. 1. If such a figure is correct, one conclusion is clear: the more clones in a plantation, the safer it will be.

Clonal forestry offers many genetic and management advantages (Ehrenberg, 1977; Libby, 1977; Lindgren, 1977; Heybroek, 1978), and thus many tree-improvement programs are exploring the clonal option (Wilcox et al., 1976; Kleinschmit & Schmit, 1977; Lepistö, 1977; Roulund, 1977; Werner, 1977; Libby & McCutchan, 1978; Rauter, 1979). In considering clonal forest-

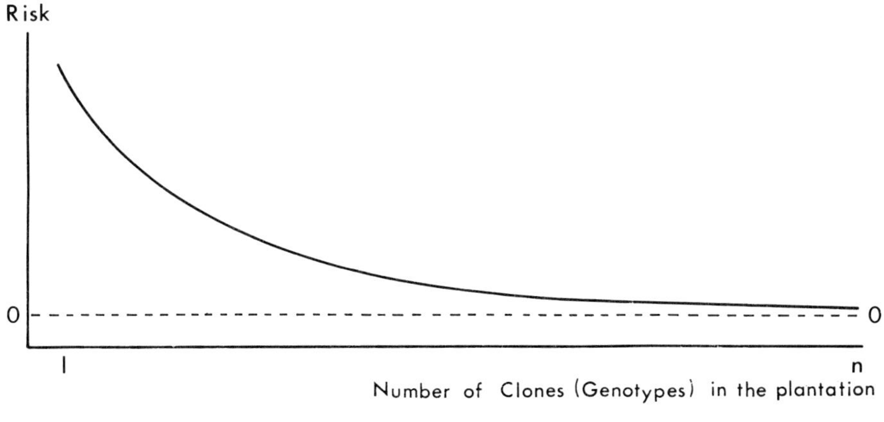

Figure 1. Schematic relationship between genetic variation and risk (after Kleinschmit, 1979).

ry, managers are asking: 'What is safe enough?' and 'How many clones provides that level of safety?'.

At this point, the central problem seems to be a balancing of the gains obtainable with one, or few, or many clones against the risks associated with that number of clones. It is frequently assumed that both gain and risk are maximum with 1 clone, and minimum in a seedling forest. This paper will explore both ends of Fig. 1, and will suggest that both ends may be of a different shape than as shown.

AN APPROACH TO THE PROBLEM

Disasters certainly happen, even in the safest of multi-species seedling forests (Bormann & Likens, 1979). Diseases such as blister-rust have done enormous damage to seedling forests of western white pine, and chestnut blight has driven the American chestnut to virtual extinction. It is unlikely that any clonal strategy we might devise would have protected such forests or species against such disasters. Let us, therefore, consider such general disasters as outside the topic of this paper (Burdon, 1977).

Some damage and mortality are both normal and acceptable in most forests grown as renewable resources. It is up to forest managers to decide what level of damage and mortality is acceptable, and to plan plantation spacings and thinnings accordingly (Appendix Comment 2). What managers wish to guard against is exceeding some unacceptable level of loss, rather than attempting to minimize the loss per se.

AN EXAMPLE

Let us first take a simple case (Appendix Comment 3). As a thinking framework in this example, and in the remainder of this paper, let us consider the composition of plantations of 2 ha or more in size. Consider some single physical event that, if it occurs, seriously damages or kills some proportion of the trees in the general population (say 10 %), but does little or no damage to the remaining 90 %, which are not susceptible to it. Management has previously decided that continuation of a plantation with losses greater than 25 % would not be economic.

If such an event occurs in a plantation of genetically-diverse seedlings, drawn from the general population, about 10 % of them are badly damaged or killed. The other 90 % are sufficiently well-formed and healthy so that they are acceptable within the management plan, and management of the plantation continues.

Given that there has been no selection affecting resistance to this event, there is a 10 % chance that the genotype in a single-clone plantation is susceptible. If it is not susceptible, most or all of the trees sustain little or no damage from this event. If it is susceptible, then most of the trees are killed or seriously damaged. For susceptible plantations, this is a disaster, and monoclonal plantations are (in this case) more likely to sustain unacceptable losses than are seedling plantations.

In a 2-clone mixture, if the clones are not related, the probability that both are susceptible to this event is 0.01. However, one or the other of the 2 clones is susceptible in 18 % of such mixtures, and if so, about 50 % of the plantation will be killed or seriously damaged if the event occurs (Appendix Comment 4). (Since many live trees are still on the site, but undamaged trees are not in sufficient density to be effectively managed, management may view 50 % damage and mortality as a greater disaster than if all trees in the plantation had been killed.) A 3-clone mixture is even worse, with a 0.27 probability of more than 25 % of the trees being damaged or killed by this one event. (Furthermore, the second and third clones in the mixture are probably susceptible to some events that little affect the first clone, thus further increasing the chance of unacceptable loss in a mixture of 2 or 3 clones, compared to a monoclonal plantation of any one of them.)

Now consider a plantation that is a mixture of many small clones. (For instance, each clone may have 10 members dispersed in the plantation, and thus there are 1/10th the number of genotypes in such a plantation as in a seedling plantation of the same size.) If the clones came from a broad genetic base, and their selection was not correlated with susceptibility to the event, then the occurrence of the damaging event produces similar results to those in seedling plantations. About 10 % of the trees are killed or badly damaged, and about 90 % suffer little or no damage. Management can continue.

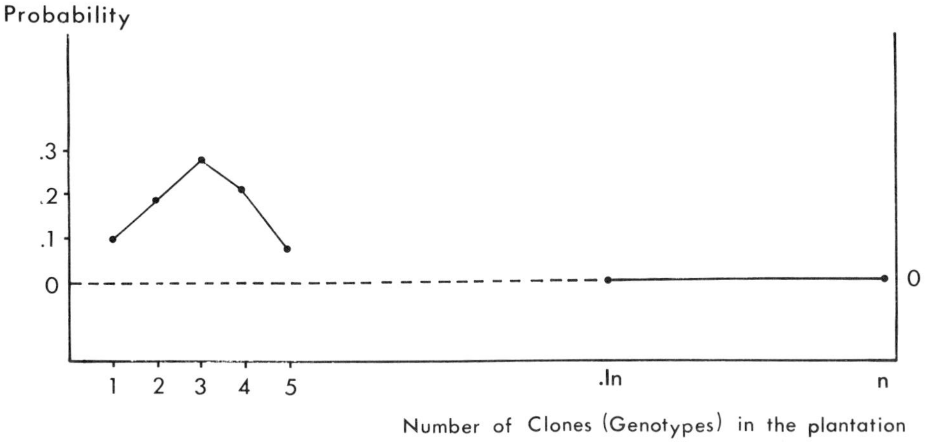

Figure 2. Probability of unacceptable loss, given a single physical event to which 10 % of the tree genotypes are susceptible, and a maximum acceptable loss level of 25 %. Plotted only for 1-5, $n/10$, and n clones.

Fig. 2 plots the probabilities of losses exceeding 25 % from such a 10 %-probability physical event, for the 5 points we have explored, and for mixtures of 4 and 5 clones. Under the concept of 'acceptable loss', Fig. 2 departs from Fig. 1 at both ends. Under the conditions described above, there is no increase in risk of unacceptable loss, compared to a seedling plantation, associated with using a large number of clones. Furthermore, mixing a very small number of clones is more likely to result in unacceptable loss than is a monoclonal plantation. Both of these results, if general, have important implications for the practice of clonal forestry.

GENERALIZATION FOR PHYSICAL EVENTS

Binomial algebra was used to calculate the probabilities in the above example. It allows only 2 classes. Surely there will be some genotypes that sustain intermediate levels of damage, or that are seriously damaged in some circumstances but not in others. Fig. 3 presents 4 sets of curves for different susceptibility levels in the general population, and for 6-8 levels of unacceptable loss. These curves where chosen to illustrate several generalizations derived from many such curves. They were generated by 1 000 to 10 000 random numbers drawn from each of various sets of Weibull distributions (Johnson & Kotz, 1970). (Tore Skrøppa, of NISK, greatly aided in programming and statistical interpretation.)

The following interpretations were used when analyzing the probability curves generated by samples from the Weibull distributions: A sampled clone with value 1.0 meant that all members of that clone were killed or seriously damaged; 0.0 meant that all members were undamaged, or were only damaged to some acceptable degree; 0.3 meant that 30 % of the members were killed or seriously damaged, but the remaining 70 % were damaged to some acceptable degree or were undamaged. (A sample value of 0.3 was not, however, interpreted to mean that the clone sustained a 30 % loss in value or in growth ... Appendix Comment 5.) 'Seriously damaged' meant that, because of the damage, the tree would be removed at the next thinning, or that management would consider such a tree no longer likely to contribute to an economic harvest. 'Acceptable damage' meant that the tree could be economically managed to rotation, and that if such a tree were removed in a pre-harvest thinning, that decision would be based on spacing and other considerations in addition to considering the damage sustained. In short, the Weibull distributions and samples from them allowed us to generate percentages of trees per clone and per plantation that would still be considered economically manageable after some damaging event(s).

The analyses of the Weibull samples included distributions that are normal, skewed, and (by truncating and summing both tails) bimodal. The mean value in each distribution is defined as the Risk to a Random Genotype in that circumstance, hereafter called *RRG*. Repeated samples of 1, 2, 3 or more clones were drawn from each distribution and tested against various levels of Maximum Acceptable Loss, namely the highest loss level that management has planned for or is willing to accept, hereafter called *MAL*. The following generalizations seem possible and appropriate:

1. At least 3 factors strongly influence the probability of having a manageable plantation after some damaging event(s): (a) the shape of the distribution of damage sustained by random genotypes (clones) in the population; (b) the *RRG*; and (c) the relationship of *MAL* and *RRG*, i.e., *MAL* - *RRG* or perhaps *MAL/RRG*.

2. If *MAL* is less than *RRG*, then the probability of exceeding *MAL*, if the event(s) occur, is 1.0 in plantations of seedlings or of mixtures of many clones. Monoclonal plantings provide the best strategy in this situation. Note that the probability of even monoclonal plantings exceeding the *MAL* is generally higher than the *RRG* (it equals *RRG* if the population is a mixture of 100 % resistant and 100 % susceptible genotypes, with no clones sustaining only partial loss, as in Fig. 2). See the upper portions of the graphs in Fig. 3.

3. If *MAL* - *RRG* is about 0.2 (or more) ... or if *MAL/RRG* is about 1.7 (or more) ... i.e., if the maximum acceptable loss level is comfortably higher than the risk, then mixtures of 7-15 clones (or fewer) will provide a probability of less than 5 % of exceeding the *MAL* level. See the lower portions of the graphs in Fig. 3. In these circumstances, such mixtures are safer than monoclonal plantations, and almost as safe as mixtures of many clones or of seedlings.

4. When *RRG* is only a little less than *MAL*, larger numbers of clones in mixture are necessary to reduce to a negligible amount the probability of the plantation sustaining unacceptable loss. See the central portions of the graphs in Fig. 3. However, this apparent mathematical reduction with large numbers of genotypes may be misleading. *RRG* near *MAL* is a marginal situation, and if such damaging event(s) occur, the decision to terminate or continue management is very close in plantations of seedlings or of many clones in mixture. A mosaic of small monoclonal plantations may be the better option in such cases, as the risks to individual clones tend to be more dispersed, both below and above the *MAL* level. While some or many may be destroyed, monoclonal plantings of the clones that prove to be resistant will remain available for continued effective management.

5. If there is little difference between clones in average damage sustained (characterized by a distribution with a single mode and a small variance), then clonal strategies are not very effective. They do appear simple. If *MAL* is greater than *RRG* by even a modest amount, then mixtures of 5-10 clones are about as safe as mixtures of a large number of clones or of seedling plantations. If *RRG* is similar to or greater than *MAL*, then monoclonal plantings offer the best chance of having some economically manageable plantations if the event(s) occur.

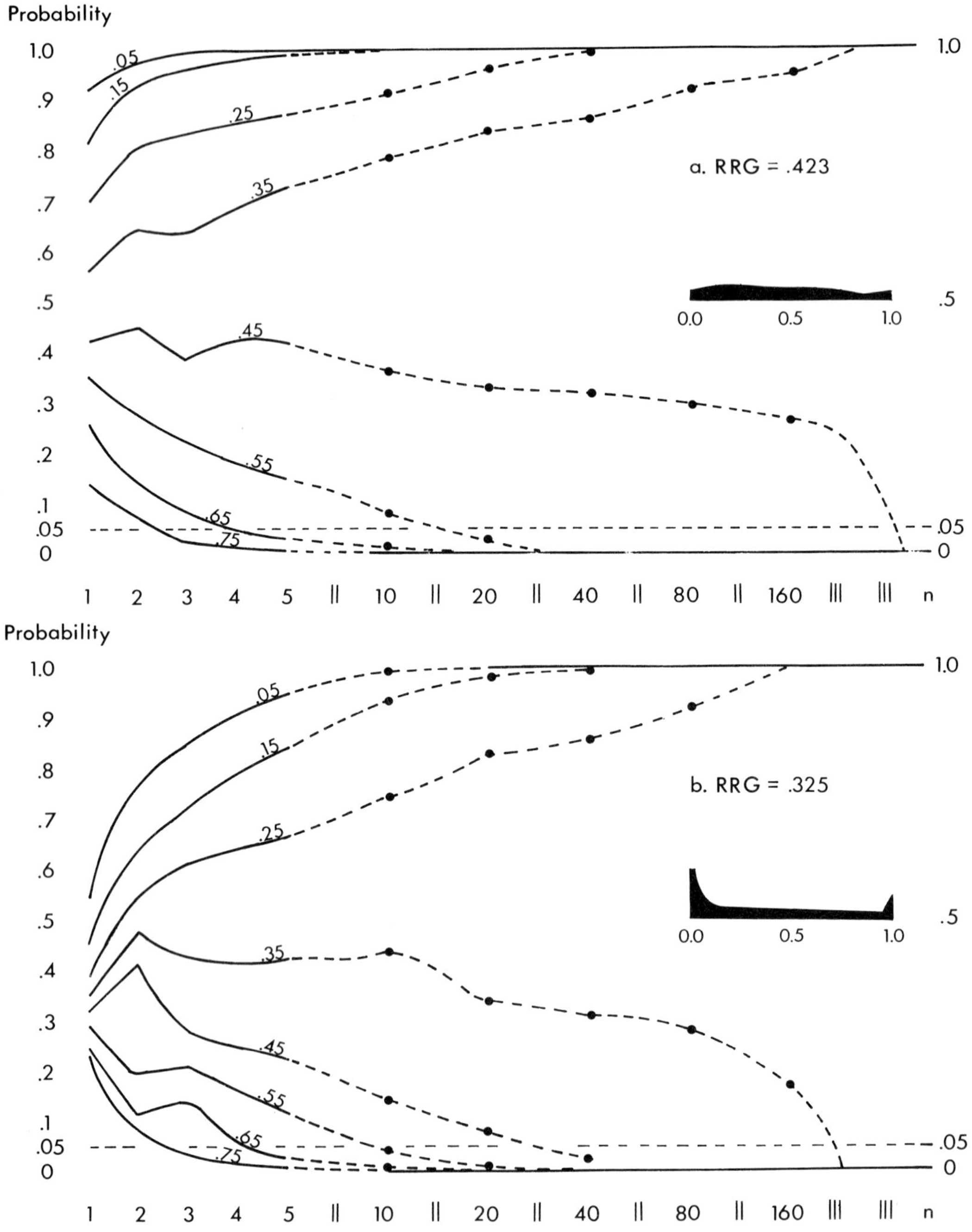

Figure 3. Curves showing probability of unacceptable loss, given a single physical event to which 0.423, 0.325, 0.152, and 0.044 of the genotypes in the general population are susceptible (*RRG*). The curves indicate the probability of plantations exceeding maximum acceptable loss (*MAL*) levels of 5 %, 15 %, 25 %, 35 %, 45 %, 55 %, and, for the higher *RRG*s, 65 % and

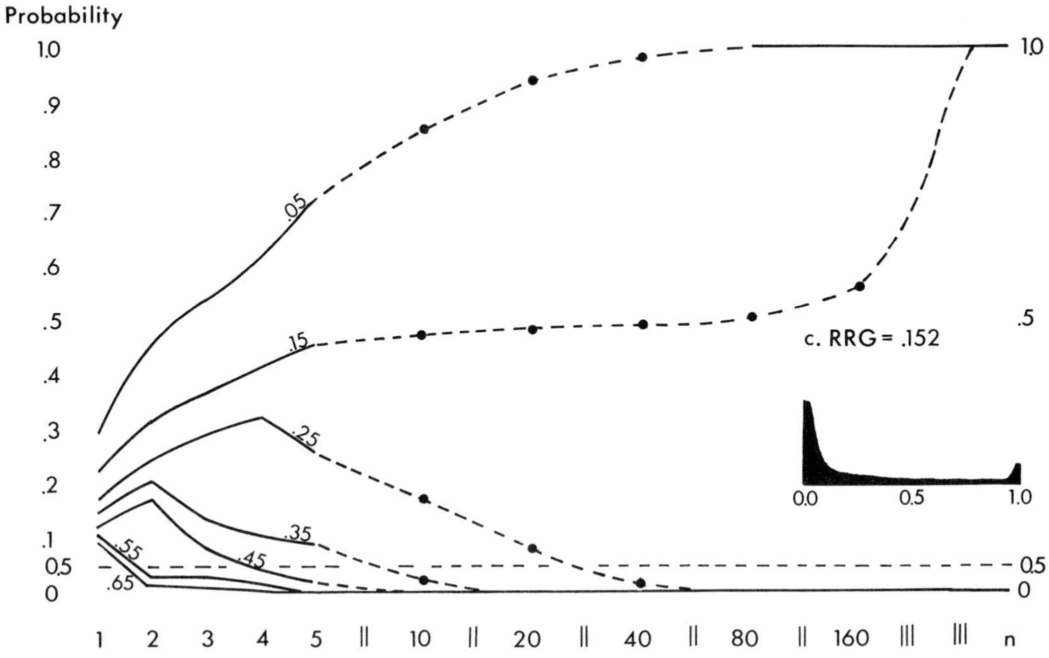

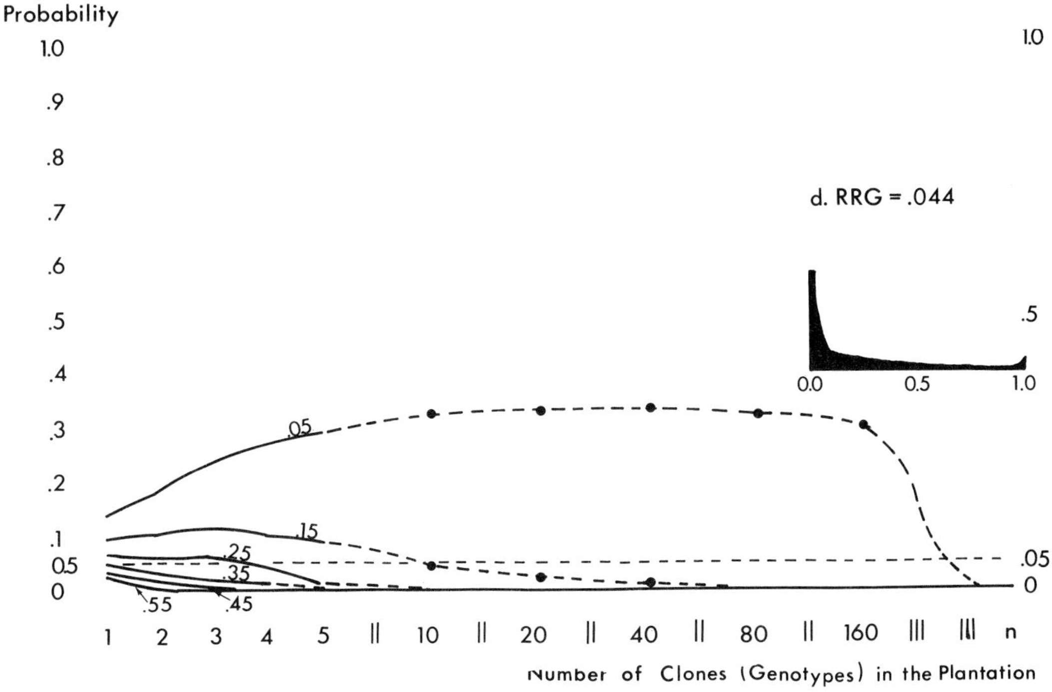

75 %. These are typical of the curves produced with numbers drawn at random from various Weibull distributions of clonal damage levels. The particular distributions producing each of these 4 sets of curves are indicated on the right side of each graph.

6. When most clones are either highly resistant or highly susceptible to an event (the distribution is bimodal), and *MAL* is greater than *RRG*, then mixtures of 2 (and sometimes 3 or more) clones are often more likely to sustain damage exceeding the *MAL* than are monoclonal plantations. (Recall that a second event intensifies this effect.) The closer *RRG* is to *MAL*, the greater is the number of clones whose mixture has a higher probability of exceeding the *MAL* than a monoclonal plantation drawn from the same group of clones. However, if *MAL* is greater then 0.5 (i.e., 50 % sustainable loss) and also exceeds *RRG*, then monoclonal plantations are always the worst option, and an increase in number of clones in mixture leads to a fairly continuous decrease in the probability of exceeding the *MAL*. See lower curves in the graphs in Fig. 3.

7. At higher *RRG* (greater than about 0.35) it takes mixtures of more clones to reduce the probability of exceeding the *MAL* to a negligible level than it does at lower *RRG* (less than about 0.25).

It seems likely that selection of well-adapted clones will lower the average risk to such genotypes below *RRG*. But this effect may be smaller than one might hope, due to the uncertain and complex nature of the risks. It is possible that ecological gradients in risk will affect regional strategies. Taking species diversity as a cue (Goodman, 1975; Burdon, 1977), *RRG* appears lower at high latitudes and increases toward the tropics. The major contributors to this increasing *RRG* appear to be biotic (Janzen, 1970).

BIOTIC EVENTS

Biotic events can be more complicated than physical events (Harper, 1977). For instance, Day (1974) listed 19 variables that affect the colonization and subsequent spread of a pest, and which may result in an epidemic. Underlying at least some of the development that follows is a conceptual model of genetic influences on host resistance to colonization and susceptibility to damage, and on pest host-range, virulence, and aggressiveness, that includes several alleles at each of several loci (not necessarily of equal importance) for each characteristic (Brown, 1975; Groth & Person, 1977; Marshall, 1977; Griffiths, 1978).

Given the background of all these variables, and their much larger number of possible interactions, Fig. 4 attempts to schematically and simply present 2 important processes, the colonization and subsequent genetic adaption of a pest to a single clone. Using the same probabilities for colonizing ability as for the above simple physical example, there is a 90 % chance of the clone being resistant (colonization fails) and a 10 % chance of it being susceptible. However, unlike the physical event, if the insect or disease successfully colonizes a plantation, its subsequent generations may evolve, and in more than one way (Barrett, 1978).

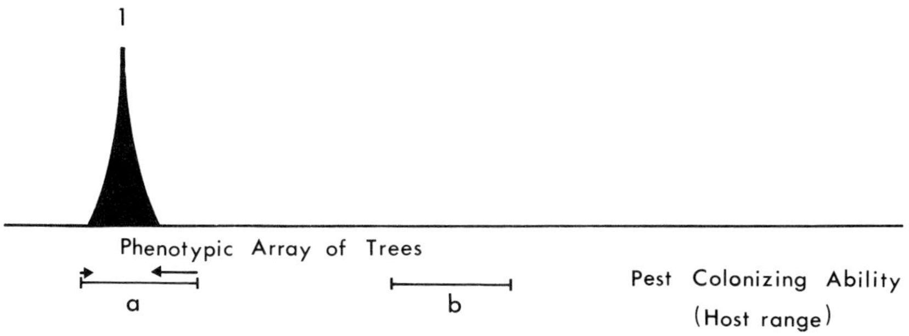

Figure 4. The top portion of the figure represents the phenotypes of clone 1 in a monoclonal plantation, arrayed with respect to susceptibility to a particular pest. If the colonizing ability of a pest includes the ability to colonize clone 1, it will become established and may adapt increasingly precisely to that single host (pest *a*) (Wolfe, 1978). However, if the invading pest has the host-range of '*b*', it will not establish in a monoclonal plantation of clone 1. (It should be noted that the phenotypic array of the host population is probably not the same for pests '*a*' and '*b*', particularly if they belong to different species.).

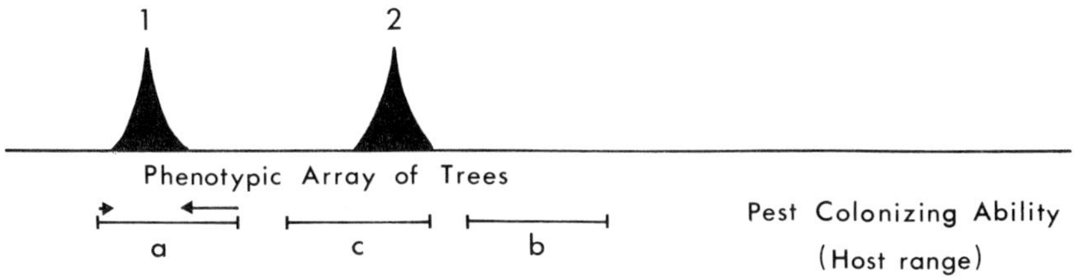

Figure 5. It is possible that clone 2 will be increasingly safe from attack by pest *a*, as pest *a* adapts more precisely to the interplanted clone 1 (Appendix Comment 6). Clones 1 and 2 will both escape colonization by pest *b*. But pest *c* could colonize and damage clone 2 if it invades the plantation, thus raising the overall probability of damage exceeding 25 % in the 2-clone mixture.

If the pest colonizes a large monoclonal plantation, it may adapt increasingly precisely to that clone. This may create an epidemic more severe than would have occurred (or that might not have occurred at all) in a genetically-diverse seedling population (Day, 1974). Thus, it appears that a large monoclonal plantation is more likely to sustain serious damage, compared to a seedling plantation, during biotic events than during physical events.

Using the same probabilities in a 2-clone mixture, 90 % of the time the insect or disease preadapted to colonize one clone will not be preadapted to colonize the second. Furthermore, as the pest adapts increasingly precisely to the one clone, it may be less able to colonize, or will be less virulent (or aggressive) on, the second clone (Fig. 5) (Edmunds & Alstad, 1978) (Appendix Comment 6). However, following the same logic as with physical events, the risk of unacceptable loss (in this example, 25 % or more) is greater with 2- or 3-clone mixtures than with a monoclonal plantation. And, as indicated in Fig. 5, a second clone is probably susceptible to pests that cannot successfully attack the first clone, further raising the overall risk of an unacceptable loss in a 2-clone mixture.

In a seedling plantation, the adaptation of an insect or disease population following successful colonization may follow a different course than that described above for 1 clone or in a mixture of very few clones. While there are no 2 genotypes identical in most seedling plantations of forest trees, there are many that are very similar. These form a genetic continuum within the plantation population. If the pest has successfully colonized 10 % of the trees, the pest population is unlikely to adapt narrowly to only 1 of those genotypes, but will itself probably maintain substantial genetic diversity (Day, 1974). It is possible (even likely) that over several of its generations it will continue to adapt to phenotypes similar to those just successfully attacked. Thus, in each generation, some members of such a pest population can successfully attack trees (genotypes) resistant to previous generations of the pest (Marshall & Pryor, 1978). Before the epidemic runs its course, it may infest and significantly damage or kill many more than the 10 % of the seedling-origin trees initially susceptible to it, and thus exceed higher *MAL* levels (Fig. 6).

The result will be little different in a plantation consisting of many clones, each cloned only a few times. This produces the relationship on the right side of Fig. 7.

Thus, 2 rather different genetic possibilities are suggested with a short-generation pest. With 1 or only a few clones in a large plantation, the pest may adapt increasingly closely to the clone it successfully colo-

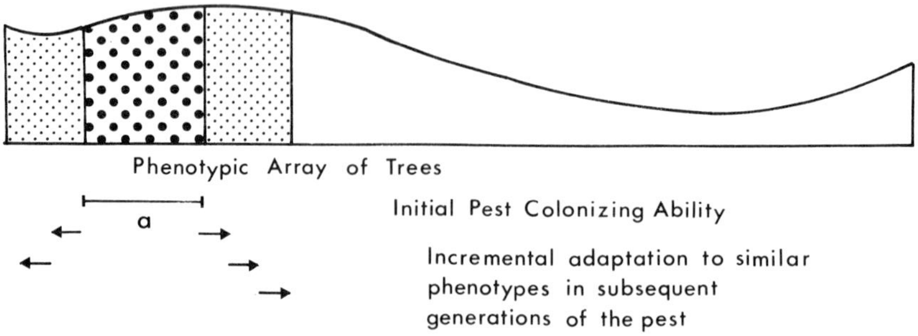

Figure 6. Colonization of 10 % of a population of seedling origin, followed by incremental adaptation and spread of the pest to initially-resistant trees (Wolfe, 1978).

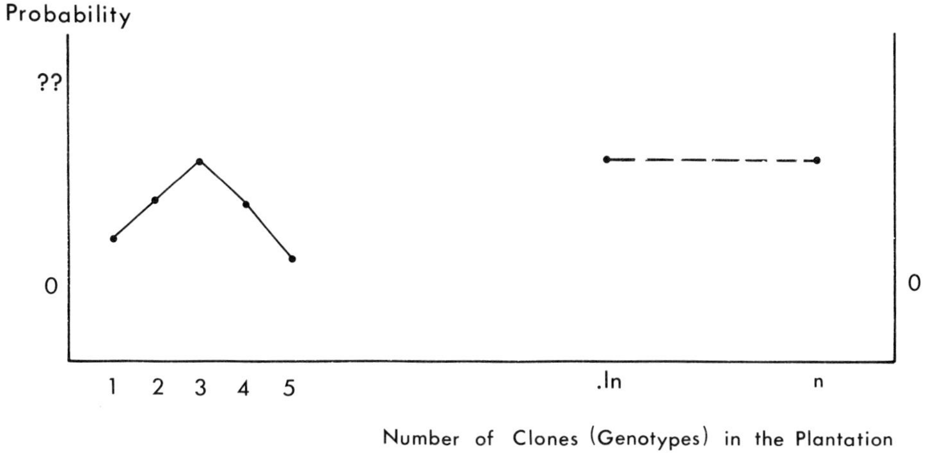

Figure 7. Relative probabilities of unacceptable loss, given a sustained biotic event to which 10 % of the tree genotypes are initially susceptible, continued adaptation by the pest, and a maximum acceptable loss level of 25 %. Because of the 2-step nature of biotic events, colonization followed by adaptation, probabilities cannot be accurately identified with our simple assumptions.

nized, with an increase in virulence on that clone and a reduced likelihood that it can colonize initially resistant clones (Wood, 1980). With very many clones or a seedling population, the pest may maintain and even increase its genetic variability, such that it can adapt to successfully attack previously resistant trees, although perhaps with reduced virulence as the epidemic spreads (Groth, 1976; Wolfe & Barrett, 1977; Edmunds & Alstad, 1978; Gould, 1979).

This suggests an interesting possibility. There may be mixed plantations with enough clones to acceptably spread general risk, but with sufficient genetic differences among the clones so that cross-adaptation of a narrowly-adapted pest is unlikely to continue to many of the other clones in the plantation. The lower number of clones in such a mixture is influenced by the *MAL* decided by management. If protection against cross-adaptation is attainable, the upper number would be influenced by the number of clones that can be included in a plantation and still maintain genetic dissimilarities among them sufficient to impede continuing cross-adaptation by a narrowly-adapted colonizing pest (Fig. 8).

The theoretical models of Marshall & Pryor (1979) give some hope that this is on the right track. Their theory is developed for homozygous lines in the host (most forest-tree clones will be highly heterozygous), and they particularly wish to guard against the evolution of a 'super race' of pest, able to attack all lines. Their several models use different assumptions on gene action, and achieve some generality - i.e., it appears better to have few lines, all differing in many alleles, than to have many lines, each differing by only a few alleles from some of the other lines in the mixture. The somewhat different models of Groth & Person (1977) also support this view.

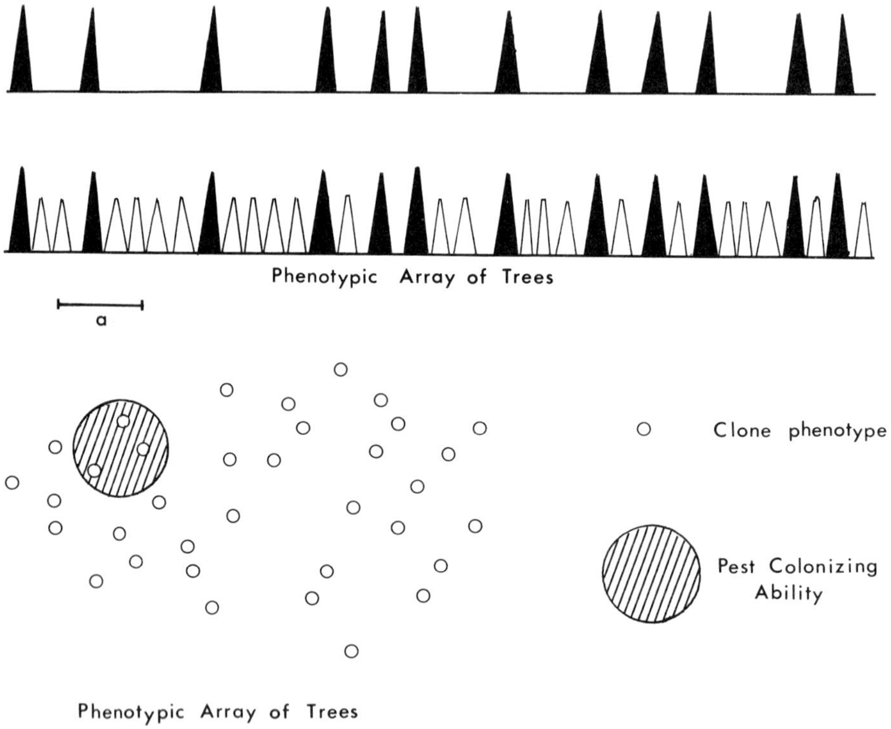

Figure 8. (top) Pest *a* would be able to colonize between 0 and 3 clones (depending on where in the host phenotypic array it is initially adapted) if only 12 clones are mixed, as in the upper plantation. The genetic distances between most of these 12 clones appear great enough so that little cross-adaptation of this narrowly-adapted pest is likely to occur. (middle) If the plantation mixture includes 23 additional clones, then colonization by pest *a* seems almost certain if it invades, and substantial cross-adaptation seems possible. (bottom) The clone phenotypic arrays and pest colonizing abilities may be more realistically shown in more than 1 dimension. The above 2-dimensional illustration allows greater distances between the 35 clones in the mixture, with less apparent probability of pest cross-adaptation, even though its initial host-range includes about 10 % of the population.

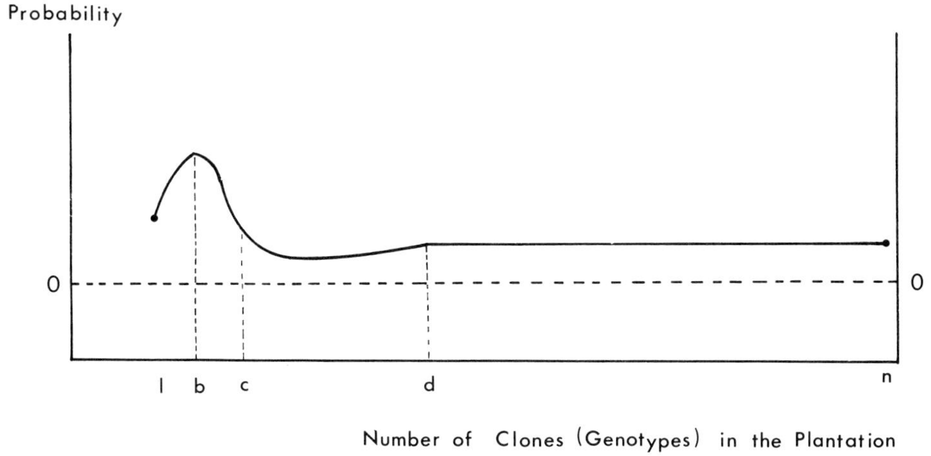

Figure 9. Probability of unacceptable loss, given a combination of physical and biotic events of various kinds, and a maximum acceptable loss level that is not too low. Region *1* - *b* is characterized by an increasing probability that an unacceptable proportion of the plantation will be killed or seriously damaged, as the number of clones in mixture is increased from *1* to *b*. The size and shape of region *b* - *c* is sensitive to management's judgement concerning the proportion of the plantation that is designed to allow damage, mortality, or eventually be thinned. Region *c* - *d* is where some protection against cross-adaptation by dynamic pest populations may be afforded by mixtures of relatively few dissimilar clones. Region *d* - *n* is essentially flat, offering about the same safety and risk as a population of *n* seedlings.

INTEGRATION OF PHYSICAL AND BIOTIC EVENTS

Without specifically identifying numbers other than *1* and *n*, a graph in the general shape of Fig. 9 may be predicted for a mixture of unknown damaging events. If the gain maximum is not at 1 clone (Appendix Comment 1), but at or near the number associated with the minimum risk of exceeding acceptable loss, then decision-making becomes easy.

SOME SUGGESTED GUIDELINES

The relationships developed in Figures 3 and 9 now allow a few guidelines to be considered.

1. Management may choose *MAL* levels that are very low, where most trees planted are expected to be harvested. When such *MAL* levels are applied to genera with known biotic or physical vulnerabilities, such as *Populus* (Schreiner, 1971), i.e., when the reasonable expectation of *RRG* is

higher than the *MAL*, then monoclonal plantings appear to be the best strategy. It seems prudent to use several different clones in a mosaic of monoclonal plantings, so that an entire enterprise is not incapacitated by one or a few damaging events (Kolster, 1978).

2. The higher the acceptable loss level that management is willing to assume, the more flexibility it will have in choosing a 'safe' number of clones. This argues for planting a relatively high number of trees per unit area, with a need to thin several times during development of the plantation. In the absence of significant mortality or damage, it allows the forester to select and favor those trees growing best. If serious damage or mortality occurs, thinnings will remove damaged and malformed trees, and adjust spacing to favor the best of the survivors.

3. Several factors argue for mixtures of relatively few clones. These include: the possibility of mixing highly-selected well-known complementary clones to increase unit-area productivity; easier and more efficient nursery management; and the possibility of reducing cross-adaptation of narrowly-adapted pests following colonization. A mosaic of different mixtures in a region may have additional beneficial effects with respect to the first and third of these factors (Ehrenberg, 1977; Libby, 1977).

4. Much plant-breeding work attempts to create crop mixtures based on known virulence and resistance genes in the host and in specific pests (Wolfe, 1978). This, particularly if the multiline is bred for agronomic uniformity, has a tendency to reduce background variability in the crop with respect to other unanticipated events or epidemics (Day, 1974; Wolfe & Barrett, 1977; Marshall, 1977). With the long generation and rotation times of forest trees, it appears wise to avoid a strategy that concentrates on breeding for resistance to a specific set of known problems. It may be better to develop a strategy that includes protection of the plantation against unknown problems (Allard & Hansche, 1964).

5. Heybroek (1978) properly noted that it should be possible to maintain a higher proportion of unusual recombinants and divergent genotypes clonally than can be repeatedly produced in populations of sexually recombined seedlings. As a strategy for dealing with unidentified possible disasters, strict pedigree control from a broad genetic base appears to be a good approach, with the added restriction that all clones be well-adapted to the site (Kleinschmit, 1979). One could construct rules against more than a certain number or proportion of clones from the same family being in the same clonal mixture.

6. If there is validity in the above lines of reasoning, it would be useful if laws and guidelines affecting clonal forestry would avoid prescribing or encouraging either too many or too few clones per plantation (Marshall, 1977; Wolfe, 1978; Fowler, 1979/80). As an early guess, 2 or 3 clones in mixture is too few, and in many cases is worse than a monoclonal plantation. Numbers near 30 or larger substantially erode the possible

advantages in point (3) above. Only 1 clone per family should be allowed in mixtures of fewer than about 30 clones.

Finally, a mixture of clones from 2 or more species should reduce pest cross-adaption not only between the species, but within each of the species, since fewer clones from each species can be used.

ACKNOWLEDGEMENTS

Consultations on this topic with K. Aastveit, R. Burdon, J. Dietrichson, L. Fins, A. Franclet, V. Guzina, D. Harry, J. Harper, I. Herpka, H. Heybroek, M. Jošt, M. Karlman, J. Kleinschmit, R. Koster, A. Krstinić, H. Muhs, S. Papadopol, H. & F. Roll-Hansen, T. Skrøppa, P. Spieth, H. Weisgerber, D. Zagory, and B. Zobel are acknowledged and appreciated. The manuscript was prepared while at The Department of Forest Genetics and Dendrology, Forestry Faculty, Zagreb, Yugoslavia; and at the Norwegian Forest Research Institute, Ås, Norway. Access to the NISK computer and library was particularly helpful.

APPENDIX COMMENTS

1. There are at least 3 alternative concepts bearing on the question of productivity per unit area (Harper, 1977). At issue is the nature of biotic interactions between neighboring trees. One alternative is that some 'best' clone produces more per unit area in pure monoclonal culture than does any mixture of clones (Kolster, 1978). A second is that genetically identical neighbors make similar demands on the site at about the same times; but that neighboring trees with complementary genotypes interfere which each other less, and use the site better than the best of them in a monoclonal planting (Allard & Hansche, 1964; Challinor, 1968; Wolfe, 1978; Kleinschmit, 1979). Included in this second alternative is a possible interaction between the plants and the host-specific insects and diseases that damage them (Chilvers & Brittain, 1972; Burdon & Chilvers, 1974; Marshall, 1977). It should be noted that much of the evidence now available on the relative performance of mixtures and their components in pure plantings is drawn from agronomic crops, selected to grow as pure plantings (Marshall, 1977), rather than from populations of plants (such as trees) that grow and have evolved in genetically diverse populations (Harper, 1977). The third alternative is that productivity per unit area is constant; that one tree's gain is a neighboring tree's loss; and that, if the trees are adapted to the site, about all a tree breeder can do is change the allocation of excess photosynthate within selected trees. The second alternative is appealing, and suggests that on productivity grounds alone, a mixture of a few well-understood excellent complementary clones will be

more productive than a monoclonal planting of the best of them. At square spacing, the minimum mixture with no identical neighbors is of 5 clones; at hexagonal spacing, of 7.

2. Levels of acceptable loss may be different at different stages of a plantation's development. For example, in a young plantation, management may accept a higher percentage of damage or mortality before thinning than following thinning. Later in the rotation, when some of the dead or damaged trees can be effectively salvaged, a higher percentage of loss may again be acceptable.

3. More than 1 damaging or potentially damaging event may be expected during the life of a plantation. Some of the examples and calculations in this paper take this into account. Others simplify the problem by summing 2 or more independent events over the genotypes they affect. However, one event may change the probability of a second event occurring, or change the severity of its damage. This seems equally true for plantations of seedlings and of clones, and therefore, is largely ignored in the development of this paper.

4. Even with a physical event, the presence of a neighboring genotype may change the probabilities of damage. For instance, a windfirm clone may be damaged by falling members of a clone susceptible to windthrow. Or, the damage in a drought-susceptible clone may be intensified by neighboring drought-resistant trees if they extract scarce water from the soil with greater effectiveness.

5. When interpreting curves such as those in Fig. 3, one may consider loss in value or growth, rather than percentage of manageable trees. Such losses may be general in a clone or plantation. For instance, a defoliating caterpillar or leaf rust may reduce average growth 30 %, with individual-tree losses in growth varying from 20 % to 40 %, and with no trees killed by that pest or free of attack and damage. A clonal strategy seems of little help in such cases. If the levels of growth loss that do or do not allow continued management can be identified for each tree by foresters on the ground, then this can be treated as in the main paper.

6. In grains, perhaps because of the size, shape and short lifespan of individual plants, there is evidence of protection of susceptible plants by interplanted resistant lines (Browning & Frey, 1969; Burdon & Whitbread, 1979), which usually includes a reduction of susceptible-host density (Burdon & Chilvers, 1977), both effects leading to reduced rates of pathogen spread. Such protection effects are influenced by the number and percentage of resistant lines and plants (see, for instance, the manipulative experiments of Luthra & Rao, 1979). However, Heybroek (1978) argued that the large size of tree crowns and long rotation times would greatly reduce such an effect. Within-tree variability must also be considered. A successful attack on susceptible portions of a tree (say, shade leaves) may allow the disease or insect to build population levels that can adapt to or

overwhelm more resistant parts of the tree. Adaptation across physiological states within a single tree, or among members of a clone occupying different local environments, is very much like adaptation among similar genotypes. Finally, if the population size of a pest population attacking a susceptible clone becomes large, it may generate mutants, unusual recombinants, or overwhelming numbers, that then successfully colonize previously-resistant neighbors (Browning & Frey, 1969; Dietrichson, 1969; Day, 1974; Groth & Person, 1977; Barrett, 1978; Edmunds & Alstad, 1978).

REFERENCES

Allard, R.W. & P.E. Hansche, 1964. Some parameters of population variability and their implications in plant breeding. Advances in Agronomy 16: 281-325.

Barrett, J.A., 1978. A model of epidemic development in variety mixtures. In: P.R. Scott & A. Bainbridge (Eds): Plant disease epidemiology. Blackwell Scientific Publications, Oxford. p. 129-137.

Bormann, F.H. & G.E. Likens, 1979. Catastrophic disturbance and the steady state in northern hardwood forests. American Scientist 67: 660-669.

Brown, J.F., 1975. Factors affecting the relative ability of strains of fungal pathogens to survive in populations. Journal of the Australian Institute of Agronomical Science 41: 3-11.

Browning, J.A. & K.J. Frey, 1969. Multiline cultivars as a means of disease control. Annual Review of Phytopathology 7: 355-382.

Burdon, J.J., 1977. Mechanisms of disease control in heterogenous plant populations - An ecologists view. In: P.R. Scott & A. Bainbridge (Eds): Plant disease epidemiology. Blackwell Scientific Publications, Oxford. p. 193-200.

Burdon, J.J. & G.A. Chilvers, 1974. Fungal and insect parasites contributing to niche differentiation in mixed species stands of eucalypt saplings. Australian Journal of Botany 22: 103-114.

Burdon, J.J. & G.A. Chilvers, 1977. Controlled environment experiments on epidemic rates of barley mildew in different mixtures of barley and wheat. Oecologia 28: 141-146.

Burdon, J.J. & R. Whitbread, 1979. Rates of increase of barley mildew in mixed stands of barley and wheat. Journal of Applied Ecology 16: 253-258.

Challinor, D., 1968. Alteration of surface soil characteristics by four tree species. Ecology 49: 286-290.

Chilvers, G.A. & E.G. Brittain, 1972. Plant competition mediated by host-specific parasites - A simple model. Australian Journal of Biological Science 25: 749-756.

Day, P.R., 1974. Genetics of host-parasite interaction. W.H. Freeman & Co., San Francisco. 238 p.

Dietrichson, J., 1969. Provenance and resistance to Scleroderris lagerbergii Gremmen. Norwegian Forest Research Institute, Vollebekk 92: 398-410.

Edmunds, G.F., Jr. & D.N. Alstad, 1978. Coevolution in insect herbivores and conifers. Science 188: 941-945.

Ehrenberg, C., 1977. Genetic consequences of clonal forestry. Genetika 9: 159-171.

Eliasson, L.A., A. Dunberg, G. Eriksson & K.R. Samuelson, 1977. Vegetative propagation of forest trees - Physiology and practice. Institute of Forest Improvement and Department of Forest Genetics, Swedish College of Forestry, Uppsala. 159 p.

Fowler, C., 1979/1980. Plant patenting. Coevolution Quarterly. Winter 79/80: 34-38.

Goodman, D., 1975. The theory of diversity-stability relationships in ecology. Quarterly Review of Biology 50: 237-266.

Gould, F., 1979. Rapid host-range evolution in a population of the phytophagous mite Tetranychus urticae. Evolution 33: 791-802.

Griffiths, E., 1978. Plant disease epidemiology - Retrospect and prospect. In: P.R. Scott & A. Bainbridge (Eds): Plant disease epidemiology. Blackwell Scientific Publications, Oxford. p. 3-9.

Groth, J.V., 1976. Multilines and 'super races': A simple model. Phytopathology 66: 937-939.

Groth, J.V. & C.O. Person, 1977. Genetic interdependence of host and parasite in epidemics. Annals of the New York Academy of Sciences 287: 97-106.

Harper, J.L., 1977. Population biology of plants. Academic Press, London. 892 p.

Herpka, I. & V. Guzina, 1979. The present state of black poplar improvement in Yugoslavia. (Današnje stanje oplemenjivanja crnih topola kod NAS.) J. Gračan (Ed.): Proceedings of Annual Meeting of Yugoslavian Forestry Genetics, Jastrebarsko (in press).
Heybroek, H.M., 1978. Primary considerations: Multiplication and genetic diversity. Unasylva 30: 27-33.
Janzen, D.H., 1970. Herbivores and the number of tree species in tropical forests. American Naturalist 104: 501-528.
Johnson, N.L. & S. Kotz, 1970. Continuous univariate distributions. Volume I, Chapter 20. Houghton Mifflin, Boston. 300 p.
Kleinschmit, J., 1979. Limitations for restriction of genetic variation. Silvae Genetica 28: 61-67.
Kleinschmit, J. & J. Schmit, 1977. Experiences with Picea abies cutting propagation in Germany and problems connected with large scale application. In: Eliasson et al. (Eds): Vegetative propagation of forest trees -Physiology and practice. Institute of Forest Improvement and Department of Forest Genetics, Swedish College of Forestry, Uppsala. p. 65-86.
Kolster, H.W., 1978. Menging van klonen in populieren beplantingen I & II. (Mixture of poplar clones.) Populier 15(1): 3-11, (2): 27-32.
Lepistö, M., 1977. Vegetative propagation by cuttings of Picea abies in Finland. In: Eliasson et al. (Eds): Vegetative propagation of forest trees - Physiology and practice. Institute of Forest Improvement and Department of Forest Genetics, Swedish College of Forestry, Uppsala. p. 87-95.
Libby, W.J., 1977. Rooted cuttings in production forests. In: Proceedings of the 14th Southern Forest Tree Improvement Conference, Gainsville. p. 13-19.
Libby, W.J. & B.G. McCutchan, 1978. 'Taming' the redwood. American Forests 84(8): 18-23, 37-39.
Lindgren, D., 1977. Possible advantages and risks connected with vegetative propagation for reforestation. In: Eliasson et al. (Eds): Vegetative propagation of forest trees - Physiology and practice. Institute of Forest Improvement and Department of Forest Genetics, Swedish College of Forestry, Uppsala. p. 9-16.
Luthra, J.K. & M.V. Rao, 1979. Multiline cultivars - How their resistance influences leaf-rust diseases in wheat. Euphytica 28: 137-144.
Marshall, D.R., 1977. The advantages and hazards of genetic homogeneity. Annals of the New York Academy of Sciences 287: 1-20.
Marshall, D.R. & A.J. Pryor, 1978. Multiline varieties and disease control.I. The 'dirty crop' approach with each component carrying a unique single resistance gene. Theoretical and Applied Genetics 51: 177-184.
Marshall, D.R. & A.J. Pryor, 1979. Multiline varieties and disease control II. The 'dirty crop' approach with components carrying two or more genes for resistance. Euphytica 28: 145-159.
Rauter, R.M., 1979. Spruce cutting propagation in Canada. In: J. Kleinschmit (Ed.): IUFRO Joint Meeting of Working Parties on Norway Spruce Provenances and Norway Spruce Breeding. Lower Saxony Research Institute, Escherode. p. 158-167.
Roulund, H., 1977. Vegetative propagation of forest trees at the Arboretum in Hørsholm, Denmark. In: Eliasson et al. (Eds): Vegetative propagation of forest trees - Physiology and practice. Institute of Forest Improvement and Department of Forest Genetics, Swedish College of Forestry, Uppsala. p. 103-128.
Schreiner, E.J., 1971. Genetics of eastern cottonwood. USDA Forest Service Research Paper WO-11. 19 p.
Scott, P.R. & A. Bainbridge, 1978. Plant disease epidemiology. Blackwell Scientific Publications, Oxford.
Singer, M., 1979. The third forest. New West 17 (Dec.): 43-48.
Werner, M., 1977. Vegetative propagation by cuttings of Picea abies in Sweden. In: Eliasson et al. (Eds): Vegetative propagation of forest trees - Physiology and practice. Institute of Forest Improvement and Department of Forest Genetics, Swedish College of Forestry, Uppsala. p. 97-102.
Wilcox, M.D., I.J. Thulin & T.G. Vincent, 1976. Selection of Pinus radiata clones in New Zealand for planting from cuttings. New Zealand Journal of Forestry 21: 239-247.
Wolfe, M.S., 1978. Some practical implications of the use of cereal variety mixtures. In: P.R. Scott & A. Bainbridge (Eds): Plant disease epidemiology. Blackwell Scientific Publications, Oxford. p. 201-207.
Wolfe, M.S. & J.A. Barrett, 1977. Population genetics of powdery mildew epidemics. Annals of the New York Academy of Sciences 287: 151-163.
Wood, T.K., 1980. Divergence in the Enchenopa binotata complex effected by host plant adaptation. Evolution 34: 147-160.

Balance in indigenous plant populations

A. Segal[1], J. Manisterski[1], J.A. Browning[2], G. Fischbeck[3] & I. Wahl[1]

ABSTRACT

Studies in various parts of the world have shown that in undisturbed natural ecosystems, diseases of indigenous plant populations caused by indigenous parasites are seldom destructive. The prolonged host-parasite coevolution associated with reciprocal selection pressure culminates in a state of balanced polymorphism. The 'protection of indigenousness' is based on genetic diversity of host and parasites adapted to a wide range of biotic and abiotic parameters in the ecosystem.

Israel is located in the center of origin and genetic diversification of *Hordeum spontaneum* and *Avena sterilis*, the putative ancestors of their cultivated counterparts. *H. spontaneum* is annually attacked by *Erysiphe graminis hordei* and *Puccinia hordei*. *A. sterilis* is affected every year by *P. coronata avenae* and *P. graminis avenae*. Populations of the four fungi represent broad spectra of virulence in which strains combining numerous virugenes prevail.

In the natural defense structures, the following mechanisms participate: (i) conventional resistance accompanied by hypersensitivity, (ii) slow rusting or slow mildewing, (iii) tolerance, and/or (iv) escape. Their proportions in the population vary with the host, parasite, and environment. For example, the first mechanism is not uncommon in profiles of defense operating against crown rust and powdery mildew, but it is virtually absent on *Ornithogalum* plants stricken by the gametophytic stage of *P. hordei*.

The diverse genetic patchwork of natural plant populations minimizes the danger of preferential selection pressure on the

1. Faculty of Life Sciences, Tel Aviv University, Tel Aviv, Israel.
2. Department of Plant Pathology, Iowa State University, Ames, Iowa 50011, USA.
3. Lehrstuhl für Pflanzenbau und Pflanzenzüchtung, Technische Universität München, 8050 Freising-Weihenstephan, BRD.

parasite populations and enables the establishment of a wide gamut of fungus strains. Urediospore production in the plant populations is markedly reduced and the disease is thus mitigated.

The relevance of knowledge about natural ecosystems to genetic disease control in agroecosystems is indicated.

INTRODUCTION

Natural, undisturbed ecosystems are characterized by biological balance. Plant populations, beneficial and harmful microorganisms, and abiotic environment are enmeshed in a complex system maintaining dynamic equilibrium regulated by homeostatic mechanisms.

'Plant disease occurs wherever plants grow' (Zadoks & Schein, 1979). Plant diseases in indigenous plant communities are a natural phenomenon, and do not indicate 'that some aspect of the biological balance is not in equilibrium' (Baker & Cook, 1974). They are considered by some researchers as subsystems in the ecosystem (Zadoks & Schein, 1979). We adopted this concept in this study. The disease seldom attains damaging proportions. This is particularly true of centers of host-parasite coevolution where both constituents of the subsystem reach a state of balanced polymorphism as a result of a long-lasting reciprocal selection pressure. Such an equilibrium is labeled by Hoff & McDonald (1972) 'balanced symbiosis'.

Mechanisms regulating diseases in natural ecosystems have attracted considerable attention. According to Person & Sidhu (1971), information for effective managing of 'man-guided' systems of parasitism 'can come only through studies of naturally occurring systems'.

STUDIES OUTSIDE ISRAEL

Zhukovsky (1959) analyzed the equilibria between wild relatives of cultivated crops and their parasites in the common homeland. As a rule, the hosts are not immune. Instead, they are endowed with 'field resistance' or tolerance. The fungus parasite attacks only some plant parts, eliciting limited necrosis and reduced sporulation. This has enabled host survival over millenia along with the parasite. The coevolution has given rise to new and more effective resistance diversity matched by new races of the parasite, referred to also as 'aggressive' races. Other examples of homeostasis in natural ecosystems outside Israel were discussed by Segal et al. (1980).

Balanced host-parasite interaction has been ascertained in natural forests. According to Dinus (1974), such forests, 'though not immune to catastrophe', are buffered by their 'ordered diversity against outbreaks of pests and other destructive agents'. Bingham et al. (1971), dealing with disease resistance in forests, emphasized that 'in the balanced systems of gene centers, rampaging epidemics and immunity from disease are absent and both host and parasite maintain their reproductive capacity'. Consequently, studies of the stabilized gene pools as models are recommended. A detailed analysis of mechanisms mitigating diseases in natural forests was presented by Schmidt (1978). He stressed that epidemics caused by indigenous pathogens are not rare in these ecosystems. 'A failure to distinguish between disease incidence and disease loss has contributed to the myth that epidemics are rare in natural forests'. Nevertheless, epidemics caused by indigenous pathogens are usually limited in time and space. They are mollified by mechanisms collectively named 'ecosystem disease resistance', or 'functional diversity' of the host, pathogen, climatic, edaphic, and biotic environmental parameters.

STUDIES IN ISRAEL

Our investigations were mainly concerned with homeostasis in subsystems involving the wild barley *Hordeum spontaneum* C. Koch and wild hexaploid oats *Avena sterilis* L. and their respective fungus parasites. Both species are indigenous to Israel, which is a part of their centers of origin and genetic diversification. Both species are distributed countrywide and consist of numerous ecotypes. *H. spontaneum* is annually attacked by *Erysiphe graminis* (DC.) Merat *hordei* Em. Marchal and *Puccinia hordei* Otth. *P. coronata* Cda. *avenae* Frazer & Ledingham and *P. graminis* Pers. *avenae* Eriks. & E. Henn. are incident on *A. sterilis* every year. The 4 parasites have coevolved with their hosts from remote antiquity. *P. hordei* and *P. coronata avenae* complete their life cycles on the respective indigenous hosts, *Ornithogalum* spp. and *Rhamnus* spp. Hence, the host-parasite subsystems implicating the 2 fungi are complex and comprise the following biological components. (i) *H. spontaneum* - *P. hordei* - *Ornithogalum* spp. (ii) *A. sterilis* - *P. coronata avenae* - *Rhamnus* spp. Both subsystems are well balanced.

Most of the parasites have broad spectra of alternative hosts belonging to a number of species or even genera.

The broad gamut of variation in virulence is attributed to genetic recombination that takes place in the common and functional cleistothecia of *E. graminis hordei*, and on the alternate hosts of the crown rust and barley leaf rust fungi. *P. graminis avenae* has no alternate host in Israel. Its variation is ascribed to somatic hybridization on the wide range of receptive grasses (Anikster & Wahl, 1979). Populations of all organisms are characterized by the prevalence of strains combining a wide array of virugenes.

The investigated plants were sampled with the transect method at 1-2 m intervals, regardless of their disease reaction in nature. Their seed was planted in test nurseries in the order of disposition of the parents in the original setting. The nurseries were periodically inoculated with cultures of specific parasites collected across the country.

Four components regulating the protection structures were identified, namely (i) conventional resistance associated with hypersensitivity and a low type of infection, (ii) slow rusting or slow mildewing, (iii) escape - avoidance, and/or (iv) tolerance. The first 2 factors will be discussed in greater detail; our information on escape and tolerance is rather limited. For the sake of clarity, the 4 diseases are classified into 2 groups based on the prevalence of hypersensitivity in defense structures. Hypersensitivity is relatively common in the powdery mildew and leaf rust of barley as well as in oat crown rust, but rare in oat stem rust and practically lacking in barley leaf rust on the alternate *Ornithogalum* hosts.

The host-parasite subsystems of the first group are represented in our discussion by oat crown rust. Other associations were analyzed by Segal et al. (1980).

Subsystem: Avena sterilis - Puccinia coronata avenae - Rhamnus spp.

Murphy et al. (1967) explained the evolution of homeostasis in this subsystem with this statement: 'A natural balance ... appears to have been established between *A. sterilis*, crown rust, and *R. palaestina*, in which *A. sterilis*, although infected, produces seed of good quality. Various levels of resistance and tolerance to crown rust have apparently resulted from natural selection under conditions of regular and heavy crown rust infection and a relatively high level of outcrossing in *A. sterilis*'. Notably, the fungus inhabits in nature grasses of several genera (Dinoor, 1967).

The defense structures in natural communities of *A. sterilis* have embraced the 4 aforementioned mechanisms.

Conventional resistance was present in all investigated populations. Its frequency varied with the region of origin, time of the season, and age of the plant. It operated throughout the whole life of some accessions, while in others a low type of infection was limited to some growth stages. Conceivably, restricted duration of hypersensitive reactions offers sufficient protection to the host and receptive substrate to the parasite, facilitating their balanced coexistence.

Slow crown rusting is characterized by markedly reduced disease progress, low infectability of the host, and diminished sporulation. Tolerance occurs in *A. sterilis* in Israel and the USA (Segal et al., 1980).

In climatically adverse regions, *A. sterilis* is only slightly affected or escapes the disease for a longer period by early ripening. In the partially escaping hosts, the duration of development of very susceptible growth phases is relatively short.

The more common defense structures in the *A. sterilis* communities are distinguished by prominence of plants with infection type 3 and low to moderate infection severity, while in about 30 % of the plants infection type 2 is associated with infection severity of 5-25 %. All investigated populations encompassed fast crown rusters. Conceivably, some of the fast rusters were protected by tolerance. There was no visible indication that protection has developed at a cost of fitness in the host.

The effect of the defense structures in *A. sterilis* populations on the race composition of *P. coronata avenae* was compared with that of Iowa Multiline I-78, and its recurrent *A. sativa* L. host represented by a mixture of cultivars C649 and Clintford. In that multiline, resistance is imparted by genes that had been extracted from *A. sterilis* of Israeli origin. At the locations Hasolelim and En Dor, the multiline and its recurrent host were grown in separate plots in proximity to *R. palaestina*. Uredia were collected from all 3 oat populations and their racial identity determined in the greenhouse with the aid of standard and supplemental differential cultivars at the seedling stage. The results (Table 1) demonstrate that race group 276-264, which predominates in Israel, also prevailed on Multiline I-78, its recurrent hosts, and *A. sterilis*. However, the heterogenic genetic patchwork of *A. sterilis* expedited the establishment of almost twice as many races as in the other oat plots. Furthermore, the *A. sterilis* genes incorporated in the wild oat populations and in I-78 stands seemed to dilute the concentration of the dangerous race group 276 & 264. This race group has constantly prevailed countrywide for many years. It combines a broad spectrum of virugenes and renders ineffective most of the known hexaploid sources of resistance to *P. coronata avenae*. Hence, it comes close to the conceptual super-race. The lush stands of *A. sterilis*

Table 1. Frequency of occurrence of *Puccinia coronata avenae* races in field nurseries at HaSolelim and En Dor on plants of *Avena sterilis*, *A. sativa* mixed check: (C 649 + Clintford) and Iowa Multiline I-78. Urediospore isolates were randomly sampled at HaSolelim in 1978 and 1979, and at En Dor in 1978.

Host	HaSolelim			En Dor		
	number of tested isolates	number of races	number of isolates, race group 276 + 264	number of tested isolates	number of races	number of isolates, race group 276 + 264
A. sterilis	235	37	62	79	19	22
Mixed check	224	21	153	53	11	37
Multiline I-78	200	20	65	30	10	10

permanently exposed to race group 276-264 suggest that the defense mechanisms operating in the natural ecosystem effectively protect *A. sterilis* against parasite strains that constitute a near super-race as far as hexaploid oats are concerned. This situation appears to dispel the lingering fears of the potential hazards that may arise with the evolution of super-races.

The intensity of crown rust infection was assessed by determining the amount of urediospores collected on rotorod spore traps 9 times during the season in plots sown to the following oat accessions: (i) the 'standard' susceptible cultivar Markton, (ii) Iowa Multiline I-78, (iii) Iowa Multiline M-73, (iv) Iowa Multiline E-77, (v) a mixture of cultivars C-649 and Clintford, recurrent parents and agronomic checks to both Multilines I-78 and M-73, (vi) cultivar C-237-89-IV, which served as the agronomic check for Multiline E-77, and (vii) 5 *A. sterilis* populations sown separately. Each population consisted of components whose parents were sampled with the transect method at one of the locations: Guara, Bet Nehemia, Ginnaton, Hagor, Rishon LeZion. Annual development of crown rust on *A. sterilis* at Rishon LeZion is more intensive than in the other 4 locations, reflecting a more favourable environment and/or a more susceptible sub-population of the host. The results (Fig. 1) show that urediospore production in all Iowa multilines and in the *A. sterilis* plots is distinctly lower than on Markton and all recurrent hosts. Thus, results with multilines in the USA (Browning, 1974) and in Israel are corroborated by results from indigenous populations in Israel (Browning et al., page 371 in this book). The *A. sterilis* populations constitute to some extent 'natural multilines'. Such studies enable quantitative estimation of defense structures operating in natural ecosystems.

Subsystem: Avena sterilis - Puccinia graminis avenae

The disease development season is shorter for the stem rust than for the crown rust, and very brief in arid regions where the host virtually escapes its impact. Conventional resistance, which protects the plant throughout the whole life of the plant, is very scarce. It occasionally operates at some growth stages. Numerous plants harbored uredia of susceptible and resistant classes on the same stems, sheaths, or blades. This pattern of reactions is characteristic of 'regional resistance' (Segal et al., 1980). Browning (1974) considered 'regional resistance' as a means of keeping stem rust in balance. As mentioned above, Zhukovsky (1959) found this phenomenon to be common in centers of host-parasite coevolution. Defense of the slow-rusting type against stem rust is widespread in *A. sterilis*. It was discovered by Dr. H.C. Murphy. This type of protection is manifested in diminished infectability of the host, retarded progress of the disease, and reduced mycelium growth in the host, while uredia signify a susceptible reaction. Slow rusting has proven to be stable and effective

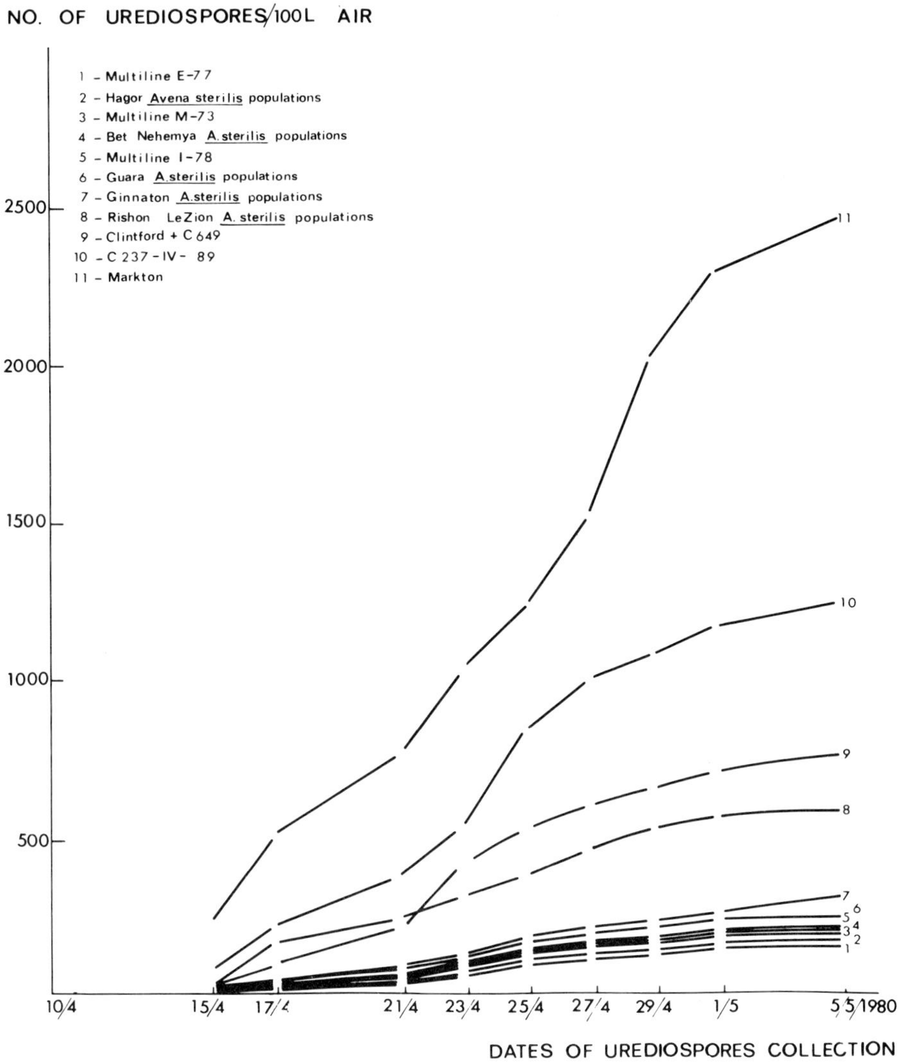

Figure 1. Cumulative urediospore production of crown rust on tested accessions.

against many races.

Defense structures against stem rust in populations of *A. sterilis* have been described by Segal et al. (1980).

The relatively brief annual association of *A. sterilis* with the parasite during the season seems to ward off serious epidemics. The host-parasite-environment interaction in Israel is less well balanced than in the case of crown rust.

Races of *P. graminis avenae* with broad spectra of virulence have been consistently prevalent over many years. Studies by Gerechter-Amitai (1973) demonstrated that such races parasitize plants of 71 species belonging to 36 genera of Gramineae.

Subsystem: Hordeum spontaneum - Puccinia hordei - Ornithogalum spp.

Puccinia hordei is widespread on *H. spontaneum*. The fungus completes its life cycle in nature on indigenous *Ornithogalum* species. The alternate host is indispensable for the perpetuation of the parasite. The precocity of teliospore formation restricts to some extent the propagation of *P. hordei* during the growing season, but enhances its association with *Ornithogalum* plants, since the spores germinate readily. The balance of the fungus with the main and alternate hosts and the excellent coordination in the development of the components in the *H. spontaneum - P. hordei - Ornithogalum* sub-system is a result of coevolution well adapted to the semi-arid environment. The restricted range of alternative hosts supports the hypothesis (Anikster & Wahl, 1979) that the fungus is of a relatively recent origin and short phylogenetic history. There is an important difference in the mechanisms balancing the fungus with the main and alternate hosts. Anikster (unpublished) has shown that *H. spontaneum* is protected against *P. hordei* by resistance accompanied by hypersensitive reactions and slow rusting. In contrast, hypersensitivity is extremely rare on *Ornithogalum* plants harboring pycnia and aecia of the parasite. Presumably, other types of defense operate in the alternate host.

In artificial inoculation greenhouse tests, *P. hordei* alternates with *Dipcadi erythraeum* Webb & Bert., and *Leopoldia eburnea* Eig & Feinbr. (Anikster, unpublished). However, plants of neither species rust in the natural arid habitats. They seem to escape the disease which is unable to develop in dry regions.

CONCLUDING REMARKS

In natural ecosystems, diseases of undisturbed indigenous plant populations caused by indigenous parasites are seldom destructive. Both constituents of the host-parasite couplet have undergone long-lasting coevolution that culminates in a state of balanced dynamic polymorphism. They represent a subsystem that possesses attributes of its own, but that maintains at the same time interdependent relations with other biotic and abiotic components of the whole ecosystem (Zadoks & Schein, 1979). The equilibrium and coexistence in the subsystem are regulated by homeostatic mechanisms inherent in the nature of the host, parasite, and the environment. The stabilizing factors that dampen excessive perturbations are collectively manifested as 'ecosystem disease resistance' (Schmidt, 1978) or 'protection of indigenousness' (Browning, 1974). They operate by engendering 'functional diversity' (Schmidt, 1978) or 'patterned diversity' (Dinus, 1974). Some of these components were listed by Browning (1974) and Schmidt (1978, Table 3).

Israel is located in the center of origin and genetic diversification of *Avena sterilis* and *Hordeum spontaneum*, the putative progenitors of their cultivated counterparts. They are annually attacked by indigenous obligate parasites, but the ensuing diseases never become very damaging. Most of the causal agents inhabit a wide range of alternative hosts. *Puccinia hordei* and *P. coronata avenae* complete their life cycles on indigenous alternate hosts.

Studies on natural defense structures were conducted by inoculating fungus cultures collected countrywide to plant populations reconstructed in test plots. The reconstruction was attained by growing the entries according to disposition of their parents in natural ecosystems. These entries were sampled with the transect technique. This method synthesizes a multiplicity of protection patterns and sorts out the components and processes acting in natural habitats.

Four components of defense were identified: (i) conventional resistance, (ii) slow rusting or slow mildewing, (iii) tolerance, and (iv) escape. The profiles of their interaction at a given site are stable, and depend on the genetics of the host and parasite, their age, ontogenetic stage of the rust fungus, and environment of the original ecosystem as well as of the site of the test. The fundamentally different reaction of *H. spontaneum* to the sporophytic stage of *P. hordei* as compared with that of *Ornithogalum* plants to the gametophytic stage of the fungus reveals important aspects of the host-parasite balance. Alternative hosts enhance the parasitic fitness of the fungi by augmenting their survivability and reproductivity. At the same time they increase the buffering capacity of plant populations by maximizing their genetic heterogeneity.

Studies of native plant populations with the transect method provide a better understanding of the individual constituents of the defense structure and the patterns of their integration; quantitation of spore yields and determination of race composition in the parasite illuminate the protective action of the balanced ecosystem as a complex entity. To paraphrase the statement of Baker & Cook (1974): while man may never understand all the interactions in the natural ecosystem, the end result - biological balance - can, however, be appraised and exploited. Significantly, defense afforded by this balance seems to minimize the damaging effect of near superraces such as the race group 276 & 264 of *P. coronata avenae*.

Knowledge about natural ecosystems offers valuable implications for management of disease in agroecosystems (Browning, 1974) and for understanding of plant epidemiology (Apple, 1977). 'The determination of the contrasting conditions under which the same pathogen may be either destructive or benign will be a major task of plant pathologists in the future' (Yarwood, 1967).

ACKNOWLEDGEMENTS

Our studies were supported by research grants of the United States - Israel Binational Science Foundation (BSF), Jerusalem Israel; the Deutsche Forschungsgemeinschaft (DFG), Federal Republic of Germany; and the German - Israel Fund for International Research and Development (GIFRID).

REFERENCES

Anikster, Y. & I. Wahl, 1979. Coevolution of the rust fungi on Gramineae and Liliaceae and their hosts. Annual Review of Phytopathology 17: 367-403.

Apple, J.L., 1977. The theory of disease management. In: J.G. Horsfall & E.B. Cowling (Eds): Plant disease 1. Academic Press, New York/London. p. 79-101.

Baker, K.F. & R.J. Cook, 1974. Biological control of plant pathogens. W.H. Freeman and Company, San Francisco. 433 p.

Bingham, R.T., R.J. Hoff & G.I. McDonald, 1971. Disease resistance in forest trees. Annual Review of Phytopathology 9: 433-452.

Browning, J.A., 1974. Relevance of knowledge about natural ecosystems to development of pest management programs for agro-ecosystems. Proceedings of the American Phytopathological Society, St. Paul, Minnesota 1: 191-199.

Dinoor, A., 1967. The role of cultivated and wild plants in the life cycle of Puccinia coronata Cda. var. avenae F. & L. and in the disease cycle of oat crown rust in Israel. PhD thesis. Hebrew University, Jerusalem. 373 p. (Hebrew, with English summary).

Dinus, R.J., 1974. Knowledge about natural ecosystems as a guide to disease control in managed forests. Proceedings of the American Phytopathological Society, St. Paul, Minnesota 1: 184-190.

Gerechter-Amitai, Z.K., 1973. Stem rust, Puccinia graminis Pers. on cultivated and wild grasses in Israel. PhD thesis. Hebrew University, Jerusalem. 420 p. (Hebrew, with English summary).

Hoff, R.J. & G.I. McDonald, 1972. Stem rusts of conifers and the balance of nature. In: R.T. Bingham, R.J. Hoff & G.I. McDonald (Eds): Biology of rust resistance in forest trees: Proceedings of a NATO-IUFRO Advanced Study Institute. Miscellaneous Publications no. 1221. U.S. Department of Agriculture, Washington, D.C., Forest Service, 1972. p. 525-536.

Murphy, H.C., I. Wahl, A. Dinoor, J.D. Miller, D.D. Morey, H.H. Luke, D. Sechler & L. Reyes, 1967. Resistance to crown rust and soilborne mosaic virus in Avena sterilis. Plant Disease Reporter 51: 120-124.

Person, C. & G. Sidhu, 1971. Genetics of host-parasite interrelationships. In: Mutation breeding for disease resistance. International Atomic Energy Agency, Vienna, 1971. p. 31-38.

Schmidt, R.A., 1978. Diseases in forest ecosystems: The importance of functional diversity. In: J.G. Horsfall & E.B. Cowling (Eds): Plant disease 2. Academic Press, New York/London. p. 287-315.

Segal, A., J. Manisterski, G. Fischbeck & I. Wahl, 1980. How plant populations defend themselves in natural ecosystems. In: J.G. Horsfall & E.B. Cowling (Eds): Plant disease 5. Academic Press, New York/London. p. 75-102.

Yarwood, C.E., 1967. Response to parasites. Annual Review of Plant Physiology 18: 419-438.

Zadoks, J.C. & R.D. Schein, 1979. Epidemiology and plant disease management. Oxford University Press. New York, Oxford. 427 p.

Zhukovsky, P.M., 1959. Vzaimootnoshennia mezdu khozainom i gribnym parazitom na ikh rodine i vne ee. (Interrelations between host and fungus parasite in their origin and beyond it). Vestnik Sel' skokhoziaïstvennoi Nauki. Moskow. 4 (6): 25-34 (Russian, with English summary).

Extrapolation of genetic and epidemiologic concepts from indigenous ecosystems to agroecosystems

J.A. Browning[1], J. Manistersky[2], A. Segal[2], G. Fischbeck[3] & I. Wahl[2]

ABSTRACT

In geographical areas where both host and pathogen are indigenous, and with time and space for genetic maneuvering, diverse host and pathogen populations have coevolved to a dynamically balanced polymorphism. The end result of this coevolution is carried as genetic information in each organism. Learning and applying these results to agroecosystems can help build superior cultivars, population structures, and cultural systems. Individual plants may be protected by the host's basic system of resistance to the pathogen's basic system of pathogenicity. The host population (not individual plants) may be protected by the incompatibility-compatibility system. Both systems function epidemiologically as components of ecosystem-buffering by slowing the rate of pathogen development and dispersal. Incompatibility genes have been found to be highly effective when deployed in diverse, indigenous populations. Such genes have also been dramatically effective in agroecosystems when similarly deployed against highly epidemic pathogens.

INTRODUCTION

Man learns very reluctantly, if at all, from history. In the area of crop protection, for example, following the United States' 1970 southern corn leaf blight pandemic, Ullstrup (1972) wrote that 'the first and most important lesson to be learned is that never again should a major cultivated species be molded into such uniformity that it is so universally vulnerable to attack by a pathogen, an insect, or environmental stress. *Diversity must be maintained in both the genetic and cytoplasmic constitution of*

1. Department of Plant Pathology, Iowa State University, Ames, Iowa 50011, USA.
2. Faculty of Life Sciences, Tel Aviv University, Tel Aviv, Israel.
3. Lehrstuhl für Pflanzenbau und Pflanzenzüchtung, Technische Universität, München, 8050 Freising-Weihenstephan, BRD.

all important crop species'. (Ullstrup's emphasis.) Yet in the United States today, corn and several other crops are more homogeneous (Day, 1977) than even in 1970 when homogeneity was recognized as 'the crux of genetic vulnerability' (NAS, 1972) to disease. Never mind that the 1970 pandemic could have been avoided had man learned from previous, equally poignant, lessons (Browning, 1972).

The basic lesson that probably would have prevented all of the major plant disease pandemics is one that could have been learned from wild ecosystems. That lesson is that coevolution of hosts and pathogens in the region where both are indigenous has resulted in a state of dynamic balance, and that both groups of organisms obviously carry genetic information that conditions the behavior that results in the maintenance of the dynamic balance characteristic of indigenous ecosystems. For example, the wild oat, *Avena sterilis* L., in Israel 'knows' to maintain a certain variety of different reaction types in the population in a given environment; susceptible genotypes are not eliminated from the population and resistant genotypes do not dominate it. Similarly, the oat crown rust fungus, *Puccinia coronata* Corda, 'knows' to sporulate only at a certain rate on wild *A. sterilis* to maintain itself in the population and neither dominate (and risk destroying its host) nor risk being eliminated - an evolutionary tight wire for an obligate parasite. When the pathogen finds itself removed to a susceptible, homogeneous cultivated population, the same rate of sporulation that maintains the fungus in the wild population could destroy both the host and its obligate parasite, with favorable weather conditions, if the cultivated host were not maintained artificially by man. Man, who manipulates pathogens via their hosts (Johnson, 1961), must ferret out how host and pathogen effect dynamic balance in the wild population. What resistance mechanisms or population structures are the products of this coevolution and are responsible for the dynamic balance observed? Can such knowledge help in building superior genotypes, population structures, and management practices so as to stabilize agro- and forestry ecosystems and prevent future epidemics?

The types of studies reported by Segal et al. (page 361 in this book) give important clues. The Israeli populations of wild progenitors of small grains and their pathogens are especially instructive because they are the same biologic species (Harlan & Zohary, 1966) and are attacked by the same array of pathogens as their respective counterparts in agriculture. Hence, anything learned from them can be interpreted immediately in terms of the accumulated literature and applied without delay to building superior disease-management systems for small grains and other crops (Browning, 1974). In this paper, we will study the coevolution of hosts and pathogens and extrapolate the principles learned to agroecosystems, first in theory and then in practice.

Host range - A source of theory gone awry

Most contemporary pest-management practices are based on theory developed from studies of agricultural systems. The concepts of 'formae speciales' and 'race' reflect specialization on crop species and cultivars, respectively, of interest to man because of his agriculture. Ever since formae speciales and race concepts were formulated, agricultural scientists have come to expect great specificity and narrow host ranges in the obligately parasitic rust and many other fungi. For example, *P. graminis tritici* commonly is considered to attack primarily wheat, and *P. graminis avenae* primarily oats, while race 1 might be virulent on cultivar A but not B, and race 2 the opposite. Flor's (1956) gene-for-gene hypothesis placed specificity on a solid genetic footing and, expressed in pathogenicity formulae, eventually enabled isolates of the pathogen population to be described more completely (Browder et al., 1980). All of this concentrated on agricultural systems and paid minimal attention to host-parasite interactions outside a farmer's field.

But agriculture is, first and foremost, biology. Biological theory of how host and pathogen coevolved and of how to prevent epidemics (genetic imbalance in favor of the pathogen) should be obtainable from indigenous populations where all components have genetic maneuvering room, interact without disease usually being very severe, and yet still result in dynamic balance. Comparing 'host range' in an indigenous population with that considered normal to an agroecosystem is an instructive example.

P. graminis avenae race 2 is, by definition, virulent on only one standard oat stem rust differential cultivar, Jostrain, which carries gene *Pg*-3 (Stewart & Roberts, 1970). In a wild ecosystem in Israel, however, Gerechter-Amitai (1973) showed that *P. graminis avenae*, i.e. strains of *P. graminis* that can parasitize oats, had a very wide host range, being virulent on 66 species in 34 genera in the field and 82 species in 41 genera in the greenhouse. One single-pustule isolate (a genetic clone) of *P. graminis avenae* race 2 even sporulated on 80 species of wild grasses in the greenhouse, 69 of which were rated susceptible! Gerechter-Amitai (1973) obtained similar data for *P. graminis tritici* and *P. striiformis* Westend., as did Eshed & Wahl (1970) for *Erysiphe graminis* DC. and Eshed & Dinoor (1980) for *P. coronata* and their respective hosts in Israel.

Theory must be developed to explain such a vast discrepancy between an indigenous ecosystem and what man has experienced, and come to expect, in an agroecosystem. The discrepancy suggests that the difference between a 'host' and a 'non-host', or a 'pathogen' and a 'non-pathogen', may be one of definition. If man can cross members of a wild species with a cultivated host species that includes some susceptible cultivars, both share a common gene pool and are defined as hosts; individual plants are said to carry genes for 'host immunity/resistance'. If man cannot cross them, then the immune/resistant species is a non-host and any *R* genes for immunity/resistance are not available to man with present technology. Thus, genes that condition mating compatibility may be more important to our agricentric concept of host and non-host than genes for immunity/resistance or susceptibility as the terms imply.

A proposed model of host-parasite coevolution

Reasoning that host and non-host immunity/resistance could be conditioned by the same genes, Browning (1980) developed a model (Fig. 1) of host-parasite coevolution and interaction. Starting from non-host immunity (Fig. 1, top) and non-pathogen avirulence (bottom) that he reasoned characterized progenitor species (just as immunity/resistance and avirulence characterize the vast majority of higher plants and pathogens, respectively, today; susceptibility and virulence remain the rare exceptions), the model leads to the basic system of host-pathogen incompatibility (left) and

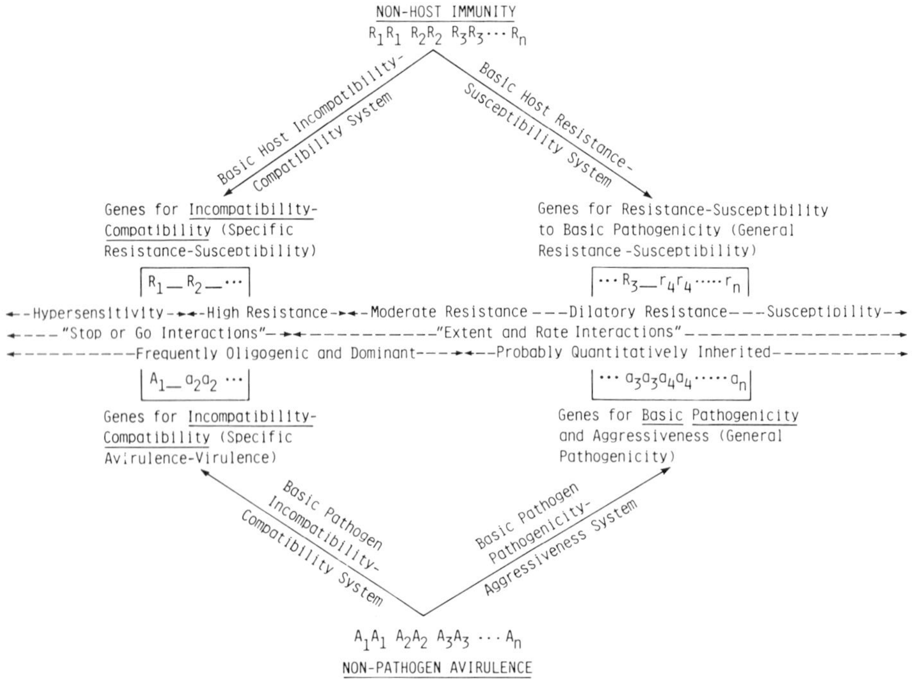

Figure 1. Proposed scheme of host-pathogen coevolution from non-host immunity and non-pathogen avirulence to the 2 basic systems of host-pathogen interaction; (left) the basic host-pathogen incompatibility-compatibility system; and (right) the basic pathogen pathogenicity system and its counterpart, the basic system of host resistance/susceptibility to basic pathogenicity. (From Browning, 1980.)

(right), the basic system of pathogen pathogenicity and of its counterpart, host resistance or susceptibility to basic pathogenicity. Thus, Fig. 1 has a double dichotomy: 2 organisms that become genetically interlocked, and 2 basic systems that govern their interaction. In Fig. 1, 'non-host immunity' and 'non-pathogen avirulence' describe coexisting populations of progenitor organisms before the onset of parasitism and pathogenicity. Following the onset of parasitism and with the continuation of coevolution, the 2 eventually come together in the horizontal center lines that describe contemporary agricultural systems. Doubtless the left-right dichotomy is the same one that Vanderplank (1968) sensed when he described vertical and horizontal resistance in contemporary agricultural systems.

Essentially, the host-pathogen incompatibility-compatibility system determines whether the pathogen can 'unlock the door' of the host. By signaling 'stop or go' interactions (Person & Mayo, 1977) through a series of 'switching points' (Heath, 1974), the incompatibility-compatibility system either allows the pathogen to 'unlock the door' and proceed through the incompatibility barrier to the next system or signals the end of the line. If the signal is 'stop', incompatibility may express itself in hypersensitivity. If 'go', the second basic system takes over; it determines the amount that the 'door' is opened and the rate and ultimate extent that the 'room' is occupied. In it the pathogen expresses its basic pathogenicity/aggressiveness and the host responds with its basic resistance/susceptibility to basic pathogenicity.

The first system frequently is controlled oligogenically; the second frequently is polygenic in inheritance. The first system frequently is under gene-for-gene control; the second system may be. As greater resistance is epistatic to lesser resistance (Flor, 1956), so the first system, which usually conditions the greater resistance and controls the incompatibility barrier to the second system, doubtless can be considered epistatic to the second.

The theoretical model (Fig. 1) suggests that similar *R* and *A* genes, with similar modes of action, are involved in host-pathogen and non-host-non-pathogen interactions. And they could be the same genes! Probably host *R* genes are carry-over genes indistinguishable in nature and agriculture from non-host *R* genes. The model was suggested from Gerechter-Amitai's (1973) host range study; however, it is supported by ultrastructure studies (Heath, 1974, 1977), gene-action models (Day, 1974; Callow, 1977), and recognition-non-recognition theory (Callow, 1977; Sequeira, 1978). An attractive feature of the incompatibility-compatibility system is that it seems similar to the universal biological phenomenon of recognition-non-recognition by which like recognizes like and rejects unlike (Callow, 1977).

Epidemiologic role of the incompatibility genes

In agroecosystems, single incompatibility genes frequently have been used in an attempt to protect field crops from highly epidemic pathogens, sometimes over millions of hectares. Use of incompatibility genes to protect the plant, or over large areas, is foreign to nature. In natural populations, incompatibility genes do not serve to protect the plant, but to protect the population by keeping the pathogen from becoming too aggressive (Parlevliet, 1980). They protect the population by reducing the dispersal efficiency of the pathogen among the genetically diverse host plants. Epidemiologically, this is measured the same as reduced aggressiveness in the pathogen. Also, they provide a finely tuned genetic feed-back mechanism to help maintain balance. It is the role of the second basic system - the basic host resistance-susceptibility system - to protect the individual plant. The second basic system accomplishes this by slowing the rate of development of the pathogen in the plant and, therefore, contributes also to slowing its development in the population. We will illustrate these points in the Section called 'Frequency of incompatibility genes needed to protect cultivars in agroecosystems'.

EXTRAPOLATIONS IN PRACTICE

Frequency of incompatibility genes in an indigenous population

Transect studies in Israel have shown that nature uses many different types of host reaction seemingly at random in dynamically balanced populations. These were presented and discussed by Segal et al. (1980). Here we will consider primarily the frequency and spatial arrangement of plants carrying incompatibility genes for it is from the supposed 'failure' of incompatibility genes due to man's mismanagement that most resistance problems have arisen, and it is wise management of incompatibility genes that can enable man to most quickly apply knowledge gained from studies of indigenous ecosystems.

In certain populations of wild barley, *Hordeum spontaneum* K. Koch, attacked by the powdery mildew fungus, *Erysiphe graminis*, the second basic system contributed very little protection; the protection that these populations enjoyed was due largely to the presence of genes for incompatibility. When seed was collected at 1-meter intervals along a transect without regard to the disease reaction of the mother plant, as little as 1/3 of the plants had resistance characteristics of incompatibility in subsequent tests with the pathogen. See, e.g., data for Transect 207 (Segal et al., 1980, Fig. 1) and Transect 217 (Wahl et al., 1978, Fig. 2). Similar data were reviewed for *P. coronata* on *A. sterilis* by Browning (1974) and Segal et al. (1980). Thus, it seems that in a given environment, an indigenous population can be, and is, adequately protected and stabilized chiefly by genes for incompatibility if ca. 1/3 of the plants carry effective genes for incompatibility.

Frequency of incompatibility genes needed to protect cultivars in agroecosystems

Can knowledge of the frequency of incompatibility genes that protect an indigenous population be extrapolated in practice to an agroecosystem? Yes. Since 1957, the Iowa Agriculture Experiment Station has developed multiline cultivars of oats for protection against crown rust (Frey et al., 1977). Thirteen multiline cultivars in 2 maturity classes, early and midseason, have been released. The Multiline E (for early) Series is protected by both resistance systems, and the two systems augment each other very nicely in this cultivar. But the Multiline M (for midseason) Series is especially interesting, for it is protected from crown rust only by incompatibility genes. These multilines have been tested against crown rust in extensive tests in Iowa, which has a short disease season, and on the Texas Coastal Plain, where the disease season is longer and more severe. In both locations, dramatic protection from the pathogen resulted when 1/3 to 1/2 of the plants carried resistance to one component or another of the pathogen population. The susceptible pure line recurrent parent was killed prematurely by crown rust in the same experiments (Browning, 1974).

The Iowa multilines were tested in Israel also, in an epidemiology experiment that included several transect-derived populations of *A. sterilis*. For these experiments, a special Iowa multiline was composited, Multiline I (for Israel) -78. Multiline I-78 is very similar to Multiline M-73 except that it was composited equally of 10 isolines each of which carried a different gene for incompatibility to crown rust from *A. sterilis* so that it would be completely relevant to the Israeli pathogen population. The results, shown in Fig. 2, also were described by Segal et al. (1981). Markton is a pure line universal suscept. C237-89IV, a sister to the recurrent parent in the early multiline, carries no gene for incompatibility but carries considerable protection from the second basic resistance system characteristic of the early multiline series. The 'Clintford+C649 mixed check' consisted of C649, recurrent parent to the M-series multilines, and Clintford, the very similar recurrent parent that was used in breeding some of the isolines used in Multiline I-78. This mixed check was protected by 2 incompatibility genes from donor cultivars Bond and Landhafer, but both of them seemed to be completely overcome by strains of the pathogen in the experiment. The Clintford + C649 check has no known genes that condition the second basic resistance system for protection from crown rust. This characterizes the background of Multiline M-73 and Multiline I-78 as well as that of the mixed check.

Thus, the disease progress curves for Multiline M-73 and Multiline I-78 in Fig. 2 show the results of rust buildup where the host populations were protected solely by genes for incompatibility deployed in diverse popula-

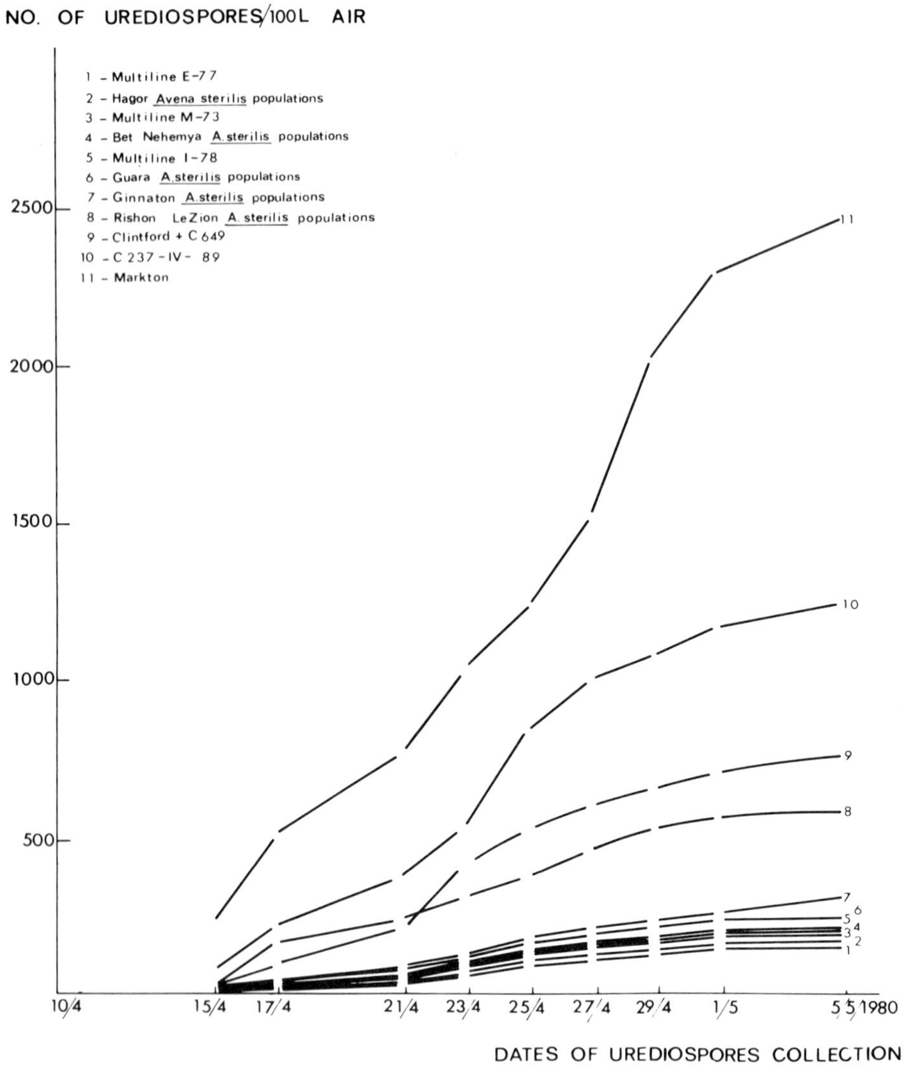

Figure 2. Disease progress curves for *Puccinia coronata* on *Avena sativa* pure line checks and multiline cultivars, and on 5 *Avena sterilis* transect-derived populations. Bet Dagan, Israel, 1980. (From Segal et al., page 361 in this book)

tions, much as in nature. Susceptible-type uredia were easy to find in the populations, but damaging increase of the pathogen was negligable, as in Iowa and South Texas. Curves for the other cultivars or populations show effects of both systems of genetic protection and of other components of ecosystem protection.

Curves (Fig. 2) for the Iowa multilines and for the 5 *A. sterilis* populations show a continuum, regardless of whether genetic protection was conditioned exclusively by the incompatibility-compatibility system (as in Multiline M-73 and Multiline I-78), the basic host resistance-susceptibility system, or both, as in most other populations. This illustrates that 'nature uses primarily a single type of epidemiologic resistance - dilatory resistance - to protect populations, but it uses many different genetic systems and population structures to achieve it' (Browning, 1980). (See also Browning (1980) Fig. 2 for another illustration of this principle.) Others, especially Wolfe & Barrett (1980), also have used incompatibility genes effectively to buffer against highly epidemic pathogens.

REFERENCES

Browder, L.E., F.L. Lyon & M.G. Eversmeyer, 1980. Races, pathogenicity phenotypes, and type cultures of plant pathogens. Phytopathology 70: 581-583.

Browning, J.A., 1972. Corn, wheat, rice, man: Endangered species. Journal of Environmental Quality 1: 209-211.

Browning, J.A., 1974. Relevance of knowledge about natural ecosystems to development of pest management programs for agro-ecosystems. Proceedings of the American Phytopathological Society 1: 191-199.

Browning, J.A., 1980. Genetic protective mechanisms of plant-pathogen populations: Their coevolution and use in breeding for resistance. In: M.K. Harris (Ed.): Biology and breeding for resistance to arthropods and pathogens in agricultural plants. Texas Agricultural Experiment Station, Miscellaneous Publication 1451. p. 52-75.

Callow, J.A., 1977. Recognition, resistance and the role of plant lectins in host-parasite interactions. Advances in Botanical Research 4: 1-49.

Day, P.R., 1974. Genetics of Host-Parasite Interaction. W.H. Freeman, San Francisco. 238 p.

Day, P.R., 1977. The genetic basis of epidemics in agriculture. Annals of the New York Academy of Sciences 287: 1-400.

Eshed, N. & A. Dinoor, 1980. Genetics of pathogenicity in Puccinia coronata: Pathogenic specialization at the host genus level. Phytopathology 70: 1042-1046.

Eshed, N. & I. Wahl, 1970. Host ranges and interrelations of Erysiphe graminis hordei, E. graminis tritici, and E. graminis avenae. Phytopathology 60: 628-634.

Flor, H.H., 1956. The complementary genic systems in flax and flax rust. Advances in Genetics 8: 29-54.

Frey, K.J., J.A. Browning & M.D. Simons, 1977. Management systems for host genes to control disease loss. Annals of the New York Academy of Sciences 287: 255-274.

Gerechter-Amitai, Z.K., 1973. Stem rust, Puccinia graminis Pers. on cultivated and wild grasses in Israel. PhD thesis. Hebrew University, Jerusalem. 420 p. (Hebrew, with English summary).

Harlan, J.R. & D. Zohary, 1966. Distribution of wild wheats and barley. Science 153: 1074-1080.

Heath, M.C., 1974. Light and electron microscope studies of the interactions of hosts and non-host plants with cowpea rust - Uromyces phaseoli var. vignae. Physiological Plant Pathology 4: 403-414.

Heath, M.C., 1977. A comparative study of non-host interactions with rust fungi. Physiological Plant Pathology 10: 73-88.

Johnson, T., 1961. Man-guided evolution in plant rusts. Science 133: 357-362.

National Academy of Sciences, 1972. Genetic vulnerability of major crops. Washington, D.C. 307 p.

Parlevliet, J.E., 1980. Disease resistance in plants and its consequences for plant breeding. In: K.J. Frey (Ed.): Plant breeding II. Iowa State University Press Ames. (in press).

Person, C. & G.M.E. Mayo, 1977. Genetic limitations on models of specific interactions between a host and its parasite. Canadian Journal of Botany 52: 1339-1347.

Segal, A., J. Manisterski, G. Fischbeck & I. Wahl, 1980. How plant populations defend themselves in natural ecosystems. In: J.G. Horsfall & E.B. Cowling (Eds): Plant disease V. How plants defend themselves. Academic Press, New York. p. 75-102.

Sequeira, L., 1978. Lectins and their role in host-pathogen specificity. Annual Review

of Phytopathology 16: 453-481.
Stewart, D.M. & B.J. Roberts, 1970. Identifying races of Puccinia graminis f. sp. avenae. A modified international system. U.S. Department of Agriculture Technical Bulletin 1416. 23 p.
Ullstrup, A.J., 1972. The impacts of the southern corn leaf blight epidemics of 1970-1971. Annual Review of Phytopathology 10: 37-50.
Vanderplank, J.E., 1968. Disease resistance in plants. Academic Press, New York. 206 p.
Wahl, I., N. Eshed, A. Segal & Z. Sobel, 1978. Significance of wild relatives of small grains and other wild grasses in cereal powdery mildews. In: D.M. Spencer (Ed.): The powdery mildews. Academic Press, New York. p. 83-100.
Wolfe, M.S. & J.A. Barrett, 1980. Can we lead the pathogen astray? Plant Disease 64: 148-155.

New problems, trends and visions in agricultural resistance breeding

Vilhelm Umaerus

Swedish University of Agricultural Sciences, Department of Plant and Forest Protection, P.O. Box 7044, S-750 07 Uppsala, Sweden

ABSTRACT

Breeding for resistance will continue to be the number one defense against plant diseases and pests. Genetic uniformity creates large problems and diversified defense against pests and diseases is advocated. Race-specific resistance should be used with care and more attention should be given to non race-specific resistance. The use of induced resistance opens up new perspectives.

INTRODUCTION

After having listened to the contributions at this workshop concerned with the difficulties encountered in resistance breeding I could make it easy for myself by saying: breed for resistance but avoid genetic uniformity and do not trust the *R*-genes. But reality is usually much more complicated. It has been said that breeding is the art of throwing away; it is also the art of compromises and so is agricultural resistance breeding. But sometimes you can not compromise. There is no future in a potato you can not eat.

Let me first state that breeding for resistance has become a powerful tool in agriculture and will probably increase in importance. Since the turn of the century, when plant breeding started, we have seen so far only the 'top of the iceberg' concerning the incorporation of genes for resistance in acceptable cultivars.

Clearly disease resistant plants are the number one defense against plant diseases and pests, and the use of resistant cultivars appears to be one of our most important measures in modern plant cultivation to combat the attack of plant parasites. The use of plant resistance presents little hazard to the environment and presents to the growers a 'built in' plant protection, which without risks and extra costs or efforts is provided with the seed. In many cases resistance is the only means of control, e.g. for wart disease of potato. You are all aware that chemicals are not available for the control of plant diseases caused by viruses. In other cases resistance is the only economically feasible control method because other methods are too expensive.

Furthermore, resistance breeding is a good investment for the society. A recent calculation has shown a net return of 1 in 100, which can be compared with the farmer's net return in the use of chemicals which is about 1 in 4 (Sundell, 1979). The drawback is that plant breeding is a slow process and investments usually do not pay back early enough.

What lessons from the past have been learned from this process which can be of value for similar efforts in silviculture? The much discussed use of clonal propagation immediately leads to one dominating issue: genetic uniformity and its influence on disease resistance.

GENETIC UNIFORMITY

Uniformly high-yielding crops grown in dense stands in an intensive agriculture are also uniformly vulnerable to any parasite with epidemic potentials. We have learned that the hard way, and I will give 2 classic examples. A new oat variety, Victoria, was released in Iowa in 1942. Victoria was resistant to all races of crown rust, and also resistant to stem rust and to smuts. It became very popular and by 1945 together with some derivative varieties covered 97 % of the total oat acreage. In 1946 a seedling blight caused by *Helminthosporium victoriae* Mee. & Mur. appeared which in 2 years reached epidemic proportions. Victoria was particularly susceptible to this fungal disease which previously had only been known as a minor parasite on wild grasses. Unluckily, the unforeseen reason was a pleiotropic effect of the dominant gene Pc-2 for crown rust resistance. The following years the Iowa acreage planted with Victoria and its derivatives was reduced; eventually the susceptible varieties were replaced by other resistant varieties and Victoria blight ceased to be a problem.

The southern leaf blight on corn which reached epidemic proportions in 1970 is a repeat of the Victoria blight history, but this time a uniform corn monoculture had developed with the introduction of hybrid corn based on cytoplasmic male sterility. The most dependable source for cytoplasmic male sterility was derived from plants found in Texas and by 1969 this so-called Tms cytoplasm had been introduced into up to 90 % of the field corn grown throughout the United States. That year reports appeared that Tms lines were unusually susceptible to leaf blight and in 1970 the hitherto passive leaf blight reached epidemic proportions. It appeared first in Florida and then rapidly spread west and then north, until by September it had been reported throughout the eastern half of the United States. The epidemic was caused by race T of *Helminthosporium maydis* Nis. & Miy. Later investigations proved that race T existed earlier, but up to 1969 only race O was generally known and this race did not differentiate between Tms and normal cytoplasm. In 1971 seed producers went back to normal cytoplasm with resistance to races O and T, and hybrids were produced by mechanical detasseling.

Modern breeding techniques may increase genetic vulnerability in other ways, too. Male-sterile barley and wheat in hybrid seed production fields are especially susceptible to ergot (*Claviceps purpurea* (Fr.) Tul.). The disease infects the flowers at the time of pollination. If flowering is cleistogamous infection is rare. For hybrid production flowers remain open a long time, greatly increasing their chances of becoming infected.

Hybrid sugar beet and sorghum also run a great risk because of narrow sources of cytoplasmic male sterility. The list can be longer. Any trait that man introduces without assurance of a varying background of other genes is risky, and it certainly is not easy to predict what special risk may be encountered.

Breeding for disease resistance should introduce to the crop important genetic variation designed to prevent epidemics. The traditional approach employs a succession of monogenic resistant varieties, where each one is replaced as it fails. This relieves genetic uniformity by creating genetic discontinuity in time. An alternative is to create discontinuity in space by deliberately introducing variation by growing mixed populations. Another approach is the use of non race-specific resistance, which rarely provides complete control but slows down the infection rate of an epidemic disease.

RACE-SPECIFIC RESISTANCE

The useful life of a race-specific resistance gene deployed against a crop parasite is determined by the frequency of the allele for virulence in the parasite population. This, in turn, is determined by the mutation rate to virulence, and by the fitness of the virulent mutant in competition with avirulent non-mutants and other organisms.

Race-specific resistance is usually controlled by one gene, or a few genes whose individual effects are readily detected. Thus race-specific resistance is easy to work with. Testing is usually done on seedlings or small plants, which allows a considerable saving in space and time. Selection is usually performed under the severe conditions of a greenhouse or laboratory test, where specific resistance almost always is strongly expressed as an 'either-or' reaction favouring a mass-selection technique. It is understandable that most breeders in the past have favoured breeding for race-specific resistance, which also in the farmer's fields showed up as complete protection.

The limitation of race-specific resistance for diseases with epidemic potentials are well known for the same reasons as previously discussed under genetic uniformity - the system is very vulnerable. The vulnerability is greater than expected because during the process of breeding for race-specific resistance a loss of genes for non race-specific resistance usually under polygenic control can occur, or as Day (1974) expresses it: 'Plant breeders stripped oligogenic resistance of associated protective polygenic effects and exposed the genes one at a time in agriculture on a tremendous scale, forseeing the parasite to respond in time'.

This may give the impression that race-specific resistance is of little long-term practical value, which, of course, is a generalization which does not hold; the picture is not that black and white.

Soil-borne parasites, e.g. cystnematodes, or the agent of wart disease of potato, have a slow spread, only 1 generation per vegetation period and host-crop, and long-distance transport is rare. It is easy for the plant breeder to be one step ahead with introduction of new resistance genes provided predictions can be made early enough which virulence gene the parasite will develop. Race or pathotype surveys are still of high importance.

Contrary to the obligate parasites, non-obligate plant pathogens have often been effectively controlled by the introduction of a single gene for resistance and physiologic specialization has not been a problem.

The obligate parasites are different. The more widely a strong gene is used, the less it can be relied on for protection; the crop selects the race that may ultimately destroy it.

One example where race-specific resistance has not been successfully used is late blight of potato. The host plants are polyploid and clonally propagated, and are consequently difficult to breed by the method of crossing an established variety with a source of resistance and of screening thousands of seedlings for agronomic characters that match the susceptible parent. This means that rapid production of a succession of good agronomic varieties carrying different *R*-genes is impractical.

In the case of resistance to rust or powdery mildew in cereals this strategy may be successful, provided a sufficient source of resistance genes exists. However, the combination of even a few genes buys time, since virulent mutants that arise independently must be combined, or else they must arise simultanously or sequentially before infection can occur.

The most successful strategy for use of race-specific resistance in epidemic diseases is the creation of discontinuity in space by growing mixed populations.

The multiline variety approach has been carried furthest by the oat breeders in Iowa (Browning & Frey, 1969). The use of mixed varieties has the same motive, and is less sophisticated and more flexible to meet the need of protection against more than one disease.

Regional deployment of resistance genes is another way of creating discontinuity in space and has been best exemplified along the *Puccinia* path in USA, also described by Browning et al. (1969).

NON RACE-SPECIFIC RESISTANCE

The pros and cons of non race-specific resistance have been discussed by Persson et al. (page 318 in this book). Successful use of non race-specific resistance is resistance to late blight in potato (Umaerus, 1970), common rust (*Puccinia sorghi* Schweinitz) in corn (Hooker, 1969), yellow rust (*Puccinia striiformis* Westendorf) in wheat (Lupton & Johnson, 1970). Efforts are now being made also to look for non race-specific resistance to other diseases in cereals.

The value of non race-specific resistance has been emphasized strongly by Wahl in his lecture on studies of natural ecosystems (Segal et al., page 361 in this book). My vision is that future breeding for resistance against epidemic diseases should first secure a high level of non race-specific resistance and as a second objective introduce race-specific resistance, not the reverse. Improvements must and can be made concerning selection methods as has been demonstrated in component analysis of resistance to late blight in potato (Umaerus, 1970) facilitating both early selection on young seedlings and screening for a discrete character: plants with a long minimum accession period. Measures also must be taken to overcome the cover effect of known or unknown *R*-genes by using proper races for inoculation or using test grounds with the highest heterogeneity for virulence genes possible.

INDUCED RESISTANCE

If I now have the privilege to give a vision, may I draw attention to the paper of Kuć presented at the IX International Congress of Plant Protection 1979 in which he remarks that 'modern agriculture can not wait for the 'evolutionary' development of disease-resistant plants that may be effective for survival but prove totally ineffective economically in the degree or persistance of resistance'. Evidence is now rapidly accumulating that resistance can be induced in plants treated with cultivar-non-pathogenic races of pathogens, avirulent forms of pathogens, non-pathogens, which protects against disease caused by subsequent infection by pathogens. The intriguing point is, and this is so difficult to comprehend, that all plants, even the susceptible ones, have the capacity of resistance. Browning et al. (page 371 in this book) indicated that resistance is the normal reaction, susceptibility is the exception. Can we use the resistance which obviously prevents non-pathogens from establishing a relationship with a host plant? How do we release the signal that commits cells of even susceptible cultivars to resistance? With that question solved the perspectives of disease control will be enormous.

REFERENCES

Browning, J.A. & K.J. Frey, 1969. Multiline cultivars as a means of disease control. Ann. Rev. Phytopathol. 7: 355-382.

Browning, J.A., M.D. Simons, K.J. Frey & H.C. Murphy, 1969. Regional deployment for conservation of oat crown-rust resistance genes. In: J.A. Browning (Ed.): Disease consequences of intensive and extensive culture of field crops. Iowa Agr. Exp. Sta. Spec. Rep. 64. p. 49-56.

Day, P.R., 1974. Genetics of host-parasite interaction. W.H. Freeman & Co. San Francisco.

Hooker, A.L., 1969. Widely based resistance to rust in corn. In: J.A. Browning (Ed.): Disease consequences of intensive and extensive culture of field crops. Iowa Agr. Exp. Sta. Spec. Rep. 64: 28-34.

Kuć, J., 1979. Biochemicals and plant protection. IX International Congress of Plant Protection. Proceedings: Opening Session and Plenary Session Symposium. p. 21-26.

Lupton, F.G.H. & R. Johnson, 1970. Breeding for mature-plant resistance to yellow rust in wheat. Ann. Appl. Biol. 66: 137-143.

Sundell, B., 1979. Pests, diseases and weeds in agriculture. Report 2: Economic evaluation of different control methods. Swedish University of Agricultural Sciences. Department of economics and statistics Report 151.

Umaerus, V., 1970. Studies of field resistance to Phytophthora infestans. 5. Mechanisms of resistance and applications to potato breeding. Z. Pflanzenzüchtung 63: 1-23.

Breeding of pines for resistance to the wood nematode, Bursaphelenchus lignicolus

Kihachiro Ohba

Forestry and Forest Products Research Institute, P.O. Box 16, Tsukuba Norin Kenkyu, Danchi - nai, Ibaraki 305, Japan

ABSTRACT

A project of breeding pines (*Pinus densiflora* and *Pinus thunbergii*) for resistance to the wood nematode, *Bursaphelenchus lignicolus*, was started in southwestern Japan in 1978. To develop the schemes for the project, a pilot test and a research project were performed with the cooperation of the Forestry and Forest Products Research Institute, the National Forest Tree Breeding Institutes, and the Regional Forest Office. In these undertakings, practical techniques for candidate pine selection, mass cultivation of the wood nematode, artificial inoculation, propagation of plant materials to be tested, and testing the pathogenicity of nematodes were developed. In this project, about 25 000 candidate pines are expected to be tested for their resistance to the wood nematode using their grafted clones in a 5-year plan. A second inoculation test will use the grafted clones which succeeded in the first test, and draw comparisons with check seedlings of *Pinus taeda*.

INTRODUCTION

In Japan, there are 2 two-needled pine species having wide natural distributions, *Pinus densiflora* Sieb. & Zucc. (Japanese red pine) and *P. thunbergii* Parl. (Japanese black pine). They are important in forestry, and about 200 000 ha of forest land are in plantations of these species. The pine wood has multiple uses, such as material for housing, pulp, and paper. Moreover, these species grow in less fertile soils and have protective effects in reducing soil erosion and in aiding water conservation in this country.

Extensive mortality of the pines has prevailed in southwestern Japan since the 1940's. The cause of their death has been attributed for a long time to many kinds of pine bark beetles. In 1971, Kiyohara & Tokushige found that the epidemic occurrence of pine death (pine wilt) resulted from a species of wood nematode, *Bursaphelenchus* sp., and the nematode was named *Bursaphelenchus lignicolus* Mamiya et Kiyohara (1972). With cooperative work of many scientists in different fields, it was confirmed that there was a vector insect species, *Monochamus alternatus* Hope, by Morimoto & Iwasaki (1972).

A STRATEGY OF PINE BREEDING

A presentation of breeding aims in a certain breeding district for the pines (Table 1) indicates the primary aims are expected to be solved by genetic improvement. Working toward general aims would be useless unless the specific one, resistance to the wood nematode, were solved. So, it is very important to have insight into factors which have destructive effects on survival, bole traits, and growth of individual trees - kinds, distribution, and overlapping of the factors - and to determine economic tech-

Table 1. Breeding aims in a certain breeding district. Tree species: *Pinus densiflora* and *Pinus thunbergii*. District: Lowland of Kyushu, lower than 500-600 m above sea level.

Specific aims		General aims	
primary aims[a]	secondary aims[b]	primary aims[a]	secondary aims[b]
Disease: *Bursaphelenchus lignicolus* Mamiya et Kiyohara	Disease: *Lophodermium pinastri* Schrad. Insects: *Petrova cristata* Walsingham *Matsucoccus* spp. *Dendrolimus spectabilis* Butler *Thecodiplosis japonensis* Uchida et Inoue Others	Fast growth Persistent growth Straight bole Non-twisted bole Narrow crown with small branches	Others

a. Aims for which genetic improvement is of primary importance.
b. Aims for which various procedures such as silvicultural tending, uses of natural enemies, chemical control and genetic improvement are feasible.

niques to overcome these factors. Priority of some counter-measures for these factors will be judged in the following order.

1. Factors resulting in the death of the individual.
2. Factors resulting in damage to the bole.
3. Factors resulting in growth reduction.

The counter-measures seem to be as follows:

1. Use the forest land for purposes other than productive forest management.
2. Use the forest land for forestry purposes with some counter-measures alone or in combination:

- silvicultural tending,
- use of natural enemies,
- use of chemical control,
- genetic improvement,
- others (change to other tree species, etc.).

A PILOT TEST OF BREEDING PINES FOR RESISTANCE TO THE WOOD NEMATODE

After the discovery of this nematode, a pilot test of breeding pines for resistance to the wood nematode was started in cooperative work among several institutions such as the Forestry and Forest Products Research Institute, Kyushu Branch Station, the Kyushu Forest Tree Breeding Institute, and the Kumamoto Regional Forest Office. Candidate pines for resistance breeding were selected and propagated by grafting and by using seeds to test their nematode resistance. Meanwhile, in the Kyushu Forest Tree Breeding Institute, a research project was started with the cooperation of 2 other national forest tree breeding institutes to develop a practical breeding system for resistance to the wood nematode. This led to the adoption of the following techniques:

(1) Grafts are used as the main testing materials

At present, we have no information on the genetic constitutions of the pines which might have resistance to the nematode. They might have dominant gene(s) and/or polygenes. As the pines are wind pollinating species, it is quite probable that a resistance test with grafted plants raised from scions taken from the candidate pines will produce more resistant plants than a test with seedlings from open pollinated seeds of the candidates. Using *Pinus taeda* L. which is almost immune to the nematode, as reported by Kiyohara & Tokushige (1971) and Ibaraki et al. (1978a), experiments were made to test resistance to the nematode after exchanging stocks and scions between Japanese black pine and taeda-pine. Taeda-pines grafted on stocks of Japanese black pines showed sufficient resistance to the wood nematode after artificial inoculation as reported by Toda et al. (1977).

(2) The bark-peel-inoculation for practical use

After trials of several kinds of techniques for artificial inoculation of the nematode, the bark-peel-inoculation method was adopted for practical use (Nishimura et al., 1977). Techniques of artificial cultivation and inoculation of the nematode had already been established by scientists at the Pathology Laboratory in the Forestry and Forest Products Research Institute, Kyushu Branch Station, and described in a manual (1974). The bark-peel-inoculation was developed by simplifying conventional techniques. In bark-peel-inoculation, the bark of the main shoot is peeled with a knife at a location about 20 cm above the grafted point for about a 5 cm length. After making the peeled surface rough with a small saw, 1 ml of a water suspension of the nematode containing ca. 10 000 nematodes is dropped on this surface with a micropipette and no cover is used. It is very simple and easy and about 500 grafts can be inoculated by a 2-man team in 1 day.

(3) Mass production of the nematode

Techniques for artificial cultivation of the nematode were established at the Pathology Laboratory (1974). The nematode can be cultivated on a fungus, *Botrytis cinerea* Pers., on potato agar medium. Mass production of the nematode was made possible by the use of petridishes for cultivation vessels. It is expected that about 0.5 million nematodes per petridish can be raised in 2 weeks.

(4) Control of the pathogenicity of the nematode

It is very important to clarify the diversity of pathogenicity among nematode strains, not only for nematology but also for resistance breeding. This makes it possible to use adequate strains of the nematode for the test of resistance of pines by artificial inoculation. A very wide variation in pathogenicity of the nematode was found by Kiyohara et al. (1977) and Ibaraki et al. (1978b). Research on this problem has been continued since the beginning of this research project. Nematode strains which have low, medium, and high pathogenicity are kept in the institutes for use in resistance testing.

(5) Artificial inoculation of pine grafts in a glass-house

Artificial inoculations commonly have been made around the 20th of July. But this time is usually at the end of the rainy season in southwestern Japan. Conditions of humid and cool weather result in a higher survival of pines after inoculation. The optimum period for artificial inoculation is expected to be from late July to early August in the field nursery. If the inoculation were made in a glass-house, we could expect to have a more severe test for resistance under high temperature and limited water irrigation. Moreover, we could combine the techniques of selective use of nematode strains having varied pathogenicity with environmental control during the test.

(6) Safe and easy collection of the scions from candidate pines

The candidate pines should be 30-50 years of age or more. Because they will be tall, there was a need to develop safe techniques to collect scions from these pines. Several techniques and tools for climbing trees were tested and it is recommended that common and new safety tools be used in combination.

(7) Selection of the candidate pines

There are still several surviving trees left in many spots where severe damage by the nematode had occurred. Good growth and straight stem form were added as criteria for selection of candidate pines for wood production.

In the course of this pilot test, about 10 pines showing high resistance to the nematode were found, giving us bright prospects for the coming breeding project (Ohba et al., 1977; Ibaraki et al., 1978a, 1978c).

SURVEY OF THE NUMBER OF SURVIVING PINES IN AN AREA WHERE SEVERE DAMAGE BY THE NEMATODE HAD OCCURRED

A typical progression of the death in pine stands suffering from the nematode is as follows: in the first year a small number of pines died, but mortality increased exponentially until almost all pines died within a few years. The owners of damaged pine stands used to fell all pines, including healthy ones. So in a large area, especially at lower elevations in Kyushu and Shikoku, pines largely disappeared except for protected pine forests. The latter were sprayed with insecticides every year because of their public importance.

In areas where mass death of pines had occurred, a survey was made to obtain a rough estimate of surviving pines that might be considered as candidates for resistance breeding. Following directions given by the Forestry Agency, 14 organizations made the survey under the following limitations: (a) pine stands specifically surviving in the severely damaged area should be included, (b) at altitudes lower than 100-200 m, (c) within a distance of 4 km from the sea-shore.

About 34 000 pines were counted consisting of ca. 6 000 from stands of under 1 % survival, 19 000 where survival was 1.1-5.0 %, and 9 600 pines in stands with 5.1-10 % survival.

After the pilot test and the survey of surviving pines in the proposed pine breeding area, a practical project was planned and started in 1978 by the Forestry Agency, Ministry of Agriculture, Forestry and Fisheries.

(1) Purpose of the project

The aim is to breed pines (red pine and black pine) resistant to the wood nematode in southwestern Japan (Fig. 1). Concerning the degree of resistance, pines having a resistance equal to or above those of *Pinus rigida* Mill. and *Pinus taeda* L. are expected to be selected.

(2) Organizations responsible for this project

The Forestry Agency made the final plan to conduct these breeding activities. Three national forest tree breeding institutes and 14 prefectural organizations will participate in this project with a 5-year plan.

(3) Tree species to be bred and the breeding district

Pinus densiflora and *Pinus thunbergii* are the objective species. Two

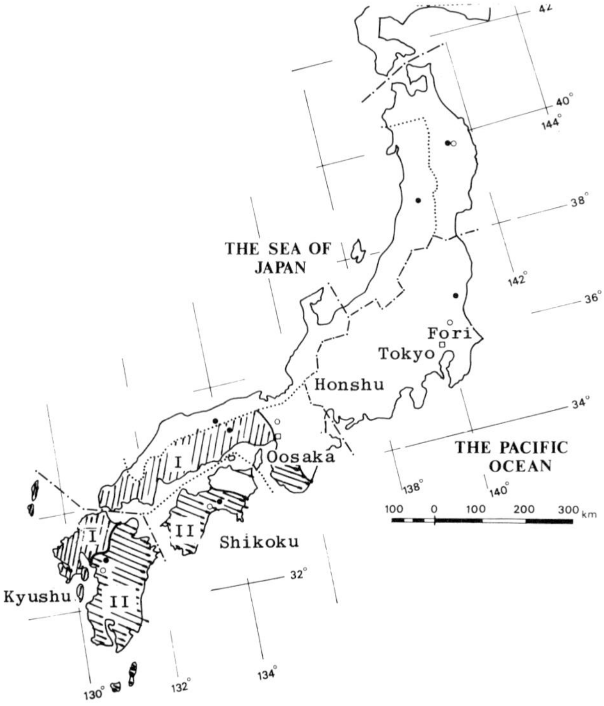

Figure 1. Two breeding districts (hatched areas) were set for the breeding project with the pine wood nematode. o, FORI, Forestry and Forest Products Research Institute; •, National Forest Tree Breeding Institute.

Table 2. The 5-year plan of pine breeding with the wood nematode.

	Fiscal year (April to March)				
	1978	1979	1980	1981	1982
Selection of candidate pines	o	o	o		
Scion collection	o	o	o		
Grafting	o	o	o		
Raising grafts		o	o	o	
The first inoculation test			o	o	o

breeding districts are set: I, north Kyushu and Setouti; and II, south Kyushu, Shikoku, and Kinki (Fig. 1).

(4) A 5-year plan of the project

The project is operated under a 5-year plan, with yearly plans such as selection of the candidate pines, grafting, and resistance tests with artificial inoculation (Table 2).

(5) Number of candidate pines which will be tested in each organization

A total of 25 010 candidate pines expected to be selected (*P. densiflora*: 11 035 and *P. thunbergii*: 13 975) is alloted to 17 organizations.

(6) Breeding scheme

A breeding scheme of mass selection was adopted as developed by the research and survey made in preceding years.

- The candidate pines will be selected from surviving pines in stands having less than 10 % survival (preferably 1 % or less). From these candidate pines, scions will be taken and grafts will be raised.
- Ten grafts from each candidate pine will be inoculated with the nematode by each organization as the first test. The nematodes will be cultivated at the 3 national forest tree breeding institutes, and will be distributed to each organization.
- As a check of the degree of resistance, 2-year-old seedlings of *Pinus taeda* will be treated the same as the grafts. Success of the candidate pines will be judged by the survival rate (including grafts with partial death) and the rate of healthy grafts in combination.
- An inoculation test will be made twice, the first by 17 organizations under the guidance of the national forest tree breeding institutes. The second inoculation test will be made at 3 national forest tree breeding institutes, using those candidate pines which were successful in the first

test. In the second test, 20 grafts for each candidate pine will be tested. After these inoculation tests, pines resistant to the wood nematode will be selected.
- The artificial inoculation will be made on potted grafts in a glass-house.
- Strains of the nematode which had been tested for their pathogenicity in the preceding year at 3 forest tree breeding institutes will be used selectively. Moderately pathogenic strains are planned for use in the first test to avoid over-kill of the grafts. In the second test, strains with a high pathogenicity will be used.
- After the selection of resistant pines, a new programme will be set to produce bred materials resistant to the wood nematode. Some of the resistant pines will be propagated by grafting and the grafts will be planted for emergency use in sea-shore plantations, national parks, and other places. Seed orchards will be established with the grafts concurrent with research on the heredity of resistance to the wood nematode.

(7) Seedling test

Beside the test of grafted candidate pines, seedlings will be raised from open-pollinated seeds taken from some of the candidate pines. An inoculation test on these seedlings will be continued to provide information on the genetic behavior of resistance. As reported by Ibaraki et al. (1978c), seedling-progenies of some candidate pines in the pilot test showed very high resistance.

CONCLUSION

A constructive pathway for the project of breeding pines for resistance to the wood nematode has been briefly described. The result of the first inoculation test on grafts is expected this fall (1980), and it will appear in the next 2 years. It is expected that a sufficient number of resistant pines will be secured in this project.

Meanwhile, we have deep concern for the following points related to the project and future work after the project:
- To select as many pines resistant to the wood nematode as possible.
- To clarify the hereditary nature of the resistance.
- To clarify the mechanism of resistance.
- To combine through artificial crossing the different genes responsible for resistance in individuals for use as seed-parents.
- To clarify the provenance variation of pathogenicity of the nematode strains.
- To make experimental plantations with grafts and seedlings derived from open-pollinated seeds of the resistant pines.
- To establish seed orchards with the resistant pines.

ACKNOWLEDGEMENTS

To the former directors, Mr. T. Niga and Mr. C. Ibaraki, and to all members of the Kyushu Forest Tree Breeding Institute; to the former head of the Silviculture Divison of the Forestry and Forest Products Research Insitute, Dr. R. Toda, and the present head of the Division, Mr. K. Hachiya; and to Dr. N. Ohyama in the Kyushu Branch Station of the Institute, I express my heartfelt gratitude for their guidance and cooperation.

REFERENCES

Forestry Agency, Ministry of Agriculture, Forestry and Fisheries, Japan, 1978. [Directions for the breeding project of pines resistant to the wood nematode] (in Jap.).

Ibaraki, C., K. Ohba, H. Yamate, T. Ritsusen, T. Toda, K. Nishimura & K. Matsunaga, 1978a. [Graft-tests of the candidate pines for pine wood nematode resistance] (in Jap.). Nihon Ringakukai, Kyushu-sibu Kenkyu-ronbun-syu 31: 59-60.

Ibaraki, C., K. Ohba, T. Toda, H. Hashimoto & T. Kiyohara, 1978b. [Different pathogenicity of 23 strains of the nematode, Bursaphelenchus lignicolus to seedlings of Pinus thunbergii] (in Jap.). Nihon Ringakukai, Kyushu-sibu Kenkyu-ronbun-syu 31: 211-212.

Ibaraki, C., K. Ohba, T. Ritsusen, K. Nishimura, T. Toda, N. Ohyama, T. Takagi & K. Kawanobe, 1978c. [Seedling tests of the candidate pines for pine wood nematode resistance] (in Jap.). Nihon Ringakukai, Kyushu-sibu Kenkyu ronbun-syu 31: 61-62.

Kiyohara, T., H. Hashimoto, K. Ohba & K. Nishimura, 1977. [Pathogenicity of four isolates of Bursaphelenchus lignicolus to plus tree progenies of Pinus densiflora and Pinus thunbergii] (in Jap.). Transactions of the 88th Meeting of the Japanese Forestry Society. p. 329-330.

Kiyohara, T. & Y. Tokushige, 1971. [Inoculation experiments of a nematode, Bursaphelenchus sp. onto pine trees] (in Jap.). Journal of the Japanese Forestry Society 53: 210-218.

Mamiya, Y. & T. Kiyohara, 1972. Description of Bursaphelenchus lignicolus n. sp. (Nematoda Aphelenchoididae) from pine wood and histopathology of nematode-infected trees. Nematologia 18: 120-124.

Morimoto, K. & A. Iwasaki, 1972. [Role of Monochamus alternatus (Coleoptera: Cerambycidae) as a vector of Bursaphelenchus lignicolus (Nematoda: Aphelenchoididae)] (in Jap.). Journal of the Japanese Forestry Society 54: 177-183.

Nishimura, K., K. Ohba, T. Ritsusen, K. Matsunaga & M. Imamura, 1977. [Development of artificial inoculation techniques for pine breeding with the pine wood nematode] (in Jap.). Nihon Ringakukai, Kyushu-sibu Kenkyu-ronbun-syu 30: 61-62.

Ohba, K., K. Nishimura, T. Toda, H. Yamate & N. Ohyama, 1977. [Artificial inoculation tests of the candidate-grafts of Pinus thunbergii for the pine wood nematode] (in Jap.). Nihon Ringakukai, Kyushu-sibu Kenkyu-ronbun-syu 30: 71-72.

Pathology Laboratory, Forestry and Forest Products Research Institute, Kyushu Branch Station, 1974. [A manual of cultivation and artificial inoculation of the pine wood nematode, Bursaphelenchus lignicolus] (in Jap.). p. 1-5.

Toda, T., T. Ritsusen & H. Yamamoto, 1977. [Resistance tests of pine-grafts and of waterpotted pine branches] (in Jap.). Nihon Ringakukai, Kyushu-sibu Kenkyu-ronbun-syu 30: 65-66.

The Swiss Endothia-resistance breeding program

G. Bazzigher

Swiss Federal Institute of Forestry Research, CH-8903 Birmensdorf, Switzerland

ABSTRACT

In the fight against the chestnut blight in Switzerland a selection program was started to find resistant individuals in native (*Castanea sativa*) and Asian (*C. crenata*) populations. The greatest number possible of blight-resistant trees form the base population for reforestation.

INTRODUCTION

Since the beginning of the 20th century chestnut blight caused by *Endothia parasitica* (Murr.) And. & And., which comes from eastern Asia, has in less than 50 years killed the American chestnut (*Castanea dentata* (Marsh.) Borkh.) in its natural range in eastern United States to such an extent that only small remnant stands remain. About 35-40 years ago this fungal disease was also discovered in various European stands; in 1947 it was identified in the Canton of Tessin in southern Switzerland. Alarming descriptions of the progress of the epidemic in the USA prompted forestry authorities to take measures against the disease and its consequences. In order to save the threatened chestnut, a selection program was carried out by the Swiss Federal Institute for Forestry Research, with the goal of selecting (using suitable testing methods) as many resistant individuals as possible from the native and Asian chestnut populations. These selected trees were to form the basic material for future reforestations. In order to carry out this selection and numerous expensive experiments, nurseries were provided in the Canton of Tessin where we could raise several thousand chestnut seedlings each year for infection and selection experiments.

It rapidly became clear that the epidemic in our European chestnut (*Castanea sativa* Mill.) progressed much more slowly and was weaker than in the American chestnut. No doubt the European chestnut is more blight-resistant than its American relative, and this higher resistance justifies a more favorable prognosis for the future of our chestnut species. If it could be proven that higher resistance is genetically controlled, there would be the possibility, through systematic selection of resistant individuals, to improve the chances for survival of the species. It would be even more efficient to simultaneously cross the European chestnut with suitable resistant Asian chestnuts, especially the Japanese chestnut (*Castanea crenata* S. & Z.).

TIMETABLE

In retrospect the progress of the selection project can be divided into 3 phases:

1951-1960: Propagation and preparation of plant material and development of inoculation and test methods. During this period the main part of the preliminary phytopathological studies were also carried out. In addition, the method for vegetative propagation of chestnut was developed (Bazzigher, 1968). In 1960 an important step was taken with further development of inoculation and test methods.

1960-1972: Period of intensified systematic selection. The effect of the new test method was studied (Bazzigher & Schmid, 1962), and we learned how the selections should be made in the future. Several spontaneous hybrids, especially hybrids with the early-flowering *C. crenata*, were included in the program. Today there are about 40 000 trees from this selection work. For technical reasons, only part of these trees could be propagated vegetatively.

1964-1982: Period of checking and additional selection of blight-resistant trees. The total selected material of blight-resistant chestnuts now had to be tested in plantations. In the mid-1960s we had decided to build up these plantations systematically as gene banks and to observe and measure this plant material further. These data, which we collect from each tree 3, 10, and possibly also 20 years after planting, give us differentiated possibilities for evaluating the results of selection. Thus we can make additional selections that satisfy the special requirements for reforestations.

PREPARATORY PHYTOPATHOLOGICAL STUDIES

Predisposition and resistance

We can only reach our goal of saving the threatened chestnut through comprehensive knowledge about the biology of both pathogen and host. This knowledge must be obtained from experimental work (Bazzigher & Schmid, 1962). Results that allowed us to draw conclusions about disease resistance and predisposition were especially interesting. Resistance, as defined by Gäumann (1951), is considered here to be the genetically controlled ability to restrict disease development; while predisposition is a temporary state of susceptibility in the plant, caused by environmental effects and within the limits of the genetic variation.

SEASONAL VARIATION IN PREDISPOSITION

In inoculation experiments we determined that the seasonal variation in predisposition followed the rhythmical changes in the vegetative condition of the host. Optimum growth of the pathogen in host tissue takes place during the warmest summer months (Fig. 1). The best time for infection in resistance tests, therefore, is mid-May (Bazzigher & Schmid, 1962). For the resistance tests it is important, as has been shown in our experiments, that only results are compared from experiments that have been performed at the same time and under as similar climatic conditions as possible.

EFFECT OF WOUND AGE ON INFECTION AND DEVELOPMENT OF SYMPTOMS

The pathogen causing chestnut blight is parasitic in wounds, i.e., for infection the fungus depends on wounds in the bark of the host. A wound heals after a certain time and the tissue is regenerated. If a wound parasite like *E. parasitica* cannot enter the wounded tissue early enough, the infection fails, or succeeds but is greatly delayed. With older wounds, i.e., with increasing development of callus tissue, the initial conditions for the pathogen are poorer. In an experiment on wound age, e.g., the growth rate of the pathogen in 2-day-old wounds was reduced to less than half that found in fresh wounds. If such a tree recovers, it is not due to greater resistance but rather to unfavourable initial conditions for the fungus. Such sources of errors also must be excluded from tests on resistance of chestnut to blight or must be accounted for when the results are interpreted.

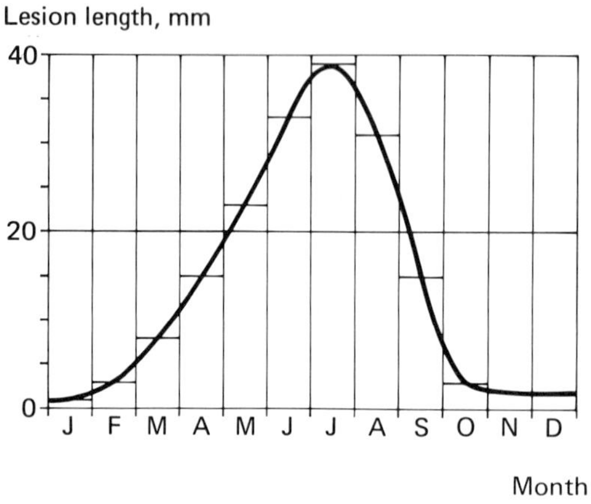

Figure 1. Average monthly growth of *E. parasitica* in chestnut bark in nature. Inoculation experiments done in 1960-1961.

DIFFERENTIAL WOUND HEALING: A CAUSE OF VARIATION IN SUSCEPTIBILITY OF DIFFERENT AGED SHOOTS ON YOUNG CHESTNUTS

In various experiments (e.g., Bazzigher & Schmid, 1962) considerable variation in resistance was found within the tree itself. We thought, as was shown in an experiment, that this was due to different rates of healing for wounds in different parts of the tree. At 2-day intervals the upper and lower ends of 1-, 2-, and 3-year-old shoots on five 7-year-old chestnuts were wounded. The wounds were made and covered with adhesive tape in the same manner as in our ordinary inoculation method. In this experiment we were thus able to study the wound healing in 0-, 2-, 4- to 12-day-old wounds. The amount of callus tissue in microscopic sections from each wound was used as a measure of wound healing. The results indicated that there is a rapid reduction in healing capacity from the youngest shoots to the older ones. The older the shoot the more time it required for the wound to heal. This is, however, the case only for 1 to 3-year-old shoots; for older wounds these differences are reduced. From this experiment we learned that for comparative studies the inoculation must be made on comparable spots on shoots of the same age.

ONTOGENETIC VARIATION IN PREDISPOSITION

The ontogenetic variation in predisposition is of great significance for interpretation of inoculation tests. Within the range of variation in genetic resistance, chestnut trees are maximally susceptible up to age 6. Susceptibility is reduced with increasing age, but increases again in older trees. These changes in resistance due to age are classified in Fig. 2. Up to age 4 chestnut trees are extremely susceptible. This is expressed by a special preference of the pathogen for the cambium. The growth rate in this tissue is up to 3 times faster than that in the outer bark tissue. At age 4-6 the lesion close to the cambium is about as large as in the outer bark. We have therefore carried out our inoculation tests with trees of this age

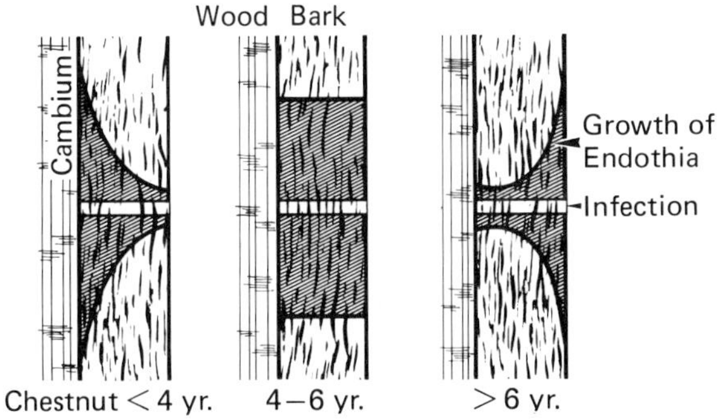

Figure 2. Ontogenetic changes in predisposition reflected in growth of *E. parasitica* in bark tissue of chestnut trees of different ages.

(error-free measurements of lesion size). The trees become increasingly resistant after 6 years. Growth of the pathogen is increasingly concentrated in the outer bark (a tendency towards exclusion of the pathogen from the bark, a healing process).

Inoculum and inoculation methods

PATHOGENICITY OF THE FUNGUS

Usually there are no difficulties in finding isolates with sufficient pathogenicity. Since, however, within a pathogen population a fairly large variation is to be expected, the suitability of an isolate for inoculation tests must be tested in pathogenicity tests (Fig. 3). In an inoculation test 33 isolates from chestnuts around Bellinzona (the Canton of Tessin) were compared. Host mortality and lesion length were taken as measures of pathogenicity.

STORING LIVE ISOLATES OF THE PATHOGEN

The best method for storage of live isolates has proven to be freeze

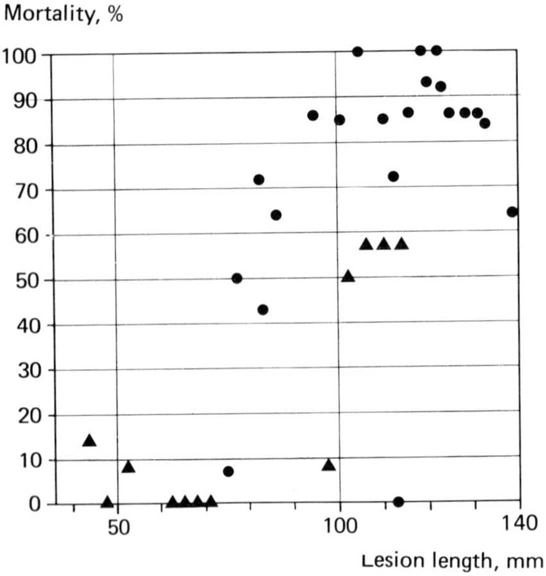

Figure 3. Pathogenicity test of 33 *Endothia* isolates (462 inoculations on 5-year-old *C. sativa*). Inoculation: June 15, 1976. Evaluation: lesion length Sept. 16, 1976, mortality Sept. 7, 1977. ● = virulent isolates, ▲ = isolates with transmissible hypovirulence. Transmissible hypovirulence (hypovirulence contagieuse, Grente, 1975) is an epidemic disease on *E. parasitica* that reduces its pathogenicity but not its vitality. The hypovirulence is controlled by a genetic factor in the cytoplasm of the fungus.

drying (without prefreezing) and subsequent storage in vacuum-sealed ampuls. No loss of vitality and pathogenicity has been found. Method: small pieces of nylon mesh (pore size 215 μm) were cut into smaller strips (3 × 10 mm), sterilized, and stored in absolute ethanol. Dry strips were dipped in a suspension of conidia and placed on malt agar in Petri dishes (1.2 % Difco Malt Extract + 2.0 % Difco Bacto Agar) and incubated at 24 °C for about a month until pycnidia formed. The nylon strips then were carefully removed from the agar and each enclosed in an ampul and dried in 10^{-3} Torr. The ampuls were then melted shut in a vacuum, labelled, and stored below 20 °C.

INOCULATION METHOD

The inoculation method and its statistical tests have already been described in detail (Bazzigher & Schmid, 1962).

Production of inoculum: culture flasks (2.5 l) each with 500 ml KYG-nutrient solution (Knop's solution + 2 % glucose + 0.5 % Difco Yeast Extract) were inoculated with a spore suspension of *E. parasitica* and then rotated horizontally at 20 °C for 6 days. Small balls of mycelium formed; their size can be controlled by varying the numbers of spores used for inoculating the nutrient solution. When pigment starts to form on the balls, the mycelium is separated by sterile filtration and washed in sterile water. The excess liquid is decanted and this inoculum (shelf life: 1 week, possibly longer) is then stored in plastic bottles (100 ml) in a refrigerator (+ 2 °C).

Inoculation: the test seedlings are wounded in the stem with a special instrument. The bark is removed as a small round plate (diameter 0.5 cm) so that the cambium is exposed. Inoculum is put into this wound with a spatula. The inoculation spot is closed with Scotch adhesive tape to protect it from contamination and dessication. It is best to make the inoculations in the second half of May. Growth rate of the pathogen is determined by periodically measuring the length of the lesion.

This evaluation, using growth rate of the pathogen in the host tissue, has been successful (Bazzigher & Schmid, 1962). Resistance to blight, i.e., the inborn ability to survive infection, is partly correlated with resistance to spread of the pathogen. The extent to which resistance occurs in clones and populations can be estimated by percentage mortality in relation to duration of symptoms. Stem diameter at the inoculation site must also be considered.

SELECTION OF BLIGHT-RESISTANT CHESTNUTS

In our program we select blight-resistant individuals from different species of *Castanea* and their hybrids. The hybrids are, except for special crossing experiments, formed spontaneously. Free pollination occurring in our conditions most likely gives rise to considerable inbreeding.

Every year chestnut seed is collected from trees selected for seed-collection (266 provenances) and their progeny. The seed is sown and raised in various experimental nurseries in the Canton of Tessin. We test each 4- or 5-year-old tree in individual inoculation tests. During our selection program we have tested more than 120 000 plants. During the last 10 years we have included in this program a large number of hybrids, especially with the Japanese chestnut (*C. crenata*), which is considerably more resistant than the European chestnut. *C. crenata* is especially well-suited for crossing with our chestnut since it flowers precociously, a trait controlled by dominance, making possible a very short generation time.

Chestnuts that survive the inoculation test are preserved as clones (vegetatively propagated through layering or grafting) or as individuals by transplanting them to gene banks.

Every bank has its own number, and the trees are numbered consecutively. Using these 2 numbers the location of each tree is unmistakably identified; each chestnut tree is tagged with its number and mapped. In addition, these numbers and information on parent (♀) tree, generation (F_1, F_2, etc.) and vegetative propagation are filed on punch cards (Parent Tree Card for Chestnut Selection) for computers.

In a breeding and selection program such as we have carried out for the last 25 years, a long initiation period is needed before a larger produc-

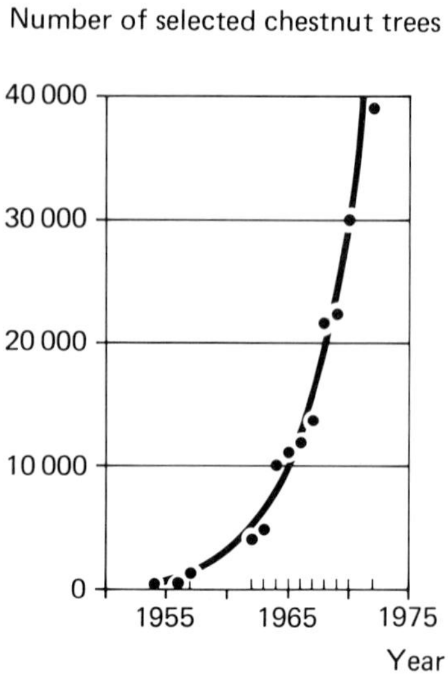

Figure 4. In 1962 improved testing methods and intensification of the selection program resulted in an increase in annual production of blight-resistant chestnut trees, which could be checked in gene banks. Ordinate: number of blight-resistant chestnut trees in the gene banks; abscissa: year of planting.

tion of resistant plants is possible (Fig. 4). Today the ca. 40 000 selected and registered trees in the gene banks include trees with medium to high blight resistance.

The gene banks serve not only for checking the selected material but also for additional selection procedures. In these banks the trees are controlled and measured according to phytopathological, phenological, and growth characteristics 3, 10, and 20 years after planting. These data are also punched on the cards and are necessary for the second-generation selection. With these we hope to be able to show the different blight resistance in the parent tree, progeny, and lines. The evaluation is possible after completion of the second measurement (10 years after planting, i.e., in 1982) in all gene banks.

Because the testing methods were too inexact for the selection work, it was necessary to improve them considerably. In addition, reasons for variation in infection and symptom development became apparent, prompting additional phytopathological studies. Only then was it possible to interpret the test results.

Several thousand chestnut seedlings were raised annually and inoculated with the pathogen at age 4-5 years. Surviving or only slightly infected trees were transplanted as clones or single trees into gene banks in order to facilitate checking and second-generation selection. Over 120 000 trees have gone through inoculation tests. About 40 000 selected trees have been outplanted and registered individually. For additional selection, measurements and controls will be carried out 3, 10, and 20 years after planting.

REFERENCES

Bazzigher, G., 1968. Die Selektion Endothia-resistenter Kastanien und ihre Vermehrung. Schweiz. Beitr. zur Dendrologie Nr. 16-18. p. 29-38.

Bazzigher, G. & P. Schmid, 1962. Methodik zur Prüfung der Endothia-Resistenz bei Kastanien. Phytopath. Zeits. 45: 169-189.

Gäumann, E., 1951. Pflanzliche Infektionslehre. Verlag Birkhäuser, Basel. 681 p.

Grente, J., 1975. La lutte biologique contre le chancre de châtaignier par "Hypovirulence contagieuse". Annales de Phytopathologie 1: 216-218.

Engineering blister rust-resistant western white pine

G.I. McDonald & R.J. Hoff

Plant pathologist and plant geneticist, respectively, Intermountain Forest and Range Experiment Station, USDA Forest Service; Forestry Sciences Laboratory, Moscow, Idaho, USA

ABSTRACT

A method of measurement of white pine blister rust hazard is described. Two illustrations of application are discussed. The first is aimed at reduction of severity of selection imposed on the white pine population for improvement in rust resistance by matching hazard and resistance. The second application is directed toward increasing efficiency of selection for phenotypic resistance through a more exact knowledge of rust intensity. Data are provided to show long term field performance of western white pine developed from light to severe levels of selection pressure and planted on low, medium, and high hazard sites.

INTRODUCTION

White pine blister rust, caused by *Cronartium ribicola* J.C. Fischer, has been a major forest disease problem in North America since its introduction into the Northeastern United States early in this century. The disease began killing large numbers of trees in the extensive western white pine forests of Oregon, Washington, and Idaho about 1920. By 1950, development of rust resistant pines was underway (Bingham et al., 1960) and, by 1972, the first commercial plantations of resistance material were established in northern Idaho. Many kinds of resistance mechanisms have been defined and possible use of some of them in integrated rust management programs determined (McDonald, 1979; Kinloch, page 119 of this book). The materials in these first plantations were derived from a limited number of parents (Hoff & McDonald, 1980). In the meantime, the desirability of a broadened genetic base has become evident. The increased base would provide for future improvement in various cultural traits, serve as a source of variation in resistance to other pests and environmental challenges, and the additional genetic breadth would facilitate development of integrated rust management (Dinus, page 452 of this book). Lastly, maximum breadth is necessary to wage a successful campaign against the inevitable pathogenic variability of the rust (Powers, page 236 of this book).

The quickest and least expensive way to increase genetic breadth is to accept reduced levels of resistance that are also mechanism non-specific. In addition, why expend resources, both genetic and monetary, in the pursuit of non-essential levels of resistance and, at the same time, run the risk of increasing the selection pressure on the rust? Therefore, use of rust resistance in the environmentally variable Pacific Northwest, to manage a highly variable rust, will require maximum flexibility in kinds of resistance and methods of development. The objective of this paper is to outline a breeding and rust management program under development by the USDA Forest Service, to obtain this flexibility by specification of hazard at parent tree selection sites and by matching resistance in created populations (percent of population and mechanisms incorporated) to expected hazard on a site-by-site basis. We want site-specific specification of both selection and performance levels (resistance engineering).

VARIATION OF RUST INTENSITY

Sites were known to vary in rust intensity early in the North American history of blister rust. Many programs have been developed to capatalize on this variation. These range from the long-standing ribes eradication program (Ketcham et al., 1968) to Van Arsdel's (1972) climatic hazard zones. We know that factors such as pathogen genetics, host genetics (pine and ribes), weather, soils, and canyon physiography work together to produce the observed variation (McDonald et al., 1980). Table 1 shows that such variation exists in northern Idaho, how it is manifest, and 2 of its major causes, ribes species and ribes plant distribution. Given that rust hazard varies, how does one obtain a useful site specific measure. Since stands vary in age, stocking density, growth rate, and years of exposure and origin (natural or planted), a parameter independent of these variables was needed.

Infection rate has much appeal in this regard. Infection rate expressed as proportion or percentage of trees infected/unit time can be determined from actual proportions of infected trees or from infections/tree/unit time. Both measures would produce the same infection rate if spore distribution was uniform. We know it is not (Van Arsdel, 1972) and we know that ribes bushes are not distributed uniformly (Fracker & Brischle, 1944). It has been suggested that the difference in curves erected by both measures from data obtained in the same stand is an index to the amount of ribes bush clumping in that stand (Fracker, 1936). Such an index could be useful, but our concern now is the site-specific measurement of rust hazard. Both rates can be calculated from field data or they can be predicted in the absence of ribes, pine, or rust by a computer simulation that integrates

Table 1. Blister rust infection in 3 plantations of susceptible western white pine located on the Priest River Experimental Forest (PR), Bonner County, Idaho, U.S.A., in the Merry Creek resistance outplanting near Clarkia, Idaho, and the Canal Gulch natural stand near Pierce, Idaho.

Site	Seasons of exposure	Number of trees	Percent infected	Percent dead rust	Actual infection rate[a]	Ribes population (bushes/acre)[b]	
						R.v.	R.l.
PR low elevation	8	1277	11.4	2.4	0.015	0	1
PR mid elevation	8	916	97.7	88.8	0.471	1500	10
PR high elevation	8	1196	91.5	6.2	0.308	37	263
Merry Creek	5	204	92.0	66.2	0.505	400	0
Canal Gulch	10	825	55.0	20.1	0.080	NA	NA

a. Actual infection rate = $\frac{1}{t_z}\left[\log e \frac{1}{1 - x_2}\right]$

b. R.v. = *Ribes viscossissimum* Pursh., R.l. = *Ribes lacustre* (Pers.) Pour.

various aspects of ribes, pine, and rust biology with weather and topographic characteristics specific to a site (McDonald et al., 1980). Our interest in this discussion is focused on the measurement and use of actual infection rates for the 'simple interest' type of epidemics of which white pine blister rust on pine is a classic example.

The rate calculated from infections per tree will be called the theoretical infection rate (R) and the equation (Vanderplank, 1975) for its calculation was derived as follows. The theoretical proportion infected (Y) is:

$$Y = 1 - \frac{1}{e^m}$$

where:

m = infection intensity expressed as average infections per tree and is a product of a rate, cankers/tree/year (r_c), and period of exposure in years.

The simple interest infection rate (r_s) derived by Zadoks & Schein (1979) is given by

$$r_s = \frac{1}{t_2 - t_1} \log_e \frac{1}{1 - x_2} - \log_e \frac{1}{1 - x_1} \qquad (1)$$

where:

t_1 = time of first observation,
t_2 = time of second observation,
x_1 = proportion infected at time 1,
x_2 = proportion infected at time 2.

For blister rust $t_1 = 0$ and $x_1 = 0$. Thus,

$$r_s = \frac{1}{t_2} \log_e \frac{1}{1 - x_2}$$

If $x_2 = 1 - \frac{1}{e^m} = y$, then

$$R = \frac{1}{t_2} \log_e \frac{1}{e^{-r_c t_2}}$$

$$= \frac{1}{t_2} \log_e r_c t_2 = \frac{1}{t_2} r_c t_2 = r_c$$

and R = theoretical proportion infected under uniform spore distribution = r_c.

The equation used to calculate actual infection rate (r) was derived from equation [1] as follows. Since t_1 and $x_1 = 0$ in our discussion, then

$$r = \frac{1}{t_2} \log_e \frac{1}{1 - x_2} \qquad (2)$$

Infection curves can be calculated from each rate (Zadoks & Schein, 1979)

$$Y = 1 - e^{-Rt_2} \qquad (3)$$

and

$$X = 1 - e^{-rt_2} \qquad (4)$$

where:

Y = proportion infected at t_2 calculated from average cankers per tree,

X = proportion infected at t_2 calculated from proportion of stand infected.

Since the variable distribution of ribes bushes, as well as the interaction of other factors, causes uneven distribution of spores, the actual progress curve will fall short of the theoretical progress curve, but it should not exceed it. Regression of r on R for 30 stands on the St. Joe National Forest was $r = 0.18 R + 0.026$ (data from Region I, USDA Forest Service). For r, the regression was based on 644 trees per stand and for R, on 84 trees per stand. The stands varied from 9 to 44 years of age and, in this case, exposure time also. The correlation coefficient was $r = 0.72$ ($p < 0.01$, 28 d.f.). After 5 years exposure at the Merry Creek Plantation, OP controls (Table 1) showed $R = 0.48$ and $r = 0.505$. The difference is probably not significant. In any event, the larger r could be attributed to variations associated with a short exposure time and small trees.

Actual infection rate can be estimated by recording numbers of healthy and cankered stems and number of whorls for each stem. In white pine, the number of whorls is an excellent indicator of exposure time because one whorl is produced each year. So the data can be sorted by the number of whorls and r calculated for each class that containes 10 or more trees. The estimated rate for a stand is the average of the rates of whorl classes that contains more than 10 trees.

The parameter of most significance to the forest manager is mortality rate. Blister rust-infected white pine is killed when the canker girdles the main stem. Relatively small amounts of damage result from either non-girdling cankers or branch cankers and typically there is no growth reduction until just prior to death. It is also important to keep in mind that there is a 5- to 50-year period between infection and completion of the girdle, depending on age of the tree at time of infection and resistance mechanisms present. We have not yet accumulated sufficient data to correctly establish the relationship between infection rate and mortality rate for different classes of resistance material under various degrees of hazard. We do, however, have some preliminary findings and short-term comparisons (Table 1). Some preliminary simulations (described by McDonald et al., 1980) also graphically illustrate this relationship. In the meantime, infection rates will serve to illustrate some points.

GENE MANAGEMENT

Knowledge about infection rates can be used to assist in both development and use of resistant populations. A major problem facing those selecting phenotypically resistant trees is the degree of confidence one can have in a given tree's status. Is it resistant or is it an escape? For example, a stand that averaged 7.5 cankers/tree after 30 years exposure would be expected to be 99.94 % infected under ideal conditions of spore distribution (calculated from Equation 3), or about 6 trees/10 000 would be expected to be canker free due to chance. The theoretical rate (R) is based on the $\bar{x}$ number of cankers/tree for the stand (m) divided by the years of exposure. The estimated actual infection rate according to our regression is equal to 0.18 R + 0.026. So the 7.5 cankers/tree should yield our actual rate (r) of 0.071. This in turn (Equation 4), gives 88.12 % infection, or about 1 canker-free tree in 10. Not much confidence here and a lot of resources could be wasted in progeny testing susceptible trees.

If susceptible trees in selection areas averaged even 50 cankers/tree after 30 years then the expected actual infection rate would be 0.33/year. This means that, after 30 years, one should expect 99.995 % of the stand to be infected, or 1 tree in 20 000 should be canker-free due to chance. At an infection level of 100 cankers/tree after 30 years, the actual rate should be 0.626 and the expected ratio of canker-free to infected trees would be 1 tree in 143 million. Such specification of selection intensities would be most helpful. The original Forest Service program in Idaho found 1 canker-free tree in 10 000 - after 30 years exposure that produced about 100 cankers/tree (R.T. Bingham, personal communication). Many trees selected under these levels of disease transmitted high levels of resistance as we shall see. One would not be too confident in selecting for canker-free white pines from stands growing in northern Idaho and exhibiting much less than 50 cankers/tree after 30 years of exposure.

Determination of infection rates also has potential for the application of resistance to rust management. Various extensive and intensive options of gene management are available. Extensive options include selection of leave trees for natural regeneration (Hoff & McDonald, 1977), creation of natural selection seed production orchards (Hoff et al., 1976), and open-pollinated (OP) seed from untested phenotypically resistant trees.

Intensive options include OP seed from tested parents, F_1 crosses from tested parents, F_2 seed from tested F_1 parents, screened seedlings from OP tested parents, and a generalized immune population.

The usefulness of resistance-hazard alignment (REHAL) is shown by the performance of the different classes of materials in a series of field tests (Steinhoff, 1971) and an updated comparison of infection rates (Table 2) on these same materials.

These infection rates are an average of 19-, 21-, and 23-year exposure for the Priest River, Deception Creek, and Emerald Creek plantations and 5 years exposure from Merry Creek. Several trends are evident. Performance of nearly all classes of pine population is consistent from site to site. Only OP controls at Priest River do not conform. These rates also provide information about the effectiveness of progeny testing white pine. The populations composed of crosses between non-general combining ability (non-GCA × non-GCA) and GCA × non-GCA should be given little weight in this analysis

Table 2. Actual infection rates (proportion of stand/year) of susceptible to resistant western white pine populations after 5 to 23 years of field exposure to *Cronartium ribicola* in northern Idaho, USA.

Pine population	Plantation				Average number of trees inspected
	Merry Creek 5 yrs	Priest River 19, 21, 23 yrs	Deception Creek 19, 21, 23 yrs	Emerald Creek 19, 21, 23 yrs	
Wild seedlings	-	0.095	0.115	0.230	200
Controls	0.505	-	-	-	204
OP controls[a]	-	0.056	0.089	0.095	102
OP non-GCA[b]	-	0.061	0.065	0.071	122
OP GCA	-	0.032	0.037	0.049	219
Non-GCA × non-GCA	-	0.021	0.024	0.035	38
GCA × non-GCA	-	0.025	0.024	0.030	255
GCA × GCA	0.127	0.018	0.019	0.031	240
GCA × $GCAF_1$	0.096	-	-	-	200
$GCAF_1$ × $GCAF_1$	0.063	-	-	-	175

a. Open pollinated seed from surviving trees located in intensely infected areas (100 cankers/tree).
b. GCA = general combining ability (phenotypically resistant tree progeny tested for ability to transmit resistance).

because only 4 non-GCA parents were involved and most of these 4 were borderline GCA. There seem to be significant increments of resistance between OP non-GCA and OP GCA, and between OP GCA and higher levels of selection. Also, significant gain appears to be associated with OP controls and OP non-GCA (about 25 parents) over wild seedlings. The OP controls consisted of open-pollinated seed from trees surviving 30 years of intense exposure to rust levels that produced 100 to 500 cankers/tree; so one would expect some increase in resistance. There is probably a significant reduction in infection rate with production of the F_2 population, as the Merry Creek rate shows and as would be expected from other analyses (Hoff et al., 1973). In conclusion, the findings clearly show that engineering of blister rust resistance of western white pine is feasible and should provide the basis for an effective integrated rust management program.

REFERENCES

Bingham, R.T., A.E. Squillace & J.W. Wright, 1960. Breeding blister rust resistant western white pine, II. First results of progeny tests including preliminary estimates of heritability and rate of improvement. Silvae Genetica 9: 33-41.

Fracker, S.B., 1936. Progressive intensification of uncontrolled plant-disease outbreaks. Journal of Economic Entomology 29: 923-940.

Fracker, S.B. & H.A. Brischle, 1944. Measuring the local distribution of ribes. Ecology 25: 283-303.

Hoff, R.J. & G.I. McDonald, 1977. Selecting western white pine leave-trees. USDA Forest Service Research Note INT-218. Intermountain Forest and Range Experiment Station, Ogden, Utah. 4 p.

Hoff, R.J. & G.I. McDonald, 1980. Improving rust-resistant strains of inland western white pine. USDA Forest Service Research Paper INT-245. Intermountain Forest and Range Experiment Station, Ogden, Utah. 13 p.

Hoff, R.J., G.I. McDonald & R.T. Bingham, 1973. Resistance to Cronartium ribicola in Pinus monticola: structure and gain of resistance in the second generation. USDA Forest Service Research Note INT-178. Intermountain Forest and Range Experiment Station, Ogden, Utah. 8 p.

Hoff, R.J., G.I. McDonald & R.T. Bingham, 1976. Mass selection for blister rust resistance: a method for natural regeneration of western white pine. USDA Forest Service Research Note INT-202. Intermountain Forest and Range Experiment Station, Ogden, Utah. 11 p.

Ketcham, D.E., C.A. Wellner & S.S. Evans, Jr., 1968. Western white pine management programs realigned on Northern Rocky Mountain National Forests. Journal of Forestry 66: 329-332.

McDonald, G.I., 1979. Resistance of western white pine to blister rust: A foundation for integrated control. USDA Forest Service Research Note INT-252. Intermountain Forest and Range Experiment Station, Ogden, Utah.

McDonald, G.I., R.J. Hoff & W.R. Wykoff, 1980. Computer simulation of white pine blister rust epidemic: model formulation. USDA Forest Service Research Paper INT-258. Intermountain Forest and Range Experiment Station, Ogden, Utah. 136 p.

Steinhoff, R.J., 1971. Field levels of infection of progenies of western white pines selected for blister rust resistance. USDA Forest Service Research Note INT-146. Intermountain Forest and Range Experiment Station, Ogden, Utah. 4 p.

Van Arsdel, E.P., 1972. Environment in relation to white pine blister rust infection. In: R.T. Bingham, R.J. Hoff & G.I. McDonald (Eds): Biology of rust resistance in forest trees. USDA Forest Service Miscellaneous Publication 1221. Washington, D.C. p. 479-493.

Vanderplank, J.E., 1975. Principles of plant infection. Academic Press, New York. 215 p.

Zadoks, J.E. & R.D. Schein, 1979. Epidemiology and plant disease management. Oxford University Press. New York. 427 p.

The international IUFRO experiment on resistance of white pines to blister rust (Cronartium ribicola). The French trial

C. Delatour[1] & Y. Birot[2]

The paper gives a first report of a French trial devoted to the international IUFRO experiment on resistance of white pines to *Cronartium ribicola* Fischer ex Rabenh.

In this test are 154 seed lots belonging to 14 species. The test seedlings in each species represented either bulk collections (provenances) or families of various origins. These families consisted of (a) F_1 progenies from open pollination of single-tree selections, (b) F_1 progenies from controlled crosses between 2 selected wild parents, and (c) F_2 progenies from controlled crosses between 2 selected F_1 parents. In the nursery (lining out stage), interplanted *Ribes nigrum* L. (Wellington clone) plants were used as a source of rust inoculum after artificial inoculation by aecidiospores collected from older infected eastern white pines. The design in the nursery was a full randomized block design with plots of 10 trees per row, 5 replications, and spacing between trees of 0.6 m × 0.2 m. One *Ribes* plant was interspersed in each second row between each 20 seedlings in that row. Only fast-growing populations or species (*P. monticola* Dougl., *P. strobus* L., *P. strobiformis* Engelmann, *P. balfouriana* Grev. & Balf., *P. flexilis* James, *P. lambertiana* Dougl., *P. armandi* Franch., *P. griffithii* McClelland, *P. koraiensis* Sieb. & Zucc.) were planted out as (1-2) seedlings near Orléans in an incomplete block design with 20 trees per plot, 3 replications and a spacing of 3 m × 1.5 m. *Ribes* plants were also interspersed: one row for every 16 rows of pines. The slowest growing (mostly *P. albicaulis* Engelm., *P. parviflora* Sieb. & Zucc., *P. cembra* L., *P. aristata* Engelm., *P. sibirica* Rupr.) populations were maintained in the nursery. Three years after natural inoculation of pines in the nursery bark symptoms only of rust were assessed on the 5-year-old seedlings.

When wild populations were considered, species from Europe and Asia (the native habitats of the rust) proved to be less susceptible to rust than the American ones with some exceptions (Table 1). Comparisons of rust

1. CNRF Laboratoire de Pathologie Forestière Champenoux, 54280 Seichamps, France.
2. CRFO Station d'Amélioration des Arbres Forestiers Ardon, 45160 Olivet, France.

Table 1. Frequency of rust-free seedlings in bulk collections of white pines.

Geographic group	Species	Number of seed lots	Rust-free (%)
European species	*P. cembra* L.	12	100
	P. peuce Grisebach	7	80-100
Asian species	*P. sibirica* Rupr.	2	100
	P. parviflora Sieb. & Zucc.	3	100
	P. koraiensis Sieb. & Zucc.	7	91-100
	P. griffithii McClelland	10	33- 93
	P. armandi Franch.	1	42
American species	*P. aristata* Engelm.	1	98
	P. albicaulis Engelm.	5	0- 60
	P. flexilis James	17	0-100
	P. strobiformis Engelm.	9	0- 60
	P. lambertiana Dougl.	5	4- 36
	P. strobus L.	24	0- 12
	P. monticola Dougl.	5	0- 6

infection in the collections representing *P. monticola*, *P. lambertiana*, and *P. strobus* showed some discrepancies but also some agreements in their susceptibility in France and the United States (Table 2). According to a cursory analysis of these results 2 different features appear: resistance ratings in USA and in France are either quite similar of very different; these erratic performances of the progenies are related to the selection areas where the original, parental selections are found. According to Bingham's opinion (personal communication) it appears that the pathogenic races of *C. ribicola* found both in France and Idaho are similar, but different from those found in coastal Washington and Oregon.

More detailed information on the rust-pine system is to be expected in the future, when results from other cooperating countries (Finland, Western Germany, Japan, South Korea) are brought together.

Table 2. Comparison of response to nursery inoculation in France and United States of progenies of white pines selected for rust resistance[a].

Seed lot No	Rust free percentage in France	Rust free percentage in USA, Idaho[b]
P. lambertiana		
41 C (F1)	0 - 1 - 4	10 - 70*
41 B (F1)	0 - 2 - 3	10 - 75*
41 A (F1)	39 - 48 - 55	50 - 100*
P. monticola		
43 B (F1)	0 - 4 - 7	60*
43 E (F1)	2 - 7 - 17	60 - 70 - 80*
43 C (F1)	17 - 31	60*
43 A (F1)	68	70
43 F (F1)	13 - 27 - 51	20 - 35 - 60
43 D (F1)	61 - 69	50 - 55 - 65
47 A (F1)	0 - 25 - 60	35
46 H (F2)	37	65
46 E (F2)	44	60
46 B (F2)	45	55
46 F (F2)	56	65
46 G (F2)	60	70
46 D (F2)	62	80
46 A (F2)	63	70
46 C (F2)	76	85

a. Some seed lots comprised several separate progenies; in this case the range of progeny means and grand mean are given.

b. * Discrepancy between French and American performance.

Relative blister rust resistance of native and introduced white pines in Romania

Ioan Blada

Statiunea Experimentala Forestiera, Padurea Verde, Timisoara, Romania

In Romania there are 7 species of white pine, among which *Pinus cembra* L. is the only native, all the others being introduced. Among the introduced pines, *P. strobus* L. (though economically important) is susceptible to *Cronartium ribicola* J.C. Fisch. ex Rabenh. whereas *P. griffithii* McClell. and *P. peuce* Griseb. are resistant to the same parasite (Patton, 1966; Bingham, 1972; Søegaard, 1972). *C. ribicola* was first noticed in Romania in 1934 (Georgescu et al., 1957), and severe attacks have occurred since 1973 (Blada, 1978).

Because of the importance of *P. strobus* and of the potential danger the pathogen represents, a genetic resistance improvement program was started in 1977. The program includes both intra- and interspecific hybridization and has for a final objective the establishment of seed orchards composed of selections with high combining ability for resistance.

Results of the first step of the program are as follows:

1. At present *C. ribicola* is spread over all the country with the exception of the mountain regions, with a higher frequency on *Ribes nigrum* L. (92 % attacked populations) than on *P. strobus* (51 % attacked young populations).
2. The degree of infection varies from 'low' to 'very high' on young populations of *P. strobus* whereas the mature populations are free from *C. ribicola* even if they are in the proximity of some centers of infection. But, when introduced in the seed orchards, the families of selections and the clones from such populations have been attacked.
3. Except for *P. strobus* and *P. monticola* Dougl., all the introduced pines (*P. peuce*, *P. griffithii*, *P. flexilis* James, *P. koraiensis* Sieb. & Zucc.) are still free from blister rust.
4. Both natural populations of *Ribes* (e.g. *R. alpinum* L., *R. petraeum* Wulf.) and the native *P. cembra* that grow in the same natural mountain habitat are free from *C. ribicola*. This fact (and others) do not support the hypothesis of the existence of a gene center for *C. ribicola* in the Romanian Carpathians.
5. Because some attacked plantations of *R. nigrum* are farther than 100 km from any white pine, it is presumed that the pathogen is either transmitted

from *Ribes* to *Ribes*, or that it has the ability to form aecia on hosts other than the white pines. In Romania, Minoiu (1974) found aecia of a rust that he identified as *C. ribicola* on *Senecio vernalis* W. et K. (a common plant with Eurasian habitat). Then, after artificial inoculation with these aeciospores he obtained uredia and telia on *R. nigrum*.

REFERENCES

Bingham, R.T., 1972. Taxonomy, crossability, and relative blister rust resistance of 5-needled white pines. In: R.T. Bingham, R.J. Hoff & G.I. McDonald (Eds): Biology of rust resistance in forest trees. USDA Forest Service. Miscellaneous publication No. 1221. p. 271-278.

Blada, I., 1978. Ameliorarea pe cale genetica a rezistentei la boli şi insecte a speciilor de raşinoase. Manuscris ICAS, 30 p. (unpublished).

Georgescu, C. et al., 1957. Bolile şi dăunătorii pădurilor. EAS, Bucureşti. 638 p.

Minoiu, N., 1974. Bolile şi daunatorii coacazului, p. 201. In: Suta, V. et al. (Eds): Protectia pomilor şi arbuştilor fructiferi. Ceres, Bucureşti. 287 p.

Patton, R.F., 1966. Interspecific hybridization in breeding for white pine blister rust resistance. In: H.D. Gerhold, E.J. Schreiner, R.E. McDermott & J.A. Winieski (Eds): Breeding pest-resistant trees. Pergamon Press, Oxford. p. 367-376.

Søegaard, B., 1972. Relative blister rust resistance of native and introduced white pine species in Europe. In: R.T. Bingham, R.J. Hoff & G.I. McDonald (Eds): Biology of rust resistance in forest trees. USDA Forest Service. Miscellaneous publication No. 1221. p. 233-239.

Developing fusiform-resistant trees in the south-eastern United States

Bruce Zobel

Professor Emeritus, School of Forest Resources, North Carolina State University, Raleigh, North Carolina, USA

ABSTRACT

In the N.C. State Cooperative Tree Improvement Program, we have been emphasizing fusiform resistance for the past 24 years. Results have been good! Several approaches appear to be successful. The following are suggested: (1) Make full use of species differences, using species resistant or quite resistant to fusiform rust. (2) Within the limits of adaptability, use better sources within the preferred species. (3) Select resistant individuals within suitable species: choose resistant individuals from very heavily infected stands; choose resistant individuals based on phenotypic selection, followed by progeny testing; after testing, establish specialty disease resistance orchards; establish 'two-clone' orchards of parents with special good combining and specific combining ability for disease resistance. (4) Produce hybrids, such as *P. taeda* × *P. echinata*, when feasible. (5) Control the alternate host oaks as much as possible.

We have a rule that if rust infection is no greater than 15 %, we watch it closely but take no special action. If it is greater than 25 %, we then take one or all of the 5 steps listed above. Despite the complexity of the rust and the differing kinds of resistance in the host, gains in fusiform resistance have been remarkably good. Resistant material is currently being planted on at least 100 000 acres each year. It is obvious that fusiform rust resistance breeding is the most important result of the tree improvement activities in the south-eastern United States.

INTRODUCTION

The most destructive forest disease on pines in the south-eastern United States is fusiform rust (*Cronartium fusiforme* Hedgc. & Hunt ex

Cumm.). Losses from fusiform rust are difficult to assess but one thing is certain - the effects of this disease are very important. In the symposium on fusiform rust in 1977 (Anon.), Phelps estimated that there was an annual loss of 562 million board feet (ca. 1.3 million m^3) of sawtimber and 194 million cubic feet (ca. 5.5 million m^3) of growing stock. Powers et al. (1974) estimated the loss in timber to be $ 28 000 000 in 1972, with a subsequent loss of $ 150 000 000 in finished wood products. Of perhaps greater importance is the fact that rust is spreading. Holley & Veal (1977) state 'Its damaging effects are intensifying and the timber resource being lost to rust is becoming more valuable'. This disease is a problem that will not go away and which has a major impact on forestry operations and on the total economy in the southern United States.

Fusiform rust attacks several species of pine but primary damage is to loblolly pine (*Pinus taeda* L.) and slash pine (*P. elliottii* Engelm.); species such as *P. palustris* Mill. (longleaf pine) and *P. serotina* Michx. (pond pine) are subject to damage but on a much lesser scale than the first two mentioned. Of great importance, a couple of pine species native to the south-eastern United States are resistant to the disease; *P. echinata* Mill. (shortleaf pine) and *P. virginiana* Mill. (Virginia pine) are not susceptible to *C. fusiforme* although they are susceptible to related Cronartiums.

The history of fusiform resistance breeding in the South has been of special interest since I have been directly involved for nearly 30 years. Although there were some exceptions, in the early years pathologists assured me that there would be little benefit from breeding for fusiform rust resistance. Their idea was that the variation among trees and stands was chance or a direct response to local environmental variations. Further, I was led to believe there was no variability in the rust organism itself. As a result we were strongly urged not to waste time including fusiform rust resistance in our applied programs.

Based on early studies and observations, it became apparent that there was indeed a large component of resistance to fusiform rust in the southern pines. Especially good data, including quantitative inheritance patterns, was obtained from the very large Heritability Study sponsored by N.C. State University, the International Paper Company, the National Science Foundation, and the National Institutes of Health (Stonecypher, 1966; Blair, 1970; Kinloch, 1968).

THE INCREASED INCIDENCE OF FUSIFORM RUST

Fusiform rust has always been present in the southern pines but is has become progressively worse in recent years. When natural regeneration is closely observed, rust is quite evident but the sick trees are usually overtopped and killed through competition with the healthy trees. Since natural stands are established at several thousands per acre, the diseased trees are lost. The movement toward plantations does result in a greater prevalence of fusiform rust for several reasons:

1. Especially in the past, seed were purchased from anyone by quantity; cone collectors took cones from trees with the heaviest crops. Pines heavily infected with fusiform rust usually have heavy cone crops. Since susceptibility to the disease has a reasonably strong inheritance, the proportion of trees inherently more susceptible to fusiform rust is generally greater from nursery-grown seedlings than within the population in general.
2. Seedlings are protected by sprays in the nursery; this results in very susceptible individuals being grown to plantable size and then being field-planted. Because of site preparation and planting at wide spacing, the infected trees are given the chance to survive and grow rather than be killed by competition in dense natural stands.
3. Site preparation and fertilization result in an increase in infection (Miller, 1977; Gilmore & Livingston, 1958). I am convinced that in some instances where rust infection is bad, the added loss from fusiform rust more than offsets the added growth and quality expected from better site preparation and fertilization.
4. Although fusiform infection in the nursery is controllable, it sometimes is not done correctly and diseased seedlings have been shipped and outplanted throughout the Southeast.
5. Fire suppression has resulted in a strong buildup of oaks, the alternate host for the rust. This is an important reason for an increase in rust (Squillace & Wilhite, 1977).
6. The widespread planting of the very susceptible slash and loblolly pines is a strong reason for increased rust. Mixed pine-hardwoods, longleaf pine and hardwood stands are being site-prepared and planted to the susceptible species.
7. Some bad 'offsite' planting has resulted in heavy rust infection. When loblolly pine is planted in the extreme southern areas and when slash pine has been planted north of its range, it is very susceptible (Zobel & Zoerb, 1977). The latter has been particularly important in the past because slash pine was considered the 'wonder tree' and it was extensively planted to the north of its suitable sites, with disastrous results.

Fusiform rust, which was formerly not considered important or was a nuisance at best, has now become the most serious forest pest in the Southeast. Without control of some sort, whether silvicultural or through resistance breeding, many millions of acres of otherwise prime forest lands will be economically marginal or submarginal for pine production. Because of the difficulty of control of fusiform rust through forest management methods, the really only viable alternative is resistance breeding. Luckily the resistance component is strong (Blair, 1970; Kinloch, 1968).

Currently the very rapid spread of rust has slowed because of good attention to planting the proper geographic source and to good nursery practices. But in previous years there had been a grand mixing of both pine and rust genotypes. In the past it was common to grow seedlings in a nursery and then to send them wherever needed throughout the South. Since at the same time there was much nursery infection, different races of rust and pines were scattered throughout the South. How much the mixing will affect future breeding programs is not known, but it probably will cause added difficulty in an already difficult breeding program.

THE SELECTION AND BREEDING APPROACH WITHIN SPECIES

When the applied industrial programs were started in 1951 we knew essentially nothing about fusiform resistance inheritance patterns. Luckily, considerable fundamental research has been done, so much is now known about the disease (Jewell et al., 1962; Miller et al., 1976; Dwinell, 1973; Powers, 1975). In addition, a few persons such as Siggers (1947), Czabator (1971), and Wakeley (1954) were concerned many years ago about the potential of fusiform rust. But little action was taken relative to disease resistance in large plantation programs until the 1950's.

Our early formula was simple - 'Do not use any tree for a parent that has rust whatsoever', whether it be a limb or bole gall. Such rigor resulted in the rejection of many otherwise beautiful trees for seed orchard use and we encountered considerable criticism for being so critical, along with the comment 'After all, you don't even know if resistance to fusiform rust is inherited'.

It is hard to assess absolute gains from our early policy of rejecting all diseased trees. Based on inheritance patterns (Kinloch, 1968), we have gained considerably. But an 'error' was made. We combined trees selected from areas of low incidence of the disease with those parents selected from heavily diseased stands. When progeny-testing was done in areas of reasonably heavy infection (resistance is not assessed if the incidence of rust is less than 15 %), a number of the parents in the seed orchards from the lightly infected stands were quite susceptible to fusiform rust. The method of intensive selection where no infected trees were used and all trees were progeny-tested resulted in good gains in disease resistance. Over 1/3 of the planting of 500 000 acres each year by members of our Cooperative is in 'hot' fusiform areas. Plantations from seed orchard trees, which now supply all the seed needed for the 500 000 000 trees planted annually, confirm the better fusiform rust resistance.

In a series of tests, Goddard et al. (1975) found progeny of disease-free trees from heavily infected stands to have only half as much infection as those from lightly infected stands. When tested in 4 different areas, the following percentages of trees with rust were obtained:

Test location	From heavily infected stands	From lightly infected stands
Test 1	61	85
Test 2	36	63
Test 3	26	47
Test 4	6	14

There still are foresters who question the wisdom of using only disease-free parents, but I strongly advocate doing this. The gain of up to 35 % (Blair, 1970) when added to the 20 % or more gain from roguing based on progeny testing (Kinloch & Kelman, 1965) is well worth while. Based on progeny test results, a number of specialty disease resistance orchards have been established by several of the forest industries in the Southeast. Seedlings from these are planted in fusiform rust hot spots. Considerable quantities of seed are now available, enabling a profitable forest enterprise in areas where disease had previously been so bad that suitable tree crops could not be grown.

When the components of variance from the Heritability Study were used, gain predictions were made as to the improvement expected from rust-resistance orchards (Zobel et al., 1971). All parents used were rust-free, and gains represent improvement from mass selection combined with progeny testing. Using a selection intensity of the best 14 of 240 families tested, the predicted gain varied from 37 % to 55 % of the mean values, depending upon the year of planting.

Several persons are now making economic analyses of the monetary gains to be achieved, both from the standpoint of growth (Porterfield et al., 1975) and from the effects on wood properties (Veal et al., 1974). Light (30 %), medium (30 % - 60 %), and heavy (60 % +) severity class galls of loblolly pine were pulped by the kraft process. Compared with non-infected stemwood, Veal found the 3 classes yielded 4, 10 and 23 % less pulp and required 4, 8 and 20 % more alkali. Translated to an acreage basis, the heavily infected stands would yield 3 % less pulp per acre of plantation. The galled wood did produce more extractives. Porterfield (1973) assessed losses in volume for stands with varying degrees of infection. For the moderately infected level, the growth loss was 13 %; for the heavy level it was 24 %. If reduction in values for lumber and plywood are added, losses from fusiform rust infection are large indeed.

If I were to again start a tree improvement program I still would not use any diseased trees directly in the production seed orchard; I would, however, modify methods formerly used as follows:

1. Seeds from outstanding trees that have one or few small limb galls would be collected. Testing can be done quickly and efficiently in the United States Forest Service Fusiform Test Center in Asheville, North

Carolina (Phelps, 1977; Powers et al., page 427 in this book). The new methods being worked out (Walkinshaw et al., 1980) should be much more efficient than those used in the past; good greenhouse-to-field resistance correlations are being obtained.

2. I would be more particular in separating disease-resistance and normal seed orchards. The economics of a separation has been well documented by Porterfield et al. (1975), who showed that an organization with lands that are both rust-free and heavily rust-infected would gain considerably by having separate orchards. For the lightly infected areas, rust would not be a criterion of selection for the orchard, allowing greater gain in other characteristics.

Excellent studies have been established and more are underway to clarify the inheritance of rust resistance and gains from breeding. For example, the N.C. State University - Industry Cooperative established a series of disease diallel test plantings. The most resistant trees then known (plus a few susceptible parents) were control-pollinated, using a 22-tree half-diallel. The crosses were planted in areas selected as fusiform rust hot spots. The plantations are now 5 years old; readings of rust resistance have been made and results will be published soon. Indications are that the original choice of resistant clones was very effective. Sixteen of the 22 parents included in the half-diallel (selection of parents was based on limited testing) have exhibited outstanding resistance to rust under rigorous testing. The potential for gain appears to be extremely bright and promising.

Perhaps the most solid information on rust resistance was obtained from the North Carolina State University - International Paper Company Heritability Study at Bainbridge, Georgia. Results from all the tests from this area are too numerous to list but the more important after 20 years were:

1. Resistance to rust was very variable among the progeny of the 280 parent trees used in the study. On the worst rust sites, infection percent by family varied from 17 % to 100 %, with galls per tree varying from 0.5 to 10.8.
2. There were great differences in infection rate in plantings made in different years and on different sites. Infection on the different sites ranged from 32 % to 80 %; however, family rankings were quite stable on the 2 major sites tested.
3. Heritabilities of rust resistance were quite variable, depending upon years and families involved, varying from $h^2 = 0.04$ to $h^2 = 0.85$. Most values were in the area of $h^2 = 0.30$. It is clear there is sufficient inheritance of resistance to make good gains from resistance breeding (Zobel et al., 1971). Considerable of the genetic component of rust resistance is due to additive variance, as shown by the heritabilities, but in some instances non-additive variance was important.
4. Different families respond differently to rust over time. Some were

essentially destroyed while others seemed to be able to survive, grow and 'live with the rust'; a few families seem to outgrow and heal over the rust infection.

5. Little genotype-×-environment interaction was found.
6. Open-pollinated and control-pollinated tests gave similar rankings for the families tested.

One major conclusion from all the studies already assessed is that although breeding for fusiform rust resistance is profitable, it is difficult, time-consuming and expensive. Despite the mass of new information available, with more being reported all the time, the more fusiform resistance is studied the clearer it is how little we really know about the genetics of the host, of the pathogen, and their interactions. The problem is complicated because of the magnitude of variability in both host and pathogen. Variability in the host is very well documented; a good example is one small group of clones in a company seed orchard in Georgia. Infection by family in the seed orchard varied from 0 % to 92 % and there was a reasonably strong correlation between parental and progeny performance. However, the same clones in different orchards sometimes produce progeny with differing infection rates (Powers & Zobel, 1978), probably indicating the differing mix of parents within the different orchards.

A major approach we have used successfully, and one I recommend for the future, is to breed for resistance within a species despite the complexity of host-parasite-environmental relationships.

OTHER METHODS TO REDUCE FUSIFORM RUST

There are other useful methods to help in fusiform rust control that must be used if proper results are to be obtained. These are from use of (1) geographic variation within species, (2) hybridization, (3) different species, (4) control of the alternate host through silvicultural practices.

Geographic variation

One important difference in rust susceptibility that was early observed is related to source of seed within a species. *P. taeda* from East Texas and from the northern extremes of its range is much more resistant than loblolly from the center of the range (Wells, 1966). Trees from the Livingston Parish area in Louisiana are noted for their good growth and relatively high disease resistance (Wells & Switzer, 1975). The use of seed from such known resistance areas has become widespread and has become a standard and successful practice along the Gulf Coast area. There are dangers in moving resistant sources too far, however, and some losses have occurred (Zobel & Zoerb, 1977). Use of a proper, proven superior geographic source is the easiest and best way to obtain quick and large improvements in fusiform resistance in the southern pines.

Hybridization

Usually a hybrid has characteristics intermediate between its parents but this is not always true. For example, although the hybrids between the susceptible slash and resistant shortleaf pine is intermediate in susceptibility, the loblolly × shortleaf hybrid appears to be more resistant (Florence & Hicks, 1976; LaFarge & Kraus, 1975). The moderately infected longleaf × susceptible loblolly pine appears to very susceptible. No matter which crosses are made, the characteristics of the particular parents being used to make the cross largely determine the performance of any one hybrid. The possibility of use and value of hybrids such as the loblolly × shortleaf cross for operational planting is considerable. If efficient methods to make the hybrids can be developed, their use in 'hot spots' would be of great value.

Use of different species

The southern pines vary from the very fusiform-resistant shortleaf and Virginia pines to the very susceptible slash and loblolly. There are other southern pine species that can be used on proper sites that have only a low susceptibility to rust (Powers, 1975). Many of the heaviest losses are in slash pine grown on the deep sand sites that are marginal for wood production; sometimes the stands are totally destroyed by rust and must be abandoned. On such sites, longleaf pine can be used (Wakeley, 1968) or sand pine (*P. clausa* (Chapm.) Vasey L.); especially the Choctawhatchee variety which is immune to fusiform rust (Burns, 1972) grows remarkably well. Sand pine is rapidly developing into a major species on the deep sands. Shortleaf pine is resistant to fusiform rust; it is classed as a slow-starting species but genetically improved trees grow quite well and are nearly competitive with loblolly pine on some sites. During the past years I had expected shortleaf pine to be more widely used but this has not happened - the bias against it by most foresters is too strong.

Control of the alternate host and good silvicultural practices

Fusiform rust has oak (*Quercus* sp.) as the alternate host. Although oaks are found widespread throughout the South, they have become much more numerous with the strong practice of fire control. Recent studies such as those by Squillace & Wilhite (1977) have shown that there is a relationship between the incidence of fusiform rust and the abundance of oak, especially water oak (*Q. nigra* L.). The authors consider that the presence of a high proportion of oak indicates hazardous rust sites. Unfortunately, practical methods of controlling oak to the degree required are not available. Control burning does reduce oak, but cannot be done before the most susceptible period in a pine plantation has passed.

REFERENCES

Anon., 1977. South. For. Dis. and Insect Res. Coun. Management of Fusiform Rust in Southern Pines. Symp. Proc. (In cooperation with U.S. Forest Service, N.C. State Univ. and Univ. of Fla.) 163 p.

Blair, R.L., 1970. Quantitative inheritance of resistance to fusiform rust in loblolly pine. Ph. D. Thesis, N.C. State Univ., Raleigh. 87 p.

Burns, R.M., 1972. Management of sand pine plantations. Proc., Sand Pine Symp. p. 124-133.

Czabator, F.J., 1971. Fusiform rust of southern pines - a critical review. USDA For. Ser. Res. Pap. SO-66. South. For. Expt. Sta. 39 p.

Dwinell, L.D., 1973. Basidiospore cast from telia of Cronartium fusiforme on half-sib families of northern red and water oaks. Phytopath. 63: 441.

Florence, L.Z. & R.R. Hicks, 1976. Fusiform rust resistance of progeny from natural loblolly-×-shortleaf pine hybrids. Proc., Twenty-Third NE For. Tree Impr. Conf. p. 10-19.

Gilmore, A.R. & K.W. Livingston, 1958. Cultivation and fertilizing a slash pine plantation: effects on volume and fusiform rust. Jour. For. 56: 481-483.

Goddard, R.E., R.A. Schmidt & F. Vande Linde, 1975. Immediate gains in fusiform rust resistance in slash pine from rogued seed production areas in severely infected plantations. Proc., Thirteenth South. For. Tree Impr. Conf., Raleigh, N.C. p. 197-203.

Holley, D.L. & M.A. Veal, 1977. Economic impact of fusiform rust. Symp., Management of Fusiform Rust in Southern Pines. South. For. Dis. and Insect Res. Coun., Gainesville, Fla. p. 39-50.

Jewell, F.F., R.P. True & S.L. Mallett, 1962. Histology of Cronartium fusiforme in slash pine seedlings. Phytopath. 52: 850-858.

Kinloch, B.B., 1968. Genetic variation in susceptibility to fusiform rust in loblolly pine. Ph. D. Thesis, N.C. State Univ., Raleigh, N.C. 47 p.

Kinloch, B.B. & A. Kelman, 1965. Relative susceptibility to fusiform rust of progeny lines from rust-infected and non-infected loblolly pines. Plant Dis. Reporter 49: 872-874.

LaFarge, T. & J.F. Kraus, 1975. Resistance of shortleaf-×-loblolly pine hybrids to inoculation with fusiform rust. Proc., Thirteenth South. For. Tree Impr. Conf. p. 113-118.

Miller, T., 1977. Fusiform rust management strategy in concept: site preparation. Symp., Management of Fusiform Rust in Southern Pines. South. For. Dis. and Insect Res. Coun., Gainesville, Fla. p. 110-115.

Miller, T., E.B. Cowling, H.R. Powers & T.E. Blalock, 1976. Types of resistance and compatibility in slash pine seedlings infected by Cronartium fusiforme. Phytopath. 66: 1229-1235.

Phelps, W.R., 1977. Incidence and distribution of fusiform rust. Symp., Management of Fusiform Rust in Southern Pines. South. For. Dis. and Insect Res. Coun., Gainesville, Fla. p. 25-31.

Phelps, W.R., 1977. Screening center for fusiform rust. For. Farmer 36: 11-14.

Porterfield, R.L., 1973. Predicted and potential gains from tree improvement programs - A goal-programming analysis of program efficiency. Ph. D. Thesis, Yale Univ. 240 p.

Porterfield, R.L., B.J. Zobel & F.T. Ledig, 1975. Evaluating the efficiency of tree improvement programs. Sil. Gen. 24: 33-44.

Powers, H.R., 1975. Relative susceptibility of five southern pines to Cronartium fusiforme. Plant Dis. Rep. 59: 312-314.

Powers, H.R., J.P. McClure, H.A. Knight & G.F. Dutrow, 1974. Incidence and financial impact of fusiform rust in the South. Jour. For. 7: 398-401.

Powers, H.R. & B.J. Zobel, 1978. Progeny of specific loblolly pine clones vary in fusiform rust resistance according to seed orchard of origin. For. Sci. 24: 227-230.

Siggers, P.V., 1947. Temperature requirements for germination of spores of Cronartium fusiforme. Phytopath. 37: 855-864.

Squillace, A.E. & L.P. Wilhite, 1977. Symp., Management of Fusiform Rust in Southern Pines. South. For. Dis. and Insect Res. Coun., Gainesville, Fla. p. 59-70.

Stonecypher, R.W., 1966. Estimates of genetic and environmental variances and covariances in a natural population of loblolly pine (P. taeda L.). Tech. Bull. No. 5, Southlands Expt. For., International Paper Co., Bainbridge, Ga. 128 p.

Veal, M.A., R.L. Blair, J.B. Jett & W.T. McKean, 1974. Impact of fusiform rust on pulping properties of young loblolly and slash pines. TAPPI Biol. Conf., Seattle,

Wash. p. 47-58.
Wakeley, P.C., 1954. Planting the southern pines. USDA For. Ser. Agri. Mono. 18. 233 p.
Wakeley, P.C., 1968. Rust susceptibility of longleaf pine associated with brown spot resistance and early commencement of height growth. For. Sci. 14: 323-324.
Walkinshaw, C.H., T.R. Dell & S.D. Hubbard, 1980. Predicting field performance of slash pine families from inoculated greenhouse seedlings. USFS South. For. Expt. Sta. Res. Pap. SO-160 (in press).
Wells, O.O., 1966. Variation in rust resistance among pine seed sources and species in a Mississippi planting. For. Sci. 12: 461-463.
Wells, O.O. & G.L. Switzer, 1975. Selecting populations of loblolly pine for rust resistance and fast growth. Proc., Thirteenth South. For. Tree Impr. Conf. p. 37-44.
Zobel, B., R. Blair & M. Zoerb, 1971. Using research data - disease resistance. Journ. For. 69: 486-489.
Zobel, B.J. & M. Zoerb, 1977. Reducing fusiform rust in plantations through control of the seed source. Symp., Management of Fusiform Rust in Southern Pines. South. For. Dis. and Insect Res. Coun., Gainesville, Fla. p. 98-109.

Testing for resistance to fusiform rust of pine

H.R. Powers, Jr.[1], Susan D. Hubbard[2] & Robert L. Anderson[3]

ABSTRACT

An automated system for artificially inoculating large numbers of pine seedlings under standardized conditions has been developed and is now in use at the Resistance Screening Center at Asheville, North Carolina. This is the only center in the world at which seedlings can be tested for resistance to a forest tree disease on a routine, commercially operational basis. The Center has evaluated the relative levels of resistance to fusiform rust of progeny of over 2000 clones from seed orchards of many pulp and paper companies and states from across the southern United States.

A reliable procedure for evaluating the relative resistance of host families and varieties is a key element in any disease control program utilizing genetic resistance. Field progeny tests are usually carried out for this purpose, but there are some inherent problems in using such tests for some forest tree diseases. In the case of fusiform rust, caused by *Cronartium quercuum* (Berk.) Miyabe ex Shirai f. sp. *fusiforme*, heavy infection does not take place every year. This means progeny tests must be maintained for several years. Field plantings are also expensive to install and maintain. Artificial inoculation tests carried out on 6- to 8-week-old seedlings can provide information to supplement progeny test data. Artificial inoculation is relatively inexpensive and provides data in a shorter period of time than progeny testing.

With fusiform rust of southern pines, the need for artificial inoculation techniques became apparent in the late 1950's, when variation in resistance to fusiform rust was detected among slash pine (*Pinus elliottii*

1. Chief Plant Pathologist, Southeastern Forest Experiment Station, USDA, Forestry Sciences Laboratory, Athens, Georgia 30602, USA
2 and 3. Resistance Screening Center Manager and Supervisory Plant Pathologist, Southeastern Area State and Private Forestry, USDA, Asheville, North Carolina 28803, USA

Engelm. var. *elliottii*) selections (Barber et al., 1957). The first system devised for this purpose (Jewell, 1960) used telia-bearing oak leaves suspended over pine seedlings inside a double tent kept moist with water sprays. In many tests this procedure worked very well, but in some instances there were difficulties, such as light infection in one area of the tent and very heavy infection in another. One of the biggest problems was that inoculations were limited to something less than 2 months in the spring of the year, because oak leaves with telia could be collected in nature only at that time. This is a tremendous limitation for a large-scale screening program.

In the early 1970's the U.S. Forest Service forest disease research project at Athens, Georgia, was asked to develop the technology for large-scale screening with standardized numbers of spores under controlled conditions. The goal was a test that would be repeatable and would provide similar results in any year. Fortunately, the Athens group was able to adapt many of the procedures utilized in cereal rust disease programs to the needs of the fusiform rust program. Many people are not aware of the debt that we owe the cereal rust researchers in the development of many of our current procedures. For example, the vacuum-dry spore storage process, now widely used on aeciospores of the fusiform rust organism, was developed for long-term storage of urediospores of cereal rusts. The cyclone spore collectors we use for aeciospore collection were also developed by the wheat disease researchers. Our progress with fusiform rust owes much to earlier workers in the field of plant pathology.

Dr. Thomas Miller made a key innovation in the development of the new inoculation system for fusiform rust. He was able to concentrate large numbers of basidiospores of the fusiform rust organism by use of Millipore filters (Miller, 1970). He found that these spores could be stored for some time on the filter pads. Also important was the adaptation of the Coulter electronic particle counter to standardize spore density for inoculations (Dwinell, 1973). The first atomized spray procedure was developed by Matthews & Rowan (1972), utilizing the techniques for spore concentration and standardization. This is now known as the concentrated basidiospore spray (CBS) inoculation system. This system involves the following steps:

1. Aeciospores representative of specified areas are collected (Fig. 1).
2. These spores are processed by screening, drying, and sealing under vacuum.
3. The spores are used to inoculate seedlings of northern red oak (*Quercus rubra* L.) in order to produce abundant telial columns on the undersides of the large juvenile leaves.
4. Oak leaves are suspended over acidified water at 20 °C to promote the germination of teliospores and the discharge of sporidia onto the water surface (Fig. 2).
5. Sporidia are poured into a series of Millipore filters and collected on

Figure 1. Aecidiospore collection in a natural forest stand.

Figure 2. Inducement of teliospore germination and collection of basidiospores.

Figure 3. Inoculation of pine seedlings with basidiospore suspension.

Figure 4. 8-month old slash pine seedlings with gall development (inoculated at 6 weeks).

the filter pads. The pads with the spores are washed into a solution, and the solution is diluted to the desired spore density by use of the electronic particle counter.

6. The resulting spore suspension is applied by spray nozzles onto 6- to 8-week-old seedlings (Fig. 3).

7. Response of the host is assessed 6 to 9 months after inoculation, depending on the tree species (Fig. 4).

All of these procedures, plus many other individual steps, were described in a procedural manual (Matthews et al., 1971) and turned over to Forest Insect and Disease Management in the U.S. Forest Service Southeastern Area State and Private Forestry. That organization's assignment was to use this basic technology to develop a large-scale, operational screening center which could handle several hundred thousand seedlings annually. The CBS system was found to be effective for such an operation (Laird & Phelps, 1975), and the Resistance Screening Center was established at Bent Creek, N.C. Its assignment was to screen seed lots for all tree improvement programs concerned with selecting and breeding for fusiform rust resistance. Many innovative improvements were made in the basic system in order to test large numbers of seedlings under standardized conditions. Hand-held sprayers were replaced with spray nozzles mounted over a conveyor belt, so the seedlings were exposed to the spray inoculum for a precise number of seconds as they moved along the conveyor. Continual refinement of the inoculation system has resulted in consistent distribution of uniform droplets of inoculum on the pine seedlings. This has been achieved through experimentation with various nozzles and air pressures, and use of peristaltic pumps.

The Resistance Screening Center routinely collects aeciospores from loblolly (*P. taeda* L.) and slash pine in 30 two-three county areas from Texas to North Carolina. Samples of 30 galls in 3 locations are mixed from each area. The client selects the source area of the spores with which his seed lots are to be tested, usually to coincide with the breeding region of the families involved.

Resistant pine selections are either open-pollinated or full-sib families, and there is much genetic variation within a family. Therefore, resistance to fusiform rust is relative, not absolute. Performance of each seed lot is presented to the user in comparison to standard resistant and susceptible checks (Table 1). Results also include an analysis of variance and a Duncan's multiple range test on seed lot means as aids to interpretation of the data. These analyses have shown the repeatability of family means over replications to be quite high. As in field progeny tests, results for a seed lot are most reliably compared to those for other lots within the same test. The data derived from these tests, however, can be used in conjunction with information coming from other sources, usually field progeny tests.

Table 1. Information provided users of the Resistance Screening Center regarding the results with their seedlots.

Infection rank	Seedlot No.	Test mean (% galled)	Duncan's multiple range on seedlot test means
Check	Resistant (Liv. Parish)	40	.
1	Seedlot 1	52	.
2	Seedlot 2	54	..
3	Seedlot 3	57	...
3	Seedlot 4	57	...
3	Seedlot 5	57	...
6	Seedlot 6	60	....
7	Seedlot 7	61	...
7	Seedlot 8	61	...
9	Seedlot 9	63	...
10	Seedlot 10	64	...
10	Seedlot 11	64	...
10	Seedlot 12	64	...
Check	Susceptible (11-23)	64	...
13	Seedlot 13	65	...
14	Seedlot 14	67	..
15	Seedlot 15	70	.
	All seedlots	60	

A key question about artificial inoculation is the correlation between its results and the performance of the same materials in the field. Observations to date suggest that the results may relate better to field performance for slash than for loblolly pines. With both species, however, resistant materials can be easily identified. The highly susceptible materials in the greenhouse are usually highly susceptible in field plantings. Problems usually involve families rated as intermediate in resistance. This seems to be particularly true for loblolly pine; we have found some loblolly families which appeared to be relatively susceptible in inoculation tests, but were moderately resistant under field conditions. However, we have never found the reverse to be true; that is, we have never rated a family as resistant in greenhouse tests and found it to be susceptible in field plantings. Thus, the CBS testing procedure is quite rigorous, and estimates of overall resistance of the seed lots are conservative.

Methods for observing and interpreting CBS results are being refined.

Preliminary data suggest that the predictability of greenhouse results can be improved by looking at several symptom types rather than only at the presence or absence of a gall (Walkinshaw et al., 1980). These refinements appear to improve the correlation between greenhouse and field results.

No artificial inoculation system, whether on late blight of potatoes, wheat stem rust, or any other disease, is anything but an artificial system. Perfect correlations with field results cannot be expected. In fact, field results are quite variable, and progeny tests in 2 field locations seldom agree completely. Results from the Resistance Screening Center are meant to give additional information to help in the evaluation of specific families as to their resistance to rust.

The CBS inoculation system is being used very effectively at the Italian Forest Disease Center in Florence, Italy, in work on pine blister rust caused by *Cronartium flaccidum* (Alb. & Schw.) Wint. Since maritime pine (*Pinus pinaster* Ait.), and Italian stone pine (*P. pinea* L.) differ considerably in susceptibility, scientists in Florence use different spore concentrations for inoculating these 2 species. A much lower spore dosage is applied to Italian stone pine, which is the more susceptible species (Raddi, 1976).

During the first several years of operation the Resistance Screening Center has evaluated progeny of over 2000 clones from state and industry seed orchards in the high rust hazard area. These results have helped tree improvement programs across the South in evaluating the material in their seed orchards. Many current tests are of second generation selections, which should have considerably improved rust resistance. Preliminary tests have already been carried out on the feasibility of evaluating pines for resistance to needle spot diseases and to white pine blister rust. Work on these two areas may be initiated in the future.

In addition to evaluating resistance of seed lots from the various tree improvement cooperatives, the Screening Center inoculates and evaluates research materials for various universities and research centers across the South. It is relatively easy for the Screening Center to set up and carry out an inoculation on a large number of seedlings. Several research studies have already been carried out under special arrangements, and many more such cooperative ventures probably will be undertaken in the future.

REFERENCES

Barber, J.C., K.W. Dorman & E. Bauer, 1957. Slash pine progeny tests indicate genetic variation in resistance to rust. U.S. Department of Agriculture Forest Service, Southeastern Forest Experiment Station Research Notes 104. Asheville, North Carolina. 2 p.

Dwinell, L.D., 1973. The relationship of the oak host to basidiospore cast of Cronartium fusiforme. Phytopathology 63: 441 (abstract).

Jewell, F.F., 1960. New pine hosts for southern fusiform rust. Plant Disease Reporter 44: 673.

Laird, P.P. & W.R. Phelps, 1975. A rapid method for mass screening of loblolly pine and slash pine seedlings for resistance to fusiform rust. Plant Disease Reporter 59: 238-242.
Matthews, F.R. & S.J. Rowan, 1972. An improved method for large-scale inoculations of pine and oak with Cronartium fusiforme. Plant Disease Reporter 56: 931-934.
Matthews, F.R., S.J. Rowan, T. Miller, L.D. Dwinell & H.J. Duncan, 1971. Procedures used in fusiform rust research. Office Report, U.S. Department of Agriculture Forest Service. Athens, Georgia. 16 p.
Miller, T., 1970. Inoculation of slash pine seedlings with stored basidiospores of Cronartium fusiforme. Phytopathology 60: 1173-1174.
Raddi, P., 1976. Indagine sui tipi di macchie fogliari e sul loro significato in discendenze di Pinus pinaster infette da Cronartium flaccidum. Rivista di Patologia Vegetale 12: 91-98.
Walkinshaw, C.H., T.R. Dell & S.D. Hubbard, 1980. Predicting field performance of slash pine families from inoculated greenhouse seedlings. U.S. Department of Agriculture Forest Service Research Paper SO-160. Southern Forest Experiment Station, New Orleans, Louisiana. 6 p.

Italian studies on resistance to pine blister rust (Cronartium flaccidum)

P. Raddi & A. Ragazzi

Centro di studio per la patologia delle specie legnose montane, CNR Instituto di Patologia e Zoologia forestale e agraria, Firenze, Italia

ABSTRACT

This presentation provides a summary of studies of the biology of *C. flaccidum* and points out some problems which we meet in the breeding program for blister rust resistance. The biology studies deal with the possibility of storage of aeciospores, basidiospore production in controlled environments, and the mode of penetration of pine needles by *C. flaccidum*. We describe the breeding program for blister rust resistant pines and discuss the results obtained.

INTRODUCTION

Blister rust of 2-needle pines is caused by *Cronartium flaccidum* (Alb. & Schw.) Wint. and is widespread in several countries of Europe within the ranges of Scots pine (*Pinus sylvestris* L.), maritime pine (*P. pinaster* Ait.), Austrian pine, Villetta Barrea and Laricio pine (*P. nigra* Arnold), Italian stone pine (*P. pinea* L.), Swiss mountain pine (*P. mugo* Turra), Aleppo pine (*P. halepensis* Mill.), and Brutia pine (*P. brutia* Tenn.). In central and southern Italy in the past 20 years epidemic outbreaks of this disease have caused severe damage in plantations and in nurseries of 2-needle pine species (Moriondo, 1975).

The primary objectives of this presentation are to provide a summary of studies of the biology of *C. flaccidum* and to point out some problems which we meet in the breeding program for blister rust resistance in Italy.

C. flaccidum is heteroecious with one part of the life cycle on the leaves of white swallowwort (*Vincetoxicum officinale* Moench), which is the most frequent in Italy, *Gentiana* spp., *Grammatocarpus* spp., *Impatiens* spp., *Loasa* spp., *Nemesia* spp., *Paeonia* spp., *Pedicularis* spp., *Tropaeolum* spp., *Verbena* spp., and the alternate generation on pine stems and branches. The most frequent pine hosts are: Italian stone pine, maritime pine, Brutia pine, Austrian pine, Villetta Barrea pine, Laricio pine, Swiss mountain pine, and Aleppo pine. Other species of exotic pines with the exception of ponderosa pine seem to have a very high degree of resistance to blister rust after artificial inoculation (Raddi & Fagnani, 1978a).

Infection occurs through cotyledons, primary and secondary needles and possibly tender stem tissue. The first signs of infection are yellow or red needle spots. These first become visible 8-10 weeks after inoculation, but are very evident after 6-8 months. Infection occurs on the needle, and through the branches the fungus then invades the stem. Diseased seedlings usually die within 2 or 3 years after inoculation, when the cambium surrounding the canker is killed. Pines are susceptible from the cotyledonary stage until they become old, but infections decrease as tree age increases and the progress of the disease is slowed. The greatest mortality is in pine plantations up to 4-5 years of age.

The wide range of host species suggests that the fungus evolved simultaneously with its host. In central Italy the main hosts of blister rust are *P. pinea*, *P. pinaster*, and *P. nigra*, each with different climatic requirements. Sometimes in the overlapping zones of the habitats of 2 of these pine species one is found badly damaged by rust and the other perfectly healthy. It has been demonstrated (Mittempergher & Raddi, 1977) that in central Italy blister rust exists in different environments as populations of diverse pathogenicity, each of which is best adapted to infect the pine species prevalent in its own habitat. This has an important bearing on the work of selection for blister rust resistance. Our research work deals with: (1) some aspects of fungus biology and methodology connected with the breeding program, and (2) the breeding for blister rust resistant pines.

SOME ASPECTS OF FUNGUS BIOLOGY AND METHODOLOGY

The goal of such studies is to standardize the production of basidiospores of *C. flaccidum* and to provide a better understanding of the infection processes and relation and expression of host resistance in individual pine selections. The use of inoculum collections from several areas insures a good representation of fungal sources. In central Italy it is difficult to collect telia-bearing leaves of white swallowwort from several localities at the same time because of its different stages of vegetative growth.

Research is addressed to the culture of white swallowwort plants in the greenhouse or in a controlled-environment growth chamber and to the harvest and preservation of aeciospores and basidiospores. Aeciospores are collected from pine cankers with a cyclone spore collector during the warmest hours of clear days. They are prepared immediately for long-time storage and vacuum packed in aluminium foil at 5 °C. With this vacuum storage technique the germination capacity suffered a reduction of 25 % after 1 year of storage. Aeciospores can be stored for 90 days when lyophilized and kept in a container full of gaseous nitrogen. In this case a steady increase in germinability was noted during the first month of storage, but it decreased rapidly after 90 days (Ragazzi, 1978).

The storage possibility of aeciospores has permitted inoculation of the white swallowwort anytime during the year. But there are some problems concerning telia production when white swallowwort is artificially cultured in controlled environments. The under surface of young leaves of white swallowwort is inoculated with a water suspension of aeciospores. Usually abundant urediospores were produced on leaves, and the neighbouring plants then became infected by the urediospores. The presence of telial columns was never noted. In field experiments white swallowwort plants, inoculated each month of the year, always gave urediospores but only inoculations made in late spring and summer developed urediospores and well-distributed crops of telia. For these reasons, till now artificial inoculations of pine necessarily have been confined to the period from July to September, when temperatures are conducive to good telia formation on white swallowwort.

Another aspect of the studies on the biology deals with the mode of penetration of pine needles by *C. flaccidum* to determine whether non-spotted seedlings have escaped infection or whether the absence of needle spots may indicate another mechanism of resistance. In fact, the open-pollinated families of maritime pine with the highest percentages of non-spotted seedlings also tended to have the highest number of 'spotted-only' seedlings, i.e., seedlings which are able to stop the fungus mycelium in the needle tissue. The variability of this character among the families seems to indicate the existence of mechanisms which stop the pathogen in the first stages of the infection process by influencing the way *C. flaccidum* mycelium behaves and grows in the needle tissue of resistant maritime pines.

A fluorescent labelling technique was used for observation of basidiospores of *C. flaccidum* on the needles of 5 species (maritime pine, Austrian pine, Villetta Barrea pine, Laricio pine, and Scots pine). Basidiospores on the needles were labelled by soaking the needles for 10 minutes in a solution of Calcofluor (0.05 ml on 100 ml of distilled water). Observations of basidiospore behavior on the needle surfaces of the above-mentioned pine species pointed out the existence of significant differences for the following traits (generally, Scots pine gave the lowest values):

- percentage of basidiospore germination after 40 hours: 44 to 93,
- beginning of germination: 5 to 20 hours (Scots pine),
- germ tube length after 40 hours: 1-10 μm to 200-400 μm,
- percentage of branching germ tubes: 13 to 71,
- percentage of germ tubes that penetrated between the guard cells: 7 to 64.

The mechanism of resistance that is being expressed in 'spotted-only' seedlings is being investigated by microscopic examination of mycelial growth in needles from susceptible and resistant maritime pines, but results are not yet available.

The nursery-bed inoculation gave good results for identifying resistance mechanisms and the most resistant families of maritime pine. Nevertheless, variation in infection intensity across a large test was associated with the location of the seedlings in the nursery beds. It created relatively high variances among the replicates. To reduce this variability the concentrated basidiospore spray system, which was developed by U.S. Forest Service scientists at the Forestry Sciences Laboratory, Athens, Georgia, was tested on a large scale. The objective was to determine the inoculum-density levels capable of causing successful infection on Italian stone pine and maritime pine seedlings which had been transplanted into planter trays and then were inoculated at about 100 days of age.

On the basis of the results (Raddi & Fagnani, 1978b) we can suggest a concentration of 12 500 basidiospores/ml or less for the large scale inoculation of Italian stone pine and of 75 000 basidiospore/ml for maritime pine. These densities caused an average number of needle spots per plant of 18.4 ± 1.4, an infection level that gave us the most reliable results in producing 'spotted-only' seedlings.

BREEDING FOR BLISTER RUST RESISTANT PINES

Research started in 1973 by selecting parent trees free from rust infections in areas where the incidence of disease was more than 80 %. Progenies from open pollination of 40 Italian stone pines and 33 maritime pines were transplanted into nursery beds. Fifteen naturally infected leaves of white swallowwort bearing telia of *C. flaccidum* were attached to cellophane tape and placed 20 cm above the test seedlings in each row across the nursery bed. The seedlings were inoculated during their second growing season.

Italian stone pine seedlings inoculated in the primary needle stage were infected without any evidence of resistance (Table 1). Maritime pine was less susceptible than Italian stone pine: significant differences were found among families for the trait 'seedlings with needle spots without fungus fruiting bodies on the stem' (Raddi et al., 1979).

The breeding project of this Center is based on the following methods:

1. Inoculation of maritime pine progenies with a controlled mechanical spray inoculation system providing a known and uniform quantity of inoculum in a given time (conveyor speed is controlled at 10 m/min and volume passing through both spray nozzles is standardized at 7 ml of spore suspension per tray).
2. For each family, 2 sets of 100 seedlings each are inoculated at about 100 days of age, in each of 2 runs 1 month apart.
3. All seedlings are inspected for the presence of needle spots and spermagonia. Seedlings with spermagonia on the stem are removed because none of them recovered in previous tests.

Table 1. Results of inoculation of Italian stone pine (*Pinus pinea*) and maritime pine (*Pinus pinaster*) seedlings in nursery beds with the blister rust fungus, *Cronartium flaccidum*. The number of families were 40 for Italian stone pine and 33 for maritime pine. Analyses of variances were computed on the percentages transformed to arcsin. F_f, F_r, F_{fr}, and F_d are F values referring to the effects due to the families, dates of inoculation, their interaction, and to the type of seedling distribution in the nursery beds, respectively. ** significance at 0.01 level, ns not significant.

Seedlings	*Pinus pinea*			*Pinus pinaster*						
	number	percentage	seedlings/ family	number	percentage	seedlings/ family	F_f	F_r	F_{fr}	F_d
inoculated	5985	100.0	148 to 150	4587	100.0	124 to 143				
needle spotted	5981	99.9	148 to 150	3824	83.4	84 to 134	**	**	ns	ns
with fungus fruiting bodies	5440	91.0	125 to 143	2399	52.3	37 to 91				
alive 4 years after inoculation:										
'spotted-only'	none			645	14.0	8 to 41		**	ns	ns
'with cankers'	practically none			710	15.5	9 to 33				

4. 'Spotted-only' seedlings are transplanted from the trays into nursery beds 18 months after the inoculation to obtain furter information on their behavior towards the blister rust.

5. Nonspotted seedlings, which we assume were able to stop the pathogen in the first steps of the infection process or are escapes, are transplanted into nursery beds and inoculated again by placing telia-bearing leaves of white swallowwort over them.

6. All 'spotted-only' seedlings and those exposed to 2 inoculations without fungus fruiting bodies on the stem having formed and alive 4 years after the inoculation are transplanted into plantations located in areas of severe incidence of the blister rust. If it is necessary, artificial rust epidemics are created by planting white swallowwort plants adjacent to the plantation and inoculating these bushes with aeciospores of *C. flaccidum* from different areas. The most promising seedlings in these plantations are then subjected to another screening test, which takes into consideration rust resistance and also seedling growth rate and form.

7. Undesirable trees, including all rust susceptible trees, are removed at regular intervals, thereby automatically converting the plantations into potential seedling seed orchards.

Our current status of knowledge about the biology of *C. flaccidum* and the mechanisms of rust resistance is sufficient to obtain a generally good level of resistance to blister rust. Every effort of this breeding program is addressed toward providing a broad genetic base of resistance in the parental trees of maritime pine.

REFERENCES

Mittempergher, L. & P. Raddi, 1977. Variation of diverse sources of Cronartium flaccidum. European Journal of Forest Pathology 7: 93-98.

Moriondo, F., 1975. Caratteristiche epidemiche della ruggine vescicolosa del pino, Cronartium flaccidum (Alb. et Schw.) Wint. in Italia. (Epidemic characteristics of pine blister rust by C. flaccidum in Italy.) Annali dell'Academia Italia Scienze forestali 24: 331-406.

Raddi, P. & A. Fagnani, 1978a. Relative susceptibility to blister rust caused by Cronartium flaccidum of several species of pine. European Journal of Forest Pathology 8: 58-61.

Raddi, P. & A. Fagnani, 1978b. Miglioramento genetico del pino per la resistenza alla ruggine vescicolosa: controllo ed efficacia della densità di inoculo. (Blister rust resistance in maritime and Italian stone pine: effects of the inoculum density level in resistance testing.) Phytopathologia mediterranea 17: 8-13.

Raddi, P., L. Mittempergher & F. Moriondo, 1979. Testing of Pinus pinea and P. pinaster progenies for resistance to Cronartium flaccidum. Phytopathology 69: 679-681.

Ragazzi, A., 1978. Conservazione e germinazione di ecidioconidi di Cronartium flaccidum (Alb. et Schw.) Wint. (Storage and germination of aeciospores of C. flaccidum.) Rivista di Patologia Vegetale 14: 17-28.

Lophodermella sulcigena in clones and progenies of scots pine in Finland

Risto Jalkanen

Finnish Forest Research Institute, Rovaniemi Research Station, Eteläranta 55, SF-96300 Rovaniemi, Finland

ABSTRACT

The incidence of gray needle cast (*Lophodermella sulcigena*) was examined in progeny trials and forest tree seed orchards of Scots pine (*Pinus sylvestris*). Whenever needle cast was present in a progeny test, all progenies were diseased at least slightly. There were highly significant differences in susceptibility to gray needle cast between the 30 progenies studied. The further the progeny was transferred to the cultivation location, the higher was its resistance. However, the progenies from north of the cultivation location were healthier than the ones south of it. More than half of the clones in the seed orchards were completely healthy, the differences among the clones being considerably greater than in the progeny trial. Both healthy and heavily infected trees were found within diseased progenies or clones.

INTRODUCTION

Epidemics of gray needle cast (*Lophodermella sulcigena*) (Rostr.) v. Höhn) have occurred every 10-15 years in Finland, (from the collections of prof. Kujala, the Finnish Forest Research Institute). The present extensive epidemic began in 1976 and is still continuing. The situation is most serious in even-aged 10 to 20-year-old pine plantations growing on fertile sites, where the disease has spread throughout entire stands. In natural stands almost 100 % of the trees are healthy.

Gray needle cast infects growing needles of Scots pine in June-July. A completely infected tree loses all of its current year needles. Mature hysterothecia develop within one year in the infected needles. Disseminating spores will again infect growing needles. The disease does not kill a tree, it only causes severe growth losses.

The fungus is also common in field experiments, and has been found in seed orchards, too.

Field trial no. 232 is composed of 32 plus pine progenies with open pollinated seed. From these, 11 subtrials were established in different parts of Finland in 1966 (Fig. 1). Each subtrial consists of 6 blocks. At the start of the trial, there were 25 seedlings of the same progeny in each plot. The number of progenies included in the different subtrials varies between 18-29.

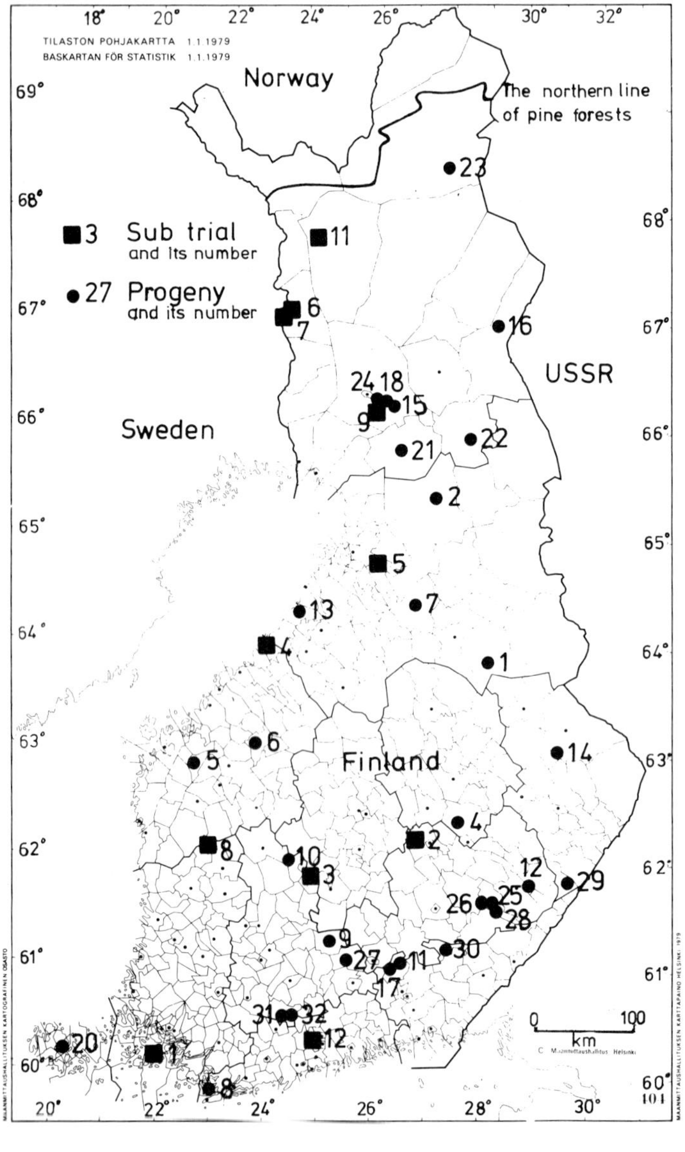

Figure 1. Subtrials and progenies in field trial no. 232.

Gray needle cast was found in 4 of the subtrials covering 30 of the progenies being tested. In 2 of them the epidemic began in 1978 and in the other 2 in 1979. The stands were inventoried during the year the epidemic started. Two seed orchards about 20 years old were also examined. Both the field trial and the seed orchards were established by the Department of Forest Genetics of the Finnish Forest Research Institute.

The progenies and clones are subdivided into 3 groups according to their geographical origin (Oskarsson, 1972): E refers to South Finland (to approximately N 62°), K to Central Finland (between N 62°-64°), and P to North Finland (north from N 64°).

The amount of current year needles destroyed by *L. sulcigena* in each tree was estimated by eye using the following classification:

Class	Needles destroyed (%)	Class average
0	none	0.0
1	1- 33	16.5
2	34- 66	49.0
3	67-100	83.5

Using the class averages, the degree of infection for each progeny in each subtrial could be calculated. When the progenies in different subtrials were compared with each other, the degree of infection of each progeny was divided by the average for the subtrial and then multiplied by 100. Thus the average of both the subtrials and of the whole material was 100. This value is called the infection index. When a progeny is more infected than the average, its infection index is more than 100.

RESULTS

Progenies

All the infected subtrials were situated in Central Finland. The southernmost and northernmost subtrials were free from *L. sulcigena*. Although the degree of infection varied considerably in different subtrials, the fungus was found throughout all the stands (Table 1). The progenies differed ($P < 0.01$) from each other in all subtrials but subtrial 8. Although the progenies in each subtrial could be ranked according to degree of infection, the within-progeny variance was high.

The infection index of the most heavily infected progeny (no. 26) was 169 and that of the healthiest (no. 7) was 40. Infection index between subtrials differed only within 3 progenies ($P < 0.01$). The healthiest subtrial always deviated from the average (nos. 30, 14, 16) (Fig. 2).

North-south transfer of the progeny to a cultivation location had the effect that the further the distance which the progeny was transferred,

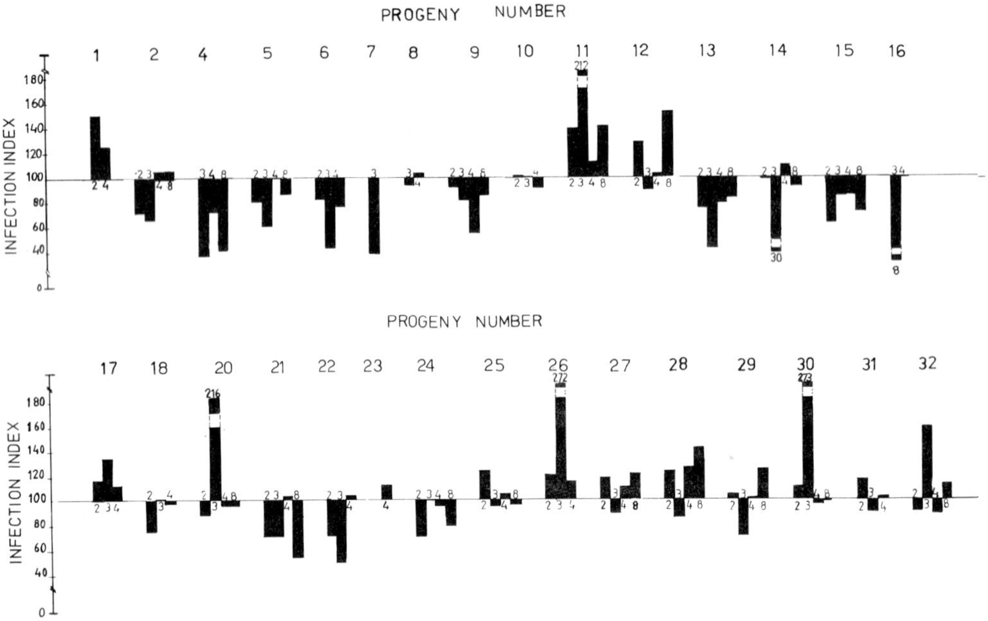

Figure 2. Infection index of progenies in subtrials 2, 3, 4, and 8.

Table 1. Degree of infection in subtrials.

Subtrial No.	Trees in different infection classes				Infection degree (%)
	healthy	1	2	3	
232/2	16	1605	323	143	27.3
232/3	3155	508	20	9	2.8
232/4	84	1048	744	976	46.9
232/8	786	1125	152	155	17.7

either south or north, the greater its resistance to *L. sulcigena* (Table 2). However, the progenies originating from north of the cultivation location were on the average healthier than the ones from south of it. The infection indices were 79 for the former and 115 for the latter (t-value = 6.61***).

By observing the zone division, progenies from South Finland (E) were the most severely infected. There were no differences between K- and P-plus tree progenies although they were clearly healthier than the E-trees

(Table 3). No differences were found in subtrial no. 4, perhaps due to its location on the southern border of the P-area, over 200 km north of the other subtrials.

Table 2. The effect of north-south and south-north transfers on the degree of infection by *L. sulcigena* as indicated by the infection index. F-value = 5.21***.

Transfer to the planting area expressed in degrees of latitude (L)[a]	Infection index
L < 1	110
1 ≧ L < 2	112
2 ≧ L < 3	90
3 ≧ L < 4	85
4 ≧ L < 5	84
L ≧ 5	8
Mean	100

a. L means the distance of the provenance from the cultivation location. One L is about 110 km. Thus all progenies the origin of which is less than 110 km from the cultivation location belong to class 1.

Table 3. The effect of progeny zone on the amount of *L. sulcigena* in different subtrials as indicated by the infection index.

Zone, N. Lat.	Subtrial number				Average
	232/2	232/3	232/4	232/8	
E 60° - 62°	113	137	101	117	117
Ķ 62° - 64°	104	45	98	75	82
P 64° - 69°	72	63	98	81	79
Mean	100	100	100	100	100
Infection degree (%)	27.3	2.8	46.9	17.7	
F-value	22.94***	10.72***	0.31	4.66***	20.98***

Clones

Differences between the clones in the tree orchards were very much greater than the differences between the progenies in the field trial. Over 50 % of all the clones were completely healthy and about 36 % of all the grafts were uninfected. Completely healthy trees were very rare in the field trials. However, the most heavily infected clone had lost 68 % of its current year needles. This is more than the values observed on any of the progenies in the field trials. There was only 1 clone in which all the grafts were infected. In the other seed orchard only 18 grafts had been attacked by needle cast. All of them belonged to the same clone. No statistical differences between clones could be tested because the layout of the seed orchards was not for testing.

DISCUSSION

Susceptibility to *L. sulcigena* was dependent on both the original geographical location and on the cultivation location. The farther from its origin, the less infected was the progeny. Resistance increased when trees had been moved either southwards or northwards, but more greatly when moved northwards. The more northerly its origin, the greater is the resistance of Scots pine to *Phacidium infestans* Karst. (Björkman, 1949). It has also been demonstrated that *L. sulcigena* causes less damage to northern progenies (Lagerberg, 1910). He suspected that the reason for this lies in the changes in the growth rhythm of trees at the southern cultivation location.

The growth rhythm of transferred progenies may deviate from the local progenies to such an extent that infection cannot spread to its maximum extent while spores are in the air. The structure of the needle, and especially the formation of wax, may play an important role in protecting against infection (Campbell, 1972a, 1972b). The resistance of Scots pine to *L. sulcigena* cannot have developed in the same way as its resistance to *P. infestans*. As *L. sulcigena* occurs only sporadically and is not fatal, the selection pressure directed at the progenies is small or absent. However, a persistent epidemic in dense stands helps less infected individuals to become dominant. *P. infestans* infects Scots pine every winter in North Finland. The resistant individuals maintain their vitality and infected parts die off.

If resistant or less infected trees are desired for forest tree breeding, it would be worth while directing selection at the healthy individuals of the healthiest progenies. Clones suitable for forest tree breeding can only be selected by excluding the most diseased clones. Clones infected by gray needle cast could be eliminated when roguing the seed orchards. However, it is quite a different matter when resistance to other pathogens is examined. Since the aim of forest tree breeding is to find trees which grow as fast as possible, progenies cannot be transferred from very far north.

Therefore local progenies, as resistant as possible, should be found.

REFERENCES

Björkman, E., 1949. Studier över snöskyttesvampens (Phacidium infestans Karst.) biologi samt metoder för snöskyttets bekämpande. Meddelanden från Statens Skogsforskningsinstitut 37(2): 1-128.

Campbell, R., 1972a. Electron microscopy of the epidermis and cuticle of the needles of Pinus nigra var. maritima in relation to infection by Lophodermella sulcigena. Annals of Botany 36: 307-314.

Campbell, R., 1972b. Electron microscopy of the development of the needles of Pinus nigra var. maritima. Annals of Botany 36: 711-720.

Lagerberg, T., 1910. Om gråbarrsjukan hos tallen, dess orsak och verkningar. Meddelanden från Statens Skogsförsökanstalt 7: 127-174.

Oskarsson, O., 1972. Suomalaiset plusmännyt ja pluskuuset. Finnish plus trees of Scots pine and Norway spruce. Folia Forestalia 150: 1-138.

Resistance trials of scotch pine clones in the Latvian SSR

I. Baumanis, D. Pirāgs & Z. Spalviņš

Latvian Scientific Research Institute for Forestry Problems, 229021, Riga Region, 111 Riga str., Salaspils, Latvian SSR, USSR

In the last 2 to 3 decades countries practicing intensive forest management have seen widespread improvements in forest seed management, and particularly in the establishment of seed orchards.

In the Latvian SSR forest seed orchards of the first generation (1006.8 ha) have been established, about 80 % of them being Scotch pine (*Pinus sylvestris* L.). We are now working toward the establishment of second generation orchards. At present we are evaluating the existing clonal material for periodicity and intensity of fruiting, increment, wood quality, plant survival and resistance.

The analysis of insect resistance traits in Scotch pine in experimental areas showed that selection of trees for a certain degree of increased wood and resin productivity increased resistance to *Aradus cinnamomeus* Panz. Resin productivity and pine shoot moth (*Rhyacionia duplana* Hb.) attack were not correlated, but the progeny having the highest resin productivity was resistant to both kinds of insects. Because of the great variability of these traits objective evaluation of the correlation between resin productivity and resistance requires that a larger number of clones be tested.

Under the conditions in Latvia pine plantations are more seriously damaged by needlecast (pathogen: *Lophodermium pinastri* (Schrad. ex Hook.) Chev.). Under conditions favouring disease incidence considerable mortality and growth reduction have occurred. Because chemical control in forest plantations is not very efficient, selection for resistance is the only effective method of eliminating damage from needlecast.

Family comparisons of needlecast resistance were made using deviations of distributions of disease ratings from a normal distribution. The mode of class frequencies for families with the highest amount of infection ranged between 0.5-0.9 (1 = completely healthy). A rather good correlation was found between amount of infection and the distribution of phenotypes in families of different seed years. In the younger age classes of 2-9 years the amount of infection in families was higher than in older age classes, but their ranking in distribution of phenotypic resistance was upheld. The distribution of resistant phenotypes changes with age, but in a population of 9-year-old progenies resistant phenotypes are distributed more or less according to the curve for a normal distribution. The amount of infection was negatively correlated with annual increment ($r = -0.32$ to -0.97), fast-growing phenotypes being the least infected. Thus, we conclude that individuals in pine populations differ in their degree of susceptibility to needle-cast. Completely resistant (immune) phenotypes have not been discovered.

A parental analysis of sibs shows that the introduction of a resistant partner into crosses in all combinations increases the resistance of progenies and decreases the frequency of susceptible phenotypes. Consequently, the selection of resistant parent trees both for open and controlled crosses increases the resistance of progenies, and especially in optimum crossing combinations. Selection for resistant phenotypes increased the juvenile growth of progeny by 23-71 %, depending on the history of infection and progeny age.

The predicted genetic effect of selection within the population for resistance to needlecast was 18 %.

Impact of western gall rust in natural stands of lodgepole pine

Owe Martinsson

Department of Silviculture, The Swedish University of Agricultural Sciences, S-901 83 Umeå, Sweden

The influence of western gall rust (*Endocronartium harknessii* Y. Hiratsuka) on growth of lodgepole pine (*Pinus contorta* Dougl.) was investigated in 3 natural stands in central and southern British Columbia.

Tree height, diameter at breast height and radial increment were measured in 3 stands, Baldy Hughes, Punchaw and Chilliwack. In the 2 younger and denser stands the height above ground of the oldest stem gall was measured. The following calculations were done:

- Annual radial growth was measured on increment cores and stem discs. Geometric means of the width of the first 5 and last 5 annual rings were calculated separately for trees with and without one or more stem galls.
- The relation between height and diameter of trees with and without stem galls.
- The relation between total tree height and height above ground of the oldest stem gall (Fig. 1).

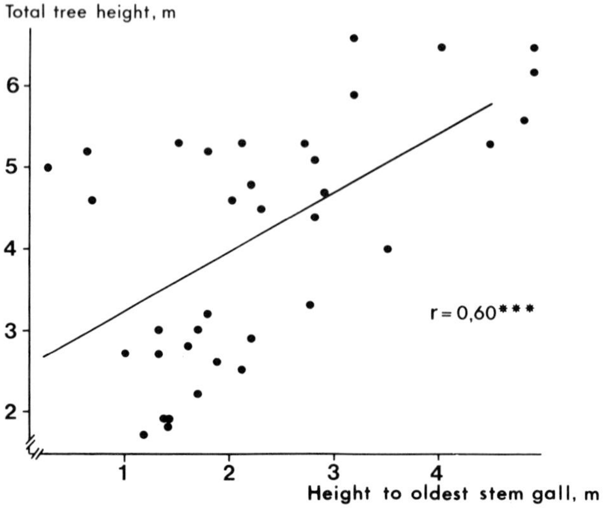

Figure 1. Relation between tree age at the time of attack (= the height above ground of the oldest stem gall) and total tree height.

Table 1. Means of 4 variables of trees with and without stem gall in 3 stands of lodgepole pine. * = difference between the two groups significant at p ≦ 0.05, ** = at p ≦ 0.01 and *** = at p ≦ 0.001; n.s. = not significant.

Stand	Age (years)	Tree height (m)*		Stem diameter (cm)**		Average radial growth first 5 years (mm)***		Average radial growth last 5 years (mm)$^{n.s.}$	
		with	without	with	without	with	without	with	without
Baldy Hughes	36	9.66	10.48	7.8	8.2	0.65	0.86	1.20	1.10
Punchaw	22	5.29	5.78	3.2	3.6	1.10	1.19	0.42	0.48
Chilliwack	32	2.55	2.89	2.5	2.9	0.54	0.61	0.34	0.33

The mean radial growth for the first 5 years was lower among infected than non-infected trees (Table 1).

No significant difference was found between the 2 groups of trees in growth for the last 5-year period. Measurements of radial growth were made, however, in the stem below gall formation. The mean stem diameter and the mean total tree height were smaller for attacked trees than for non-attacked trees. The relation between height and diameter shows that in the 2 younger and closer stocked stands, Punchaw and Chilliwack, diseased trees of a certain diameter are shorter than healthy trees of the same diameter. In the Baldy Hughes stand diseased trees of a certain diameter were taller than healthy trees of the same diameter. This may indicate that in natural stands of lodgepole pine, attacks of western gall rust have the greatest influence on height growth in young stands, whereas in older and more open stands the negative influence of gall rust infection is more on diameter growth. If the tree survives, gall formation does not seem to decrease radial stem growth below the gall. The slower height growth of attacked young trees results in a reduced ability to compete for light and space.

In these stands trees which were attacked by western gall rust had a slower growth rate before the attack than those which were not attacked.

Integrating resistance breeding and tree improvement

Ronald J. Dinus

International Paper Company, Western Forest Research Center, Lebanon, Oregon 97355, USA

ABSTRACT

Usable levels of pest resistance occur in many tree species. Breeding for resistance can complicate the improvement process and efforts should proceed only after thorough assessment of costs and benefits. Too little emphasis limits utility of improved material; too much reduces overall gains. Where special resistance trials are needed, they should fulfill several purposes, including isolation of material for future breeding and characterization of the nature and extent of resistance. Frequent reports of additive inheritance suggest that resistance can be manipulated the same as other traits and that long-term effectiveness can be had. Individual resistance reactions, their inheritance patterns and epidemiological consequences should be understood before widespread usage. Resistance should not be used alone, but should be deployed on an as-needed basis with other measures in comprehensive control programs. This seems best accomplished via organized teams of specialists, balanced as to discipline, and with extension capabilities.

INTRODUCTION

Breeding forest trees has a short history compared to that of agricultural crops. Even where manipulated for centuries, trees remain but few generations removed from their wild ancestors (Libby, 1973). Their relatively undisturbed status and considerable genetic variability provide a unique opportunity to improve productivity. Despite the newness of tree improvement, substantive programs abound, progress has been made, and pest resistance has been an important consideration.

That integration need be addressed seems more a function of human institutions than of hosts or pests. Scientific and technical personnel tend naturally to pursue matters of individual interest and concern. Like others, forest scientists specialize, gather in societies of common interest, and are organized into discrete disciplines in the workplace. Such trends often inhibit, if not prevent, meaningful collaboration. Forests and plantations, however, are complex, and pest problems are unlikely to be resolved by unitary action.

This paper seeks to explain how tree improvement and resistance breeding can and must be integrated. Improvement practices are described and related to methods for increasing resistance. Problem areas are noted, and solutions are suggested. Lastly, a few thoughts on program organization and management are offered.

THE IMPROVEMENT PROCESS

Genetic variation

Tree breeders use genetic variation at several levels. Variation among species is considered in choosing the species best-suited to site and product requirements, and may also be exploited via interspecific hybridization. Geographic variation often occurs within species as trees from different physiographic regions evolve differently in response to varying selection pressures. Genetic differences may also arise among populations or stands within regions. Lastly, individual trees vary, often greatly. The relative proportion and pattern of variation at each level govern the scope and nature of improvement programs.

Selection

Selection is the identification of desirable trees for use as parents. Criteria necessarily vary with species and objectives. Trees are long-lived and multi-purpose crops, and many traits must be considered. Two important factors affecting selection efficiency are selection differential - the difference between the mean of selected trees and that of the population, and heritability - the degree to which a trait is governed by heredity. Genetic gain, the product of the two, can be enhanced by increasing either. Differentials can be raised by increasing numbers of trees examined, and selecting only a few. That for any one trait will decline, however, as more traits are considered on a set number of trees. Traits affected greatly by the environment have lower heritabilities than those influenced to lesser extents. And, heritability will be highest when environmental variability is low. Thus, heritability can be raised and useful genotypes selected more reliably by lessening environmental variability. Commonly used approaches are the comparison tree and base-line methods (Ledig, 1974).

Selecting for many traits is complicated by differences in variability and heritability, correlations among traits, and value. Independent culling, one solution, uses genetic and economic data to set levels of size or quality for each trait. Trees having less than the desired rating for any trait are rejected, even if otherwise superior. The approach is somewhat subjective, but works reasonably well.

Selection indices have been used in efforts to optimize gain. Decisions are weighted by mathematical integration of heritabilities, genetic correlations, and expected economic return (Bridgewater & Stonecypher, 1979). Selection indices have many advantages, but wider usage awaits more and better data (Faulkner, 1979).

Tandem selection, selection for one trait per generation, is considered impractical because of long generation intervals (Faulkner, 1979). The method could be useful when one trait severely limits culture and/or improvement, but large populations would be necessary to insure improvement over generations.

Indirect selection, selection for one trait through correlation with another, has not been used frequently (Faulkner, 1979). Such correlations are not often known and can be had only by observing offspring of known parentage at meaningful ages. The method could be useful, where one trait can be assessed easily or early and another can be evaluated only with difficulty or time.

Production of improved material

Improved seed is secured by grouping selected trees in seed orchards (Faulkner, 1975). Most orchards are established with grafts or rooted cuttings arranged to maximize crossing and minimize inbreeding. Locations are chosen to favor production and minimize contamination from external pollen sources. Cultural and protective measures are used to hasten and increase production. Designs often provide for increasing gain by removal of materials later proven undesirable.

Seedling orchards generally yield less improvement than clonal orchards, but are useful alternatives in species that produce seed early or are difficult to graft. A major drawback is the need to establish them on sites typical of where improved seed will be used. Such situations do not always favor seed production and are often subject to pollen contamination.

Vegetative propagation is a fast way to multiply improved material. A major advantage is that all desirable genes can be captured and replicated over space and time. Propagation, however, becomes more difficult as trees age (Faulkner, 1979), and multiplication is either an uncertain proposition or individuals are selected before desirability is known. Age and crown position also affect propagule growth and development. Cuttings from a tree old enough to be evaluated may have the form of that individual, but grow more slowly than genetically similar seedlings. The many benefits will be reaped only as such problems are solved.

Testing and evaluation

Improved material must often be released before gains are known or provisions are made for future selection. Sooner or later, however, tests are established with clonal lines or open or control-pollinated offspring. One purpose is to monitor progress and obtain genetic information. Another purpose, often compatible with the above, is to validate parental quality and eliminate poor individuals. Such tests are most useful where heritabilities are low or special needs occur (Libby, 1973; Kellison & Dinus, 1977). A third purpose, perhaps the most important, is to select material for use in succeeding generations. Such tests are more important to the future than those designed to upgrade seed orchards, but the latter can pay handsome returns. Tests compatible with both ends can and should be used. Testing, regardless of purpose, is costly, and separate trials for all purposes can seldom be afforded.

Hybridization

Interspecific hybrids generally are most useful outside natural ranges of the parents or where desirable combinations of traits cannot otherwise be obtained. Long generation intervals, crossing problems, and low seed yields presently limit efficiency. Useful materials, however, could emerge from hybridization efforts, especially in advanced generations and when vegetative propagation is feasible. Greatest utility may be in increasing genetic variation.

BREEDING FOR RESISTANCE

Decision and emphasis

Choosing to improve resistance is often not a decision per se as productivity would otherwise be lost. Most times, however, the decision concerns emphasis rather than being a matter of yes or no. Such decisions are not easy as a spectrum of possibilities exists. Situations involving introduced trees or innocuous pests are one extreme, and require minor though positive action - pest-free parents, large breeding populations, and flexible programs. Breeders should also lend expertise to those prescribing silvicultural practices, applying other control measures, and monitoring pest impact.

Other situations are more complex. Including resistance impacts every phase of improvement. Only by weighing benefits against losses can breeders determine how much to emphasize resistance with respect to other traits. Sound decisions require answers to such questions as: what pests are important; which is most important; where is the pest most damaging; how much loss occurs; what loss can be tolerated; how much can loss be reduced by other measures; how much loss will resistance prevent; what combination of measures is most effective; and do returns justify resistance breeding? Such concerns are a few of the many that must be addressed before ongoing programs are altered or new programs are started.

Too little emphasis can cause reduced yield. Too much can have the same effect by reducing gain in other traits. Maximum economic benefit can be had only by giving each trait its rightful emphasis. Breeders should seek just enough resistance to reduce losses to acceptable levels, other control measures considered. Seeking unnecessarily high degrees of resistance may also jeopardize future progress by reducing variation in other traits, including resistance to other pests, and by placing undue selection pressure on the pest in question.

Sources of resistance

Usable levels of heritable resistance to a variety of pests have been found in most important genera. Some species and seed sources are more resistant than others, and such variation can be exploited by substituting resistant but adapted materials for susceptible ones. Useful resistance can also be found at the population or stand level, especially in the wake of severe infestations. Selecting for growth and form among pest-free residuals can provide resistant seed and material. Individual tree variation in most species is such that resistance can be increased by judicious selection and breeding, without losing adaptability or introducing undesirable traits. When otherwise unavailable, resistance may be transferred from related species. Breeders seeking to integrate interspecific hybridization efforts into improvement programs nevertheless must be prepared to face problems with susceptibility to other pests, undesirable traits or poor adaptability, repeated and protracted testing, and multiplication of improved materials.

Little, if any of the foregoing is new or startling. The point to be made is that information required to choose among and exploit various options is seldom available when needed. As a result, reaction or crisis management becomes necessary. Delaying a program to gather all information is senseless, but having advance information on the amounts and kinds of resistance as well as where to find it would be beneficial.

Testing and selection for resistance

Selection for resistance requires that resistant material be identified with ease and certainty. Individual populations or trees must be given equal and high probabilities of being attacked. Such conditions do not often occur in routine species, provenance, and progeny tests, but data and material from them should be used wherever possible. Special resistance trials, however, are often needed. Given the expense of even routine testing, resistance trials must fulfill several purposes and provide all information needed by breeders. Foremost among purposes should be isolation of material for future breeding, with attention given to discerning nature as well as extent of resistance.

Field trials are best as large populations can be tested, resistance can be examined where it will be used, and yields can be contrasted. Efficiency may be increased by modifying environments or provoking artificial epidemics. Tests in standard environments with known pest concentrations and sources can save time and enhance precision, but may cause selection for only a few or simply-inherited forms of resistance. Pest numbers and sources should therefore be varied to maximize information about the nature and strength of resistance. Utility must be confirmed in areas where resistant material will be planted.

Nature of resistance

Previous remarks about the nature of resistance require explanation, especially in view of doubts about durability of resistance. Some would argue that pests will evolve faster than trees can be bred. Pests are variable, but risks occasioned by such variability and widespread planting of resistant trees are not easily determined.

Experience with self-pollinated annual crops has not been reassuring as breeders have often used vertical resistance - a form typically controlled by one or a few dominant genes. Though exceptions occur, wide usage imposes selective pressures, and pest biotypes able to circumvent the resistance become prevalent. New resistance genes or some means of combining and deploying old genes must then be found. Breeders of other crops, such as corn, have found more lasting protection in horizontal resistance - a form that does not provoke great selective pressure, and is conditioned by many genes, each having small effects. Such resistance is additively inherited, and not so difficult to use as is often presumed.

Viewed from another perspective, vertical and horizontal resistance are extreme forms at either end of a continuum, as professor J.C. Zadoks pointed out in his address to this workshop. Many forms can be expected, each having an inheritance pattern and epidemiological consequences determined by its position in the continuum. Analyzing host/pest systems may reveal the position of each and host and pest characteristics can then be used to decide the best forms of resistance to use. The foregoing suggests that tree breeders should not rely solely on vertical resistance, and that forms approaching horizontal resistance or a combination would be best. Resistance in some trees appears to be additively inherited, so breeders are already working at the appropriate end of the continuum and longlasting resistance is within reach. Tree breeders routinely work with additive traits, and should not have difficulty including resistance.

For most species, however, information on inheritance patterns and epidemiological consequences is woefully inadequate. Difficulties also stem from measurement methods, as the end result of resistance is monitored and underlying reactions are often overlooked. Overall effects may seem quantitative, but some components may be quantitative and others qualitative. Better information will enable breeders to assemble more durable products, through directed combination of resistance forms. Similar considerations apply to detection and use of tolerance - that capacity of a host to endure severe attack without serious yield loss. Evidence from annual crops suggests significant heritability of tolerance and its independence from resistance. Tolerance occurs in forest trees (Griggs & Dinus, 1977), but little more is known.

Much research is needed, especially on measuring resistance and tolerance, quantifying inheritance patterns, and evaluating epidemiological consequences. Clarifying these issues will enable breeders to proceed more confidently and convincingly. Helping with such concerns need not involve research on physiological or histological details of resistance mechanisms. Knowing the how and why of resistance will facilitate continued progress, but such work can be deferred until more important questions are answered. Hastening release of a first usable product, however preliminary, is more important.

Attention must also be drawn to the dangerous notion that resistance is an all or none proposition. Horizontal resistance can provide only partial protection. Vertical resistance may have utility, mainly as a supplement to other forms, and provided its risks are considered. Thus, resistance should be used in combination with other measures on an as-needed basis and within the framework of a comprehensive pest control program.

PROGRAM ORGANIZATION AND MANAGEMENT

Serious pest problems seem best resolved by a formal team approach. Clear goals must be set at the start, and one individual charged with choosing and coordinating the approach. Personnel must be chosen carefully and made to feel as partners rather than rivals. Nothing inhibits collaboration more than specialization, and individual objectives must be set such that workers understand the program and what each is to contribute. Objectives should also foster inter-dependencies. Incentives should be proportional to team accomplishments as well as individual performance. Personnel are best concentrated at one central location to heighten interaction and lessen costs. Funding must be commensurate with problem severity and stable across time. The team should include essential disciplines, without being unwieldy in size. Quantitative sciences should be emphasized. Talent needed only on occasion can be borrowed. For less urgent problems, a less formal approach may suffice, but the same principles should be followed.

Research, development, and application, like resistance, are continuous. The continuum is a circle, however, and slowing or ignoring one affects the others. Scientists may not desire or be able to translate theory into application, but are obliged, as team members, to communicate with co-workers in ways that will facilitate their work and engender feedback. Such linkages should be formed within the team at the outset, by including an extension specialist. Well-balanced teams are apt to move from start to application with efficiency and dispatch.

REFERENCES

Bridgewater, F.E. & R.W. Stonecypher, 1979. Index selection of volume and straightness in a loblolly pine population. In: Proceedings of the 15th Southern Forest Tree Improvement Conference, 1979. p. 132-139.

Faulkner, R., 1975. Seed orchards. Forestry Commission Bulletin No. 54. Her Majesty's Stationery Office, London. 149 p.

Faulkner, R., 1979. The exploitation of genetic variation by selection and breeding. In: E.D. Ford, D.C. Malcom & J. Atterson (Eds): The ecology of even-aged forest plantations. Proceedings of the Meeting of Division I, IUFRO, Edinburgh, 1978. p. 99-117.

Griggs, M.M. & R.J. Dinus, 1977. Patterns of fusiform rust increase and their implications for selection and breeding. In: Proceedings of the 14th Southern Forest Tree Improvement Conference, Gainesville, Florida, 1977. p. 43-53.

Kellison, R.C. & R.J. Dinus, 1977. Recent advances in genetic improvement of Pinus taeda and Pinus elliottii in the southern United States. In: Proceedings of the Third World Consultation on Forest Tree Breeding, Canberra, Australia, Paper FO-FTB-77-3/2, Vol. 2. p. 475-488.

Ledig, F.T., 1974. An analysis of methods for the selection of trees in wild stands. Forest Science 20: 2-16.

Libby, W.J., 1973. Domestication strategies for forest trees. Canadian Journal of Forest Research 3: 265-276.

Progeny testing of Douglas-fir seedlings for resistance to western spruce budworm

G.I. McDonald

Plant Pathologist, Intermountain Forest and Range Experiment Station, USDA Forest Service, Ogden, Utah, 84401; located at Intermountain Station's Forestry Sciences Laboratory, Moscow, Idaho, USA

ABSTRACT

Second instar larvae of *Choristoneura occidentalis* successfully mined needles and buds and fed on current and year-old needles of 1-year-old *Pseudotsugae menziesii* after inoculation into large cages. The introduced budworm population appeared to behave in a normal manner including mating, oviposition, and hatching. Percent of stems debarked and number of egg masses per tree were both significantly correlated with tree height. Larval feeding was too heavy to obtain a reliable measure of either seed tree or stand related indication of larval food preference.

INTRODUCTION

Resistance of many agricultural crop plants to various insects is well known (Kogan, 1975), and resistance to various insects in forest tree species has been studied for many years (Hanover, 1976). Foresters are especially attracted to resistance because it generally does not require repeated attention and is easily incorporated into integrated pest management plans. Another consideration for localities having extensive wild forests is the preservation of coevolved balances as these forests come under more intensive management.

A pest of great importance to the conifer forests of North America is spruce budworm - various species of *Choristoneura*. Inland Douglas-fir (*Pseudotsuga menziesii* var. *glauca* (Beissn.) Franco) of the Northern Rocky Mountains of western North America is particularly susceptible to attack by *C. occidentalis* Freeman (Lepidoptera: Tortricidae), the western spruce budworm. In 1975, we began investigating the Douglas-fir budworm system to elucidate the nature of this interaction.

A tour of several stands subjected to extremely heavy defoliation from past outbreaks or currently in outbreaks quickly demonstrated ample phenotypic variation on which to base studies (McDonald, 1981). Two questions that seemed most important were: is there a genetic component in this variation; and, if so, how are the resistance mechanisms expressed and inherited? A preliminary progeny test of Douglas-fir indicated the presence of a genetic component (McDonald, 1979). Our first priority has been the development of progeny test procedures. The objective of this paper is to describe the procedure developed and to present the results of our first full-scale progeny test.

MATERIALS AND METHODS

Douglas-fir

The host material consisted of 42 open pollinated seedlings from each of 7 mother trees from 7 stands to give 2058 seedlings. Both stands and mother trees were selected without regard to budworm. Seedlings were container grown for one full season in plastic tubes in a mixture of forest soil, sand, and peat moss. They were overwintered in these tubes and then placed in a 13 °C cooler before bud break at the beginning of their second growing season in order to synchronize bud break with larval development. Their average height was 101 mm.

Budworm

Hibernating 2nd instar larvae were collected on stem sections of Douglas-fir, western larch, and ponderosa pine growing on the University of Montana's Lubrecht Forest located 25 miles east of Missoula, Montana. Larval populations were sampled prior to the main collection. On May 1, 1979, enough stem length was collected to provide an estimated 5000 larvae and transported back to Moscow, Idaho. On May 2, the stem sections were placed in light-proof containers and traps affixed to catch the larvae as they broke hibernation and migrated to the lighted traps. On May 3 and 4, the 2058 seedlings were removed from the cooler and placed in the test bed.

Test procedure

Two cages were constructed to provide 1044 planting spots each. A 91.4 cm × 304.8 cm sheet of plywood was drilled with 1044 2.54 cm diameter holes on a 5.08 cm spacing. The holes were arranged into 58 rows with 18 holes each. This floor was then installed in a 91.4 cm × 304.5 cm × 30.5 cm cage made of 32 mesh (32 strands/2.54 cm) nylon screen. The top was totally removable and was held in place by screendoor hooks. This cage was then suspended over a metal tray 91.4 cm × 304.8 cm × 15.25 cm. Periodic filling of the tray watered the seedlings from below. Each hole was designated by row and column. A computer was used to randomize the 42 seedlings from each tree-stand combination into 2 completely randomized single-tree-plot blocks (21 seedlings/stand mother tree combination).

As the larvae emerged from their hibernation sites, May 4 to May 11, and migrated into the traps, they were collected in small plastic petri-dishes, about 30 to the dish. These dishes were then placed in a regular fashion among the seedlings to give about 1649 larvae in block 1 and 1665 larvae in block 2.

The larvae were allowed to crawl from the dishes onto the foliage of the test seedlings where they developed through the various larval instars to the adult moths. Damage and oviposition data were collected August 1 to 14 and pupal cases were counted.

Data

The plan called for measurement of length of total foliated and defoliated branch and stem length of 1979 growth on each seedling to obtain a ratio of defoliated/total foliated length for each seedling. This plan could not be followed because 1600 larvae/block overwhelmed the seedlings and completely consumed the 1979 growth and most of the 1978 needles as well. We recorded a subjective estimate of percent defoliation of the 1978 needles for each seedling. Because this foliage was consumed before larval development was completed, many larvae consumed strips of bark. Consequently, we recorded the presence or absence of debarked stem and branch as an alternative to measuring defoliation. In addition, we removed all seedlings from block 1 to keep for 1 year to check survival after complete defoliation and then put a fresh seedling of the same family in every other hole to give a 5.08 cm × 10.16 cm spacing. The replacement operation was done after pupation and before adult emergence. After egg laying was complete, we recorded the number of egg masses/seedling. Analysis of variance and regression were completed according to Snedecor (1956).

RESULTS

Defoliation

The budworm population overwhelmed the host even though only 595 of 1649 larvae formed pupae in block 1 and 834 of 1665 in block 2. Also, the larvae tended to congregate on seedlings around the dishes; thus, most needle mining occurred on these seedlings. No new foliage or old buds survived, and defoliation of the previous year's needles averaged 98 % in block 1 and 97 % in block 2. The regression of average percent defoliation of 1-year-old needles on average family height after 1979 growth yielded an r of 0.30 (n.s., 47 d.f.). We completed an analysis of variance (Table 1). Both stand and family show some significance (Table 2).

Debarking

Study of the summarized data indicated a possible correlation between percent of stems/family debarked and the average height of a family after growth. Such a regression was run using mean family height as determined by

Table 1. Keyout of factorial-hierarchal analysis of variance used on budworm defoliation data. Assume blocks fixed; stands and families variable.

Source	d.f.	Expected mean square
Blocks	a-1	$\sigma^2 + r\sigma^2_{AC,B} + rc\sigma^2_{AB} + rcb\sigma^2_{A}$
Stands	b-1	$\sigma^2 + ra\sigma^2_{C,B,} + rca\sigma^2_{B}$
Families in stands	b(c-1)	$\sigma^2 + ra\sigma^2_{C,B}$
Blocks × stands	(a-1) (b-1)	$\sigma^2 + r\sigma^2_{AC,B} + rc\sigma^2_{AB}$
Blocks × families in stands	b(a-1) (c-1)	$\sigma^2 + r\sigma^2_{AC,B}$
Within families	abc(r-1)	σ^2

Table 2. Analysis of variance of percent of budworm defoliation of year-old needles on 1-year-old Douglas-fir seedlings.

Source	d.f.	MS.	F	P>F
Blocks	1	559.44		
Stands	6	191.37	2.00	0.086
Families in stands	42	95.45	1.50	0.022
Blocks × stands	6	69.36		
Blocks × families in stands	42	40.43		
Within families	1960	63.76		

the average height in millimeters of the seedlings placed into block 1 after larval feeding to supply oviposition sites. The r was 0.66 ($p < 0.05$; 47 d.f.) (% debarked = 100 [0.29 tree height - 14.4]).

Egg mass number

The tallest family averaged 206 mm in height and had 2.36 egg masses/seedling or 0.0115 masses/1 mm of height. The shortest family was 89 mm tall after growth and had 0.69 masses/seedling or 0.0078 masses/1 mm of seedling height. These data are based on about 10 seedlings/family. A regression of egg masses on seedling height showed r = 0.69 ($p < 0.05$; 47 d.f.). Analysis of variance of egg masses/1 mm of seedling height showed no significant sources of variation (egg masses/seedling = 0.0153 seedling height -0.84).

Survival of defoliated trees

There appeared to be a definite relationship between stands and prob-

Table 3. Survival of Douglas-fir seedlings after being defoliated by budworm in their second growing season, in relation to height before regrowth in their third year.

Stand	Number dead	Number living	Proportion dead	Height (mm)
1	5	142	0.03	108
2	5	142	0.03	134
3	10	137	0.07	88
4	13	134	0.09	86
5	18	129	0.12	69
6	11	136	0.07	121
7	5	142	0.03	88

d.f. = 6; X^2 = 126.29; P of larger X^2 = 0.0005

ability of death after defoliation, but again, much of the observed variation can be associated with seedling height (Table 3). In any event, only 7 % of the completely defoliated seedlings died.

CONCLUSIONS AND DISCUSSION

Five obvious conclusions are:

- seedling height must be removed from any measure of oviposition behavior and is also an important component of the debarking data;
- percent defoliation of individual seedlings might be analyzable without removal of seedling height;
- larvae dispersed, mined needles and buds, and consumed foliage in a normal manner;
- adults emerged, mated, and oviposited to provide workable oviposition data; and
- the test procedure is workable and should provide good data after proper insect levels and distribution are worked out.

The results from this first progeny test provided little specific information about the genetics of the budworm/Douglas-fir interaction, but they did show that such testing can be conducted and reliable feeding and oviposition preference data collected. The method should be adaptable to progeny testing of hibernacular site selection, pheromone chemistry, predator and parasite effectiveness, host and pest phenologic-synchronization, and bud and needle mining behavior.

REFERENCES

Hanover, J.W., 1976. Physiology of tree resistance to insects. Annual Review Entomology 20: 75-95.

Kogan, M., 1975. Plant resistance in pest management. In: R.L. Metcalf & W.H. Luckman (Eds): Introduction to insect pest management, John Wiley and Sons, New York. p. 103-146.

McDonald, G.I., 1979. Resistance of Douglas-fir to western spruce budworm: results of a preliminary progeny test. USDA Forest Service Research Note INT-264. Intermountain Forest and Range Experiment Station, Ogden, Utah. 7 p.

McDonald, G.I., 1981. Differential defoliation of neighboring Douglas-fir trees by western spruce budworm. USDA Forest Service Research Note INT-306. Intermountain Forest and Range Experiment Station, Ogden, Utah. 33 p.

Snedecor, G.W., 1956. Statistical methods. Iowa State University Press, American. 534 p.

Testing larch clones for Adelges laricis resistance

I. Blada

Statiunea Experimentala Forestiera, Padurea Verde, Timisoara, Romania

ABSTRACT

One hundred twenty-one phenotypically resistant trees were selected in 20 populations of *Larix decidua* and in 1 of *L. leptolepis*, extensively attacked by *Adelges laricis*. Clones of the candidates were infested in the nursery and then planted in 3 different locations in order to test their resistance to *A. laricis* under natural conditions of infestation. After 9 years of testing we drew the conclusion that among the clones of *L. decidua*, there were highly significant differences, and that the resistance to *A. laricis* is a continuously varying character controlled by polygenes. Broad-sense heritability coefficients for the 3 locations were as follows: 0.41, 0.68, 0.77 and the expected genetic gains 7.9 %, 14.1 %, 15.1 %, respectively. The clones of *L. leptolepis* were more resistant ($p < 0.05$) than those of *L. decidua*. Finally, 34 clones were selected for use in the improvement program.

INTRODUCTION

Larix decidua Mill., a native species in the Alps and the Carpathians, is important for its high wood production, its wide ecological amplitude and its high resistance to injurious factors. Nevertheless, some insects, *Adelges laricis* Vall. included, cause a decrease in vigor of attacked trees and increment losses.

The study of resistance to insects as well as the use in culture of genetically improved material as a way of reducing losses were recommended about 30 years ago (Painter, 1951). Two factors underline the importance of genetic resistance to injurious parasites: the inefficiency of the classical control methods (Søegaard, 1964) and the necessity for avoiding environmental pollution with pesticides (Gerhold, 1970).

The variability in resistance of larch to *A. laricis* has been mentioned by various authors. Thus, Larsen (1953) found numerous lice on *L. decidua*, and only a few on *L. leptolepis* (S. & Z.) Gord. In a test on 25 European larch clones, Eidmann (1966) proved that there was one clone of relatively low susceptibility, while the others had variable resistance. Considering trees exposed to natural infection, Sindelar & Hochmut (1967) found highly significant differences in resistance among different provenances of European larch. They also confirmed that resistance to *A. laricis* of Japanese larch was higher than that of European larch. Investigations in Romania on attacked populations proved the variability of individual phenotypic resistance to the same parasite (Blada, 1977).

In order to avoid losses, in 1969 a program of improvement of genetic resistance to *A. laricis* was initiated in Romania making use of intra- and interspecific hybridization methods. The final objective of this program was to create seed orchards of trees having general combining ability for resistance to *A. laricis*.

This paper deals with the results of a clonal test that was carried out in order to find new sources of resistance.

MATERIALS AND METHODS

One hundred twenty-one apparently resistant trees were selected out of 20 populations of *L. decidua* and 1 population of *L. leptolepis*, extensively attacked by *A. laricis*. Selection was possible because of the variability in phenotypic resistance in the heterogeneous populations.

Grafts on the candidates of the 2 species of larch were made on rootstocks in polyethylene bags. Infestation was induced in the nursery in the period 1970-1972 by placing the clones between highly attacked rows of larch seedlings; under these circumstances the parasite was transmitted in a natural way. In order to expose them to conditions favoring natural infestation, the clones then were planted out (in spacing 3 m × 3 m) in the following locations: Sinaia, Toplita and Huedin. Grafts of all 121 clones (111 *L. decidua* and 10 *L. leptolepis*) were planted in each location disposed in 3 blocks (replications). Each replication contained a ramet, so that the entire experimental system (replications × locations) included 9 individuals of each clone. From 1972 to 1979 resistance was estimated on the basis of percentage of attacked needles. Samples were collected at random from each ramet when the attack was at its peak; this happened in early July, when the insect was in the first larval generation (estivale). Three replications with the highest attack were included in the statistical analysis for each location and clone (not considering the year). This procedure was considered to be effective in eliminating variation in the amount of infestation in the experimental system in order to estimate the true resistance or susceptibility of each clone. Percentages were transformed into arcsin $\sqrt{\%}$ values and analyzed by means of the analysis of variance. The analysis was performed on the species both separately and together, in order to obtain a general ranking according to resistance. The significance of the differences was established by the multiple range test and finally the broad-sense heritability and the expected genetic gain (Δ G) were computed according to the following formulas (Nanson, 1967; 1970): $h^2 = s^2_V - s^2_E/s^2_V$ and $\Delta G = ih^2 \sigma_F$, where s^2_V = the variance of the clones, s^2_E = the error variance, i = the selection differential, and σ_F = phenotypic standard deviation.

RESULTS

In the 3 locations considered, there were significant differences ($p < 0.01$) among the clones of *L. decidua* in their resistance to *A. laricis* (Table 1; 3).

Among the clones there were, too, significant differences ($p < 0.01$) in their reactions to the ecological conditions in the 3 locations (Table 2; 4th column); this emphasizes the strong interaction of clone × environment, on the one hand, and parasite × environment, on the other. Nevertheless, one might conclude that resistance of the clones is under genetic control, even if their reactions to the environment were different (Table 2; 5th column).

Resistance of European larch to *A. laricis*, according to the percentage of attacked needles (continuously variable between 16 and 85 %) appeared to be a quantitative character (Fig. 1). The frequency distribution of the clones according to the percentage of attacked needles was very close to the normal curve (Fig. 1).

The amplitude of the variation in resistance of the clones of *L. leptolepis* was low (3 to 27.5 %), and the differences were not significant (Table 2).

The mean resistance of the clones of Japanese larch (19 % infection) was much higher ($p < 0.05$) than that of European larch (44.3 % infection)

Table 1. Analysis of variance of resistance of European larch clones for each of 3 experimental locations.

Source of variation	Degrees of freedom	Variance (s^2)		
		Sinaia	Toplita	Huedin
Replications	2			
Clones	110	486.4**	442.7**	461.8**
Error	220	285.7	141.9	105.6

** = $p < 0.01$

Table 2. Analysis of variance of resistance of European and Japanese larch clones for all 3 experimental locations combined.

Source of variation	Degrees of freedom	Variance (s^2)	F-test in respect to	
			s_E^2 = error variance	s_I^2 = interaction variance
European larch				
replications	6			
locations	2			
clones	110	730.6	3.27**	2.21**
clones × locations	220	330.3	1.48**	
error	660	223.2		
Japanese larch				
replications	6			
locations	2			
clones	9	280.0	1.88	1.13
clones × locations	18	247.9	1.66	
error	54	149.2		

** = $p < 0.01$

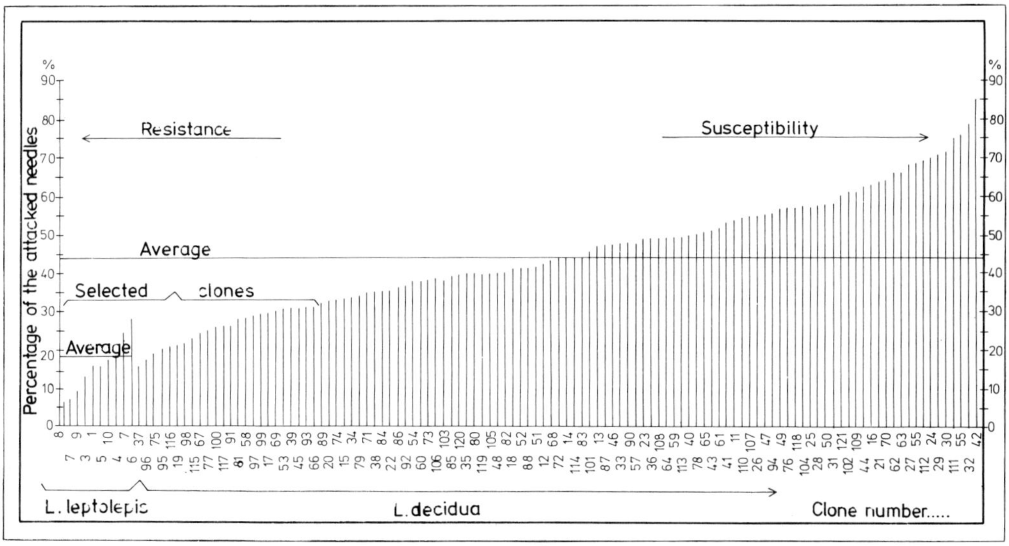

Figure 1. Ranking of the 121 *Larix* sp. clones according to the mean percentage of attacked needles.

Table 3. Analysis of variance of resistance for both European and Japanese larch clones and the combination of all locations.

Source of variation	Degrees of freedom	Variance (s^2)	F-test in respect to	
			s_E^2 = error variance	s_I^2 = interaction variance
replications	6			
locations	2			
clones	120	937.1	5.31**	2.87**
clones × locations	240	326.0	1.85**	
error	720	176.3		

** = $p < 0.01$

Table 4. Analysis of variance of resistance in comparison of clones of *L. decidua* and *L. leptolepis*.

Source of variation	Degrees of freedom	Variance (s^2)	F-test in respect to	
			s_E^2 = error variance	s_I^2 = interaction variance
replications	6			
locations	2			
species	1	1607.4	112.4**	53.2*
species × locations	2	30.2	2.1	
error	6	14.3		

* = $p < 0.05$; ** = $p < 0.01$

Table 5. Expected genetic gain (Δ G) and broad sense heritability (h^2) for *L. decidua* clones.

Location	h^2	Δ G (%)
Sinaia	0.41	7.9
Toplita	0.68	15.1
Huedin	0.77	14.1
The mean	0.62	12.4

(Table 4). *L. decidua* was characterized by an intraspecific variability much higher (see clones 11 to 121) than that of *L. leptolepis* (see clones 1 to 10), and it had some clones as resistant as the latter's (Fig. 1).

The broad-sense heritability was 0.41 at Sinaia, 0.68 at Toplita and 0.77 at Huedin (Table 5).

The expected genetic gain was 7.9 % at Sinaia, 14.1 % at Huedin and 15.1 % at Toplita (Table 5), with only the clones having up to 30 % attacked needles being considered.

Finally, 24 clones of *L. decidua* and 10 clones of *L. leptolepis* (Fig. 1) were selected as resistant and useful for the improvement program.

DISCUSSION

Because variability was continuous the hypothesis was proposed that resistance of European larch to *A. laricis* was polygenic.

The varying values of heritability and, implicitly, of the genetic gain from location to location might be attributed to the environmental conditions in the 3 locations that influenced the evolution of the parasite. The highest attack was at Toplita, whereas the lowest was at Sinaia.

Because *L. leptolepis* had a ratio of 100 % resistant clones and *L. decidua* had a ratio of only 21.6 % resistant clones, this was proof that the latter did not lack resistance genes, but only that the frequency of these was lower than that in *L. leptolepis*.

The existence of individual variability in both species of larch makes intraspecific selection possible; if hybridization follows it may lead to an accumulation of resistance genes and consequently to an extra genetic gain.

In conclusion, resistance to *A. laricis* is present and variable both at the intra- and the interspecific level which presents an opportunity for using both sources of resistance in the improvement program.

REFERENCES

Blada, I., 1977. Ameliorarea pe cale genetica a rezistenţei duglasului şi laricelui la principalele boli şi dăunători. Studii şi cercetări, I, XXXIV. p. 19-31.

Eidmann, H.H., 1966. Zur unterschiedlichen Resistenz von Lärchen gegen Läusebefall. Anzeiger für Schädlingskunde 39: 8-10.

Gerhold, H.D., 1970. Dix ans de progrès dans la production d'arbres forestiers résistants aux maladies. Unasylva 24(97-98): 37-44.

Larsen, S., 1953. Studies of diseases in clones of forest trees. Hereditas 39: 152-179.

Nanson, A., 1967. Tables de la differentielle de sélection dans la distribution normale (0,1). Biométrie-Praximétrie VIII: 40-51.

Nanson, A., 1970. L'héritabilité et le gain d'origine génétique dans quelques types d'expériences. Silvae Genetica 19: 112-120.

Painter, R., 1951. Insect resistance in crop plants. Mac Millan, New York. 520 p.

Sindelar, J. & R. Hochmut, 1967. Variability of larch (Larix sp.) invasion caused by woolly aphids Adelges laricis and Sacchiphantes viridis. XIV IUFRO-Kongress, München, III: 902-906.

Søegaard, B., 1964. Amélioration génétique de la résistance aux attaques d'insectes chez les arbres forestiers. Unasylva 18 No. 73-74: 82-88.

Variation in susceptibility of native and introduced coniferous trees to some insects of eastern Canada

W.H. Fogal, E.K. Morgenstern, P. Viidik & C.W. Yeatman

Petawawa National Forestry Institute, Canadian Forestry Service, Chalk River, Ontario KOJ 1JO, Canada

ABSTRACT

Insect-tree interactions are discussed from the viewpoint of coevolution. The host-pest interactions considered are: *Pinus strobus*, *P. banksiana*, and *Picea abies* with *Pissodes strobi*; introduced species of *Larix* with *Pristiphora erichsonii*; *Pinus strobus*, *P. banksiana*, and *P. resinosa* with *Neodiprion* and *Diprion* spp. Damage, behavioral, and ecological characteristics for each insect are outlined briefly and a review is made of variation in susceptibility of host trees.

INTRODUCTION

Susceptibility of forest trees to insect damage is a complex phenomenon intimately related to the evolution of both the host and parasite. Trees have evolved certain strategic defenses comprising a complex of biochemical, physiological, and morphological properties that enable some to resist and survive insect attack: oviposition or feeding may be prevented; the insect may be killed or its growth impaired; or the tree may be able to tolerate heavy attacks. Presumably the level of success achieved by the insect depends on the degree to which it has adapted to these defensive strategies (Dethier, 1976; Painter, 1951; Stern & Roche, 1974). We do not yet fully understand the interaction between insects and host trees or the genetics of defense mechanisms underlying insect-host preferences and selection. However, knowledge of natural variation in susceptibility within species is a prerequisite for exploiting tree defense mechanisms in forest pest management.

NATIVE PINES

White pine : White-pine weevil

Eastern white pine, *Pinus strobus* L., is the most valuable lumber species of eastern Canada and occurs in the Deciduous, Great Lakes -

St. Lawrence, Acadian, and parts of the Boreal Forest regions (Rowe, 1972). When it is planted or regenerated on open land, it is heavily and repeatedly attacked by the white-pine weevil, *Pissodes strobi* Peck, an indigenous pest that causes extreme deformation and development of multiple stems. The result of this damage is a reduction in merchantable volume and lumber value (Brace, 1971).

Selection of weevil-resistant genotypes has been one of the goals in breeding of white pine (Fowler & Heimburger, 1969; Heimburger, 1966). However, there are no conclusive data from provenance studies that demonstrate differences in susceptibility among populations. A study of heavily weeviled, 13-year-old trees in a provenance test of white pine including 27 sources from the Maritime Provinces of Canada through Ontario and Quebec to Minnesota, USA, and south to Georgia revealed no differences in damage (Garrett, 1972). In another test, of 12 sources, differences in weevil damage at age 13 were detectable, but all sources were heavily weeviled (Garrett, 1973). In yet another test, of 21 seed sources, there was no evidence of genetic resistance to weeviling independent of tree size (Pauley et al., 1955). In cage experiments, differences among provenances were found on 3-year-old trees, but they were dependent on weevil density (Soles & Gerhold, 1968). The last two experiments suggest that detectability of variation in weevil susceptibility among provenances depends upon the size of the weevil population and may be confounded by variation in tree size.

An alternative method for developing weevil resistance is interspecific hybridization. For example, hybrids of *P. strobus* with the western white pine, *P. monticola* Dougl., grow well at some sites and may be valuable in areas having heavy weevil populations because of resistance inherited from *P. monticola* (Wright, 1970; Wright & Gabriel, 1959).

Jack pine : White-pine weevil

Jack pine, *Pinus banksiana* Lamb., is a pioneer species, is intolerant of shade, and commonly regenerates after fire to form even-aged stands of high commercial value. Significant seed source variations in growth, wood quality, form, and susceptibility to insects and diseases are known, and genetic improvement in growth and form by selection and breeding within suitable seed sources is considered highly feasible (Yeatman & Teich, 1969). The species is susceptible to white-pine weevil (Baker, 1972), and genetic variation in susceptibility has been reported at the provenance level in test plantations in the Lake States (Arend et al., 1961; Batzer, 1961; King, 1971). The more susceptible provenances came from areas south and west of Lake Superior in Wisconsin and Minnesota, and the least susceptible originated in northern Minnesota. There is no clear explanation for the differences in susceptibility among provenances, and there appears to be wide variation among the more susceptible southern seed sources.

Weevil injury has also been recorded at a number of provenance test locations in Canada, but in none has the incidence been high enough to associate susceptibility with seed origin (Yeatman, pers. obs., data on file). In contrast to white pine, jack pine apparently has evolved a considerable tolerance to white-pine weevil and is not a likely host for the build-up of weevil populations of epidemic level in forests of temperate and boreal North America.

Pines : Pine sawflies

Genetic variation in growth, wood quality, form, and susceptibility to insects has already been noted for white and jack pine. Improvement is possible. Red pine, *P. resinosa* Ait., has limited natural variation, so genetic improvement is considered more difficult (Fowler & Heimburger, 1969).

Some diprionid sawflies are important pests in commercial stands of these pines: the red-headed pine sawfly, *Neodiprion lecontei* Fitch, and the European pine sawfly, *N. sertifer* Geoff., attack young plantations; the Swaine jack-pine sawfly, *N. swainei* Midd., is a threat to stands of jack pine, particularly in Quebec; and the introduced pine sawfly, *Diprion similis* Htg., attacks stands of white pine, but it is also found on jack pine and red pine of all sizes (Davidson & Prentice, 1967). Pine sawflies are just a few of many species of the family Diprionidae, which are characterized by adaptive association with a narrow range of hosts (Knerer & Atwood, 1973). Some species are apparently confined to a single host; for example, the Swaine jack-pine sawfly.

Significant geographic variation in susceptibility of jack pine to the red-headed pine sawfly has been demonstrated in provenance studies (Arend et al., 1961). Henson et al. (1970) studied susceptibility of several species and hybrids of hard pines to European pine sawfly and selected genotypes of comparatively low vulnerability by subjecting trees to a battery of tests that measured oviposition, initiation of feeding, growth, and egg production. They suggested that rate of adult strike, initial establishment, and survival of larvae could be used as criteria to select stock in a breeding program.

INTRODUCED CONIFERS

Norway spruce : White-pine weevil

Norway spruce is grown in eastern Canada as an ornamental tree and has been used in reforestation in a few localities. It is not winter-hardy in boreal Canada and is highly susceptible to attack by the white-pine weevil in temperate climates, but carefully selected provenances planted on good sites will outperform native spruces in growth and yield. Behaviour of the weevil on Norway spruce is essentially the same as it is on eastern white pine (Vandersar et al., 1977).

Norway spruce provenance and selection work at Petawawa National Forestry Institute dates back to 1924, when the first plantation of the species was established on an abandoned farm field. It soon became evident that winter hardiness and the ability to withstand weevil attacks were the principal factors limiting its suitability for use in the Ottawa Valley and other areas in eastern Canada. Observations on weeviling were made from 1963 to 1972 in a local plantation of 22 provenances from the International Norway Spruce Trial of 1938 (Viidik, pers. obs., data on file). Four provenances from northern Europe confirmed Holst's (1955) observation that narrow-crowned phenotypes are less prone than others to weevil attack and/or have a higher recovery rate. The best provenance was a local population derived from the first plantation established in 1924 and later subjected to intensive selection for winter hardiness and weevil tolerance by Dr. Carl Heimburger. This suggests that selection for weevil resistance in this introduced species is effective.

Larches : Larch sawfly

Tamarack, *Larix laricina* (Du Roi) K. Koch, the native larch of eastern and boreal North America, was an important species at the turn of the century for lumber, railroad ties, and mine timbers. However, larch sawfly, *Pristiphora erichsonii* Htg., which is thought to have been introduced from Europe, became epidemic towards the end of the 19th century and caused widespread losses as infestations advanced westward (Davidson & Prentice, 1967; Drooz, 1960; MacGillivray, 1969).

Exotic larches and their hybrids are being considered for reforestation because of their potential for rapid early growth. However, fear of sawfly has prevented their widespread use. In Europe the larch sawfly generally has not been so destructive as it has in North America (Drooz, 1960; Hunt, 1932); the exception is Britain, where both larch and the sawfly were introduced (Benson, 1950). It may be possible to select species and develop hybrids of European and Asiatic larches that will be relatively resistant to larch sawfly under Canadian conditions.

Experiments were established at Petawawa in 1960 and 1961 to compare the silvicultural potential of exotic larch species and to select superior phenotypes for further breeding. In 1977 a severe outbreak of larch sawfly in these plantations provided an opportunity to rate the seed sources for defoliation (Fogal, pers. obs., data on file). There were significant differences in defoliation, height, and volume among seed sources. There was also a tendency for fast-growing, high-yielding sources to suffer heaviest defoliation. However, the weakness of the correlation between defoliation and height was illustrated by an alpine seed source grown in Denmark for several generations before planting at Petawawa; it was among sources suffering lightest defoliation even though it was one of the fastest growing. A seed source from Poland had a relatively high yield and appeared less susceptible to sawfly than did other sources. In contrast, a hybrid, *L. eurolepis* (Sieb. & Zucc.) Gord., was high in yield but appeared to be very susceptible to sawfly. Therefore, combinations of low susceptibility and high yield may be rare but can be found.

Genys & Harman (1976) found that F_2 hybrids of Japanese, *L. leptolepis* Gord., and European, *L. decidua* Mill., larch grew most rapidly but were highly susceptible to sawfly. Among pure species, Japanese larch was most susceptible; Dahurian, *L. gmelini* (Rupr.) Litvin., and Siberian, *L. sibirica* Ledeb., larches were intermediate; whereas western larch, *L. occidentalis* Nutt., was least susceptible. Susceptibility was highly variable among geographic strains of European larch.

CONCLUSIONS

Available information suggests that there are variations in susceptibility among closely related tree species, geographic strains of a particular species, and interspecific hybrids, but more documentation is required if such information is to be effectively used in forest management practice. Use of trees that are resistant or widely adapted for survival in the presence of potentially damaging insects is a highly desirable pest-management strategy. This is especially true where other alternatives, such as direct chemical control, biological control, or stand management, fail or where risks to the environment and/or costs outweigh benefits in terms of wood and fibre saved.

The possibility for employing knowledge of variation in susceptibility in pest management strategies will depend on the host-pest relationship. In the case of an indigenous pest adapted to a range of indigenous tree species, possibilities may be limited and other alternatives more appropriate. For example, white-pine weevil damage to eastern white pine can be controlled by planting in shade (Berry & Stiell, 1976). Where a host-pest relationship is highly specific it may be possible to select trees outside the range of host acceptability. Where an indigenous tree has been shown to be particularly susceptible to injury from an exotic insect, exotic trees and hybrids that are less susceptible may be used to replace the native tree. To take advantage of desirable growth and yield characteristics of exotic trees, selection for resistance to native pests is desirable and possible in some cases.

REFERENCES

Arend, J.L., N.F. Smith, S.H. Spurr & J.W. Wright, 1961. Jack pine geographic variation - five year results from Lower Michigan tests. Michigan Academy of Sciences, Arts, and Letters 46: 219-238.
Baker, W.L., 1972. Eastern forest insects. United States Department of Agriculture Forest

Service Miscellaneous Publication 1175. 642 p.
Batzer, H.O., 1961. Jack pine from Lake States seed sources differ in susceptibility to attack by white-pine weevil. United States Department of Agriculture Forest Service, Lake States Forest Experiment Station Technical Note 595. 2 p.
Benson, R.B., 1950. An introduction to the natural history of British sawflies. Transactions of the Society for British Entomology 10: 45-142.
Berry, A.B. & W.M. Stiell, 1976. Control of white pine weevil damage through manipulation of stand climate: Preliminary results. Canadian Forestry Service Information Report PS-X-61. 8 p.
Brace, L.G., 1971. Effects of white-pine weevil damage on tree height, volume, lumber recovery, and lumber value in eastern white pine. Canadian Forestry Service Publication 1303. 33 p.
Davidson, A.G. & R.M. Prentice (Eds), 1967. Important forest insects and diseases of mutual concern to Canada, the United States and Mexico. Department of Forestry and Rural Development, Ottawa, Canada. 248 p.
Dethier, V.G., 1976. Man's plague? Insects and agriculture. The Darwin Press Inc., Princeton, New Yersey. 237 p.
Drooz, A.T., 1960. The larch sawfly. Its biology and control. United States Department of Agriculture Forest Service Technical Bulletin 1212. 52 p.
Fowler, D.P. & C.C. Heimburger, 1969. Genetic improvement of red pine and eastern white pine. The Forestry Chronicle 45: 414-420.
Garrett, P.W., 1972. Resistance of eastern white pine (Pinus strobus L.) provenances to the white-pine weevil (Pissodes strobi Peck). Silvae Genetica 21: 119-121.
Garrett, P.W., 1973. Geographic variation in resistance to white-pine weevil (Pissodes strobi) (Coleoptera: Curculionidae) by eastern white pine (Pinus strobus). Canadian Entomologist 105: 347-350.
Genys, J.B. & D.M. Harman, 1976. Variation in larch sawfly attack of different species and geographic strains of larch, exhibiting diverse growth rates. Journal of Economic Entomology 69: 573-578.
Heimburger, C., 1966. Genetic improvement for disease and insect resistance of forest trees in Canada. In: H.G. Gerhold, E.J. Schreiner, R.E. McDermott & J.A. Winieski (Eds): Breeding pest-resistant trees. Proceedings of an advanced study institute on genetic improvement for disease and insect resistance of forest trees, University Park, Pennsylvania, 1964. Pergamon Press, Oxford. p. 45-51.
Henson, W.R., L.C. O'Neil & F. Mergen, 1970. Natural variation in susceptibility of Pinus to Neodiprion sawflies as a basis for the development of a breeding scheme for resistant trees. Yale University School of Forestry Bulletin 78. 71 p.
Holst, M.J., 1955. Breeding for weevil resistance in Norway spruce. Zeitschrift für Forstgenetik und Forstpflanzenzüchtung 4: 33-37.
Hunt, S.S., 1932. European larch in the northeastern United States. Harvard Forest Bulletin 16. 45 p.
King, J.P., 1971. Pest susceptibility variation in Lake States jack pine seed sources. United States Department of Agriculture Forest Service Research Paper NC-53. 10 p.
Knerer, G. & C.E. Atwood, 1973. Diprionid sawflies: Polymorphism and speciation. Science 179: 1090-1099.
MacGillivray, H.G., 1969. Larches for reforestation and tree improvement in eastern Canada. The Forestry Chronicle 45: 440-444.
Painter, R.H., 1951. Insect resistance in crop plants. Macmillan Co., New York. 520 p.
Pauley, S.S., S.H. Spurr & F.W. Whitmore, 1955. Seed source trials of eastern white pine. Forest Science 1: 244-256.
Rowe, J.S., 1972. Forest regions of Canada. Canadian Forestry Service Publication 1300. 172 p.
Soles, R.L. & H.D. Gerhold, 1968. Caged white pine seedlings attacked by white-pine weevil, Pissodes strobi, at five population densities. Annals of the Entomological Society of America 61: 1468-1473.
Stern, K. & L. Roche, 1974. Genetics of forest ecosystems. Springer-Verlag, N.Y. 330 p.
Vandersar, T.J.D., J.H. Borden & J.A. McLean, 1977. Host preference of Pissodes strobi Peck. (Coleoptera: Curculionidae) reared from three native hosts. Journal of Chemical Ecology 3: 377-389.
Wright, J.W., 1970. Genetics of eastern white pine (Pinus strobus L.). United States Department of Agriculture Forest Service Research Paper WO-9. 16 p.
Wright, J.W. & W.J. Gabriel, 1959. Possibilities of breeding weevil-resistant white pine strains. United States Department of Agriculture Forest Service, Northeast Forest Experiment Station Paper 115. 35 p.
Yeatman, C.W. & A.H. Teich, 1969. Genetics and breeding of jack pine and lodgepole pines in Canada. The Forestry Chronicle 45: 428-433.

International bacterial canker testing programme on poplar

M. Ridé

Station de Pathologie Végétale et de Phytobactériologie, INRA, route de Saint Clement, 49000 Angers, France

ABSTRACT

On behalf of the international group of cooperators, the author presents results of inoculation with *Xanthomonas populi* on Populus clones. The same series of clones was distributed to each of 4 participating stations in Belgium, France, Great Britain and the Netherlands. In each station a local strain of the bacterium was used. Inoculations were made following the same methods and at comparable inoculation dates in September 1978, and May, June, July 1979, but adjusted to local stage of development of clones. Inoculation in September 1978 proved more stringent than inoculation in May and June 1979. These spring and early summer inoculations allow better prediction of resistance under field conditions. Strains can differ greatly in aggressiveness.

INTRODUCTION

The bacterial canker of poplar due to *Xanthomonas populi* Ridé remains a risk for poplar planting in Europe. Various programmes for testing of resistance are developing mainly in Belgium, France and the Netherlands, but also in Great Britain, and West and East Germany. Inoculation by dropping a bacterial suspension into fresh leaf scars produced by removing the leaves in September was proposed by Ridé and Steenackers and adopted by the FAO International working group on poplar diseases (1966) to test various clones and selected trees in European countries (Ridé, 1963; Steenackers, 1966; Zycha et al., 1967; Burdekin, 1972; Gremmen & Koster, 1972). This method, though useful for a first screening against bacterial canker, is very severe and does not offer any indications of the degree of field resistance that some clones demonstrate.

Moreover a curious lack of parallelism was observed between resistance test results obtained by the same method in the recent past in Belgium, France, Great Britain and the Netherlands. This led us to take into account the level of aggressiveness of bacterial strains used in the different countries and the prevailing ecological conditions.

Therefore an international cooperative programme was proposed by Koster in Wageningen 4 years ago in order to screen preselected, promising poplar clones in the 4 countries and to compare and standardize inoculation methods. Results obtained by Ridé about the response of the tree to *X. populi* in terms of inoculum dose, physiological stages of the poplar and aggressiveness of the bacterial strain constituted a base for test planning (Figures 1 and 2). The aim was to study differences in susceptibility in poplar plants simultaneously grown in each of the 4 countries from cuttings originating from the same set of stools in the Netherlands.

MATERIALS AND METHODS

Test design

The test was set up with 24 clones (see Table 3, plus NL 2066, 'Rochester'; NL 1310, *deltoides* × *nigra*, and NL 2752, '*hybrida* 275' Poland). At Hees 3 Belgian clones ('Hunnegem', Raspalje' and 'Boelare') were added.

These clones were tested simultaneously in 4 countries: at Hees (the Netherlands), Grimmingen (Belgium), La Rétuziere Angers (France) and Grange Estate (England).

Spacing was 3 m × 3 m, the number of plants per clone was 23. This number corresponds to 4 dates of inoculation on 5 plants each time; at the last 3 dates 1 plant was added and not inoculated in order to serve as control.

Times of inoculation

Inoculations were made in September 1978 and May, June and July 1979. In order to account for climatological differences between the locations in 4 countries, the inoculation dates were related to flushing stages. E.g., the May inoculation was to be made when 5-6 leaves of a clone had flushed.

For practical reasons the clones were divided into 3 groups: the first group consisted of clones flushing (with 5-6 leaves) at the same time as 'Oxford' (348); the second group at the same time as 'Robusta' (2098); the third group at the same time as clone 1266.

The inoculations in June and July were made in principle 30 and 60 days after the May inoculation.

Reasons for choosing these times of inoculation

The choice of the method of and time for inoculating resulted from the observations made in France during the past 10 years. An example is given in Table 1. From the results in the column Sept. 74 it is not possible to obtain an impression of differences in susceptibility under field condi-

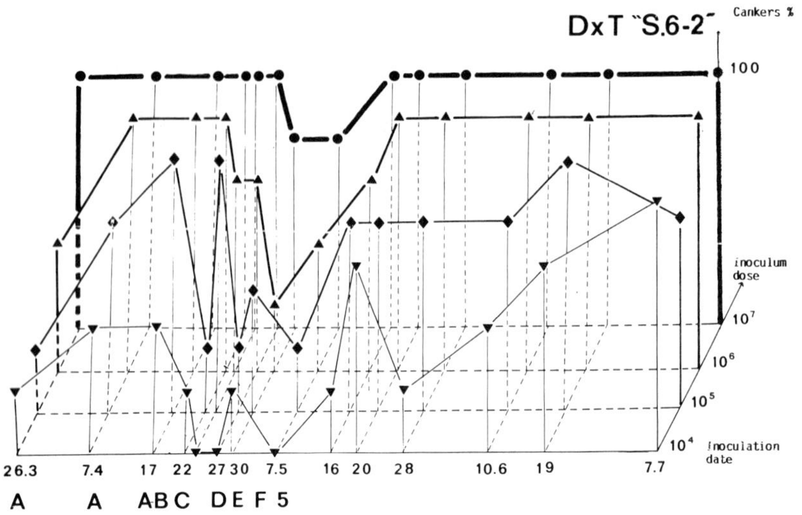
DxT 'S.6-2'
Cankers %
100
inoculum dose
10^7
10^6
10^5
10^4
inoculation date
26.3 7.4 17 22 27 30 7.5 16 20 28 10.6 19 7.7
A A ABC DEF 5

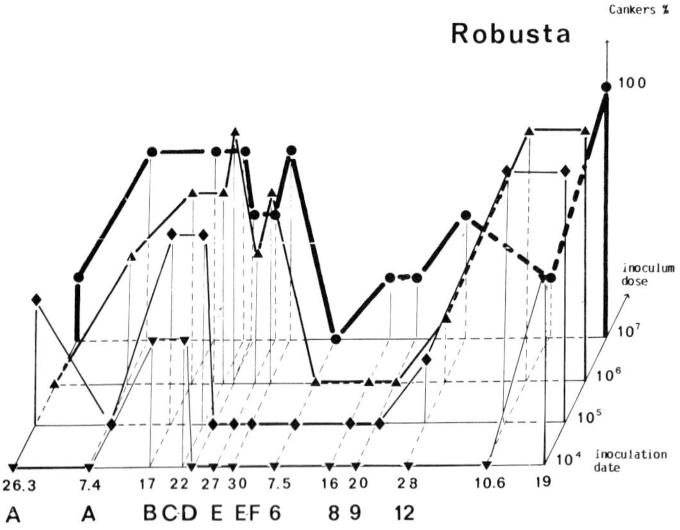
Robusta
Cankers %
100
inoculum dose
10^7
10^6
10^5
10^4
inoculation date
26.3 7.4 17 22 27 30 7.5 16 20 28 10.6 19
A A BCD E EF 6 8 9 12

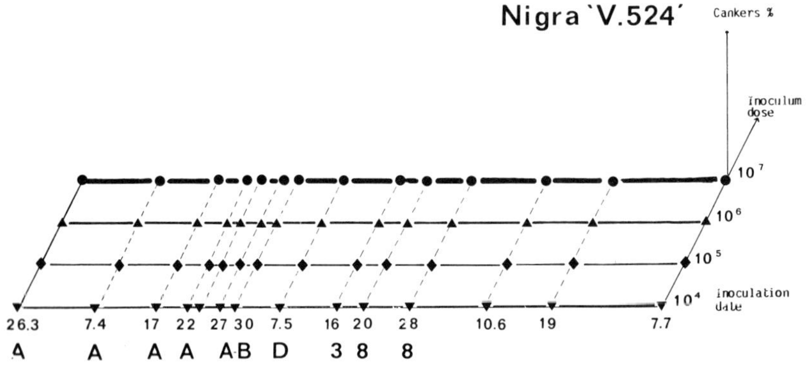
Nigra 'V.524'
Cankers %
Inoculum dose
10^7
10^6
10^5
10^4
inoculation date
26.3 7.4 17 22 27 30 7.5 16 20 28 10.6 19 7.7
A A A A AB D 3 8 8

S 620-96 S 5-3 S 239-12

Figure 2. Expression of bacterial canker on various poplar clones in 1980, subjected to inoculation in 1965. (Poplar Research Center, Geraardsbergen (Grammont) B). Clones: S 620-96: *P. deltoides* (Michigan); S 5-3: *P. deltoides* V5 (Iowa) × *P. trichocarpa* V21 (Montana); S 239-12: *P. deltoides* (Missouri) × *P. grandidentata*. Correlation between results of juvenile test and adult expression is now established.

Figure 1. Response of different poplar clones to *X. populi* in terms of inoculum dose and phenological stages. (Spring and summer inoculations.) *P. deltoides* × *trichocarpa* S.6-2; *P.* × *euramericana* cv. 'Robusta'; *P. nigra* V. 524. (after M. Ridé & S. Ridé, 1978.)

A = dormant bud
B = swelling bud
C = first leaf tip
D = half appearing leaves
E = petiole appearing but leaves still rolled up
F = leaves completely spread out
5-12 = number of preformed leaves

Table 1. Susceptibility of poplar clones to bacterial canker according to inoculation period. Angers, scored February 1978. Explanation of scores: 1 = no reaction; 6 = attack spreading to larger parts of the stem; stem often dies. 052-2 to GRM 51: *P. deltoides* clones obtained from National Center for Forest Research, Orleans (F). 70-039/110 to 3: *P. trichocarpa hastata* (Idaho) × *P. trichocarpa* (Oregon) obtained from Poplar Research Center, Geraardsbergen (B).

Clone	Inoculation			
	Sept. 74	May 75	June 75	July 75
052-2	5.6	3.1	4.2	3.8
061-3	5.3	4.1	3.3	3
234-1	5.8	4	3.1	2.3
053-2	6	3	3.2	2.3
257-1	6	3	3.1	1.8
044-3	5.5	2.7	2.7	3
257-2	5.8	2.3	3	1.3
244-2	5.3	1.8	1.8	1.3
044-2	5.2	1.6	1.8	1.1
241-1	5.2	1.5	1	1.1
063-1	5.7	1.2	.	.
253.3	6	1	1.3	1.3
Alpaga	6	2.8	2.8	2.1
Albiruch	6	2.6	3	2.7
Bergerac	5.5	1.8	2	2.3
WGN/19	5	1.5	1.2	2.1
S.335-12 × S.332-1/11	5.3	3.8	4	4.1
V8 × V11 = B/5	6	3	3.3	3.2
B/6	6	3.3	2.8	2.7
B/14	6	3.1	2.2	2.5
GRM 51	3.2	1.1	1	1.2
70-039/110	5.7	2.7	3.6	4
/108	5.7	2.5	2.3	3.7
/ 40	4.7	2.7	2.3	2.1
/ 97	5	2	2	2.3
/ 64	6	1.8	1.1	1.7
/ 98	5.2	1.3	1.5	1.8
/ 99	4.2	1.1	1	1
/ 85	2.8	1.5	2	1.6
/ 39	2.7	1.7	2.2	1.7
/ 7	3	1.3	1.5	1
/ 31	3	1	1.1	1
/ 14	2.1	1.5	2	1.5
/ 20	1.6	1.8	1.2	1
/ 90	1	1.5	1.3	1.2
/ 3	1.1	1	1.1	1

tions. Autumn inoculation has very severe consequences and shows all clones (except GRM51) to be susceptible. On the other hand, if we compare the results of the inoculations in May and June we are able to establish a ranking of clones according to their degree of susceptibility under field conditions.

Inoculation methods

On the shoot of the previous year the French incision method was used: a wedge-shaped cut with a knife into the cambium (as far as the sapwood); on the current year's shoot, fresh leaf scars were inoculated. A drop of bacterial suspension was placed on the fresh leaf scar or on the fresh wound. For further details see Table 2.

Inoculum

Local strains of the bacterium were used in each country. Strains were not mixed; their aggressiveness is being compared in Angers on 14 reference clones.

Strains were incubated at 23 °C during two days. Bacterial suspension concentration was 10^8 cells/ml. Isolates used were in:

- Belgium: M1 (isolated from *P. euramericana* cv. Regenerata, April 1978).
- France: SP (isolated from *P. euramericana* cv. Blanc de Poitou, April 1978).
- Great Britain: Rap (isolated from 'Rap', April 1976).
- The Netherlands: 115 (isolated from *P.* 'Serotina', April 1978).

Scores

The score system used is identical to the FAO working group system (scale 0 to 5) (1966) but renumbered by Gremmen & Koster (1972) to 1 to 6 for computer analysis.

Table 2. Inoculation scheme for the international test.

Date	Number	Place of inoculations
Sept. 1978	2	at 30 % and 60 % of the length of the plant
May 1979	2	at 30 % and 60 % of the length of the shoot of '78
	1	on the new (very short) shoot of '79
June 1979	2	at 30 % and 60 % of the length of the shoot of '78
	1	at 30 % of the length of the shoot of '79
July 1979	2	at 30 % and 60 % of the length of the shoot of '78
	2	at 30 % and 60 % of the length of the shoot of '79

RESULTS AND DISCUSSION

The trial will probably have to be continued for 4 to 5 years depending on the development of the disease in each country. Nevertheless, the first results were compared in 1979. These only took into account the scores of the September inoculations in Belgium, France and the Netherlands. Ranking was found to be similar through all of these first observations, probably as a result of the very comparable level of strain aggressiveness. This was contrary to previous experiments in which the 'Rap' strain proved very highly aggressive while strain 88 NL was only slightly aggressive (strain 88 NL was not used in the trials at Hees).

In 1980, results (Table 3) showed a good correlation for Dutch and French experiments, but not for the Belgian experiments in these months. After discussion in the field in Grimmingen (B), it was thought that two factors probably can explain this difference: the low nutrient level of trees in the Belgian experimental plot (Steenackers, personal communication) and, above all, the fact that in most of the spring and summer inoculations the bacterial suspension had been dropped on the outer-bark layer of the trees (cortex parenchyma) instead of into the cambium as mentioned in the experimental procedure. Outer bark contains inhibiting substances which could explain inoculation failure (Ridé & Ridé, 1978; Ridé, 1980) (Table 3).

In the future, it will be interesting to compare the results obtained by this method to those observed after spraying clones in spring with bacterial suspensions (Gremmen & de Kam, 1980).

In the experimental plot at Hees, the normal disease expression was sometimes disturbed by attacks of cambium borers (*Dendromyza carbonaria* Hendl.). The larvae of this insect can distribute *X. populi* and other bacteria along the trunk, twigs and branches but they can also by their presence induce an appearance of the stem or branch similar to that caused by bacterial canker.

ACKNOWLEDGEMENTS

This paper summarizes a field demonstration presented by Dr. and Mrs. Ridé in the nursery of the Dorschkamp institute at Hees on 19 IX 1980 on behalf of the international working group. Other members of this informal working group are R. Koster and M. de Kam (the Netherlands), M. Jilbert (United Kingdom), V. Steenackers (Belgium).

I want to express my gratitude to Mrs. Ridé (F) primarily for the design and drawing of posters, to Mr. Van Lokhorst (NL), Mr. Larsin and Remy (B) for their kind and constant assistance in this cooperative programme.

Table 3. International bacterial canker testing programme on poplar. Mean scores of inoculations, assessed in August 1980, for 21 clones. Scoring as in Table 1.

Identity of clones[a]		Netherlands				France				Belgium			
		Sept. 78	May 79	June 79	July 79	Sept. 78	May 79	June 79	July 79	Sept. 78	May 79	June 79	July 79
NL 350	*maximowiczii* × *trichocarpa* 'Androscoggin' (USA)	5.4	2.3	4.6	4.6	6	4.8	5	5	5.1	1.6	1.9	2.9
NL 1266	*trichocarpa* S3-31 (B)	2.6	2.9	2.6	4.4	5.1	4.7	4.5	5	3.1	1	2.2	2.5
NL 348	*maximowiczii* × *berolinensis* 'Oxford' (USA)	5.2	4.5	5.1	5.1	6	4	4	4.2	4.9	2.1	2.4	2.7
NL 2030	*maximowiczii* × *berolinensis* 76-56 (D)	4.2	3.5	5	4.3	5.8	4.1	4	4.2	5.7	1.5	1.9	2.7
NL 2205	*deltoides* (B) × *nigra* (I)	3	1	1.6	.	5.7	3.5	4	4.5	3.1	1	1.2	1.9
NL 1623	*deltoides* (B) × *trichocarpa* 'Barn'	3.5	1	3.7	3.2	5.6	3.2	4	4.2	3.7	1.5	1.7	2.5
NL 2118	*trichocarpa* × *tacamahaca* clone CF (GB)	5.2	2.8	3.7	4.5	6	3.6	3.2	4.1	4.5	2.9	2.1	2.5
NL 1658	*trichocarpa* (B) × *deltoides* 'Rap'	3.9	1.5	2	4.2	5.6	2.8	4	4	2.8	1.1	1.4	2.6
NL 1647	*deltoides* (B) × *trichocarpa* 'Donk'	2.7	1.1	2.8	4.1	5.1	2.8	3.5	3.6	3.1	1	1.7	2.5
NL 925	*deltoides* × *nigra* (I) 'Dorskamp'	4	2.9	4.6	3.5	5.7	2.3	3.1	3.7	4	1.1	1.6	2.1
NL 1524	*deltoides* (Illinois)	4.1	3.4	4	4.1	4.6	1.8	2.7	4	2.7	1.1	1.3	2.7
NL 2098	*deltoides* × *nigra* 'Robusta'	2.2	2.5	2.8	2.7	4.5	1.7	3	3.2	2.4	1	1.2	2.1
NL 1454	*deltoides* (Missouri)	2.6	1.7	2.6	3.4	4.1	1.7	2.3	3.6	2.2	1	1.2	1
NL 2058	*trichocarpa* × *tacamahaca* clone 32 (GB)	2	1.4	3.2	.	3.5	2	3	3	2.1	1.7	1.6	1.6
NL 1785	*deltoides* (B) × *trichocarpa*	2.7	1	2	3.9	(5)	2	3	3.2	2.7	1	1.2	1.9
NL 1255	*trichocarpa* 'Blom'	1	1	1.2	2	3.5	2.2	1.7	2.7	1	1	1.1	1.4
NL 2195	*deltoides* (B) × *nigra* (I)	1.9	1	2	1.8	2.8	1.7	2	2.5	1.4	1	1.5	1.2
NL 2043	*trichocarpa* 636-52 (D)	1.2	1.2	1.9	2.1	2	2.5	2	2	1.2	1	1.4	1.4
NL 1237	*trichocarpa* SP 127 (GB)	1.7	1.5	1.5	2.5	2.2	1.2	1.7	3	1.4	1	1	1.5
NL 1236	*trichocarpa* clone CF (GB)	1.1	1.2	3.2	2.6	1.6	1	1.7	2.2	1	1	1.3	1.7
NL 1274	*trichocarpa* 'Fritzi Pauley'	1.7	1	1.8	2	1.9	1.7	2	1.7	1	1	1	2

a. B = Belgium, NL = Netherlands, D = Germany, USA = United States of America, GB = Great Britain, I = Italy.

REFERENCES

Burdekin, D.A., 1972. Bacterial canker of poplar. Annals Applied Biology 72: 295-299.

Gremmen, J. & R. Koster, 1972. Research on poplar canker (Aplanobacter populi) in the Netherlands. European Journal of Forest Pathology 2: 116-124.

Gremmen, J. & M. de Kam, 1980. Research on poplar canker (Xanthomonas populi subsp. populi) in the Netherlands. European Journal of Forest Pathology 10: 48-53.

Ridé, M., 1963. Our present knowledge of bacterial canker caused by Aplanobacterium populi. FAO/CIP/MAL 23, Casale Monferrato (translated by Batko).

Ridé, M. & S. Ridé, 1978. Factors affecting inoculation success in woody plants. Proc. 4th Int. Conf. Plant Path. Bact. Angers, Vol. 2. p. 957-968.

Ridé, M., 1980. Interactions hôte-pathogène dans la résistance des peupliers aux infections bactériennes. Proc. FAO/IUFRO Working group on Poplar diseases - Poznan.

Steenackers, V., 1966. La sélection et la création des peupliers résistants aux diverses maladies et plus particulièrement à Aplanobacter populi Ridé. FAO/CIP Working group on Poplar diseases. Versailles.

Zycha, H., M. Ridé & H. Fröhlich, 1967. Prüfung der Krebsresistenz verschiedener Pappeln in Deutschland (unpublished data).

A coevolutionary view of resistance breeding and research[1]

Henry D. Gerhold

School of Forest Resources, Pennsylvania State University, University Park, Pennsylvania 16802, USA

ABSTRACT

Evolving relationships between research and production of disease-resistant trees are reviewed, using an analogy with symbiosis. Research developments are cited which are likely to expand the demand for resistant trees, increase the precision of manipulating genetic resistance, and improve the design of breeding systems. Resistance breeding has received increased attention in cooperative tree improvement programs, which have themselves proliferated and have also improved technology transfer. Constraints caused by economic pressures, changing pest control methods, and environmental concerns are seen as increasing the need for more resistant varieties. The coevolution of resistance breeding and research in the future holds many opportunities, and suggestions are offered for making the most of them.

INTRODUCTION

The intent of this concluding paper in the 'Workshop on Genetics of Host-Parasite Interactions in Forestry' is to provide some perspective on trends in resistance breeding and research. Two related meetings, similarly comprehensive in subject matter and international in scope, preceded this one. 'Biology of Rust Resistance in Forest Trees' was discussed in 1969 at Moscow, Idaho (Bingham et al., 1972). In 1964 a world-wide summary of 'Genetic Improvement for Disease and Insect Resistance of Forest Trees' was attempted for the first time (Gerhold et al., 1966). Other more narrowly focused meetings have been held too, many of them organized by Working Groups of the International Union of Forest Research Organizations.

1. Based partly on research supported by The Pennsylvania Agricultural Experiment Station, USDA Regional Research Project NE-27, and the US Forest Service. Journal Series Paper No. 6143.

My instructions from the Organizing Committee were to review developments since the 1964 meeting and to bring in some constructive suggestions and visions into the future. That is quite a challenge! The impressive array of papers presented at this Workshop gave us just a glimpse of some of the many accomplishments during the past 16 years. In 1969 a survey turned up over 50 resistance breeding programs in 19 countries (Gerhold, 1969), and undoubtedly the work has expanded since then. Rather than reviewing technical aspects of host-parasite interactions, I wish to comment on another type of interaction, namely the interdependency that I see between research and production aspects of improvement in pest resistance. Progress in producing more resistant trees has been dependent, of course, on research in disciplines such as genetics, pathology, entomology, and physiology. There has also been feedback from tree improvement which has enhanced research in these fields. By considering evolving relationships between research and production of resistant trees, I hope we can perceive more clearly what has contributed to progress, especially in applications of research.

Prior to 1964 the work on pest resistance of trees was mainly in the realm of research. Like organisms invading a barren landscape, investigators were exploring genetic variation in resistance and how it could be utilized to improve adaptation to hostile environments.

Today we know that resistance breeding of forest trees can be successful and economical, under appropriate circumstances and with the proper techniques. Research results have been incorporated into tree improvement programs to produce realized genetic gains in resistance, i.e. gains that earlier had only been predicted. Research and breeding programs have adapted to new knowledge and external forces, just as colonizing populations pursue adaptive strategies to exploit Nature's vacuums and accommodate its pressures. Furthermore, these complementary activities have adapted to each other.

A COEVOLUTIONARY ANALOGY

It may be useful to think of the interrelated development of research and production as analogous to the concept of coevolution, in which evolving species exert adaptive pressures on each other. The basis of the analogy is the similarity between information systems that operate within organisms and those within organizations that conduct forestry research or production activities (Table 1). Information stored in genes is equated with ideas, i.e. related facts or pieces of knowledge. Both kinds of information give rise to a series of events or specific activities. Chromosomes, cells, organs, and organisms are increasingly higher levels of biological complexity that are analogous to concepts, plans, projects, and programs of research or production organizations. Meiosis may be thought of as a re-

Table 1. Analogy of information systems in organisms and organizations.

Organisms	Organizations
genes	ideas
chromosomes	linked ideas, concepts
meiosis	rearranging ideas
mitosis	propagating ideas
cells	plans
organs	projects
individuals	programs
fitness	utility of ideas
genetic gain	technological progress

arrangement of ideas, e.g., during meetings such as this one; and mitosis represents the propagation of ideas such as in training workers. Through population genetics the analogy may be extended, and one may think about the fitness of ideas or programs as being similar to the fitness of genes or genotypes. The implied value of ideas accordingly would depend on their survival and reproduction, and perhaps also their impact on related organizations. Technological progress, like genetic gain, could be defined as a function of intensity of selection and the degree to which the best ideas are utilized.

That brings us to the coevolutionary part of the analogy. Reciprocal evolutionary changes in species that exert adaptive pressures on each other may involve predation, competition, or symbiosis. My analogy draws upon various types of symbiosis, which are classified as parasitism, commensalism, or mutualism. Parasitism works to the detriment of the host, as you well know. Commensalism benefits one partner, and does not harm the other. Mutualism enhances the ability of both species to survive and reproduce.

Parasitism could be likened to a tree production program, the host, that sustains a forestry research program, the pest (Table 2). Now, a researcher might resent being called a pest. But an industrial executive would be more apt to agree with the analogy, especially if he is preoccupied with the company ledgers. After all, research can cause quite a drain on the budget, and this type of investment will please management and stockholders only if results find practical applications that improve net earnings. Applications depend upon technology transfer, represented by the aegricorpus, a term coined by Loegering (1966). This pseudo-organism actually is the interaction of host and parasite, leading to infection of a susceptible host. Rejection of research applications, reflected by resis-

Table 2. Analogy of parasitism with forest research and production.

Parasitism	Research & production
host	tree production program
pest	forestry research program
aegricorpus	technology transfer
resistance	rejection of research results
susceptibility	acceptance of research results

Table 3. Analogy of commensalism with forestry research and production.

Commensalism	Research & production
root	production
fungus	research
mycorrhizae	research applications

tance in the host, is perhaps a natural tendency of production organizations to resist innovation and change. Susceptibility then would take on a favorable connotation, as it implies an acceptance of research results. Tree breeders might regard this as backwards, but perhaps not pathologists or entomologists. For they have been heard to exclaim more than once, 'There's a beautiful epidemic!'

Commensalism is a preferable form of symbiosis between research and production (Table 3). A mycorrhizal relationship seems apropos, for it benefits the fungus in all situations, but benefits to the host are most apparent on harsh, nutrient-deficient sites. Similarly, research goes on continually, but seems to be appreciated most during an impending calamity or when there is pressure to boost production. Commensalism may be the most appropriate analogy for most research - production relationships, especially in their initial stages of development.

An example of mutualism serves my purposes better, however (Table 4). It can illustrate an ideal relationship between a tree improvement program, represented by a fig tree, and the supportive research projects represented by wasps. They buzz about, disseminating research results (pollen), while helping to ensure the future of fig trees and their own kind. The waspish researchers are rewarded with the sweet nectar of fame and fortune, though it is usually doled out in small doses. In this intricate relationship

Table 4. Analogy of mutualism with biological research and tree improvement.

Mutualism	Tree improvement & research
fig tree	tree breeding program
wasps	research (genetics, pathology, etc.)
pollen	research results
bats, birds	liaison with industries
seeds	knowledge and genotypes

(oversimplified here), if the wasps failed to transfer the research results, the tree improvement program would not be worth a fig. Furthermore, let's not overlook the bats and birds which provide liaison with industrial tree growers. After gorging themselves with the fruit of improvement programs, they disseminate information and selected genotypes. As they deposit these seeds, they also help them to grow with a healthy dose of dung.

This analogy could be elaborated in greater detail, but that would detract from my assignment to review past developments and to peer into the future. For convenience, I shall group developments under research, tree improvement, and the external environment that influences these activities and their interactions.

DEVELOPMENTS IN RESEARCH

Most individual impacts of the great majority of research findings have not been dramatic. But as papers in this Workshop have indicated, the steady growth in knowledge about pest resistance of forest trees has been impressive. It has expanded beyond the more important diseases and insects to lesser ones, and also to viruses, nematodes, and mycoplasmas. A cumulative effect of this research and its applications has been to firmly establish resistance breeding as an accepted means of controlling certain forest diseases caused by fungi. The acceptance of this control method will certainly be extended to additional diseases as research progresses, and perhaps also to some insects and other pathogens. But it won't happen automatically, in most cases. When a piece of research is completed, further efforts will be required to engineer the applications (Shigo, 1977). This need for involvement of researchers with practitioners is so obvious it hardly seems worth mentioning, yet administrators often neglect to bring the two parties together.

There are 3 types of research whose effects appear to have great potential for influencing resistance breeding in forestry. Their effects are likely to (1) expand the need for more resistant trees, (2) increase the precision of manipulating genetic resistance, and (3) improve the strategy and design of breeding systems by modelling the outcomes of alternative improvement strategies.

An expansion of the demand for more resistant trees has been underway for some time, and is likely to accelerate. Research on vegetative propagation, provenance experiments, and exotic species trials has contributed to this trend, mainly by pioneering silvicultural alternatives which may be implemented when warranted by industrial conditions. Clonal plantations were considered a novelty until recently, except for poplars. Now there are extensive areas planted to clones of Monterey pine, eucalyptus, Norway spruce, and other species. Over half of the landscape trees grown in the United States are vegetatively propagated, and all of these species are represented by just a few clones of each. Research with exotic species and provenances is another part of a general intensification of silvicultural practices. Increased productivity is the goal, but new disease and insect problems will surely be encountered, so at least in some cases genetic improvement in resistance will be an integral part of the intensified management practices.

Several research developments some day may greatly increase the precision in manipulating genetic resistance, compared to conventional methods (Abelson, 1980; Gwynne, 1980; Kleinhofs & Behki, 1977). New tissue culture techniques and isoenzyme analyses can be used to obtain more definitive information about gene action and interaction, and thus to elucidate the genetic control of resistance mechanisms. Genotypes may be altered through cell fusion or transformation. Recombinant DNA technology may enable the splicing of individual resistance genes into chromosomes. Such techniques for more precise handling of genes can be especially significant in short-cutting tree breeding. Many years may be required to adapt such technologies to forest trees, but the research is under way.

The applicability of the gene-for-gene model to forest diseases is a finding that I believe will have more far-reaching consequences for us than any related research in recent times. It is a powerful concept that can be used not only for identifying specific resistance genes, but more importantly for predicting the outcomes of different strategies in breeding and deploying improved varieties. To attain this capability an extension of the model will be required to include fitness of resistance and virulence genes and associated genomes in various environments, and disease levels induced by various gene combinations over a period of generations. Non-specific types of resistance would also have to be accommodated and in some cases interactions with vectors. A synthesis of such modelling studies with other analyses could be used to determine the optimum design of a resistance breeding system. Several papers here have made valuable contributions to such an approach.

If analogous gene-for-gene models and more powerful selection methods could be developed for insect resistance, it might provide a great stimulus to progress in this field. Resistance research with insects seems to be generally more difficult than with diseases, presumably due to the insects' greater mobility, responses to diverse stimuli, and complex relationships with predators. That may help to explain why indirect selection for biochemical traits has been considered in many cases, though its effectiveness remains elusive after much research. Insect resistance is not now a principal goal in breeding programs, but it may be in the future.

DEVELOPMENTS IN TREE IMPROVEMENT

The great proliferation of cooperative tree improvement programs in North America (Kang, 1980) and cooperative research in Europe on elm and poplar diseases are phenomena worth nothing. Not only have new programs started, but older, established programs have increased their work on resistance breeding. There is a greater awareness of the value of pest resistance by tree breeders and an acceptance of tree improvement by foresters. Furthermore, the organizational nature of these cooperative programs helps to assure effective technology transfer, by providing close working relationships between researchers and practitioners.

Resistance testing centers may be viewed as a related organizational development, facilitating interactions between pathologists and tree breeders. Their intellectual interplay and melding of expertise may be just as important as the specialized facilities for imposing higher, more uniform selection pressures. Artificial inoculation is commonly employed for mass screening and resistance studies, but there is some concern about how closely results are correlated with resistance in nature.

Another significant development is the survival and even expansion of programs that have been challenged by more virulent strains of fungi. In each case the initial impact must have been shocking for tree breeders and sponsors alike. The fact that these programs survived may imply not only that the genetic approach to disease control is a valid one, but that confidence in it has been strengthened.

DEVELOPMENTS IN THE ENVIRONMENT

An essay on evolution would be incomplete without comments on the role of the environment, in this case meaning external influences on the coevolution of resistance breeding and research. The past 16 years has been a time of fluctuations in the economic climate, upheaval in disease and insect control methods, and drift in public concerns. These events are suggestive of major geologic changes which have altered the course of biological evolution.

Energy shortages and economic pressures have been felt by researchers, breeders, and practicing foresters alike. We feel constrained in various ways, so that it is more difficult to get our work done. A more serious consequence may be the trend toward concentration and intensification of tree growing. This is likely to lead to greater risks of disease and insect damage, and thus increase the need for more resistant varieties.

A revolution has been underway, of course, in methods for controlling insect and disease damage. Hazards of chemicals are still being discovered, and pesticide regulations are becoming ever more stringent. Exaggerated reactions of environmentalists, the news media, the public, and government agencies may be annoying, and restrictive measures may not always seem acceptable, but generally they prevail and do become accepted. Despite some misgivings, we can regard this course of action as prudent because safeguards against chemical hazards are needed, and they do stimulate the development of biological controls and resistant varieties. However, this favorable influence will be countered by another set of fears, probably held by many of the same people who object to pesticides.

Apprehensions and misapprehensions concerning genetic engineering, monoculture, endangered genotypes, and the like surely will hamper forestry practices, research, and particularly opportunities to utilize resistant varieties. Outcries have already become more strident against real or imagined hazards of genetic involvements with human intelligence, genetic diseases, abortion, sterilization, and virulent microbes. This is probably just the beginning, for there are certain to be more scientific breakthroughs in human genetics and dramatic proposals for altering our social structures. Critical attitudes surely will spill over into agriculture and forestry, which have already been criticized for hard tomatoes, monoculture, and clearcutting of forests (Popovich, 1980).

THE FUTURE OF RESISTANCE BREEDING AND RESEARCH

What does the future hold for the coevolution of resistance breeding and research? I am optimistic that the operational environment will be full of opportunities, and will be favorable if we don't neglect our educational responsibilities. The main determinants of the future probably will be those evolutionary 'ideas' generated by research and improvement programs that prove to have high 'fitness'. In other words, future directions will be set mainly by new knowledge or techniques that find widespread application in forestry practices, or whose significance is recognized and utilized by many researchers. Note that in this view, intrinsic value has little influence on 'fitness of ideas' until they have been widely disseminated and accepted. If it is true that the practical value of innovations in forestry production and research will have a strong effect on rate and direction of coevolution, then I can suggest several implications:

- We can favorably influence our professional destiny by cooperating with our coevolutionary colleagues.
- We should build stable organizations that provide effective interactions among those engaged in research, tree improvement, and forestry production.
- Resistance breeding goals should be identified early, especially in new improvement programs. Appropriate weights should be given to all important selection criteria, including measures of yield, quality, and pest resistance.
- For any threatening disease or insect a research lead should be maintained to explore genetic variation in resistance and virulence. This will make it possible to develop effective strategies for selection, breeding, and geographic deployment of improved varieties.
- Improved plants or techniques should be moved into production aggressively, despite some risks, to demonstrate their practical value.
- We should assume responsibility for assuring that genetic gains in resistance will be demonstrated and that technology transfer will occur to a sufficient degree.

In applying these suggestions, the differing needs of new programs versus firmly established programs should be recognized. A principal concern in new programs is to demonstrate the effectiveness of research or improvement methods. As a program matures, there may be a shift toward overcoming complacency about routine accomplishments, and justifying more complex procedures so that progress may continue. This workshop audience is well aware of the complex, long-term scientific work required to obtain durable types of resistance, and integrating these with environmental adaptation, improved yield, and higher quality. But are we sufficiently aware and active in educating colleagues, foresters, policy makers, and the public? If this symbiotic requirement is overlooked, we may be relegated to the evolutionary fate of the dinosaurs.

REFERENCES

Abelson, J., 1980. A revolution in biology. Science 209: 1319-1321.
Bingham, R.T., R.J. Hoff & G.I. McDonald, 1972. Biology of rust resistance in forest trees: Proceedings of a NATO-IUFRO Advanced Study Institute. USDA Forest Service Misc. Publ. No. 1221. 681 p.
Gerhold, H.D., 1969. A decade of progress in breeding disease-resistant forest trees. Unasylva 24: 37-44.
Gerhold, H.D., E.J. Schreiner, R.E. McDermott & J.A. Winieski, 1966. Breeding pest-resistant trees. Proc. NATO & NSF Advanced Study Institute. Pergamon Press. 505 p.
Gwynne, P., 1980. The cloning of Russett Burbank. Mosaic 11: 33-38.
Kang, H., 1980. Research needs in tree breeding. North American Quantitative Forest Genetics Group Workshop Proceedings, USFS Inst. Forest Genetics, Rhinelander, Wisc.
Kleinhofs, A. & R. Behki, 1977. Prospects for genome modification by non-conventional methods. Ann. Rev. Genetics 11: 79-101.
Loegering, W.Q., 1966. The relationship between host and pathogen in stem rust of wheat. Proc. 2nd Int. Wheat Genetics Symp., Hereditas Supp. 2: 167-177.
Popovich, L., 1980. Monoculture: a bugaboo revisited. Journal of Forestry 78: 487-489.
Shigo, A.L., 1977. Communication of knowledge and needs between forest researchers and practicing foresters. Northern Logger and Timber Processor 25: 7, 38.

Recommendations

Participants in the Workshop on Genetics of Host-Parasite Interactions in Forestry - 14 to 21 September, 1980, Wageningen, the Netherlands - recommend that:

1. Scientific and technical terms be used with precision and consistency.

2. International and national agencies for research and development be encouraged to acquire and train scientific and technical personnel to combat pests (includes all biotic agents that damage trees) attacking or threatening forest plantations in tropical and sub-tropical countries.

3. Systems for integrated forest pest management be designed such that they are based on genetic resistance and provide for maintenance of ecosystem stability.

4. Genetical, physiological, and epidemiological aspects of host-parasite interactions all be investigated so that mechanisms underlying resistance can be understood.

5. Legislative and regulatory agencies be alerted to the danger that well-intentioned actions to ensure excellence of clonal material may have the undesired side effect of restricting the number and variety of clones legally available for use, and that said agencies be encouraged to modify such actions, wherever warranted, so as to foster use of adequate numbers of clones in mixed or mosaic plantations.

6. Actions to conserve genetic resources contained in forest ecosystems be taken soon, that conservation be undertaken at one or more levels (alleles, co-adapted gene complexes, populations, or ecosystems), and that ecosystems of special interest be secured intact so that interacting populations can continue to co-evolve and remain available for study.

7. Procedures for international exchange and testing of agronomic host-plant germplasm be examined for applicability to forestry situations, and be modified, as necessary, to increase effectiveness in monitoring and breeding for pest resistance in forest trees.

8. Those formulating theories, models, and strategies for forest populations be encouraged to draw on genetic concepts from the broader fields of population and community biology, as well as from the narrower but

richer experience had in closely-bred agronomic systems.

9. Population genetics of major tree species and their pests be investigated so as to characterize qualitative and quantitative components of environmental and genetic variation.

10. Efforts to develop trees resistant to decay agents be increased, in view of recent, promising advances in decay research.

11. Parallel non-protected experiments be conducted to determine if resistance is lost where materials in breeding programs are given artificial protection against damaging agents.

12. Model host-parasite systems, analogous to *Drosophila* and *Escherichia coli*, be identified and developed for detailed investigations of tree-pest interactions.

13. Genetic consequences of pest control strategies be considered by forest managers since chemical and biological control measures act as agents of selection on pest populations.

14. Forest research agencies be encouraged to employ scientists equipped to study the genetics of arthropod-tree interactions.

Recommendations Committee
Wageningen
21 September, 1980

Index of hosts and parasites